ANATOMY OF THE
MONOCOTYLEDONS
X
ORCHIDACEAE

Anatomy of the Monocotyledons series

(Volumes I–VII edited by C. R. Metcalfe)

Vol. I. Gramineae. By C. R. Metcalfe, 1960.
Vol. II. Palmae. By P. B. Tomlinson, 1961.
Vol. III. Commelinales-Zingiberales. By P. B. Tomlinson, 1969.
Vol. IV. Juncales. By D. F. Cutler, 1969.
Vol. V. Cyperaceae. By C. R. Metcalfe, 1971.
Vol. VI. Dioscoreales. By E. S. Ayensu, 1972.
Vol. VII. Helobieae (Alismatidae) (including the seagrasses). By P. B. Tomlinson, 1982. ISBN 978–0–19–854502–6.
Vol. VIII. (eds M. Gregory and D. F. Cutler) Iridaceae. By Paula Rudall, 1995. ISBN 978–0–19–854504–0.
Vol. IX. (eds M. Gregory and D. F. Cutler) Acoraceae and Araceae. By Richard C. Keating, 2002. ISBN 978–0–19–854535–4.
Vol. X (eds M. Gregory and D. F. Cutler) Orchidaceae. By William Louis Stern, 2014. ISBN 978–0–19–968907–1.

Series editors: D. F. Cutler, M. Gregory and P. Rudall.

Anatomy of the Dicotyledons series

(Volumes I–III do not have editors)

Vol. I. Systematic anatomy of leaf and stem, with a brief history of the subject.
 By C. R. Metcalfe and L. Chalk, 1979. ISBN 978–0–19–854383–1.
Vol. II. Wood structure and conclusion of the general introduction.
 By C. R. Metcalfe and L. Chalk, 1983. ISBN 978–0–19–854559–0 (*v*. 1)
Vol. III. Magnoliales, Illiciales, and Laurales (*Sensu* Armen Takhtajan).
 By C. R. Metcalfe, 1987. ISBN 978–0–19–854593–4 (*v*. 3).
Vol. IV. (eds D. F. Cutler and M. Gregory) Saxifragales.
 By Hazel P. Wilkinson, R. J. Gornall, K. I. A. Al-Shammery and Mary Gregory, 1998. ISBN 978–0–19–854792–1.
Vol. I. reprinted with corrections in paperback, 1988. ISBN 978–0–19–854253–7.
Vol. II. reprinted with corrections in paperback, 1989. ISBN 978–0–19–854594–1.

Series editors: D. F. Cutler, P. Gasson, M. Gregory and D. W. Stevenson.

ANATOMY OF THE MONOCOTYLEDONS

EDITED BY

M. GREGORY and D. F. CUTLER

Honorary Research Fellows,
Jodrell Laboratory, Royal Botanic Gardens,
Kew, Richmond, UK

X. ORCHIDACEAE

WILLIAM LOUIS STERN

Emeritus Professor of Botany,
University of Florida,
Gainesville,
Florida 32611, USA

with an INTRODUCTION

by ALEC M. PRIDGEON

Sainsbury Orchid Fellow,
Royal Botanic Gardens,
Kew, Richmond, UK

OXFORD
UNIVERSITY PRESS

Great Clarendon Street, Oxford, OX2 6DP,
United Kingdom

Oxford University Press is a department of the University of Oxford.
It furthers the University's objective of excellence in research, scholarship,
and education by publishing worldwide. Oxford is a registered trade mark of
Oxford University Press in the UK and in certain other countries

© William Stern 2014

The moral rights of the authors have been asserted

First Edition published in 2014

Impression: 1

All rights reserved. No part of this publication may be reproduced, stored in
a retrieval system, or transmitted, in any form or by any means, without the
prior permission in writing of Oxford University Press, or as expressly permitted
by law, by licence or under terms agreed with the appropriate reprographics
rights organization. Enquiries concerning reproduction outside the scope of the
above should be sent to the Rights Department, Oxford University Press, at the
address above

You must not circulate this work in any other form
and you must impose this same condition on any acquirer

Published in the United States of America by Oxford University Press
198 Madison Avenue, New York, NY 10016, United States of America

British Library Cataloguing in Publication Data
Data available

Library of Congress Control Number: 60052155

ISBN 978–0–19–968907–1

Printed in Great Britain by
Bell & Bain Ltd., Glasgow

Links to third party websites are provided by Oxford in good faith and
for information only. Oxford disclaims any responsibility for the materials
contained in any third party website referenced in this work.

AUTHOR'S PREFACE

Sometime in the early 1980s I was asked by the late Charles Russell Metcalfe to prepare a compendium on orchid anatomy along the lines of the already published plant families in *Anatomy of the Monocotyledons*. Metcalfe, whom I admired, was Keeper of the Jodrell Laboratory at the Royal Botanic Gardens, Kew, and a prominent plant anatomist. He was aware that many innovations had taken place during the 50 or more years that had elapsed since a comprehensive account of orchid anatomy had appeared. Technology had advanced appreciably and newly evolved ways of understanding systematics had been applied to unravelling and answering botanical questions. Thus, the intent of this work is to provide an up-to-date analysis of the vegetative anatomy of the orchid family (Orchidaceae), to suggest any taxonomic implications of the application of anatomy to questions of systematics, and to present the results in a systematic format providing a tool for systematists, students, and others to whom orchid anatomy and systematics are relevant.

Heretofore, the only comprehensive body of work on orchid anatomy had been summarized by Hans Solereder and Fritz Jürgen Meyer in their seminal compilation of 1930 as part of the series *Systematische Anatomie der Monokotyledonen*. Solereder was Professor of Botany at the University of Erlangen, Germany, and Meyer held a similar position at the Technical Institute in Braunschweig, also in Germany. Solereder began the work and devoted more than 10 years to it. Besides compilations and interpretations of existing literature and his original observations, noted in the text by an exclamation point, '(!)', Solereder illustrated salient features. Following Solereder's death in 1920, Meyer assumed editorship, revising, updating, and completing Solereder's work. The original product was published in German, translated into English by A. Herzberg, and edited by B. Golek as part of the Israel Program for Scientific Translation. The English version was published in Jerusalem in 1969.

Descriptions in Solereder and Meyer's anatomical treatise were organ-based, however, making application of their work cumbersome for systematic studies. There were no indices for taxa nor coordinated lists of literature references; the bibliography was condensed into paragraph form and the entries were listed chronologically rather than alphabetically by author. I have expanded the earlier efforts of Solereder and Meyer and presented results in a systematic format: by incorporating taxonomic categories, listing literature and authors' names alphabetically and chronologically, providing an evaluation for each taxon described, and adding results from botanical literature and research since 1930. At the end of the introductory portion for each taxon described, I have added selected names of the chief ornamental species represented in the taxon and listed species of putative medicinal value extracted from the compilations of Lawler (1984), Gutiérrez (2010), and Singh and Duggal (2009).

It may be of interest that, prior to my study, an earlier unpublished work on the anatomy of orchids had been attempted by Edward S. Ayensu while he was associated with the Smithsonian Institution. His efforts lapsed owing to his feeling that he should put his energies into something else and to the lack of collaborators who, almost correctly, considered the anatomy to be an endless task. There are estimated to be more than 26, 000 species in the orchid family! I must confess that during my writing, at times, I was reminded of the Herculean endeavours of James. A. H. Murray, compiler and editor of the great *Oxford English Dictionary*, who spent 50 years on it! My own research began in 1983 while I was Professor and Chairman of the Department of Botany at the University of Florida, Gainesville, and continued in Florida after my 2002 retirement, studying at The Kampong, Coral Gables, and at Florida International University, Miami and North Miami.

I have chosen to follow the format developed by Metcalfe in the Monocotyledon series he initiated, including his anatomy of Gramineae in 1960 and, subsequently, anatomy of Cyperaceae in 1971. The monocot series has continued, with the latest publication in 2002 being Volume IX on Acoraceae and Araceae by Richard C. Keating.

To provide a framework for this publication I have adopted the classification utilized in *Genera Orchidacearum* (Pridgeon *et al*. 1999–2014). It is a fortunate circumstance that my work proceeds almost simultaneously with the last volumes of *Genera Orchidacearum*. Resources for my work are my original published research papers as well as publications of other botanists for groups unrepresented in my studies.

I consider it an honour and pleasure to have been chosen to undertake this wholly enjoyable task. My goals are for this treatise to enrich the body of information available on orchid anatomy, to serve as a lodestone for future studies on this, arguably, the largest of flowering plant families, to provide a source of taxonomic detritus to be picked over by systematists, and to unfold the rich fabric of orchid structure as it applies to the biology of orchid life. My hope is to reach into the distant future with information we need to sustain that which we know and to promote that which is waiting to be known about the most seductive plants in the world.

W.L.S.

ACKNOWLEDGEMENTS

I have prefaced three volumes with the names of my authorities. I have done so because it is, in my opinion, a pleasant thing and one that shows an honorable modesty to own up to those who were the means of one's achievements. . . .

Pliny, *Natural History*, Preface

It is an appealing thing for me to acknowledge the leadership and friendship of Charles Russell Metcalfe, late Keeper of Kew's Jodrell Laboratory, where, under his sponsorship, the seminal works *Anatomy of the Dicotyledons* and *Anatomy of the Monocotyledons* were conceived and produced. I will feel everlastingly honoured that he chose me to prepare this treatment of the orchids.

Every author needs an editorial companion to stand behind his/her shoulder; someone to comment objectively, suggest revisions, search for and correct inaccuracies, understand the text, provide detailed bibliographic commentary, guide the flowing and processing of words, and shadow his/her mental musings. Mary Gregory, Royal Botanic Gardens, Kew, has been that companion to me for which I will be evermore grateful. Loismay C. Abeles, my wife and advisor on English didactics, has been with me through writing the Preface, Acknowledgements, and sundry other parts of the text, bringing her skills at composition and linguistic judgement to my aid. For her wisdom in these things and for standing by me with her support I am supremely appreciative and thankful.

Alec M. Pridgeon, Royal Botanic Gardens, Kew, turned my interest from wood anatomy to orchid anatomy. I sincerely thank him for moving me in that direction and for remaining my orchid mentor these many years. Diurideae in the current work is the result of his study. For discussions on the cytology of cells in apostasiad root tubercles, I acknowledge the interest and experience of Peter E. Brandham, Royal Botanic Gardens, Kew, for confirming the presence of multinucleate cells in tissue sections of these organs and for the use of his photomicroscope to picture them. My colleague, Barbara S. Carlsward, Eastern Illinois University, acted as my anatomical barometer while she was my student exploring ideas of cellular and tissue relationships and applying cladistic methods in our joint publications. She graciously provided illustrations of Vandeae from her study of this tribe for use in this work; additionally she permitted me to use her unpublished observations on Sobralieae for this study.

Kurt Neubig, University of Florida, shared freely of his research on Sobralieae that formed the basis of the current treatment of this tribe. I appreciate his generosity.

Beate H. E. Ruetter validated for me Max Siebe's omission of root tubercles in his description of the apostasias. Heike Hettler translated into English Martin Moebius's German language monograph on orchid leaf structure, for both of which I am grateful. Ruetter and Hettler served in my laboratory at the University of Florida.

Over many years of friendship, the late Fred Fuchs, Naranja, Florida, kindly provided me with specimens of uncommon orchid species for my research. Other major sources of research materials were the Royal Botanic Gardens, Kew, through assistance there of Sandra Bell and the late Tim Lawrence. I thank Ronald O. Determann, head conservator at the Atlanta Botanical Garden, for most of the vanilla specimens I examined. It is a pleasure to acknowledge Kenneth M. Cameron, New York Botanical Garden, for specimens of rare members of Vanilleae from the South East Asian islands without whose help I would not have been able to report on the anatomy of these plants.

Words alone are often insufficient to express visual concepts. Pictures give life to words and David F. Cutler, Royal Botanic Gardens, Kew, collaborated with Oxford University Press as liaison to organize our pictures effectively, to evaluate the quality of the images, to oversee the preparation of captions, and to collaborate with Mary Gregory in assembling the manuscript. I thank him for his diligence.

I thank James M. Heaney, Florida Museum of Natural History, for permission to use unpublished material from university research on the anatomy of subtribe Polystachyinae, Beatrice N. Khayota, East African Herbarium, National Museums of Kenya, for her study of *Ansellia* and related taxa, Marlene

Rosinski, Neuenkirchen, Germany, for her research analysis of Coelogyninae and Eriinae, and the late Richard K. Baker, for leaf anatomy of Laeliinae, unpublished works from which I borrowed freely. There were those who provided me with difficult-to-find plant material whom I wish to note with gratitude. They are *Brachycorythis* from the late Joyce Stewart, *Triphora* from Susan M. Stern-Fennell, Coconut Grove, Florida, and *Calypso bulbosa* from Kenton L. Chambers, Oregon State University, Corvallis. Susan M. Stern-Fennell smoothed the way through customs at the airport in Miami for some of my field-collected specimens. Some preserved specimens came from collections of Edward S. Ayensu, formerly of the Smithsonian Institution. I have fond memories of Paul E. Stern, Ancile Gloudon, and Cicely Tobisch, who accompanied me in Jamaica while I was collecting orchids for research; for their help in the field I am grateful.

Wayne M. Morris, North Georgia College, Dahlonega, prepared most of the microscope slides of tribe Dendrobieae and reported in his doctoral dissertation the anatomy of that group. I find pleasure in acknowledging the skill of Lorraine M. McDowell, University of Florida, for preparing, processing, and photographing tissues for both scanning and transmission electron microscopy.

Initially, my research took place at the University of Florida, Gainesville, where I was a faculty member and chairman of the Department of Botany. I appreciate their generosity for providing laboratory and office space and for making available equipment and supplies for my research.

Were it not for the help and influence of David W. Lee, Florida International University, I would not have had laboratory, storage, and office space at Florida International University to continue research after my retirement. I am grateful to him and to the university administration for accommodating me and fulfilling my needs. The Kampong in Coral Gables, Florida, similarly provided space for my research activities and writing.

Permission to use the line drawings of orchid tissues was granted by the Smithsonian Institution, where I was chairman of the Department of Botany in the Museum of Natural History. These drawings were executed by staff artists under the supervision of Edward S. Ayensu. The work was supported through the Smithsonian's Walcott Research Fund to which I express my thanks for making funds available. I also thank the owners and editors of the *American Journal of Botany*, the *Botanical Journal of the Linnean Society*, the *International Journal of Plant Sciences*, *Lankesteriana*, and *Lindleyana* for permission to reproduce photographs from those publications.

For her expertise and knowledge of the intricacies of the computer, I must thank Vanessa Sandrino, whose help smoothed the passage of text from mind to print.

I would be remiss were I not to acknowledge with thanks the technical expertise and services of the librarians at the Royal Botanic Gardens, Kew, whose searches turned up many critical publications on which some of the above descriptions and discussions are based.

I owe my deepest appreciation to the late Oswald Tippo, my guardian spirit, whose wise counsel guided me safely through the meshwork of botanical idiosyncrasies, human and otherwise, and who initially sparked my interest in plant anatomy.

CONTENTS

MATERIAL AND METHODS xi

INTRODUCTION by Alec M. Pridgeon 1

DESCRIPTIONS OF VEGETATIVE ANATOMY OF ORCHIDACEAE 43
  Subfamily Apostasioideae 43
  Subfamily Cypripedioideae 49
  Subfamily Orchidoideae 52
    Tribe Diseae 52
    Tribe Diurideae by Alec M. Pridgeon 56
      Subtribe Acianthinae 57
      Subtribe Caladeniinae 58
      Subtribe Cryptostylidinae 60
      Subtribe Diuridinae 60
      Subtribe Drakaeinae 61
      Subtribe Megastylidinae 62
      Subtribe Prasophyllinae 63
      Subtribe Rhizanthellinae 64
      Subtribe Thelymitrinae 64
    Tribe Orchideae 65
    Tribe Chloraeeae 73
    Tribe Codonorchideae 73
    Tribe Cranichideae 74
  Subfamily Vanilloideae 79
    Tribe Pogonieae 79
    Tribe Vanilleae 81
  Subfamily Epidendroideae 95
    Tribe Arethuseae 95
      Subtribe Arethusinae 95
      Subtribe Coelogyninae 98
    Tribe Calypsoeae 105
    Tribe Collabieae 112
    Tribe Epidendreae 116
      Subtribe Bletiinae 116
      Subtribe Chysinae 117
      Subtribe Coeliinae 117
      Subtribe Laeliinae 118
      Subtribe Pleurothallidinae 130
      Subtribe Ponerinae 135
    Tribe Gastrodieae 136
    Tribe Malaxideae 139
    Tribe Neottieae 143
    Tribe Nervilieae 148
    Tribe Podochileae 150
    Tribe Sobralieae 157
    Tribe Triphoreae 161
    Tribe Tropidieae 164
    Tribe Xerorchideae 168

Tribe Cymbidieae	168
Subtribe Catasetinae	168
Subtribe Coeliopsidinae	172
Subtribe Cymbidiinae	173
Subtribe Cyrtopodiinae	177
Subtribe Eriopsidinae	178
Subtribe Eulophiinae	179
Subtribe Maxillariinae	184
Subtribe Oncidiinae	189
Subtribe Stanhopeinae	195
Subtribe Vargasiellinae	199
Subtribe Zygopetalinae	199
Tribe Dendrobieae	202
Tribe Vandeae	214
Subtribe Adrorhizinae	214
Subtribes Angraecinae and Aeridinae	215
Subtribe Agrostophyllinae	221
Subtribe Polystachyinae	222
TABLES OF DIAGNOSTIC CHARACTERS	226
BIBLIOGRAPHY	235
INDEX	257

MATERIAL AND METHODS

For the most part the following anatomical reports are based on original research by the author. These resulted from fluid-preserved plant parts obtained from the following sources: to a large extent from the Royal Botanic Gardens, Kew; collections from the field; R. F. Orchids, Homestead, Florida; Fuchs Orchids, Naranja, Florida; several hobby collections; conservatory collections at the University of Florida, Gainesville; personal collections of W.L.S.; and the Marie Selby Botanical Garden, Sarasota, Florida. Where fresh material was not available for study and reportage, recourse was made to already published information in different scholarly journals.

A diversity of methods was used to prepare tissues for study depending upon the results desired. Methods are outlined briefly; details are available in the sources listed. Fresh plant parts were preserved in FAA (70% ethanol, glacial acetic acid, and commercial formalin, 9.0, 0.5, 0.5 parts) and stored in 70% ethanol before further processing. For observation with the light microscope transverse and longitudinal sections of leaves and transverse sections of roots and stems were cut unembedded as thinly as possible with a Reichert Sliding Microtome, stained in Heidenhain's iron-alum haematoxylin, and counterstained with safranin. Serial sections were prepared from paraffin-embedded material, sectioned on an American Optical Company rotary microtome; paraffin ribbons were fixed to glass slides using Haupt's adhesive, and stained similarly to the unembedded tissues (Johansen 1948).

Leaf scrapings were made following Cutler's (1978) method whereby leaves were laid flat on a glass plate and tissues scraped from both leaf surfaces with a razor blade while flooded with Clorox (0.5% sodium hypochlorite) leaving only the epidermis. Epidermises were washed several times in water and stained with safranin.

Tissue sections of unembedded parts, paraffin ribbon material, and scrapings were mounted on glass slides with Canada balsam under cover glasses. Completed slides were placed on a warming table to dry and weighted to flatten the sections. Observations were made with a Nikon Optiphot microscope with an attached drawing tube. Photographs were originally taken with film and later with an Olympus BH-2 epifluorescence microscope system and a Pixera 120C digital camera. The line drawings were prepared at the Smithsonian Institution; the scale bars on the anatomical drawings represent $c.100$ µm; *this figure is approximate and should be used only as a rough guide, as are magnifications on the plates.*

For tracheary cell studies tissues were macerated in a solution of 40% acetic acid, 10% hydrogen peroxide, and 50% distilled water, thoroughly washed in distilled water and stained with toluidine blue, and cells teased apart in glycerine. Tracheary elements were examined and photographed with a Zeiss Ultraphot microscope (Thorsch and Stern 1997).

Scanning electron microscopy was carried out after tissues were fixed in osmium tetroxide, dehydrated through a graded ethanol series, coated with gold, and viewed with a Hitachi-S4000 scanning electron microscope.

Tissues destined for examination with the transmission electron microscope were treated with a solution of formaldehyde, glutaraldehyde, and calcium chloride in a cacodylate buffer, postfixed in osmium tetroxide, dehydrated in an ethanol series, and stained with uranyl acetate and Sato's lead citrate (Hayat 1981a, b). Tissues were viewed with a Hitachi HU-11E electron microscope.

Microscope slides of stained and sectioned specimens are lodged in the collection of the Cheadle Center for Biodiversity and Ecological Restoration at the University of California, Santa Barbara, California; others are in the Jodrell Laboratory anatomy slide collection, Royal Botanic Gardens, Kew, Richmond, UK. Vouchers of preserved material are housed with the Selby Botanical Garden (SEL) in Sarasota, Florida.

In descriptions of stem anatomy, the terms for circumferential sclerenchyma, 'cylinder', 'ring', 'band', and 'sheath' have been used indiscriminately and without definition to refer to this condition. Sheaths or bands imply structures that are several cells wide radially, ring or cylinder apply to structures that are one or two cells wide. All terms refer to a more or less continuous circle of sclerenchyma usually, if not always, surrounded by parenchyma and that is sometimes perceived to separate cortex and ground tissue. In the descriptive text the terms 'periclinal' and 'anticlinal' are shorthand for periclinally oriented or elongated and anticlinally oriented or elongated; stomata refers to stomatal apparatus.

INTRODUCTION

Alec M. Pridgeon
Sainsbury Orchid Fellow, Royal Botanic Gardens, Kew, Richmond, UK

SIZE AND DISTRIBUTION

Orchidaceae are among the largest families of seed plants, arguably the largest in terms of number of species. Atwood (1986) estimated the number of valid species as 19,000; Dressler (1993) raised the number to 19,500. However, in the almost 30 years since Atwood's study, hundreds of new species have been described from the tropics, chiefly in subtribe Pleurothallidinae but also in other groups, and Chase, Cameron et al. (2003) reported a new total of 24,910 based on the World Monocot Checklist compiled by Govaerts (2003). The most recent figure is 26,972 (World Checklist of Selected Plant Families 2010), whereas estimates of number of species in Asteraceae vary from 23,000 to 30,000. The number of recognized genera through volume 5 of *Genera Orchidacearum* is 619; at the end of the series with volume 6, the total number should be close to 800 (Pridgeon et al. 1999, 2001, 2003, 2005, 2009, and 2014).

Orchidaceae are also one of the most cosmopolitan of vascular plant families. Several genera such as *Dactylorhiza* and *Corallorhiza* are represented near the Arctic Circle. In the Southern Hemisphere, *Corybas* occurs on Macquarie Island, about halfway between Australia and Antarctica; the distributions of *Chloraea* and *Gavilea* extend into Tierra del Fuego, only about 700 miles from Antarctica. Most species, however, occur in the tropics as epiphytes, and especially in the cloud forests of tropical America and Asia; the Andes and New Guinea are particularly rich in species. Atwood (1986) estimated that 73% of species are epiphytes; however, taking into account the relative numbers of epiphytic to terrestrial species validly described since 1986, that percentage has risen.

HABITATS

Orchids are found in every biome type except true deserts, although a few orchids such as *Eulophia petersii* (Rchb.f.) Rchb.f. occur in arid, semi-desert habitats, and several species are known from semi-arid scrub and chaparral. At the other extreme, *Habenaria repens* Nutt. and *Spiranthes graminea* Lindl. are virtually aquatic, floating in vegetation mats (Hágsater et al. 2005), and some species are temporarily submerged during periodic flooding. Although there are no truly marine orchids, some species of *Brassavola*, *Myrmecophila*, *Dendrobium*, and other genera are epiphytic on mangroves in estuaries; many others have adapted to salt spray and soil salinity in stabilized coastal dunes (Jones 1988; Linder and Kurzweil 1999; Hágsater et al. 2005).

Terrestrial species span the elevational range of habitats, from the seashore to tepuis to elfin forests and páramo. Most occur in open forests, savannas, and grasslands at mid-elevations, with progressively fewer in both directions. Much the same is true for the distribution of terrestrials at the extremes of rainfall gradients; relatively fewer species occur in arid environments and in the wettest lowland forests such as those of the Amazon basin. Many thrive in calcareous soils, whereas others occur in neutral to slightly acid situations as in bogs, marshes, and seepage zones. Dressler (1981) estimated that roughly 25% of orchid species are exclusively terrestrials and another 5% adapted for either terrestrial or epiphytic situations, although the percentage of terrestrials has probably fallen in the last 30 years with the description of hundreds of newly discovered epiphytic species (especially those in subtribe Pleurothallidinae).

Lithophytes and epiphytes share the range of habitats of terrestrials but are more common than terrestrials at the elevational extremes and indeed can be surprisingly diverse in some lowland tropical rain forests, such as Barro Colorado Island of Panama and the Selva del Ocote and Selva Lacandona of northern Chiapas, Mexico (Gentry and Dodson 1987; Hágsater et al. 2005). They are most common, however, in mid-elevation (1000–3000 m) cloud forests, growing among other vascular epiphytes as well as bryophytes. Dunsterville (1961) famously reported at least 47 orchid species on a single tree in a Venezuelan cloud forest.

DIVERSITY

Related to the cosmopolitan nature of Orchidaceae are its remarkable morphological, ecological, and physiological adaptations within vastly different climates and life zones, chief among them adaptations for uptake of water and mineral nutrients, pollination (both allogamy and autogamy), and seed germination. In terms of both number of species and number of individuals, orchids are primarily terrestrial or lithophytic in the temperate zones and epiphytic or lithophytic in the tropics. However, many terrestrials in (for example) tribes Arethuseae, Cranichideae, Collabieae, Epidendreae, Malaxideae, Neottieae, and Sobralieae are common understorey plants in the tropics.

Diversity in size among orchids is perhaps the most extreme of the angiosperms. Entire, flowering-size plants of some species of *Platystele* and *Bulbophyllum* are of the order of a few millimetres. Indeed, some minute plants may grow on or hang from a single leaf of another angiosperm, even that of another orchid. On the other hand, *Vanilla* vines may be several metres long, and plants of *Grammatophyllum* can weigh hundreds of kilograms (Dressler 1981; Cribb 1999).

VEGETATIVE MORPHOLOGY

As might be expected in a group as large and diverse as Orchidaceae, the range in morphology of plant organs is vast and ultimately related to the two fundamental growth habits in the family. The sympodial habit is characterized by shoots arising successively from axillary buds of a rhizome. The majority of sympodial species are epiphytic and possess bifacial, simple, distichous, linear–elliptical, petiolate, conduplicate, sheathing leaves; unifoliate or bifoliate, thickened stems or 'pseudobulbs' of one or more internodes, functioning chiefly in water storage; and adventitious, velamentous, aerial or attached roots. Aerial roots are mostly cylindrical, but those attached to a substrate are dorsiventrally flattened, thereby increasing the surface area for absorption. The sympodial condition is ancestral and global in occurrence. In the monopodial habit, however, leaves arise from the same, indeterminate, apical meristem. They are sheathing, usually distichous and bifacial, and often laterally flattened. Rhizomes and pseudobulbs are absent, but roots are as above. The monopodial condition, characterizing Vandeae, is derived and almost exclusively paleotropical; in the Neotropics it has apparently arisen independently in a few taxa of *Epidendrum s.l.* (Hágsater and Soto Arenas 2004). Deviations from these generalizations are discussed below.

Leaf

Leaves are simple and generally linear–elliptical, lanceolate, spatulate, ovate, or oblanceolate. In a few species they may be cordate, sagittate (*Pachyplectron arifolium* Schltr.), or spirally twisted (*Thelymitra spiralis* (Lindl.) F.Muell.). Margins are entire, or rarely serrate as in *Triphora*, often undulate, crenate (*Townsonia*) to crenulate (e.g. *Thelyschista*), minutely dentate (*Megastylis paradoxa* (Kränzlin) N.Hallé), ciliate in some Cypripedioideae, or bulbil-producing (*Hammarbya*). At anthesis leaves may or may not be present.

Size ranges from minute scales in mycoheterotrophs and scale-like bracts in many Vanilleae (*Pogoniopsis, Cyrtosia, Erythrorchis, Galeola*) to more than a metre in length in some species of *Bulbophyllum*. At the same time, *Bulbophyllum minutissimum* (F.Muell.) F.Muell. bears true leaves less than 1 mm long. Leaflessness has arisen multiple times, not only in mycoheterotrophs (e.g. Gastrodieae, Diurideae) but also in many Vandeae such as some species of *Campylocentrum* and *Dendrophylax* (Angraecinae) and *Chiloschista* and *Taeniophyllum* (Aeridinae), in which the bulk of photosynthesis occurs in aerial roots (Carlsward et al. 2006a).

Vernation is convolute in Apostasioideae, some Cypripedioideae, most Orchidoideae, and some Epidendroideae. Convolute leaves at maturity are generally thin and either conduplicate (*Apostasia*, Orchidoideae) or plicate (e.g. *Neuwiedia, Cypripedium, Selenipedium*, Arethuseae, *Palmorchis*, Catasetinae, *Corymborkis*, Sobralieae, Collabiinae, Stanhopeinae, many Zygopetalinae). *Vanilla* leaves, however, are convolute and fleshy-thickened. Vernation is duplicate in *Paphiopedilum, Phragmipedium*, and most Epidendroideae, in which leaves are conduplicate, thin to coriaceous, and fleshy. In some Epidendroideae—Vandeae in particular—duplicate, fleshy leaves are laterally flattened instead ('equitant') in varying degrees, all the way up to becoming terete. Blades are generally bifacial (dorsiventral) but unifacial (terete) in *Luisia, Paraphalaenopsis, Thelymitra*, and many Dendrobieae and Vandeae, and terete/hollow in Prasophyllinae and *Epiblema*. Leaves of *Oberonia* are unifacial and iridiform.

Phyllotaxy is distichous in derived groups (most Epidendroideae) but spiral in primitive taxa such as Neottieae and *Diceratostele*. This spiral arrangement may take the form of a basal rosette, as in Orchidoideae, often adpressed to the substrate (*Satyrium*). Rarely, leaves are whorled at the stem apex (*Isotria*) or at a node in the lower third of the stem (*Codonorchis*). Leaves with convolute vernation are generally spirally arranged, whereas conduplicate leaves have duplicate vernation; plicate leaves may arise from either convolute or duplicate vernation (Freudenstein and Rasmussen 1999).

Articulations are present only in most Epidendroideae with the exception of some species of *Dichaea*, Pleurothallidinae, Podochileae, *Epidendrum, Notylia, Tropidia*,

Diceratostele, *Nervilia*, and Neottieae. The abscission zones generally occur either at the boundary of the leaf sheath (if present) and blade or simply at the base of the leaf if no sheath is present (Dressler 1981).

An indumentum is rare in orchid leaves, especially those of epiphytes, with some exceptions, a few of them noted here. Hairs are multicellular and reddish brown in *Trichotosia*, brown-furfuraceous at the leaf base of *Corymborkis*, woolly in some species of *Dendrobium*, *Dresslerella*, *Calanthe*, *Cypripedium*, and multicellular (glandular or not) in *Caladenia* and other genera of subtribe Caladeniinae.

Venation is more or less parallel with commissural veining in most species but reticulate in many Vanilloideae and Goodyerinae and anastomosing in Caladeniinae.

Most dorsiventral leaves are darker green abaxially, but differential coloration from localized or generalized presence of anthocyanins also occurs in many species. *Tipularia* and *Psilochilus* are green adaxially but purple abaxially. Purple leaves with contrasting veins characterize some forest-floor species ('jewel orchids') of Goodyerinae (*Anoectochilus*, *Goodyera*, *Ludisia*, *Macodes*). In addition, anthocyanin spots occur on one or both leaf surfaces in some species of *Cypripedium*, *Phaius*, *Paphiopedilum*, *Orchis*, *Dactylorhiza*, *Psychopsis*, and other genera.

When present, the leaf base is often sheathing around the stem or may be reduced to a bract. Leaves are usually petiolate, rarely sessile (e.g. *Pseudocranichis*, *Galeottiella*, *Dictyophyllaria*, Caladeniinae, Diuridinae). The petiole is variously configured. It may be dilated into a tubular, amplexicaul sheath in many Goodyerinae, inrolled and channelled in *Coelogyne*, and up to subcylindric in Stanhopeinae. *Bipinnula* and *Chloraea* (Chloraeeae) are pseudopetiolate.

Stem/rhizome

Rhizomes of sympodial orchids usually consist of the basal portions of successive shoots, each potentially with an axillary bud that gives rise to an upright shoot and from that an inflorescence, although in some species the inflorescence may arise directly from the rhizome. It is succulent in *Cypripedium*, *Zeuxine*, and many Goodyerinae and tuberous in *Epipogium* and Gastrodieae. Mycoheterotrophs often lack an aerial stem and have only a rhizome (*Corallorhiza*, *Hexalectris*).

Stems are generally slender or pseudobulbous. Slender stems are sometimes cane-like or reed-like (*Anthogonium*, *Arundina*, *Isochilus*, *Palmorchis*, *Sobralia*, *Diceratostele*, *Corymborkis*, Ponerinae), and sometimes branched (e.g. *Arundina*, *Eriaxis*, *Vanilla*). Pseudobulbs, confined to sympodial orchids and generally to epiphytic species, are variously thickened stems with a water-storage function, sometimes superposed (*Otochilus*, *Pholidota* in part). Shapes range from cylindrical to globular but are commonly clavate or fusiform. They may be hollow as in *Cleistes* and *Pogonia* and often house ant colonies (*Myrmecophila*, *Caularthron*). Terrestrial, sympodial species in Apostasioideae, Cypripedioideae, and Orchidoideae as well as many epiphytes in Epidendroideae (e.g. *Brassavola*, Pleurothallidinae) lack pseudobulbs. At the same time, there are several pseudobulbous species in Epidendroideae that can occur as either epiphytes or terrestrials. Pseudobulbs may be heteroblastic (consisting of one internode) or homoblastic (two or more internodes).

Corms are surface or subsurface storage organs consisting of several internodes, occurring primarily in Calypsoeae (e.g. *Aplectrum*, *Calypso*, *Govenia*, *Tipularia*), Arethusinae (*Anthogonium*, *Arethusa*, *Calopogon*, *Eleorchis*), and Bletiinae (*Basiphyllaea*, *Bletia*). Shapes range from fusiform to globose, and elliptical to conical.

True stem tubers (differentially swollen underground shoots with scale leaves subtending axillary buds) are rare in Orchidaceae but occur in Bletiinae (*Basiphyllaea*), Nervilieae (*Nervilia*), and Diurideae (*Rhizanthella*). There are reports of tubers in some species of *Eulophia* (Cribb 2009), but these should be investigated further.

Root

As the plant organ most closely involved in water and nutrient uptake across a broad range of microniches from semi-desert to semi-aquatic environments and from lowland rain forest to páramo habitats, the orchid root exhibits more morphological, anatomical, and physiological adaptations than the leaf or stem. Such a wide array of root types is integral in helping to explain the virtually pole-to-pole distribution of orchids as well as define the major clades in orchid phylogeny.

Absorptive roots of most epiphytic and some terrestrial orchids are bound externally by a specialized epidermis of 1–20 cell layers, termed the velamen. Interior to that is the cortex, the outermost layer(s) of which is the exodermis and the innermost layer(s) the endodermis. The innermost stele comprises the pericycle and primary xylem and phloem. Characters of each of these tissues are discussed below.

However, there are also erect, aerial (negatively geotropic) roots in some Catasetinae (*Catasetum*, *Cycnoches*), Cymbidiinae (e.g. *Ansellia*, *Grammatophyllum*, *Graphorkis*, *Cymbidium*), and Cyrtopodiinae (*Cyrtopodium*) that allow accumulation of leaf litter from above so that the plant benefits from absorbed foliar leachates.

In many genera, aerial roots are modified in morphology and anatomy when they contact a substrate. Once cylindrical roots become dorsiventral, velamen assumes a protective function dorsally and an absorptive function ventrally. Anatomical changes ventrally include formation of root hairs and fewer velamen layers with smaller cells.

Many species of Vandeae (*Microcoelia, Taeniophyllum, Campylocentrum, Chiloschista, Dendrophylax, Harrisella*) have reduced leaves or are leafless altogether, so that the major photosynthetic organ is the root. Their roots are often dorsiventral in organization; there is some experimental and anatomical evidence that the condition is genetically rather than environmentally determined (Janczewski 1885; Schimper 1888; Müller 1900; Porsch 1906; Goebel 1922; Benzing et al. 1983). Velamen is sloughed off dorsally thereby exposing the thick-walled exodermis, which offers protection from desiccation and mechanical injury. Towards the ventral face, however, velamen becomes more developed, and exodermal cell walls become thinner. Based on DNA sequence data from 112 ingroup taxa of Vandeae, leaflessness has arisen six or seven times in the tribe, up to five times in the Paleotropics and twice in the Neotropics (Carlsward et al. 2006b).

Nodular roots, or root tubercles, occur in *Apostasia* (Apostasioideae) and *Tropidia* (Tropidieae). Stern and Warcup (1994) reported that the unusual tubercles in *Apostasia* are stalked, warty growths that arise perpendicularly to the axis of roots. They bear stomata with permanently open pores and subequal guard cells but with no sign of hyphal ingress or pelotons, leading the authors to suggest that the pores facilitate gas exchange for roots immersed in moist soils.

Root tubers bearing velamen (occasionally a simple rhizodermis) characterize most species of Orchideae, Diseae, and Diurideae as well as *Epipogium, Triphora, Codonorchis, Pterostylis*, and some tropical species of *Cleistes*. They are solitary, paired, or clustered and range in shape from cylindrical or ellipsoidal to globose. New tubers form from either the stem base, apices of droppers, or stolonoid roots. Droppers arise endogenously from (1) undifferentiated cortical parenchyma of the stem base, (2) the previous dropper, or (3) the distal pole of the previously formed tuber. They grow downwards, either vertically or obliquely, and terminate in a *replacement* root tuber (Pterostylidinae, Orchideae, Diurideae, Diseae). Stolonoid roots originate in the same ways as droppers but grow mainly horizontally (often for several centimetres) and terminate in a *reproductive* tuber that results in the formation of tufts, clumps, or colonies (Pterostylidinae, Diurideae, Diseae). Tubers may also develop directly below previous tubers, forming an annually deepening chain, as in some species of *Caladenia* (Diurideae). Older tubers of Diurideae are often enclosed in a multilayered fibrous sheath or 'tunica', which presumably offers mechanical protection and prevents desiccation, whereas those of other taxa may be glabrous or hairy (Pridgeon and Chase 1995).

Finally, roots may be absent altogether in the holomycotrophic species of *Hexalectris, Gastrodia, Corallorhiza*, and other genera. In these cases, the rootstock is often rhizomatous and villose or coralloid.

REPRODUCTIVE MORPHOLOGY

The ancestral and most common inflorescence type in Orchidaceae is the raceme, produced either terminally or laterally from the stem or rhizome and opening acropetally. Panicles occur in *Erythrorchis, Galeola, Pseudovanilla*, and occasionally *Vanilla* (Vanilleae) and in some species of *Epidendrum* (Epidendreae), *Corymborkis* (Tropidieae), *Cyrtopodium* (Cyrtopodiinae), *Ansellia* (Cymbidieae), *Oncidium* (Oncidiinae), *Renanthera* (Aeridinae), *Polystachya* (Vandeae), and other genera. True spikes, in which pedicels are absent or so reduced that flowers are essentially sessile on the rachis, are rare but are known in *Corycium* (Diseae). Simple or complex cymes are known in *Vanilla* (Vanilleae) and *Lockhartia* (Oncidiinae). Solitary-flowered inflorescences occur in many groups as diverse as *Paphiopedilum* (Cypripedioideae), Vanilloideae, Diurideae, Maxillariinae, *Bulbophyllum* (Dendrobieae), Zygopetalinae, and Pleurothallidinae.

Orchid flowers are trimerous and bilaterally symmetrical. The outermost calyx is usually valvate and often brightly coloured adaxially. Sepals are generally larger than the petals and often differentiated into a dorsal sepal and lateral sepals. Lateral sepals may be connate for all or part of their length to form a synsepal or they may be variably fused to the dorsal sepal as well, forming a tubular calyx as in some species of Pleurothallidinae. In many genera (e.g. *Dendrobium, Masdevallia*), they form a spur-like mentum at the base, surrounding a hinged attachment between the labellum and column foot. Although sepals are generally glabrous, they are notably pubescent to hirsute in many species of Cypripedioideae, Diurideae, Pleurothallidinae, Eriinae, Dendrobieae, and other groups.

The corolla consists of two lateral petals and a median petal termed the labellum (lip), which is often modified as a pollinator–attractant and/or integral element of the pollination process itself. Petals are similar in size, shape, and coloration and may bear hairs, osmophores, or other glands. The labellum can be dull or brilliantly coloured, entire or trilobed, rigidly fixed or articulated to the base of the column or column foot, and may bear lamellae, calluses, glands (nectaries, elaiophores, osmophores), hairs, or any of these in combination. It can serve as a temporary pollinator trap in Cypripedioideae, Diurideae, Pleurothallidinae, and *Coryanthes* (Stanhopeinae). In many groups such as Orchideae, Diseae, Angraecinae, and Calypsoeae, the base of the labellum may be extended into a shallow or elongate, frequently nectariferous spur (although spurs are often empty in deceit flowers).

At the centre of the flower is the highly variable column (gynostemium), which represents the fusion of staminal filaments, stigmas, and styles. *Neuwiedia* (Apostasioideae) has three fertile anthers, but *Apostasia* and Cypripedioideae have only two. In all other Orchidaceae, only the median anther at the apex of the column is fertile, and the other two anthers are either absent or else present as staminodia or column wings (Dressler 1993). The position of the anther, whether erect and parallel to the axis of the column (most Orchidoideae), bending over backwards (Diseae), or bending downwards 90–120 degrees so that it rests at the column apex or ventrally at the apex (Vanilloideae, Epidendroideae), has been an important character used in orchid classification (Dressler 1993).

Pollen is shed as powdery monads in Apostasioideae, viscous monads in Cypripedioideae except *Phragmipedium* (Dressler 1993), and as masses of free monads or tetrads in Vanilloideae (Dressler 1993; Cameron 2003). In most other groups, pollen is aggregated into soft or hard pollinia composed of tetrads. Soft pollinia may be mealy (e.g. Cranichidinae, Pterostylidinae) or sectile and organized into discrete subunits known as massulae (Orchideae, Diseae, Goodyerinae, etc.). The vast majority of species of Orchidaceae produce hard pollinia, varying in number from two to eight depending on the species. Reduction series from eight to two pollinia are known in Laeliinae and Pleurothallidinae (Dressler 1993), but there is some ontogenetic support for the hypothesis of four pollinia instead of eight as the plesiomorphic state (Freudenstein and Rasmussen 1996).

Pollinia are usually shed with accessory structures and are then termed pollinaria. One of these structures is the viscidium, a sticky pad derived from the rostellum (see also FLORAL ANATOMY) that becomes directly attached to the pollinator. Connecting the viscidium to pollinia are caudicles, produced in the anther, and/or a stipe, which is a strap-like structure derived from columnar tissues other than the anther. Following the terminology of F.N. Rasmussen (1986) and Freudenstein and Rasmussen (1999), if the stipe is formed from the epidermis of the rostellum, it is a known as a tegula or tegular stipe, common throughout derived Epidendroideae. If, however, it constitutes the apex of the rostellum, it is a hamulus, exemplified by Tropidieae. A third type, the hammer stipe, occurs in *Sunipia* and *Genyorchis* (both now included in *Bulbophyllum s.l.*).

The trilobed stigma is either emergent (Vanilloideae), flat, or deeply concave (Epidendroideae; Dressler 1993). The median lobe is the largest of the three; a portion of it, termed the rostellum, plays a role in pollen transfer by (1) secreting a glue that coats the dorsum of the pollinator as it backs out of the flower and dislodges pollen or pollinia or (2) removing previously attached pollen or pollinia and directing it onto the stigmatic surface. As mentioned above, part of the rostellum may comprise the viscidium. The ovary is always inferior and pedicellate (except in *Corycium*, which essentially lacks pedicels). Placentation is mostly parietal or unilocular, but it is axile or trilocular in Apostasioideae, some Cypripedioideae, and some Vanilloideae.

In most species the labellum is uppermost in the flower bud but becomes lowermost by torsion or twisting of the pedicel and/or ovary through 180 degrees by anthesis, a process known as resupination. Non-resupinate flowers, with labellum uppermost, either do not undergo the torsion (e.g. *Calopogon*, pistillate flowers of *Catasetum*) or else twist through a full 360 degrees (e.g. *Malaxis*, *Angraecum*). In all cases the labellum plays a specific role in the pollination process such as serving as a landing platform, and its orientation is an adaptation—among many others—that often results in removal and deposition of pollinia and fertilization. Orchid pollination lies outside the scope of this volume, and the reader is referred to several titles devoted entirely or in part to the topic: Darwin (1877), van der Pijl and Dodson (1966), Dressler (1981, 1993), Williams (1982), Williams and Whitten (1983), Jones (1988), Kaiser (1993), van der Cingel (1995, 2001), and Pridgeon et al. (1999, 2001, 2003, 2005, 2009, 2014).

Fruits of all but a few orchid species are dehiscent capsules, with the exceptions found in more basal groups—fleshy berries in *Neuwiedia* (Apostasioideae), *Palmorchis* (Neottieae), and *Cyrtosia* and *Vanilla* (Vanilloideae) and indehiscent capsules in *Selenipedium* (Cypripedioideae) and *Dictyophyllaria* (Vanilloideae). Capsules may be glabrous, papillate, hairy, spiny, ridged, or beaked (the beak a remnant of the floral tube or cuniculus; Dressler 1993). The capsule usually begins to dehisce apically along the midvein of each carpel rather than between carpels; later the midvein then separates from each half-carpel, resulting in up to six valves (Dressler 1993). Exceptions to this generalized dehiscence pattern occur in Vanilloideae, Pleurothallidinae, Angraecinae, etc.

Orchid seeds are well known for their small size (0.15–6.0 mm) and sheer numbers per capsule, from 6000 to 4 million and weighing as little as a microgram (Ziegler 1981; Dressler 1993). In general, they lack endosperm and instead form mycorrhizal relationships for germination and early growth. Reinfection of mature plants by mycorrhizae also contributes to the nutritional budget of most species, terrestrials and epiphytes alike. Seeds usually have only a thin, papery testa, representing a layer of dead cells of the outer integument (Ziegler 1981). Two integuments are present in ovules of most Orchidaceae, and the unitegmic condition has been reported in only Podochileae and isolated, unrelated taxa (Clements 1999). Seed coats of *Apostasia* (Apostasioideae), *Selenipedium* (Cypripedioideae), *Palmorchis* (Neottieae), and some Vanilloideae are hard (Ziegler 1981; Molvray and Chase 1999). In addition to these features, others of systematic significance include size and shape of seeds and testa cells, sculpturing, presence of waxy deposits, and intercellular spaces or gaps (Ziegler 1981; Chase and Pippen 1988, 1990; Molvray and Chase 1999).

Embryos of Orchidaceae were first monographed by Treub (1879), followed by three significant series of papers (Swamy 1949 and earlier; Abe 1972 and earlier; Veyret 1974 and earlier). Swamy (1949) developed several embryo types based on (1) division sequence of cells from zygote to proembryo and (2) the form and development of the suspensor; he then attempted to classify Orchidaceae accordingly. Veyret (1974) noted that some taxa do not fit into Swamy's scheme primarily because his characterization of embryo types ends too early in development. She adopted the system of Souèges (1936–9), which grouped embryos into an unwieldy system of megarchetypes based on the sum of the constructive potentials of the apical and basal cells after the first transverse division of the zygote. In most angiosperms the apical cell goes on to become the embryo proper, and the basal cell develops into the ephemeral suspensor, which forces the embryo into contact with nutritive tissue of the gametophyte (Maheshwari 1950). However, endosperm does not normally form in Orchidaceae, either because the second sperm nucleus does not fuse with the polar nuclei or because the polar nuclei degenerate after double fertilization (Veyret 1974; Clements 1999). In exhaustive studies of orchid embryogensis, Clements (1995, 1999) identified 18 developmental patterns, ranging from the plesiomorphic type with a large embryo sac and minimal suspensor (*Neuwiedia*, Cypripedioideae) to the apomorphic type with a smaller embryo sac but well-developed external suspensor (e.g. Diurideae, Epidendreae, Vandeae). Although possession of a given type generally conforms to modern circumscriptions based on phylogenetic analysis of DNA sequences, the Cypripedioid pattern is present in taxa as disparate as *Apostasia*, *Bromheadia*, *Cypripedium*, and *Coelogyne*.

Following fertilization, meristematic cell divisions produce the first seedling stage termed the protocorm. Some epidermal cells elongate and develop as rhizoids, either singly or in clusters on elevated protuberances; in some species, rhizoids never develop (Rasmussen 1995). The embryonic shoot apex differentiates first, and when the embryonic adventitious root later emerges, the elongating seedling enters the mycorhizome stage, followed by further development of the rhizome with scale leaves and finally the entire sympodium (Rasmussen 1995).

MYCORRHIZA

In the absence of endosperm, mycorrhizae are critical to seed germination and nutrition of all orchids and as a result are largely responsible for the worldwide expansion, vertical distribution, and phylogeny of Orchidaceae. The vast majority of Orchidaceae are chlorophyllous and autotrophic, but 200 orchid species are recorded as being obligately non-photosynthetic and epiparasitic on nearby photosynthetic plants using basidiomycete ectomycorrhizal fungi (Bidartondo 2005). Using DNA sequence data, Molvray et al. (2000) demonstrated that mycoheterotrophy has evolved independently in different orchid lineages as the possible result of derived nutritional strategies, which affect substitution rates. Mycoheterotrophy ('saprophytism' in the older literature) occurs in several genera (e.g. *Cymbidium*, *Epipactis*, *Eulophia*, *Liparis*, *Malaxis*, *Neottia*) or encompasses entire genera (e.g. *Aphyllorchis*, *Auxopus*, *Corallorhiza*, *Epipogium*, *Galeola*, *Gastrodia*, *Hexalectris*, *Limodorum*, *Rhizanthella*).

Bernard (1902, 1903, 1904, 1909) and Burgeff (1909, 1932, 1959) were among the first to characterize the orchid–fungus relationship, beginning with infection of seeds, protocorms, and roots of adult plants and proceeding through digestion of hyphae. Their experiments showed that, without infection, seed germination and growth of the taxa studied were minimal. Bernard (1909) suggested that the symbiosis between orchid and fungus was not mutualistic but a form of controlled parasitism, whereby the orchid benefits from enhanced uptake of nutrients supplied by the mycobiont but limits the fungus to certain zones in the root and/or tuber and digests it, in the form of either individual hyphae or coiled masses of hyphae termed pelotons. Pelotons are not known to form in nature outside the orchid host. The parasitic nature of the association has been corroborated more recently by Harley and Smith (1983) and Clements (1988).

Fungal hyphae typically enter roots and protocorms through epidermal hairs, papillae, the velamen–exodermis complex (Williamson and Hadley 1970; Hadley and Williamson 1972), and rhizoids (Rasmussen 1990) and form pelotons in inner cortical cells. Simultaneously, nuclei in infected cells undergo hypertrophy, so that the parenchyma cells have 2C, 4C, and 8C nuclei and perhaps even higher (Burgeff 1932; Harley and Smith 1983). Hyphae then develop further and anastomose in one or (usually) several layers of cortical parenchyma, where they store protein, glycogen, and fat before digestion begins, initiated by the cell nucleus; reinfection may occur after digestion (Burgeff 1959; Rasmussen 1995). Infection of germinating seeds occurs through the suspensor (Bernard 1903; Burgeff 1909; Clements 1988) and/or rhizoids (Harvais and Hadley 1967; Clements 1988; Rasmussen 1990), followed by hyphal ramification and formation of pelotons as soon as 6–8 days after sowing and inoculation (Clements 1988) and persisting in the protocorm, mycorhizome, and roots. Mycorrhizal zones of infection also occur in the collar/stem, root, and root tuber of several Australian and New Zealand orchids in tribe Diurideae, in which reinfection is mediated by elaborate, multicellular trichomes with up to five tiers of cells (Pridgeon 1994a). The vast amount of literature on the subject has been summarized and supplemented by Hadley (1982) and Rasmussen (1995). Specificity of mycobiont associations with orchid hosts has also been widely investigated and reported, and the reader is referred to papers by Warcup (1973), Clements (1988), Currah et al. (1997), Otero et al. (2002), Kottke and Suárez (2009), and Suárez et al. (2009).

CYTOGENETICS

Although cytogenetics per se is beyond the remit of this volume, it is worthwhile to discuss briefly those aspects that bear on systematics and biology of Orchidaceae, namely variations in chromosome number and morphology and also genome size. In the last century, before the advent of DNA sequencing and molecular phylogenetics, special emphasis was assigned to chromosome numbers as a way to understand orchid phylogeny, and many laboratories around the world published somatic or gametic counts for hundreds of taxa (species and hybrids) throughout the family. These counts were gathered and published as chromosome indices (e.g. Duncan 1959; R.J. Moore 1970, 1971, 1972, 1973, 1974, 1977; Tanaka and Kamemoto 1974, 1984; Goldblatt 1981, 1984, 1985, 1988; D.M. Moore 1982; Goldblatt and Johnson 1990, 1991, 1994, 1996). Brandham (1999) analysed the counts by genus and suggested basic numbers from polyploid series, which in some cases are difficult to ascertain, given high levels of aneuploidy and dysploidy in the family. Indeed, Chase and Olmstead (1988) argued that aneuploidy rather than polyploidy is responsible for the variation in chromosome numbers, as shown in Oncidiinae, and Brandham (1999) noted that aneuploidy is common in cultivated, polyploid hybrids. The most common chromosome numbers reported are $2n = 38, 40, 42$, but numbers range from $2n = 10$ in *Erycina pusilla* (L.) N.H.Williams & M.W.Chase (formerly placed in the genus *Psygmorchis*) to as high as $2n = 162$ in *Catasetum* and $2n = 168$ in *Habenaria* and *Oncidium* (though rare in both cases).

The utility of chromosome number and morphology in orchid systematics is perhaps best illustrated in tribe Orchideae, comprising about 60 genera and 1800 species distributed in northern temperate and tropical areas of both hemispheres. Prior to 1997, *Orchis* (the type genus) consisted of taxa with chromosome numbers of $2n = 32, 36$ ($x = 16$ or 18) and $2n = 40$ and 42 ($x = 21$). Those taxa with $2n = 32$ or 36 have a more asymmetric karyotype than those with $2n = 40$ or 42 (D'Emerico et al. 1996; D'Emerico 2001). On the other hand, the latter often have B-chromosomes in their karyotypes, but the former do not (D'Emerico et al. 1990). In general, taxa with more asymmetric karyotypes are more highly evolved than their symmetrical counterparts (D'Emerico 2001). In companion papers, Pridgeon et al. (1997) and Bateman et al. (1997) sequenced ITS nuclear ribosomal DNA for 87 taxa of tribe Orchideae, including 21 taxa belonging to *Orchis s.l.* Parsimony analysis of DNA sequences showed that *Orchis* as then understood was polyphyletic, with a derived clade with $2n = 32, 36$ and asymmetric karyotype (including *Anacamptis pyramidalis* (L.) Rich.) and two clades with $2n = 40, 42$ and symmetric karyotypes (one including *Neotinea maculata* (Desf.) Stearn and the other including *Aceras anthropophorum* (L.) R.Br. ex W.T.Aiton). In this case, then, there was perfect congruence between chromosome number, karyotype symmetry, and DNA sequence data. Doubling the data set to 190 taxa in a later paper (Bateman et al. 2003) supported these results and allowed more transfers to *Anacamptis s.l.* and *Neotinea s.l.* from *Orchis s.l.*

Leitch et al. (2009) summarized what is known about genome size (the amount of DNA in the unreplicated gametic nucleus, expressed as the 1C value) for orchids. According to data from the Plant DNA C-values database (Leitch et al. 2009), size varies from 1C = 0.33 pg in *Trichocentrum* to 55.40 pg in *Pogonia*, but most species have small genomes with a mean of 8.50 pg with the exception of subfamily Cypripedioideae with a mean of 25.8 pg. Although there is no correlation between genome size and organismal complexity (onions have five times as much DNA as humans), a C-value of no more than 10 pg is a predictor of success in genetic fingerprinting techniques (Leitch et al. 2009). At the same time, plant species with large genome sizes (>20 pg) are obligate perennials with reduced speciation rates and prone to extinction (Vinogradov 2003, cited in Leitch et al. 2009). Among orchids in general, terrestrials have a mean 1C = 18.3 pg, whereas epiphytes have a mean 1C = 3.0 pg (Leitch et al. 2009). The authors argue that, because genome size and stomatal guard-cell size are correlated (to the extent that the former can be estimated by measuring the latter), the water stresses endured by epiphytes would seem to require small stomata to survive droughts. Moreover, species such as twig epiphytes that can grow rapidly and reach flowering size in as little as 1 year, especially in nutrient-poor sites, would benefit from smaller genomes (Chase, Hanson et al. 2005).

PHYTOCHEMISTRY

There are a few family-wide (or at least broad) reviews of different classes of compounds naturally occurring in orchids, among them flavonoids (including anthocyanins), alkaloids, phytoalexins, and floral fragrances. As such, they should be mentioned here in passing as a stimulus to additional reading.

Flavonoids

C.A. Williams (1979) surveyed 142 species of 75 orchid genera for leaf flavonoids and found that flavone C-glycosides (in 53%) and flavonols (in 37%) were the most common. Updating the taxonomy, the former characterize tropical and subtropical species of Epidendroideae except Neottieae (in 63%), whereas flavonol glycosides occur mostly in temperate species of Orchidoideae and Neottieae (in 78%). Oddly, *Aerides*, a genus of paleotropical epiphytes, accumulates flavonol glycosides instead of glycosylflavones. Other genera with uniform flavonoid profiles are *Epipactis*, *Ophrys*, *Orchis s.l.*, *Oncidium s.l.*, and *Pleurothallis s.l.* At the other extreme, *Polystachya* (26 species sampled) showed three different flavonoid patterns—flavonol, flavone C-glycoside, and xanthone. It is possible that further sampling outside of *Polystachya* would have revealed more heterogeneity in other genera, however.

Anthocyanins are present not only in orchid flowers but also in the roots of some Epidendreae species and even in the uniformly coloured leaves or differential colouring of some Goodyerinae (*Ludisia*, *Anoectochilus*, *Dossinia*, *Erythrodes*, *Goodyera*, etc.), the abaxial leaf surfaces of *Tipularia* (Calypsoeae), *Corybas* (Acianthinae), *Cryptostylis* (Diurideae), *Disa* (Diseae), etc., and the spotted leaves of some species of *Cypripedium*, *Dactylorhiza*, *Orchis*, etc. Of the 16 naturally occurring anthocyanidins, the most common listed in a survey of orchid species and hybrids by Arditti and Fisch (1977) are cyanidin, pelargonidin, petunidin, malvidin, and delphinidin.

Alkaloids

Lüning (1964, 1967) and Lawler and Slaytor (1969, 1970) conducted screens for alkaloids for more than 2044 orchid species in 281 genera (as then understood). Their results showed that 214 species in 64 genera had 0.1% or higher alkaloid content (Lüning 1974). Most species belong to modern-day concepts of Dendrobieae, Malaxideae, and Aeridinae, particularly *Dendrobium*. Until 1977, 14 alkaloids of the dendrobine type had been isolated, including dendramine, dendrine, nobilonine, dendroxine, and dendrowardine (Slaytor 1977). Others not related to dendrobine include hygrine, dendroprimine, pierardine, dendroparine, and crepidine (Lüning 1974). Several pyrrolizidine-based alkaloids had been isolated from Malaxideae: malaxin, grandifoline, paludosine, hammarbine, laburnine, nervosine, and kuramerine, to name a few. Within Aeridinae, laburnine and its acetates have been isolated from *Vanda*, and phalaenopsine and cornucervine from *Phalaenopsis* (Lüning 1964, 1967, 1974; Slaytor 1977). Alkaloids known for all genera have been reported by Nigel Veitch and Renée Grayer in *Genera Orchidacearum* (Pridgeon et al. 1999, 2001, 2003, 2005, 2009, 2014). Ethnobotanical uses of orchid species with alkaloids are discussed below.

Floral fragrances

The roles that floral fragrances have played in orchid speciation and phylogeny cannot be overestimated, as Darwin (1862, 1877) was the first to show. Although fragrances have been known and appreciated for several millennia in China and Japan, research into fragrance compositions and their selective attraction of pollinators did not begin in earnest until the middle of the twentieth century, especially with the works of Vogel (1962) and Dodson and his colleagues (Dodson 1962a,b; Dodson and Hills 1966), culminating in the best general account of orchid pollination to that point by van der Pijl and Dodson (1966). In the next 20 years, qualitative and quantitative differences in orchid fragrances were determined by gas chromatography–mass spectrometry, as summarized by Williams and Whitten (1983), and some were tested in the field on euglossine bees (Dodson et al. 1969). The most recent comprehensive accounts of what is known about orchid floral fragrances are those by Williams (1982), Gerlach and Schill (1991), and Kaiser (1993).

Osmophores and their locations in orchid flowers were discussed by Darwin (1862), Vogel (1962), van der Pijl and Dodson (1966), Williams (1982), and many others. There have been few published anatomical studies of osmophores in Orchidaceae, among them Vogel (1962), Pridgeon and Stern (1983, 1985), Stern, Curry et al. (1987), Curry et al. (1991), Teixeira et al. (2004), and Ascensão et al. (2005).

ECONOMIC USES

Lawler's (1984) treatment of ethnobotany of Orchidaceae is the most comprehensive and authoritative on the subject to date. His account detailed, by species, the phenomenal number of uses that have been made of orchids worldwide: religious items, floral emblems, food (especially salep), flavourings (especially *Vanilla*), beverages, arts and crafts, adhesives, medicines, poisons and narcotics, and even aphrodisiacs. According to Lawler, salep is a starchy mucilage (see **Root tubers**) prepared in Europe and Asia by washing, peeling, and drying tubers of Orchideae, but tubers or roots of other species have been similarly used elsewhere. It is a component in ice cream, tonics, soups, and hot beverages but has also been used for various medicinal and aphrodisiac purposes.

The commercial species of *Vanilla* are *Vanilla planifolia* Jacks. ex Andrews from Central America, *V. pompona* Schiede from Central and South America and Guadeloupe, and *V. tahitensis* J.W. Moore from Tahiti and other Pacific islands; the most important in vanilla production is *V. planifolia*, but *V. pompona* can be grown under less favourable conditions and is resistant to *Fusarium* root rot (Soto Arenas 2003). The fruit, a dehiscent capsule or slightly dehiscent berry, is first 'killed' by any of several methods (sun, oven, hot water, scarification, freezing), then cured by alternately drying and sweating to yield vanillin (Childers et al. 1959; Lawler 1984).

Orchid commerce—trade in species (wild collected or artificially propagated) and hybrids—is a multibillion dollar industry that has multiplied many times over since the discovery and application of asymbiotic germination techniques by Knudson (1922). As the result of Knudson's pioneering work, the number of valid orchid species (25,000 +) is dwarfed by the number of registered hybrids, numbering approximately 144,000 as of April 2010 with 300–400 added every month by Julian Shaw, Orchid Registrar of the Royal Horticultural Society. Thousands more hybrids have never been registered.

TAXONOMIC HISTORY AND PHYLOGENETICS

Although Linnaeus may have been the first to publish binomials in orchids in a systematic way, 69 species in seven genera in the first edition (1753) and 85 in the second edition (1762–3) of *Species Plantarum*, systematics of Orchidaceae began in earnest with the classification by Swartz (1805), who provided a key to his 28 genera, observed that some orchids have one anther and others two, and divided the monandrous orchids into three groups on the basis of the position of the anther. Robert Brown divided the 'Monandrae' into four groups using characters related to the anther and habit in his *Prodromus* (1810) of the Australian flora; later (1813) he treated 46 genera and 115 species in the fifth volume of the second edition of Aiton's (1811–13) *Hortus Kewensis*. Louis Claude Richard (1817) stressed the structure of pollinia (sectile, granulose, solid) and the column in characterizing orchids; he also introduced many terms associated with the column that we still use today—rostellum, viscidium, and gynostemium.

John Lindley (1830–40) built on Swartz's system and examined thousands of specimens for his *Genera and Species of Orchidaceous Plants*, including 394 genera and an estimated 3000 species in seven tribes. From 1852 to 1859 he expanded coverage in *Folia Orchidacea* with an additional 42 genera and 1343 species, including major monographs of *Oncidium*, *Epidendrum*, and *Pleurothallis*. But at the tribal level, he, too, relied heavily on the structure and position of the anther. George Bentham (1881) recast Lindley's classification by adding the rank of subtribe but still focused on inflorescence and anther features, especially pollinia. Bentham and Hooker (1883) proposed a system of five tribes in *Genera Plantarum*: Epidendreae (nine subtribes), Vandeae (eight), Neottieae (six), Ophrydeae (four), and Cypripedieae, including *Apostasia* and *Neuwiedia* (zero). Pfitzer (1889) took the bold step of overlaying another level of characters—vegetative characters such as vernation, number of pseudobulb internodes, and growth habit—on the conventional column features and attempted to polarize them, primitive and advanced, across the family and was probably influenced by Darwin's (1859, 1862, 1877) works. Reichenbach (1854–1900) criticized Bentham and Pfitzer for deviating too much from details of the column in their systems and relied heavily on floral characters in describing species, transferring them to different genera, and erecting or synonymizing genera. As Bentham (1881) noted, however, Reichenbach never revealed his principles of classification nor a synopsis of contrasting characters.

Rudolf Schlechter (1926) borrowed from both Bentham and Pfitzer to create a synthetic classification covering 610 genera that influenced all succeeding systems until the molecular era. It combined the floral characters used by Bentham and Hooker as well as the vegetative features introduced by Pfitzer. Schlechter divided Orchidaceae into two subfamilies based on number of anthers: (1) Diandrae, with one tribe and one subtribe to accommodate the slipper orchids, and (2) Monandrae, broken down into Division Basitonae and Division Acrotonae. Basitonae included all those species with caudicles and a viscid disc arising from the base of the anther and comprised only tribe Ophrydoideae (two subtribes), whereas Acrotonae included species with the caudicles and viscid disc arising from the apex of the anther. Acrotonae were further divided into tribe Polychondreae with soft, granular pollinia and persistent anther (26 subtribes) and tribe Kerosphaereae with waxy pollinia and a deciduous anther (50 subtribes, split among series Acranthae with terminal inflorescences and series Pleuranthae with lateral inflorescences). Dressler and Dodson (1960) later supplemented Schlechter's classification and tweaked it to bring it into conformity with the *International Code of Botanical Nomenclature*, but retaining Schlechter's scheme of only two subfamilies (Cypripedioideae and Orchidoideae). To this mix Garay (1972) added yet more types of characters such as those related to floral vascularization, embryology, and seeds and proposed a system of five subfamilies, segregating Apostasioideae from Cypripedioideae and dividing Dressler and Dodson's Orchidoideae into Orchidoideae, Neottioideae, and Epidendroideae.

Dressler (1974) revised the classification of Dressler and Dodson (1960), adding Garay's subfamily Apostasioideae. By 1981, the number of his subfamilies had risen to six with the addition of Spiranthoideae, Epidendroideae, and Vandoideae. What set this classification apart from the previous ones was its incorporation of new characters such as subsidiary cell development. Nevertheless, it still stressed the flower, the column in particular. In 1993 Dressler added even more characters to his classification after publication of new data on seeds, stegmata, and endothecial thickenings. His reasoning was cladistic in nature but without formal cladistic analysis. Problems remained, exemplified by two subtribes and two other genera labelled 'misfits and leftovers'.

Only a year later, Chase, Cameron et al. (1994) published a cladogram based on *rbcL* sequence data for 33 taxa spread across the family. It provided the first truly phylogenetic evidence that *Vanilla* and *Pogonia* did not belong to Epidendroideae, Spiranthoideae (excluding *Tropidia*, *Diceratostele*, and others) were part of Orchidoideae, and Neottieae part of Epidendroideae. Like Pfitzer and Dressler before them, they stressed the importance of analysing vegetative features and taking into account data from many sources towards producing a natural classification.

Freudenstein and Rasmussen (1999) scored 71 morphological characters using Dressler's 1993 classification and excluded 10 other characters for a variety of reasons. Their results showed substantial structure but high levels of homoplasy in the non-vandoid epidendroids. The independence of vegetative characters from floral elements as evidence was clearly shown by the position of *Tropidia* in Epidendroideae instead of its traditional placement in Spiranthoideae.

With the advent of automated DNA sequencing and improved reagents, Cameron et al. (1999) dramatically expanded the number of taxa sampled to 171 in another *rbcL* study that showed the same relationship of *Tropidia* to Epidendroideae and offered support for a subfamily Vanilloideae. At the same time it revealed no support for Spiranthoideae, Neottioideae, or Vandoideae nor for relationships among the five subfamilies along the spine of the tree. Most subtribes were monophyletic, but not so the tribes. More markers were required to help resolve these problems.

Freudenstein and Chase (2001) investigated mitochondrial DNA in Orchidaceae, sequencing an intron in the gene for NADH dehydrogenase, which has a high proportion of length mutations. Inclusion of indels in an analysis of base substitutions yielded significantly more resolved trees than use of substitutions alone. In this study, the overall signal mirrored that of Cameron et al. (1999), which means that the mitochondrial genome shows the same historical pattern as the plastid genome.

Chase, Cameron et al. (2003) based their phylogenetic classification on published evidence from morphology and multiple markers, including a three-gene dataset for the family from Cameron (2001). Freudenstein et al. (2004) expanded the combined *rbcL*/*matK* data set to encompass 173 taxa, resolving some relationships that were only weakly supported in the Cameron et al. (1999) analyses and strengthening others from those earlier studies. In the short span of a decade, a robust, natural classification of the orchids had developed, one that could serve as a framework for answering questions about orchid phylogeny, circumscription of taxa at every rank, biogeography, and character trends.

Genera Orchidacearum (Pridgeon et al. 1999, 2001, 2003, 2005, 2009, 2014) is the first comprehensive attempt to monograph the world's 800 or so orchid genera using phylogenetics and especially molecular phylogenetics. It represents a synthesis of our knowledge of the nomenclature, distribution, anatomy, palynology, cytogenetics, phytochemistry, phylogenetics, ecology, pollination, uses, and cultivation for each genus. The system followed in *Genera Orchidacearum* comprises five major lineages based mostly on the classification of Chase, Cameron et al. (2003). At the base are the two genera of Apostasioideae from tropical Asia and northern Australia. *Apostasia* and *Neuwiedia* are terrestrial, have two or three anthers, and the lip is similar to the petals. Pollen grains are shed separately as monads and are powdery. Vanilloideae, distributed worldwide, are vines or herbs with tunicate or crustose or winged seeds, different from the dust seeds of more advanced subfamilies. Pollen from the single anther is usually shed as monads. Many species of *Vanilla* are economically important. The five genera of Cypripedioideae, the slipper orchids, are widespread from the temperate zones to the tropics. They are terrestrials, epiphytes, or lithophytes with usually showy flowers having a pouched or saccate lip and two anthers. Pollen is powdery or sticky. Orchidoideae are usually terrestrials with tubers or fleshy rhizomes, spirally arranged leaves, and two or four pollinia attached by caudicles to viscidia. Epidendroideae are by far the largest subfamily with about 18,000 species in 650 genera. Most occur in the tropics but they extend from the Arctic Circle to Argentina, Tasmania, and New Zealand. Most have fleshy stems or pseudobulbs, fleshy leaves, and pollen shed as pollinia, but naturally there are many exceptions to these. All have one anther.

ORIGINS AND AFFINITIES OF ORCHIDACEAE

Suggestions for relationships based on morphology have spanned Burmanniaceae and Corsiaceae (Cronquist 1981), *Alstroemeria* (Dressler 1986), Liliales (Dahlgren et al. 1985), and Hypoxidaceae (Hutchinson 1973; Dressler 1993). The first study to resolve Dressler's question of 'Whence the orchids?' was a parsimony analysis of *rbcL* sequences of 33 orchid taxa and 62 species of other monocots by Chase et al. (1994). That and succeeding molecular studies (Chase et al. 1995, 2000; Chase 2005; Seberg et al. 2012) showed that Orchidaceae are members of lower Asparagales and sister to all other families, including Hypoxidaceae and Iridaceae.

Dating the origins of the family has been made more difficult because until 2007 there were no unquestionable orchid fossils. Schmid and Schmid (1977) reviewed fossils formerly attributed to orchids, such as the genera *Palaeorchis* Massalongo and *Protorchis* Massalongo from the Eocene, *Orchidacites* Straus from the Pliocene, and others from the Oligocene, Miocene, and Quaternary. The case for being orchidaceous was weak in all cases, and so most commentators concluded that there was no satisfactory fossil record; as Chase et al. (1994, 2005) pointed out, many commentators mistakenly used absence of fossils as evidence that orchids were recently evolved. Garay (1972), however, argued that they arose during the Cretaceous in Malaysia, although no evidence for the claim except a biogeographical hypothesis was cited, and in fact at that time Malaysia had not yet risen from the seabed (Chase 2001). Ramírez et al. (2007) reported the first credible orchid fossil—a pollinarium of *Meliorchis caribea* gen. et sp. nov. attached to the mesoscutellum of an extinct stingless bee, *Proplebeia dominicana*, recovered from Miocene amber in the Dominican Republic—that is 15–20 million years (myr) old. By applying cladistic methods to a morphological character matrix, they placed *M. caribea* in the extant subtribe Goodyerinae of subfamily Orchidoideae. Then, using the ages of other fossil monocots and *M. caribea* to calibrate a molecular phylogenetic tree of the Orchidaceae, they showed that the most recent common ancestor of extant orchids dates from the Late Cretaceous (76–84 myr ago) and that representatives of all five subfamilies were present before the end of the Cretaceous (about 65 myr ago). The first unambiguous vegetative orchid fossils are leaves of two Early Miocene taxa (*Dendrobium winikaphyllum* Conran, Bannister & Lee sp. nov. and *Earina fouldenensis* Conran, Bannister & Lee sp. nov.) from New Zealand (Conran et al. 2009). Diagnostic characters were the distinctive, raised tetrato cyclocytic stomatal subsidiary cells of *Earina* and the glandular hairs and 'ringed' guard cells of *Dendrobium*.

Chase (2001) reasoned that, given the earliest accepted monocot fossils of Arecaceae are from about 90 myr ago, lineages such as Asparagales that branched off before the palms are at least that old and probably older. He estimated that the age of lower Asparagales including Orchidaceae is then about 110 myr ago, when orchids could already have been present on all continents without the need for long-distance dispersal. Those first orchids were probably sympodial with erect, unbranched, and non-twining shoots and having spiral, plicate (or broad) leaves, velamentous roots, axillary inflorescences, trimerous flowers with inferior ovaries, powdery pollen, dry capsules, and hard seed coats (Dressler 1993; Chase

2001). They were probably understorey plants in wet tropical forests, evergreen, and mycorrhizal (Chase 2001).

FLORAL ANATOMY

Although the focus of this series is vegetative anatomy, some mention should be made of the substantial body of work on orchid floral anatomy inasmuch as the phylogeny of Orchidaceae is intimately tied to the well-known (and often unique) attractants and mechanisms involved in pollination, especially deceit pollination. Ever since Richard (1817) and Brown (1833) published some of the first hypotheses on the structure of orchid flowers, questions related to homology and function of floral parts have driven further research. Darwin (1862) in particular devoted an entire monograph to the subject, and although we now know that some of his interpretations based on vasculature were incorrect, for example (following Brown) that the labellum is a compound structure comprising 'one petal with two petaloid stamens of the outer whorl', his overall contributions to the subject are immeasurable. Since that time, and given both chronological and topical overlap, there have been essentially three phases of floral anatomy studies in Orchidaceae: comparative, ontogenetic, and now molecular.

Kumar and Manilal (1993) provided a more comprehensive history of comparative and developmental anatomy, to which the reader should refer for further information.

Comparative anatomy

Early comparative studies of floral vasculature were published by Swamy (1948), who examined 40 species in 24 genera and distinguished three distinct types based on vascular supply to the staminate and stigmatic whorls rather than the perianth: Cypripedilinae (= Cypripedioideae), Ophrydinae (= Orchidoideae), and the rest of the monandrous orchids. He showed that, contrary to the assertion by Brown and Darwin, the labellum has the same vascular supply as the rest of the perianth, and the two lateral stamens of the outer whorl are instead represented in the column. Garay (1960) added Apostasioideae to the list and noted that, like Cypripedioideae, six vascular traces enter the floral axis with an additional trace giving rise to the midrib of the bract. Rao (1969 [1970], 1974) expanded Garay's (1960) observations on floral anatomy of Apostasioideae, stressing the phylogenetic link to Cypripedioideae and refuting Garay's (1972) claim that *Apostasia* and *Neuwiedia* are not closely related to one another. St-Arnaud and Barabé (1989) updated vascularization studies in *Cypripedium*, adding three species, and found that the three differ in organizational patterns, the more elaborate of which occur in both the smallest- and largest-flowered species. Kumar and Manilal (1988) continued anatomical studies on *Apostasia* and also described floral anatomy in *Satyrium* (1985), *Paphiopedilum* (1992 [1993]), and four genera of 'saprophytic' orchids (1992)—*Anoectochilus*, *Aphyllorchis*, *Galeola*, and *Epipogium*. Freudenstein (1991) compared secondary wall thickenings in the endothecium of the anther in 210 genera spanning Orchidaceae and found four basic types. Terrestrial genera typically have regularly arranged and well-developed thickenings, whereas epiphytic taxa generally have congested, irregular, and thinner thickenings.

Another significant area of comparative anatomy is that of secretory structures, especially nectaries (both extrafloral and floral), osmophores, and elaiophores. In a survey of extrafloral nectaries of angiosperms, Zimmerman (1932) described those of *Epidendrum cochleatum* L. (= *Prosthechea cochleata* (L.) W.E.Higgins) and *Catasetum inornatum* Schltr. (= *C. ochraceum* Lindl.). Jeffrey et al. (1970) pointed out possible sites of extrafloral nectaries in *Caularthron* and *Encyclia* and speculated on their role in well-known orchid–ant associations. Nectar composition always includes glucose, fructose, sucrose, and often other sugars such as raffinose and stachyose (Ernst and Rodriguez 1984). Stpiczyńska (1997) described the anatomy and ultrastructure of nectaries in *Platanthera*. Vieira et al. (2007) studied the nectary in *Corymborkis*, located in the basal lateral portions of the labellum. Starch granules present before anthesis in the secretory parenchyma and epidermis of the nectary are hydrolysed and used as the source of sugars for nectar, which accumulates between the cuticle and outer epidermal walls before moving into the nectar chamber. More recently Bell et al. (2009) correlated nectar production, presence and size of papillae, and epidermal cell striations in labellar spurs of several genera of Orchideae.

The pioneering work of Stefan Vogel (1962, 1974) in osmophores and elaiophores was so comprehensive that virtually everything else published to date on the subjects has been a footnote to it. Among Orchidaceae, he described the location and anatomy of osmophores in *Stanhopea*, *Bulbophyllum*, *Dendrobium*, *Oncidium*, and *Phragmipedium*, several genera of Pleurothallidinae, Catasetinae, Orchideae, and Chloraeinae (Vogel 1962). Ontogenetic details of osmophores at the ultrastructural level were later published for *Restrepia* (Pridgeon and Stern 1983), *Scaphosepalum* (Pridgeon and Stern 1985), and *Stanhopea* (Stern, Curry et al. 1987). Ascensão et al. (2005) determined that the secretory cells on the labellum of *Ophrys* are papillae on the margins and distal region of the abaxial surface of the labellum, confirming Vogel's (1962) observations. Vogel (1974) also described the elaiophores in *Zygostates*, *Sigmatostalix*, and other Oncidiinae. Davies and Stpiczyńska (2009) added *Rudolfiella* (Maxillariinae) to this list.

Unusual, paired, secretory glands occur on the column of *Coryanthes*, producing an aqueous fluid that falls into the bucket-shaped epichile of the labellum and immobilizes euglossine bees so that they cannot fly out, forcing them to crawl past the anther and stigmatic region. Schnepf et al. (1983) investigated the ultrastructure of these glands and analysed the fluid, which contains low amounts of ions and sugars, a mucilage, and unknown compounds that account for the odour and taste. The authors characterized the glands as 'epithelial hydathodes'.

Ontogeny

Kurzweil (1998 [1999]) provided a thorough account of nineteenth-century studies in floral development, chiefly those related to European orchids, by Irmisch (1853), Payer (1857), Pfitzer (1888), Capeder (1898), and later on by Hirmer (1920). The modern era of ontogenetic studies of orchid flowers continued with the works by Rotor and MacDaniels (1951) on bud initiation and development in *Cattleya* and Vermeulen (1959) on the stigmas and homology of the rostellum of Orchideae and other groups.

However, it was not until the 1980s and 1990s that developmental work began in earnest and proliferated, chiefly in the laboratories of Finn Rasmussen, Hubert Kurzweil, Edward Yeung, and John Freudenstein. In a thorough light microscopy study of column (gynostemium) ontogeny, Rasmussen (1982) examined 62 genera of Neottieae, Diurideae, Tropidieae, Cranichideae, etc., in several of which he described a new type of pollinium stalk, the hamulus, which represents the recurved apex of the rostellum. This is distinguished from the tegula, which consists of the modified rostellar epidermis and sometimes one or more subjacent cell layers. Both of these types of stipes (pl. stipites) are derived from the median carpel, whereas the other kind of pollinium stalk, the caudicula or caudicle, is derived from anther tissue. The hamulus also occurs in other genera such as *Bulbophyllum* (Rasmussen 1985).

Kurzweil produced a series of four scanning electron microscopy (SEM) studies of column ontogeny, first in taxa of Epidendroideae (1987a), then Orchideae and Diurideae (1987b), Cranichideae, Tropidieae, and Neottieae (1988), and finally Cypripedioideae (1993), confirming among other observations that the rostellum is derived from the median carpel. He then turned his attention to floral morphology and ontogeny of Disinae (Kurzweil 1990) and Satyriinae (Kurzweil 1996), use of morphological and ontogenetic characters in phylogenetic studies of Diseae (Linder and Kurzweil 1990, 1996), and more recently to gynostemium morphology and ontogeny in Australian members of Diurideae and Cranichideae (Kurzweil et al. 2005). He summarized a decade's worth of observations and their phylogenetic implications in a review of floral ontogeny in orchids (Kurzweil 1998). Extending Kurzweil's (1987b) SEM study of Orchideae to Chinese representatives of Orchidinae, Luo and Chen (2000) suggested that the separation of the lateral lobes of the rostellum in these taxa as well as in *Brachycorythis* is the ancestral state, whereas the conjoining of the lobes in *Dactylorhiza* and *Orchis* is a derived character.

Yeung and co-workers utilized both light and electron microscopy to follow the ontogeny of the anther and stigma of *Epidendrum ibaguense* Kunth. Blackman and Yeung (1983) found that the caudicle begins as a mass of cells in the microsporangium, which then divide to form linear tetrads that later lay down thick secondary walls on their outer surfaces. Thin-walled cells between adjacent caudicles and between caudicle and pollinium produce a lipidic polymer and then undergo autolysis, releasing the lipidic compound that constitutes the elastic connections between adjacent caudicles and caudicles and pollinia. Yeung (1987a) found that a similar process occurs in the formation of the glue-like substance on the rostellum. Epidermal and subepidermal cells of the rostellum differentiate as the viscidium. A lipidic secretion is synthesized in the viscidium, and, as more is produced, its cells separate and then degenerate, leaving a single mass of 'glue'. Yeung (1987b) reviewed the development of pollen and accessory structures in orchids and also described (1988) the development of the stigma in *E. ibaguense*.

Two later papers also dealt with the development of stigma and transmitting tissues. Clifford and Owens (1990) described the development of the stigma and pollen-tube transmitting tract in the column and ovary of *Lemboglossum* (= *Rhychostele*). They recorded a complex cuticular layer over the stigmatic secretory zone similar to the lipidic layer (cuticle?) overlying the stigmatic surface in *Epidendrum ibaguense* (Yeung 1988). Leitão and Cortelazzo (2010) also observed a cuticle extending over the internal surface of the central canal in the transmitting tissue in *Rodriguezia*. Another similarity between these two genera of Oncidiinae is the lipidic secretion in the intercellular spaces of the stigmatic and stylar transmitting tissues, producing a 'fenestrated' or 'reticulate' appearance.

Freudenstein's developmental work on the anther answered several lingering phylogenetic questions and added several characters of systematic value. One of these questions was the origin of pollinium number. Freudenstein and Rasmussen (1996) found that—from a single, homogeneous region of meristematic tissue in each theca—two, four, or eight pollinia may result from septation or lack thereof. Without septation, two pollinia result. Partial septation yields lobed or hollow pollinia. Four pollinia, the most common condition, result from a dorsiventral, longitudinal septation in each theca (but may fuse later in some Orchidoideae). Exclusively longitudinal or both longitudinal and transverse septation produces eight pollinia.

Anther orientation (erect, reflexed, or incumbent) has long been used as a systematic character. Epidendroideae and Vanilloideae are characterized by an incumbent (bent forwards or inflexed) anther, but Freudenstein et al. (2002) showed that anther bending in Vanilloideae is caused by expansion of the connective, whereas in Epidendroideae it results either from column elongation and tipping of the mature anther (non-vandoid orchids) or from an early redirection of growth of the anther itself (Vandeae), supporting earlier observations by Kurzweil (1987a) about early bending in the latter.

Another contribution to ontogenetic study of the anther was that by Wolter and Schill (1986), treating the pollen, massulae, and pollinia of 15 species of Cypripedioideae, Orchidoideae, and Epidendroideae. Among their conclusions were that the taxa differ in presence of the endothecium, form and number of nuclei in tapetal cells, and arrangement of pollen mother cells at the onset of exine formation. The evolution of compact pollinia and absence of an endexine were said to be related to neotenic processes.

Finally, Kocyan and Endress (2001) investigated the floral development of *Apostasia* and *Neuwiedia*, the basal-most members of Orchidaceae. Unlike monandrous orchids, the lateral petals rather than the median petal (labellum) develop first. Although stamens form on only the abaxial side of the flower, vestiges of stamen primordia occur on the adaxial side. In both genera, stamens and style are basally fused.

Evolutionary developmental ('evo–devo') approaches

Floral organogenesis is now understood in terms of the A–B–C–D–E model, according to which the set of A + E genes code for formation of sepals, A + B + E genes for petals, B + C + E genes for stamens, C + E genes for carpels, and C + D + E genes for ovules. Most A–B–C–D–E homeotic genes are members of a transcription factor family that has a characteristic conserved DNA binding domain structure, termed the MADS-box (Erbar 2007). The function of these transcription factors is to activate or repress gene expression during floral development. Hsiao et al. (2011) provided an introductory look at the status of genomics, transformation technology, flowering regulation, molecular regulatory mechanisms of floral development, scent production, and colour presentation in Orchidaceae.

Xu et al. (2006) were among the first to isolate possible A, B, C, D, and E function genes in Orchidaceae using *Dendrobium crumenatum* Sw., including *AP2-*, *PI/GLO-*, *AP3/DEF-*, *AG-*, and *SEP*-like genes. They suggested that, although there is partial conservation in the B and C function genes between *Arabidopsis* and orchids, gene duplication could have led to divergence in gene expression and regulation, followed by functional divergence, to explain floral ontogeny in orchids.

Several research teams have investigated the genes involved in the transition from the vegetative phase to reproductive phase. Chen et al. (2007) isolated two *Phalaenopsis* MADS-box genes designated *ORAP11* and *ORAP13* from the *SQUA* family, both of which are strongly expressed in the flower bud and also occur in vegetative organs. Both genes occur in the apices and margins of developing petals, labellum, column, and ovule. They concluded that both genes are partially redundant and function in the process of floral transition and morphological architecture. Another set of floral meristem identifying genes are the *FLORICAULA/LEAFY* (*FLO/LFY*) homologues that are key players in floral development. Zhang et al. (2010) cloned a *FLO/LFY* homologue gene, *PhalLFY*, from a *Phalaenopsis* hybrid and analysed its expression pattern during vegetative and reproductive development. They determined that transcripts of *PhalLFY* accumulated at higher levels in the stem during the transition from vegetative to reproductive growth and especially in inflorescence and floral meristems, floral primordia, and all floral organs. Yu and Goh (2000) likewise isolated three MADS-box genes—*DOMADS1*, *DOMADS2*, and *DOMADS3*—from a *Dendrobium* hybrid and traced their expression in floral transition.

Expression of perianth development has been the subject of many studies involving B-class MADS-box genes, which as noted above play a role in petal and stamen development. Pan et al. (2011) identified 24 *APETALA* (*AP3*)-like genes and 13 *PISTILLA* (*PI*)-like genes from 11 orchid species and sorted them into four *AP3*- and two *PI*-duplicated homologues. They were able to detect several duplication events in orchid phylogeny. The *AP3* paralogous genes were expressed throughout inflorescence and floral bud development, whereas the

AP3 orthologous genes had diverse expression patterns in the various orchid species. Chang et al. (2010) investigated sepal/petal/labellum formation in *Oncidium* Gower Ramsey and characterized three paleo-*APETALA3* genes and one *PISTILLATA* gene. They found (among other conclusions) that determination of the final organ identity for the sepal/petal/labellum depended on the presence of *O.* Gower Ramsey MADS-box *gene5*.

Understanding the molecular mechanisms for floral ontogeny and identity offers the best hope for unravelling the evolution of the orchid flower and its manifold responses to pollinator pressures, ranging from scent production and mobile labella in hundreds of species to the unique columnar glands in *Coryanthes*.

SYNOPSIS OF VEGETATIVE ANATOMICAL LITERATURE

Historical and general works

The earliest known report on vegetative anatomy of Orchidaceae is probably that of the outer layer(s) of roots of *Epidendrum* by Heinrich Friedrich Link (1824): 'Interdum (in Epidendris) epidermis seu externum parenchymatis stratum albescit, ut credas apicem emergere ex interiore radicula, sed distincte vidi parenchyma tantum exsuccum evadere et inde albescere'. Although he noticed that the tissue turned white as it dried out, he was unable to ascribe any origin or function to it. Schleiden (1849) later named this outer layer the *velamen radicum* and speculated that it absorbed water vapour and other gases by virtue of its condensing ability and porosity. Secondary thickenings on the walls of velamen cells were first described by Meyen (1837), but in an overview of the general organography of terrestrial and epiphytic orchids, Link (1849–50) elaborated on these secondary thickenings and made several other significant contributions. One of these was to describe and illustrate stegmata in Orchidaceae for the first time. However, he interpreted them as projections of sclerenchyma rather than as cells lining sclerenchyma; Mettenius (1864) corrected the error and referred to them as *Deckzellen*, according to Møller and Rasmussen (1984). The second was to report multiple discrete steles in tubers of *Orchis*, each bounded by an endodermis in a common cortex; these were later misinterpreted as 'polysteles' (White 1907; Arber 1925; Sasikumar 1975b) rather than as meristeles, or individual bundles of a dictyostele. He also described the vascularization of buds of *Orchis* and columns of *Orchis sambucina* (= *Dactylorhiza sambucina* (L.) Soó) and *Maxillaria cruenta* (= *Lycaste cruenta* (Lindl.) Lindl.).

Until the end of the nineteenth century, the focus of anatomical studies was directed towards interpretation of homologies of tissues and underground organs as well as the origin and nature of velamen. Extensive debate over the nature of the tubers of subfamily Orchidoideae was attributable to their mode of development, originating as axillary shoots that produce basal extensions resembling adventitious roots (Bell 1991; Rasmussen 1995). De Candolle (1827) regarded the organs as swollen roots and named them 'tubercules' but did not examine the origin of the bud of the tuberous root. Irmisch (1850) argued that it is morphologically (and functionally) a root tuber, the product of a close connection between a shoot and the coalescence of several roots. He observed that the rhizoderm cells had spiral bands or fibres like those occurring in aerial roots of tropical orchids (i.e. velamen). However, Schleiden (1845–6) and Fabre (1856) considered the organ as simply a swollen shoot formed laterally from an axillary bud. Prillieux (1865) adopted the hypothesis of Irmisch (1850) with some modifications, namely that the tubercule is an adventitious root with a piliferous layer (velamen) externally and sheathed by a coleorhiza at its base. However, he disagreed that the tubercule represents the concrescence of several roots by pointing to the mode of formation of the divided tubercules of some *Orchis* and *Gymnadenia* (Orchidinae) species.

The issue was finally settled with Moreau's (1913) study of African Orchideae and Diseae and Arber's (1925) work on monocot morphology. Moreau argued that root tubers are formed by either the concrescence of several roots or the hypertrophy of cortical parenchyma in a single root, depending on the taxon; the root nature of the tuber is immediately recognizable by the presence of a piliferous layer (velamen) and particularly by the disposition of xylem and phloem. Arber corrected Moreau's concrescence theory by citing White's (1907) observation that tubers of *Habenaria* were monostelic at the base but 'polystelic' (= dictyostelic) at the apex. She also cited other monocotyledonous genera with root tubers, e.g. *Triglochin* L. (Juncaginaceae), *Roscoea* Sm. (Zingiberaceae), *Hemerocallis* L. (Xanthorrhoeaceae), and *Asphodelus* L. (Xanthorrhoeaceae).

Dietz (1930) surveyed underground organs of terrestrial orchids, including forms with root tubers, and mentioning in this context *Cynorkis* and *Habenaria* (Orchideae), *Thelymitra* (Diurideae), and *Pterostylis* (Cranichideae). He distinguished *Cynorkis* with several steles from *Thelymitra* with a single one, reported velamina with suberized walls in root tubers of *Thelymitra* and *Pterostylis*, and recorded the presence of fungi and the stolon in *Pterostylis*. He noted that the 'stolon' originated from the main axis, grew 20–30 cm distant from the parent plant, and developed small tubers. Along these lines Burgeff (1932) described and illustrated the development of the stolon of *Corybas* (Diurideae) and the origin of leafy buds and root tubers from stolon apices.

Later ontogenetic studies of the orchidoid stolon, more appropriately termed the 'dropper', 'sinker', or 'stolonoid root', were undertaken by Ogura (1953) and Kumazawa (1956, 1958). Ogura (1953) studied Japanese Orchideae with different types of root systems. The stoloniferous type, represented by *Platanthera japonica* Lindl., is characterized by a horizontal adventitious root with a subterminal bud that develops perpendicularly into a stem the following year. The fusiform type, as in *Platanthera nipponica* Makino (= *P. tipuloides* (L.f.) Lindl.), consists of a thickened, fusiform root growing downwards and bearing a bud at its base. The handform type (*Gymnadenia conopsea* R.Br.) is similar to the fusiform type except that the apices of the tubers bifurcate two or three times. The tuberous type (*Orchis pauciflora* Ten.) is similar in organization to the fusiform type but spherical in shape. The last type, tuberous with a stalk (*Pecteilis radiata* (Thunb.) Raf.), has spherical tubers on a distinct stalk arising from the stem base. Kumazawa (1956) investigated the development of the sinker in *Pecteilis radiata*, which originates as a lateral bud in the axils of the fourth and fifth leaves of a monopodium and comprises three distinct regions: (1) stalk, which is cauline in organization; (2) neck (between the stalk and tuber), at the base of which is a cavity bearing the terminal bud; and (3) tuber, a root with root hairs on the apical rhizoderm. Kumazawa (1958) later studied the development of the sinker in Ogura's stoloniferous type and found that, though externally root-like, it bears a subterminal bud that will become the aerial shoot the following year.

That the tubers of Orchidoideae are indeed root tubers rather than 'root–stem tuberoids' (Dressler 1981, 1993) has found support in studies on Diseae by Kurzweil et al. (1995), Diurideae by Pridgeon and Chase (1995), and Orchideae by Stern (1997a,b); in all three tribes, mature tubers have a velamen and exodermis, an endodermis with casparian strips, and a stelar configuration of alternating xylem and phloem.

Whether velamen was epidermal or extraepidermal was argued by leading anatomists soon after its discovery by Link (1824). Meyen (1837), Schleiden (1849), Chatin (1856), and Fockens (1857) viewed it as a sheath over the 'epidermis' (what we now call the exodermis), which had thick-walled long cells and thin-walled short cells; indeed, the Latinized name that Schleiden gave to it, *velamen radicum*, means 'root covering'. Chatin (1856) stated that the epidermis, the *membrane epidermoidale* (= exodermis), corresponds to the epidermis of subterranean roots but lacks stomata; above the *membrane epidermoidale* was the *enveloppe* or *peau spongieuse* (= velamen) that arises from it. In an examination of aerial roots of Orchidaceae, Liliaceae, Araceae, and Commelinaceae, Fockens (1857) wrote that he (as Schleiden before him) had indeed observed stomata in the 'epidermis'. On the other hand, Schacht (1856) and Oudemans (1861) believed that the outermost layer of velamen is the true epidermis because root hairs arise from it, and what the others were calling an epidermis was in fact a cortical layer. Oudemans (1861) termed it an 'endodermis' with alternating long and short cells, the long cells thick-walled and empty and the short cells thin-walled and nucleate. Leitgeb (1864b) undertook ontogenetic studies to resolve the dispute and correctly determined that the velamen is a multiple epidermis derived from periclinal divisions of dermatogen, later confirmed in developmental studies by Engard (1944) and Zankowski et al. (1987) among many others. In that same benchmark, influential monograph Leitgeb surveyed the root anatomy of 69 species, comparing the number of layers of velamen, secondary wall thickenings, pores, and wall thickness. The term 'endodermis' (referring to the exodermis) continued in use until the turn of the century, shortly after which Siebe (1903) and Sprenger (1904) used 'exodermis' explicitly in anatomical descriptions of Apostasioideae and Bulbophyllinae, respectively. Detailed information on literature related to developmental anatomy and functions of the velamen/exodermis complex can be found in Pridgeon (1987).

Chatin (1856) seems to have been the first to report (*passim*) the presence of *grosses saillies papilleuses* arising from the *membrane epidermoidale* of roots of *Bulbophyllum careyanum* (Hook.) Spreng. Oudemans (1861) mentioned similar, brownish *Körperchen* associated with short cells of the 'endodermis' (= exodermis) in *Sobralia liliastrum* Lindl. but, like Chatin, offered no insights into origins or function. Leitgeb (1864a,b) provided the earliest detailed descriptions of *kugelformige Körper* or spherical bodies. In the 1864(a) work he characterized them as brownish networks formed of many branching and intersecting fibres. He speculated there that, because of their porosity and association with the thin-walled cells of the 'endodermis', they function either in absorption or in condensation of water vapour from the atmosphere. In 1864(b) Leitgeb altered his ideas on their function and supposed that they form a *Wasserreservoire* and also prevent desiccation from subjacent tissues with thin-walled cells. In a detailed study of root anatomy of about 70 orchid species, Meinecke (1894) noted the presence of *Stabkörper* (rod bodies) or *Faserkörper* (fibre bodies) in many genera and determined through histochemical assays that they are lignified and suberized. Richter

(1901) referred to them as *Corpuscula Leitgebiana* in Leitgeb's honour. These outgrowths (see **Roots**) are now generally known as tilosomes (Greek *tilos*, fibre + *soma*, body).

Between about 1880 and the outset of the First World War German orchid research focused on anatomical surveys and monographs of different taxa, notably at the University of Heidelberg under the supervision of Professor Ernst Hugo Heinrich Pfitzer (1846–1906). If Leitgeb's study was the *sine qua non* for orchid root anatomy of the era, then the anatomical survey of orchid leaves by Martin Möbius (1887), dedicated to Pfitzer, was its counterpart. Möbius surveyed 193 species in 95 genera and interpreted results in the context of Pfitzer's (1889) system of classification that weighted vegetative morphology. He was one of the first to correlate structure and tissue types with ecological and climatic factors, later continued by Tominski (1905), one of his students. Leitgeb's research on root anatomy was expanded by Fellerer (1892) and especially by Meinecke (1894), who surveyed 70 species and reported number of velamen layers, hairs, wall thickenings, pores, histology, and exodermal secondary wall-thickening patterns. The other major surveys overseen by Pfitzer were those by Weltz (1897) and Hering (1900). Weltz investigated the stems, rhizomes, and/or *Luftknollen* or pseudobulbs of 130 species of monandrous sympodial orchids and found few groups with consistent, systematically useful characters. This observation still holds today, and, compared with the leaf and root, the axis is the least informative organ in orchid studies. Hering's anatomical survey of stems and inflorescence axes of monopodial orchids complemented Weltz's monograph but covered only 50 species in 17 genera. After the turn of the century Pfitzer's students produced anatomical monographs of various sympodial taxa: Siebe (1903) on *Apostasia* and *Neuwiedia*, Zörnig (1903) on Coelogyninae, Faber (1904) on *Cypripedium* and *Paphiopedilum*, Hünecke (1904) on Pleurothallidinae, and Sprenger (1904) on Bulbophyllinae. Of these, Hünecke and Zörnig were the only ones who educed systematic patterns among the observations and constructed artificial keys to the species examined.

With Pfitzer's death in 1906, the onset of the First World War, and social and economic upheavals in Germany afterwards, the impetus for anatomical studies gave way to more basic priorities. In the 1920s, however, a few works of note were published. Staudermann (1924) surveyed the hairs of monocots, among them the trichomes of genera such as *Cypripedium*, *Epipactis*, *Glossodia*, and *Renanthera*. Fuchs and Ziegenspeck (1925, 1926, 1927a–c) published exhaustive (and exhausting) comparative and ontogenetic studies of European orchids. The last major anatomical work to appear prior to the Second World War was that by Soloreder and Meyer (1930), an authoritative, illustrated synthesis of all known comparative orchid studies to that time along with original observations for almost 150 genera. Taking into account that the taxonomy has changed substantially in over 80 years, misidentifications along the way were probable, and, as vouchers were not cited nor presumably made, most of the descriptions are still valid at least to the genus level.

In the last half of the twentieth century, several family-wide surveys pinpointed systematically useful anatomical characters associated with subsidiary cells (Rosso 1966; Williams 1979; Rasmussen 1981a,b; Stern et al. 1993b), silica bodies (Møller and Rasmussen 1984; Rasmussen 1986), velamen (Pridgeon, Stern et al. 1983; Pridgeon 1987; Porembski and Barthlott 1988), and spiranthosomes (Stern, Morris et al. 1993c). Parsimony-informative vegetative characters in the phylogenetic analysis by Freudenstein and Rasmussen (1999) included presence of root tubers and velamen, type of secondary thickening in velamen, exodermal cell shape and thickening, presence of spiranthosomes, anticlinal walls of the abaxial leaf epidermis (straight or sinuous), and presence of subsidiary cells. Stern and Carlsward (2004) expanded the list of useful characters after their multiple anatomical studies of orchid taxa: (1) presence and distribution of fibre bundles and sclerenchyma in leaves and stems; (2) configuration of mesophyll; (3) sclerenchymatous or parenchymatous embedment of vascular tissue in roots; (4) wall thickenings of endodermal and pericycle cells; (5) shape and cell type of pith in roots; (6) secondary cell-wall banding in idioblasts of mesophyll and also the cortex and ground tissue of stems; (7) presence and distribution of water-storage cells in mesophyll and stem cortex; and (8) presence, position, and form of hairs on leaves and stems. More details are given below by organ.

Leaves

As mentioned above, Möbius (1887) surveyed leaves of 193 species in 95 genera for systematic purposes, mainly to evaluate the various classifications of the day such as Bentham and Hooker (1883) and Pfitzer (1889) in terms of naturalness. He also wanted to distinguish those characters that were heritable from those influenced by the climate or habitat. Perhaps not surprisingly, Möbius found closer agreement between his anatomical results and Pfitzer's system than Bentham and Hooker's classification, arguing that the morphology of the whole plant and not simply floral structure of Orchidaceae should be the basis of a natural classification. Solereder (in Solereder and Meyer 1930) studied the leaf anatomy of 73 species, mostly epiphytes, the results of which were included in their literature survey.

Hairs

Leaf hairs may be glandular or eglandular, unicellular or multicellular, superficial or sunken. The most common trichome type in Orchidaceae is the sunken hair of up to five cells, with or without a glandular apical cell (Fig. 1A,B). Pridgeon (1981a) detailed the ontogeny of glandular hairs in Pleurothallidinae, beginning with a periclinal division of a protodermal cell to produce a thin-walled, globose, apical cell with a relatively large nucleus and a subapical stalk cell with heavily cutinized lateral walls. A second periclinal division in some species results in a third, small, basal cell with thick lateral walls but thin transverse wall. During development the trichome assumes a sunken position because of continued anticlinal divisions of protoderm. The wall of the apical cell eventually ruptures, leaving a depression on the surface covered by a brown, opaque residue. After vascular tissue differentiation, two or more pitted foot cells develop at the base of the trichome (Fig. 1B). Benzing and Pridgeon (1983) showed that the hairs do not function as significant absorption agents and may instead secrete mucilage, perhaps facilitating unfolding of the lamina.

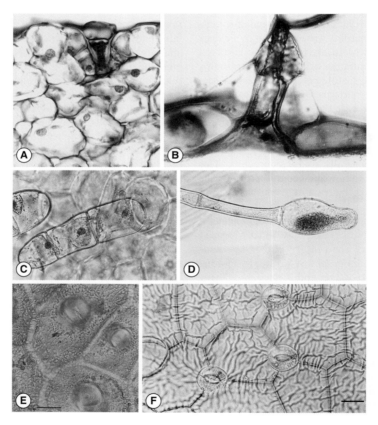

Fig. 1. Leaf trichomes and cuticle. (A) *Anathallis polygonoides*, TS immature lamina with adaxial glandular trichome, ×1900; (B) *Trichosalpinx semilunata*, TS mature lamina showing stalk cell and pitted foot cell of glandular trichome (apical cell no longer present), ×900; (C) *Cypripedium margaritaceum*, adaxial surface with multicellular eglandular trichomes, ×900; (D) *Elythranthera emarginata*, adaxial surface with multicellular glandular trichome, ×400; (E) *Drakaea glyptodon*, adaxial surface showing papillae and punctate cuticle. Scale bar = 20 μm; (F) *Arthrochilus dockrillii*, adaxial surface with anomocytic stomata and striate–anastomosing cuticle. Scale bar = 20 μm. (A) Reprinted from Pridgeon (1981a); (E,F) reprinted from Pridgeon (1994a); (B,D) photographs by A.M. Pridgeon.

Similar hairs occur in most other taxa of Epidendroideae, including Coelogyninae, Epidendreae, Zygopetalinae, Podochileae (especially Eriinae), Sobralieae, Collabieae, Triphoreae, Diceratosteleae, Tropidieae, Dendrobieae (including *Bulbophyllum*), Catasetinae, Coeliopsidinae, Cymbidiinae, Cyrtopodiinae, Eulophiinae, Maxillariinae, Stanhopeinae, Aeridinae, and Angraecinae. Conran et al. (2009) adduced as evidence (in part) the presence of these hairs in early Miocene fossil leaves to describe *Dendrobium winikaphyllum* Conran, Bannister & Lee. Hairs of any type have not been reported for Oncidiinae (Stern and Carlsward 2006) and Vanilloideae (Stern and Judd 1999, 2000).

Superficial hairs occur—but rarely—in both terrestrial and epiphytic taxa. Stern, Morris et al. (1993c) illustrated uniseriate hairs of up to 10 cells with a bulbous apical cell in *Goodyera* (Cranichideae). Unicellular but elongate and twisted hairs occur in section *Formosae* of *Dendrobium* (Dendrobieae; Yukawa et al. 1990 [1991]). Diseae leaves are mostly glabrous, but both glandular and eglandular hairs occur in *Disa*, and margins of leaves may be ciliate in Satyriinae and Coryciinae (Kurzweil et al. 1995). Bicellular trichomes occur over both leaf surfaces of *Calypso*, *Govenia*, and *Tipularia* (Calypsoeae; Stern and Carlsward 2008). In Cypripedioideae, glandular or eglandular hairs of 2–7 cells are distributed over both leaf surfaces of *Selenipedium* and most *Cypripedium* species (Fig. 1C), (Solereder and Meyer 1930; Rosso 1966; Pridgeon 1999). Hairs in Diurideae are superficial or sunken, uniseriate, multicellular, unbranched, and glandular or eglandular (Fig. 1D); *Aporostylis* is unique among Caladeniinae in having bi- or triseriate glandular trichomes (Pridgeon 1993, 1994a). Among epiphytic taxa, Solereder and Meyer (1930) described and illustrated the superficial, branched trichomes of the epiphytic *Chysis* (Chysinae), and Pridgeon (1982a) illustrated ephemeral, mucilage trichomes on the developing lamina of some Pleurothallidinae.

Epidermal papillae are known to occur in *Ceratandra* and *Disperis* (Diseae; Chesselet 1989), *Dresslerella* (Fig. 2B; Pleurothallidinae; Pridgeon and Williams 1979), *Goodyera* (Cranichideae; Staudermann 1924), *Caleana*, *Drakaea* (Fig. 1E), *Eriochilus*, and *Lyperanthus* (Caladeniinae; Pridgeon 1994b), *Acampe* (Solereder and Meyer 1930), and *Paphiopedilum* and *Phragmipedium* (Cypripedioideae; Atwood and Williams 1979) among other taxa. In describing the adaxial epidermis of the last two genera, Atwood and Williams (1979) distinguished macropapillae (single large protuberances, one per cell) from micropapillae (20 or more small projections on a single cell) and were able to find remarkable consistency in sculpturing pattern among different specimens of the same species but different patterns among species. Micropapillae were observed only in *Paphiopedilum* subgenus *Barbata* and in three of the six species of *Phragmipedium* examined. Yukawa et al. (1990 [1991]) also found them in several sections of *Dendrobium*.

Cuticle

Cuticle is smooth and conformal in most species but often ornamented in systematically useful patterns, especially in terrestrial taxa. For example, it is punctate or striate in many Diurideae (Fig. 1F; Pridgeon 1994b), punctate, striate, or reticulate in Disinae (Chesselet 1989), rugose to rugulose in many Cranichideae (Stern, Morris et al. 1993c), and striate in many sections of *Dendrobium* (Dendrobieae; Yukawa et al. 1990 [1991]). Among epiphytes, it is papillose in many Pleurothallidinae (Pridgeon 1982a) and *Isochilus* (Laeliinae; Stern and Carlsward 2009), striate in some Maxillarieae (Holtzmeier et al. 1998), granular–warty, striate, or pseudopitted in many Laeliinae (Solereder and Meyer 1930; Baker 1972), and papillose to rugose in some Pleurothallidinae (Pridgeon 1982a) and Oncidiinae (Williams 1974; Sandoval-Zapotitla and Terrazas 2001). Sprenger (1904) reported cuticular ridges in some species of *Bulbophyllum*, occasionally forming a network. Thickness varies from less than 1 µm in many terrestrial species to 25 µm in some Oncidiinae (Stern and Carlsward 2006), 27 µm in *Cymbidium finlaysonianum* Lindl. (Yukawa and Stern 2002), and an astonishing 50–60 µm in some Laeliinae (Stern and Carlsward 2009).

Epidermis

Epidermal cells are generally rectangular or polygonal and isodiametric in surface view with straight or curvilinear anticlinal walls. Sinuous (undulate) anticlinal walls have been reported, among others, in *Cypripedium* (Rosso 1966; Atwood and Williams 1979; Atwood 1984), Diseae (Chesselet 1989), Diurideae (Pridgeon 1994b), and Vanilloideae (Stern and Judd 2000). Cell walls are uniformly thickened or else the outer wall is thicker. Rarely the inner wall is thickest (e.g. *Vanilla planifolia* Jacks. ex Andrews; Stern and Judd 1999).

Adaxial cells are usually larger than abaxial cells (Fig. 2D); in extreme cases (terrestrials such as *Paphiopedilum*, *Habenaria*, *Satyrium*), the palisade-like adaxial epidermis occupies one-half to two-thirds of leaf volume (Solereder and Meyer 1930; Rosso 1966; Atwood 1984; Stern 1997a). Such enlarged cells probably function in water storage (Möbius 1887; Metzler 1924; Chesselet 1989), although no experimental evidence has ever been offered to support the claim. Larger adaxial cells above the midvein in many taxa serve as lamina expansion cells.

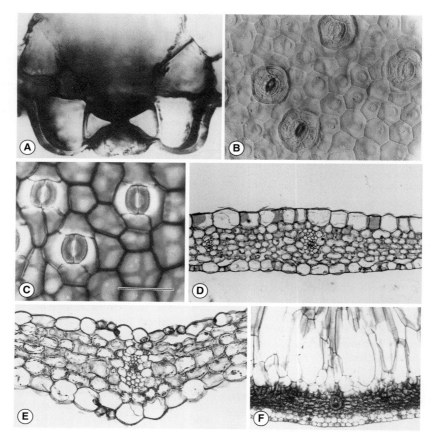

Fig. 2. Leaf. (A) *Dresslerella powellii*, leaf TS showing abaxial stoma with two subsidiary cells to either side of the guard-cell pair. Note elevation of stomatal apparatus above epidermis, ×1250; (B) *Dresslerella hispida*, abaxial epidermis with papillae and cyclocytic stomata, ×575; (C) *Ansellia africana*, leaf scraping showing tetracytic stomata. Scale bar = 50 µm; (D) *Cypripedium bardolphianum*, leaf TS depicting homogeneous mesophyll, hypostomaty, and relatively large adaxial epidermal cells, ×380; (E) *Chiloglottis reflexa*, leaf TS illustrating amphistomaty and homogeneous mesophyll, ×250; (F) *Specklinia costaricensis*, leaf TS with extensive adaxial hypodermis and 1-layered abaxial hypodermis, ×180. (A) Reprinted from Pridgeon (1982a); (C) reprinted from Stern and Judd (2002); (B,D–F) photographs by A.M. Pridgeon.

Stomata: guard cells are almost always flush with the outer surface of the epidermis, but they can be truly sunken as in *Ponera striata* Lindl. (Solereder and Meyer 1930; Baker 1972) or elevated above the surface by as much as 28 mm as in *Dresslerella* (Fig. 2A) and *Echinosepala* (Pleurothallidinae; Pridgeon 1982a). Guard-cell walls of many epiphytes are thickened and have well-developed outer cuticular ledges or walls, and walls of other epiphytes and terrestrials are generally thinner with less-developed ledges (Solereder and Meyer 1930). Outer ledges are sometimes extended into cuticular 'horns' or a 'collar' as in some Oncidiinae (Williams 1974; Stern and Carlsward 2006). Substomatal chambers are usually small relative to surrounding cells, occasionally lined with sclerenchyma (Rasmussen 1987).

Solereder and Meyer (1930) reported subsidiary cells from the literature in about 20 genera and added about 40 other genera to the list from their own observations, including an illustration of a tetracytic stoma in *Liparis*. Rosso (1966) observed subsidiary cells in four of five genera of Cypripedioideae. Nevertheless, the claim that Orchidaceae lacked subsidiary cells persisted in the literature (Stebbins and Khush 1961; Cronquist 1968; Withner et al. 1974) until Williams (1979) documented subsidiary cells in 321 species in 145 genera of Epidendroideae and Vandoideae (as then defined). He recorded their occurrence in tribe Vanilleae (but not tribe Pogonieae) of modern Vanilloideae, tribe Cranichideae (but not tribe Orchideae) of Orchidoideae, and all genera of Cypripedioideae described to that point (*Mexipedium* was not described until 1992). Rasmussen (1981a) expanded the coverage by Williams with 32 collections representing 26 species, mainly in Orchideae. A later report by Stern et al. (1993b) also clearly illustrated them in both genera of Apostasioideae. Anomocytic stomata characterize subfamily Orchidoideae (Fig. 1F) and occasional genera in Epidendroideae such as *Pleione* (Coelogyninae) and *Epipactis* and *Neottia* (Neottieae) (Solereder and Meyer 1930). Cyclocytic stomata occur in *Dresslerella* (Fig. 2B) and *Echinosepala* (Pleurothallidinae; Pridgeon 1982a) and *Earina* (Agrostophyllinae; Conran et al. 2009) among others. The tetracytic condition is most common among Epidendroideae (Fig. 2C; Rasmussen 1987). Paracytic types occur in *Dendrobium* (Morris et al. 1996) and several taxa with two lateral subsidiary cells reported by Williams (1979). Anomocytic and tetracytic types are often mixed in the same specimen. Anisocytic stomata are also common in Epidendroideae. Diacytic configurations have been reported in Cranichideae (Williams 1975; Dressler 1981; Singh 1981).

Stomata may be distributed over one or both surfaces of bifacial leaves or the entire surface of unifacial leaves. In the former, they are most often confined to the abaxial epidermis (Fig. 2D), although there are almost as many exceptions as the rule (Fig. 2E); differences can be systematically useful at lower levels of classification. Dimensions are generally larger in terrestrial species than those in epiphytic ones (Solereder and Meyer 1930). Stomata occur only on the adaxial epidermis of both *Lyperanthus* species (Caladeniinae; Pridgeon 1994a) to distinguish them from two others now segregated as the genus *Pyrorchis*.

In addition to the morphological terminology to describe the stomatal complex, there is also an ontogenetic terminology that bears no direct relationship to the former because different patterns of development may result in the same cell configuration in the mature leaf. Rasmussen (1981a,b, 1987) reviewed stomatal development in Orchidaceae, summarized here. A protodermal cell divides unequally into a meristemoid cell, which then undergoes an unequal division itself to form a new meristemoid, which usually functions directly as a guard-cell mother cell (agenous development). However, that meristemoid may also divide again to form a larger daughter cell (mesogene cell) and a smaller meristemoid (hemimesogenous development). Both types produce anomocytic stomata in Orchidinae, for example. Williams (1975) documented hemimesogenous development in several genera of Cranichideae. Williams (1975) and Rasmussen (1981a) also observed isolated perigene cells—daughter cells formed by divisions of protodermal cell abutting the guard-cell mother cell—in Orchidoideae. Rasmussen (1981b) noted that three kinds of cells may surround the stoma at maturity—agene, perigene, and mesogene cells. Singh (1981) reported perigenous development in *Cymbidium* (Cymbidiinae) and *Cirrhopetalum* (= *Bulbophyllum*, Dendrobieae) and mesoperigenous development in *Bulbophyllum*, *Dendrobium*, *Eria*, and *Pholidota*. Williams (1976) described developmental patterns in Cymbidiinae, Catasetinae, Cyrtopodiinae, Zygopetalinae, and Stanhopeinae (Cymbideae). The morphological approach to describing subsidiary cells is used in this volume simply because young, living shoots are not always available for anatomical surveys, but herbarium specimens are. However, it should be remembered that observations of mature leaves alone may obscure underlying ontogenetic pathways that, if examined more closely, could lead to different systematic conclusions.

Hypodermis

Few developmental studies have been published to show that the layers immediately subtending epidermal cells are in fact hypodermis (derived from ground meristem) rather than inner layers of a multiple epidermis (derived from protoderm). However, unpublished studies of the developing lamina in Pleurothallidinae (cited by Pridgeon 1982a) showed cell divisions only in ground meristem and no periclinal divisions of protoderm. There is no evidence for a multiple epidermis throughout Orchidaceae.

Hypodermal tissue is absent in leaves of Apostasioideae, Cypripedioideae, most Orchidoideae, and some Vanilloideae but present in many epiphytic Epidendroideae. Conduplicate-leaved taxa are more apt to have a hypodermis than plicate-leaved ones, but there are exceptions. There are rarely some terrestrials in Epidendroideae with a hypodermis, e.g. *Aplectrum* and *Tipularia* (Calypsoeae; Stern and Carlsward 2008). Among Vanilloideae, hypodermal layers occur in *Pseudovanilla*, *Clematepistephium*, *Eriaxis*, and *Vanilla* (Stern and Judd 1999, 2000).

Adaxial layers range from one to more than 10 in number (although most commonly 1–4) and often oriented anticlinally like palisade mesophyll, especially at the midrib (Figs 2F, 4D). The abaxial hypodermis is usually only one or two layers deep (Fig. 2F). Depending on the taxon, leaves may have adaxial and abaxial layers or only one or the other. Where both occur, however, hypodermis is almost always more extensive adaxially, sometimes accounting for up to 80% of leaf volume as in *Holothrix* (Orchideae; Solereder and Meyer 1930).

Composition of hypodermis is parenchyma and/or sclerenchyma, occurring as either continuous or interrupted layers and as either one or both tissues. Parenchymatous layers almost certainly function in water storage and sclerenchyma fibres in mechanical support. The former often have annular or helical thickenings that are bifringent in polarized light (Fig. 3A–D). Histochemical tests showed that the thickenings (at least in Pleurothallidinae) are cellulosic rather than suberized or lignified (Pridgeon 1982a), whereas Olatunji and Nengim (1980) reported thickenings with varying degrees of lignification in mesophyll idioblasts of African taxa. As further support for the cellulosic nature of the thickenings, hypodermal cells lacking thickenings are often bifringent as well, as in species of Laeliinae (Stern and Carlsward 2009). The thickenings in Pleurothallidinae and other taxa as reported by Pirwitz (1931) have articulations, on one side of which is a constriction or groove and on the other is a transverse dilation, which imparts a hinged or geniculate aspect (Fig. 3C; Pridgeon 1982a). Such thickenings, also found in mesophyll idioblasts (see **Mesophyll**), may thus offer mechanical support and prevent cellular collapse during periods of desiccation (Olatunji and Nengim 1980). As hypodermal cells of many genera without thickenings are frequently observed to be distorted, infolded, or even collapsed, this hypothesis is appealing.

Sclerenchyma, occurring as isolated fibres or fibre bundles, subtends the epidermis in many epiphytic taxa such as members of Laeliinae (Baker 1972; Oliveira Pires et al. 2003; Stern and Carlsward 2009), Dendrobiinae (Morris et al. 1996), Maxillariinae (Davies 1999; Stern, Judd et al. 2004), Cymbidiinae (Yukawa and Stern 2002), and Oncidiinae (Solereder and Meyer 1930; Williams 1974; Stern and Carlsward 2006).

Mesophyll

Chlorenchyma may be homogeneous (Fig. 2D,E) or differentiated into palisade and spongy mesophyll (Fig. 4A,D), sometimes even within the same genus, and so is of little systematic value. Atwood (1984) suggested that tessellation in *Paphiopedilum* (Cypripedioideae) leaves is determined by number of palisade layers, but Lawton et al. (1992) showed that it results instead from a difference in the quality of phenolic acids in mesophyll cells. Cells of spongy mesophyll often have conspicuous primary pit-fields as in Pleurothallidinae (Pridgeon 1982a) and Cymbidieae (Stern and Judd 2002; Yukawa and Stern 2002) interconnected by cellular protrusions (Solereder and Meyer 1930), giving the tissue the appearance of a lacunose network of stellately lobed or isodiametric cells.

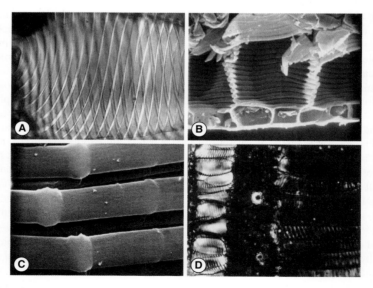

Fig. 3. Leaf TS. (A) *Phloeophila cymbula*, abaxial hypodermal cell with spiral thickenings, ×1500; (B) *Phloeophila cymbula*, abaxial epidermis and hypodermis with spiral thickenings, SEM photograph, ×625; (C) *Phloeophila cymbula*, spiral thickenings of abaxial hypodermal cell showing articulations. SEM photo, ×6000; (D) *Anathallis vasquezii*, section illuminated in polarized light to show birefringence of spiral thickenings in abaxial hypodermis, ×180. (A–C) Reprinted from Pridgeon (1982a); (D) photograph by A.M. Pridgeon.

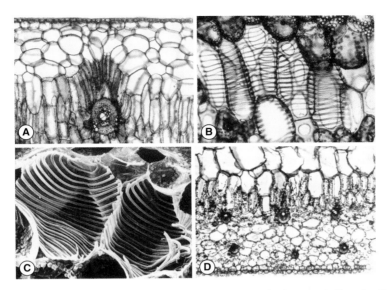

Fig. 4. Leaf TS. (A) *Echinosepala uncinata*, adaxial epidermis, hypodermis, palisade mesophyll, and collateral vascular bundle, ×180; (B) *Pleurothallopsis tubulosa*, spirally thickened idioblasts in mesophyll, ×330; (C) *Acianthera circumplexa*, spirally thickened idioblasts in mesophyll, SEM photograph, ×330; (D) *Octomeria costaricensis*, showing adaxial hypodermis, palisade and spongy mesophyll, and two series of vascular bundles, ×200. (A,B) Photographs by A.M. Pridgeon; (C,D) reprinted from Pridgeon (1982a).

Spirally or reticulately thickened idioblasts occur in mesophyll (as well as in the cortex of pseudobulbs and roots) of many epiphytic Epidendroideae and probably have arisen independently in a number of lineages (Fig. 4B,C). Within a lineage they seem to have limited systematic utility (Williams 1974; Pridgeon 1982a). The secondary thickenings are similar or identical in structure and composition to those described above. Olatunji and Nengim (1980) referred to these water-storage cells as 'tracheoidal elements' using terminology coined by Foster (1956). However, the term 'tracheoidal' (= 'like a tracheid') is ill-advised because the thickenings (1) may not be lignified like the secondary wall patterns of xylem and (2) may have articulations (unlike tracheids). As already noted (Pridgeon 1982a), spirally thickened idioblasts have no physical, ontogenetic, or functional similarity with tracheary elements and so are not 'like' tracheids in any sense.

Extravascular fibre bundles in mesophyll have been reported throughout epiphytic Epidendroideae, but especially in Oncidiinae (Solereder and Meyer 1930; Williams 1974; Sandoval-Zapotitla and Terrazas 2001; Stern and Carlsward 2006), Maxillariinae (Holtzmeier et al. 1998; Davies 1999; Stern, Judd et al. 2004), Cymbidiinae and Eulophiinae (Stern and Judd 2002; Yukawa and Stern 2002), Laeliinae (Baker 1972; Oliveira Pires et al. 2003; Stern and Carlsward 2009), Dendrobieae (Solereder and Meyer 1930; Morris et al. 1996), and Catasetinae (Stern and Judd 2001). Fibre bundles may alternate with vascular bundles or be disposed above and/or beneath them or any combination of these. Solereder and Meyer (1930) also reported unusual septate fibres or fibre-like cells with narrow lumina and arranged longitudinally in species of Aeridinae, Catasetinae, and Ponerinae.

Vasculature

Vascular bundles are collateral in all Orchidaceae studied and usually arranged in a single series. However, two or more rows of bundles are known to occur in some species of Aeridinae, Dendrobieae, Maxillariinae, Oncidiinae, Eriinae, Laeliinae, Pleurothallidinae (Fig. 4D), and Coelogyninae (Solereder and Meyer 1930; Baker 1972; Pridgeon 1982a; Pridgeon and Stern 1982; Stern, Morris et al. 1994; Holtzmeier et al. 1998; Toscano de Brito 1998; Sandoval-Zapotitla and Terrazas 2001; Stern and Carlsward 2006; Stern and Carlsward 2009). Additionally, two or more bundles of varying sizes may occur in the midrib of Maxillariinae (Möbius 1887; Holtzmeier et al. 1998; Stern, Judd et al. 2004), Coelogyninae (Zörnig 1903; Solereder and Meyer 1930), Zygopetalinae (Stern et al. 2004), and other groups. Phloem sclerenchyma is generally more developed than xylem sclerenchyma in vascular bundle caps, although both may be absent altogether in herbaceous terrestrials such as Cranichideae (Stern, Morris et al. 1993c), Diurideae (Pridgeon 1993, 1994b), Malaxideae (Solereder and Meyer 1930), and Orchideae (Möbius 1887; Solereder and Meyer 1930; Stern 1997a,b). Rarely, only xylem sclerenchyma occurs, as in *Tipularia* (Stern and Carlsward 2008). A uniseriate or biseriate sclerenchyma 'bridge' often separates xylem and phloem.

Tracheids have been reported in leaves of Cranichideae (Thorsch and Stern 1997), Apostasioideae (Fig. 13F,L; Stern et al. 1993b), Cypripedioideae (Rosso 1966), and Epidendroideae (Pridgeon 1982a; Pridgeon and Stern 1982; Sandoval-Zapotitla and Terrazas 2001), but vessel elements with scalariform perforation plates in leaves are known from only a few Epidendroideae such as *Calopogon*, *Stanhopea*, and *Huntleya* (Thorsch and Stern 1997) as well as *Epidendrum* and *Scaphyglottis* (Cheadle 1942).

Stegmata

Solereder and Meyer (1930) summarized earlier reports of stegmata with silica bodies and added new observations. More recently, Møller and Rasmussen (1984) surveyed 130 species from 105 orchid genera, and Sandoval-Zapotitla, Terrazas, and Villaseñor (2010b) surveyed Oncidiinae for them. Stegmata occur in longitudinal files lining fibre bundles and vascular bundle sheaths of all plant organs, though rarely in the root as in *Cymbidium* (Yukawa and Stern 2002) and some Maxillariinae (Holtzmeier et al. 1998; Yukawa and Stern 2002). Orchidoideae and Vanilloideae lack stegmata altogether. Stegmata with conical silica bodies occur in Apostasioideae, Cypripedioideae (rarely), Arethuseae (especially Coelogyninae), Epidendreae, Cymbidieae, and Tropideae. Conical bodies also occur in Diceratostelinae (Stern et al. 1993c) but not Triphorinae (Carlsward and Stern 2009) of Triphoreae. Stern and Carlsward (2008) reported them in leaves of two genera (*Cremastra*, *Govenia*) of Calypsoeae but not in axis organs of any genus. On the other hand, spherical silica bodies are found predominantly in Podochileae, Dendrobieae (Fig. 5A), and Vandeae. The most glaring exception is in *Eria javanica* (Sw.) Blume (Podochileae), the type species, which has conical silica bodies (Dressler and Cook 1988).

In a family-wide phylogenetic analysis of morphological and anatomical characters, Freudenstein and Rasmussen (1999) noted that the outgroup chosen (*Hypoxis*; Hypoxidaceae) lacks stegmata, and so the plesiomorphic state for silica bodies in Orchidaceae is absence. Because stegmata are always found in association with sclerenchyma, they noted that absence may be the result either of loss of the stegmata where sclerenchyma is present or of secondary loss of sclerenchyma. Conical silica bodies represent the ancestral condition as they are present in Apostasioideae, and the most parsimonious explanation for their absence in Orchidoideae and Vanilloideae is that they were lost secondarily (Møller and Rasmussen 1984; Rasmussen 1986). Spherical silica bodies probably evolved more than once (Rasmussen 1986), which has been borne out in recent molecular phylogenies (see Taxonomic history and phylogenetics). Stern, Cheadle et al. (1993b) reported conical silica bodies in the leaves of *Apostasia wallichii* R.Br. and also incipient spherical silica bodies in the stem of the same species, so that both types might occur in that same ancestral group; however, this warrants further investigation. Prychid et al. (2004) provided an excellent summary of the literature regarding stegmata in Orchidaceae as part of a broader review of monocotyledons.

Other ergastic substances

In addition to silica bodies and cruciate starch granules, the most common cellular inclusions are raphides, present as either single crystals or bundles in the epidermis and/or chlorenchyma. They may also occur in

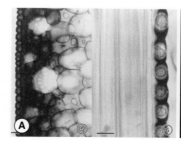 

Fig. 5. **Leaf ergastic substances.** (A) *Dendrobium albosanguineum*, leaf LS with row of stegmata containing spherical, rough-surfaced silica bodies adjacent to vascular bundle fibre, ×880. (B) *Myoxanthus trachyclamys*, flavonoid crystal, SEM photograph, ×4600. (A) Reprinted from Morris et al. (1996); (B) reprinted from Pridgeon (1982a).

longitudinally oriented mesophyll idioblasts (Fig. 6F) and often near vascular bundles and leaf margins, for example in Diseae (Kurzweil et al. 1995). Spherical crystals are present in epidermal and mesophyll cells of Oncidiinae (Sandoval-Zapotitla and Terrazas 2001; Stern and Carlsward 2006), Coelogyninae (Zörnig 1903), and other taxa (Möbius 1887). Flavonoid crystals have been reported in Laeliinae (Baker 1972, as radial crystalline masses; Oliveira Pires et al. 2003), Pleurothallidinae (Fig. 5B; Pridgeon 1982a, Pridgeon and Stern 1982, as raphides), and Oncidiinae (Sandoval-Zapotitla, Terrazas, and Villaseñor 2010b, as prisms). Druses have been observed in *Chysis* (Epidendreae) and *Dendrobium* (Dendrobieae) by Solereder and Meyer (1930), *Arundina* (Arethuseae) and *Bulbophyllum* (as *Cirrhopetalum*, Dendrobieae) by Tominski (1905), and several genera of Oncidiinae (Sandoval-Zapotitla, Terrazas, and Villaseñor 2010b).

Mucilage is much more common in orchid stems and roots, especially tubers in Orchideae (see **Root tubers**), but Möbius (1887) reported cells with a mucilage matrix in leaves of *Aerides* and *Rhynchostylis* (Aeridinae), *Aeranthes* (Angraecinae), *Pleurothallis* and *Masdevallia* (Pleurothallidinae), *Bulbophyllum* (Dendrobieae), *Brassia* (Oncidiinae), and *Brassavola* (Laeliinae), in either chlorenchyma, hypodermis, or spirally thickened idioblasts. Raphides are usually present, as in longitudinally oriented idioblasts in *Maxillaria* (Maxillariinae; Holtzmeier et al. 1998). Leaves of Cypripedioideae species also contain abundant mucilage, as anyone who has extracted DNA from those taxa can testify (e.g. Cox et al. 1997); Möbius (1887) specifically mentioned epidermal cells of *Paphiopedilum* in this regard.

Tannin droplets have been observed in the epidermis and mesophyll of many unrelated taxa (e.g. Malte 1902; Zörnig 1903; Solereder and Meyer 1930; Kurzweil et al. 1995). Malte (1902) described them as soluble in water and alcohol, usually greenish or yellowish (but sometimes colourless or reddish) and comprising at least two tannins. Malte (1902) also reported oil droplets (*Elaiosphären*) in several species of Epidendreae, Cymbidieae, Dendrobieae, and Vandeae; Pridgeon (1982a) noted them in epidermal cells of *Platystele* (Pleurothallidinae).

Fungal pelotons are usually confined to the axis and root of orchid plants, but Pridgeon (1993) also observed them in palisade mesophyll of *Caladenia* (Diurideae).

Stems

Given the wide array of stem types, the following remarks refer to aerial stems (including pseudobulbs) unless otherwise designated.

Hairs

An indumentum on axis organs is rare in epiphytes but more common in terrestrial taxa. Among terrestrials, hairs on the upper portion of the stem of *Caladenia*, *Cyanicula*, and *Elythranthera* (Caladeniinae) are multicellular, uniseriate, unbranched, and non-glandular or glandular depending on the species (Pridgeon 2001). Uniseriate, unbranched trichomes supported by a buttress of four or five epidermal cells occur in Diurideae, notably some species of Acianthinae, Caladeniinae, Drakaeinae, and Prasophyllinae. In all cases the apical cells and subjacent cortical cells are infected with fungal hyphae, but epidermal cells not associated with trichomes are rarely infected (Pridgeon 1994a). More elaborate, multiseriate trichomes with 2–5 tiers of cells are present on the stem ('collar') of many Diurideae (Fig. 6A) and also Pterostylidinae (Fig. 6B,E; Cranichideae) of Australia and New Zealand; central and peripheral cells as well as subjacent cortical parenchyma are consistently associated with hyphae, indicating the role that trichomes play in mycorrhizal reinfection of mature plants during the growing season (Ramsay et al. 1986; Pridgeon 1994b, 2001). Among other terrestrial taxa, non-glandular or glandular hairs and/or papillae are known in *Cypripedium* and *Selenipedium* (Cypripedioideae; Rosso 1966), *Disa glandulosa* Burch. ex Lindl. and some *Disperis* species (Diseae; Kurzweil et al. 1995), *Neuwiedia* (Apostasioideae; Stern, Cheadle et al. 1993b), Cypripedioideae (Rosso 1966), *Wullschlaegelia*, corm of *Calypso*, and rhizome of *Cremastra* (Calypsoeae; Stern 1999, Stern and Carlsward 2008), and *Holothrix* (Orchideae; Stern 1997a). The hairs rarely present in epiphytes are generally glandular and sunken, mirroring those in the leaf, for example in some Maxillariinae (Holtzmeier et al. 1998) and *Psilochilus* (Triphoreae; Carlsward and Stern 2009).

Cuticle

Cuticle is generally smooth and variable in thickness, but it is most developed in Epidendroideae and particularly in Laeliinae (Stern and Carlsward 2009) and Cymbidieae (Yukawa and Stern 2002; Stern and Carlsward 2006). Terrestrial taxa may have a rugose, rugulose, papillose, striate, or granulose cuticle (Stern, Morris et al. 1993c; Pridgeon 2001; Stern and Carlsward 2008, 2009). Tubercle-like elevations of the cuticle have been observed in pseudobulbs of *Coelogyne* (Coelogyninae; Zörnig 1903) and *Myrmecophila* (as *Schomburgkia*, Laeliinae; Solereder and Meyer 1930). In several genera of Aeridinae, cuticular lacunae occur in rows (Solereder and Meyer 1930).

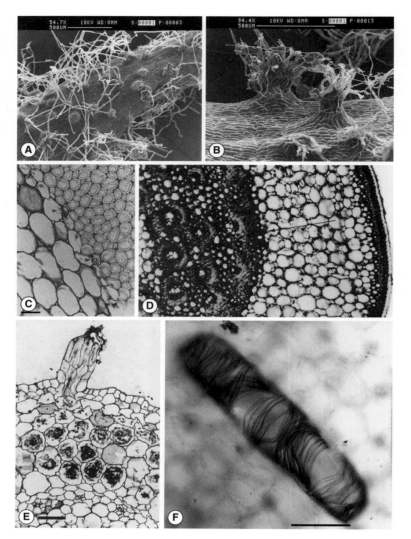

Fig. 6. Stem and leaf. (A) *Caladenia latifolia*, 'collar' of emerging shoot with multicellular trichomes at different stages of development, SEM photograph. Scale bar = 500 μm; (B) *Pterostylis nutans*, collar, SEM photograph. Scale bar = 500 μm; (C) *Neuwiedia veratrifolia*, stem TS with conical silica bodies (arrows) internal to endodermoid layer at outer margin of sclerenchymatous layer of ground tissue. Scale bar = 20 μm; (D) *Octomeria costaricensis*, stem TS with thick cuticle, lignified epidermis and subepidermis, phloic fibre sheath (arrow), and concentric rings of bundles, ×130; (E) *Pterostylis hispidula*, collar TS showing multicellular trichome and infected cortical cells. Scale bar = 100 μm; (F) *Cycnoches lehmannii*, leaf scraping, mucilaginous idioblast. Scale bar = 25 μm. (A,B,E) Reprinted from Pridgeon (1994b); (C) reprinted from Stern et al. (1993a); (D) reprinted from Pridgeon (1982a); (F) reprinted from Stern and Judd (2001).

Epidermis

Cells of the 1-layered epidermis are uniformly thick-walled, thicker only on the outer wall or both the outer and inner walls, sometimes extending to the radial walls. In rare cases, only the inner wall is thickened (e.g. Stern and Carlsward 2006). A sclerotic epidermis is known for many unrelated taxa (Solereder and Meyer 1930; Pridgeon 1982a). Pitting of the outer epidermal wall of pseudobulbs and rhizomes has been reported for, among others, Laeliinae, Aeridinae, Dendrobieae, Catasetinae,

Stanhopeinae, Oncidiinae, Maxillariinae, and Coelogyninae (Solereder and Meyer 1930; Holtzmeier et al. 1998). Stomata are generally absent or sparse, but they are more common in terrestrials than epiphytes. Among the terrestrials, they have been reported in one or more species of Coelogyninae (Solereder and Meyer 1930), Orchideae (Stern 1997a,b), Diurideae (Pridgeon 2001), Vanilloideae (Stern and Judd 1999, 2000), Apostasioideae (Stern, Cheadle et al. 1993b), Cypripedioideae (Rosso 1966), Cranichideae (Stern, Morris et al. 1993c), and Diseae (Kurzweil et al. 1995).

Hypodermis
In many taxa, mainly epiphytic Epidendroideae but also some Vanilloideae (Stern and Judd 1999, 2000), there is a 1–5-layered hypodermis. Cells may be sclerotic or possess U-shaped wall thickenings and offer mechanical support, e.g. some Pleurothallidinae (Weltz 1897; Pridgeon 1982a; Pridgeon and Stern 1982), Dendrobieae (Weltz 1897; Solereder and Meyer 1930; Morris et al. 1996), and Maxillariinae (Weltz 1897). Otherwise the cells are parenchymatous or collenchymatous, sometimes with spiral thickenings (Solereder and Meyer 1930), and probably function in water storage. Examples are *Bulbophyllum* (Dendrobieae; Sprenger 1904; Solereder and Meyer 1930) and Stanhopeinae (Krüger 1883; Stern and Whitten 1999).

Ground tissue
Ground tissue in stems, pseudobulbs, and rhizomes is either continuous throughout or is (commonly) interrupted by what has variously been termed a sclerotic sheath, sclerenchyma sheath, pericyclic ring, sclerenchymatous ring, or sclerenchyma band. If present, the sheath effectively delimits an outermost cortex and innermost pith and is often continuous with the outermost vascular bundles. In some cases, vascular and fibre bundles occur in the cortex also. Rarely, it is an endodermis with casparian strips rather than sclerenchymatous ring that separates the cortex from the vascular cylinder, as in *Holothrix* (Orchideae; Stern 1997a). Peripheral assimilatory cells are generally smaller and chlorophyllous, whereas central and inner cells are larger with fewer chloroplasts. Cruciate starch granules are present throughout. Intermixed may be relatively large water-storage idioblasts with birefringent, pleated walls and/or spiral (or pitted) thickenings (e.g. Chatin 1857; Krüger 1883; Weltz 1897; Hering 1900; Sprenger 1904; Olatunji and Nengim 1980; Morris et al. 1996; Holtzmeier et al. 1998; Stern and Whitten 1999; Stern and Judd 2001; Yukawa and Stern 2002; Stern, Judd et al. 2004; Stern and Carlsward 2006; Stern and Carlsward 2009). In addition, there may be mucilage idioblasts (e.g. Zörnig 1903; Stern and Judd 2001) and isolated fibre bundles. In some species of Diseae (Kurzweil et al. 1995) and Prasophyllinae of Diurideae (Pridgeon 2001) the innermost parenchyma cells disintegrate, leaving a central lacuna. Pseudobulbs of *Myrmecophila* and *Caularthon* (Laeliinae) are similarly hollow and inhabited by ants, but no developmental studies have been undertaken. Dannecker (1898, cited in Solereder and Meyer 1930) noted the thickened cells surrounding the central cavity and up to 10 lateral cavities as well as slit-shaped orifices at the base; Philipp (1923) suggested that the secondary cell walls are probably suberized.

Vasculature
Weltz (1897) effectively summarized the various configurations of vascular bundles in the axis: (1) absent from any sclerenchymatous sheath; (2) disposed in one or two rings continuous with that sheath; (3) embedded in the sheath and also present in the interior of the vascular cylinder; and (4) grouped in an intermittent ring of sclerenchymatous complexes. There may be 1–5 clearly defined rings (Fig. 6D) or they may appear to be scattered. Bundles are collateral, although Rosso (1966) reported amphivasal bundles in stem bases and nodes of *Cypripedium* and *Selenipedium* and throughout the stems of *Paphiopedilum* and *Phragmipedium* (Cypripedioideae). Sclerenchyma caps are generally more developed at phloem poles; xylem sclerenchyma may be absent altogether in smaller bundles. As in leaves, a sclerenchyma 'bridge' separates the vascular tissues. Tracheids are the most common conducting elements in xylem (e.g. Rosso 1966; Pridgeon 1982a; Pridgeon 2001), although vessel elements occur in Epidendroideae sampled by Thorsch and Stern (1997) and also *Vanilla* (Vanilloideae; Solereder and Meyer 1930). Phloem sieve-tubes in *Vanilla* are notably 'spacious' (Solereder and Meyer 1930).

Stegmata
(See under **Leaves**.) Stegmata in stems and rhizomes are commonly associated with sclerenchyma sheaths and xylem and/or phloem sclerenchyma in Apostasioideae (Fig. 6C), Cypripedioideae (rarely), and Epidendroideae. However, many taxa in the last lack them, e.g. Malaxideae, *Bulbophyllum*, and Calypsoeae (Solereder and Meyer 1930; Møller and Rasmussen 1984; Stern and Carlsward 2008).

Other ergastic substances

Cruciate starch granules are common in assimilatory ground tissue of all stem types. Spiranthosomes (also see **Roots**) occur in the rhizome of *Uleiorchis* (Gastrodieae; Stern 1999), stem of *Vanilla* (Vanilloideae; Stern and Judd 1999), corm of *Geodorum* (Eulophiinae; Stern and Judd 2002), pseudobulb of some species of *Eulophia* (Eulophiinae; Stern and Judd 2001), and stem of *Prescottia* (Cranichideae; Stern, Morris et al. 1993c). Raphides are the most frequently reported crystals in ground tissue, either free or bundled in idioblasts. Stern and Judd (2001) observed mucilage idioblasts in pseudobulbs of Catasetinae. Ground tissue of the 'collar' or underground stem of many Diurideae and Diseae is infected with fungal hyphae and pelotons (Ramsay et al. 1986; Kurzweil et al. 1995; Pridgeon 2001).

Rhizomes

Rhizome anatomy has received scant attention recently but is referenced frequently in the historical literature. The 1-layered epidermis in *Cypripedium* is often necrotic and sloughed off in older specimens; cell walls of the epidermis (if present) and outer cortex are cutinized or suberized (Rosso 1966). Large intercellular spaces or lacunae are present in cortical tissues of the rhizome in various epiphytic Epidendroideae, according to Moreau (1913). Outer vascular bundles occur (among others) in the cortex of *Acineta* (Stanhopeinae) and *Coelogyne* and *Pholidota* (Coelogyninae; Moreau 1913), and *Neuwiedia* (Apostasioideae; Siebe 1903). An endodermis with casparian strips was reported in rhizomes of *Cypripedium* (Fuchs and Ziegenspeck 1926; Rosso 1966), *Epipactis* (Neottieae; Solereder and Meyer 1930), some Calypsoeae (Solereder and Meyer 1930; Stern and Carlsward 2008), and *Uleiorchis* (Gastrodieae; Stern 1999). A pericycle and/or sclerenchymatous ring may be present depending on the species. Vascular bundles are collateral in all cases.

Corms

Stern and Carlsward (2008) examined the anatomy of corms of *Aplectrum, Calypso, Govenia,* and *Tipularia* (Calypsoeae). Hairs occur only in *Calypso*, and stomata only in *Govenia*. All lack an endodermis and pericycle, and all possess water-storage cells in ground tissue.

Stem tubers

True stem tubers are known only for *Basiphyllaea* (Bletiinae), *Nervilia* (Nervilieae), and *Rhizanthella* (Diurideae). Little is known about their anatomy in the first two genera except that a sclerenchymatous ring is absent and vascular bundles are arranged in a ring in *Nervilia* (Solereder and Meyer 1930). Anomalously, Dietz (1930) described the tuber of *Nervilia* as a root tuber (with velamen) instead, so new material should be examined to resolve the discrepancy. Dixon et al. (1990) investigated the anatomy and ultrastructure of the tuber of *Rhizanthella gardneri* R.S.Rogers from Western Australia. Multicellular hairs typical of many Diurideae as well as anomocytic stomata line the surface. Cortical layers are heavily infected with the endophyte. Vascular bundles are scattered in ground tissue, and sclerenchyma and starch are essentially absent. Pridgeon and Chase (1995), however, examined a specimen of the disjunct species, *Rhizanthella slateri* (Rupp) M.A.Clem. and P.J.Cribb (which occurs only in south-east Queensland and eastern New South Wales) and reported a heavily suberized cuticle and the absence of both hairs and stomata. Fungal pelotons are only sparsely distributed throughout ground tissue. Fibres and silica bodies are absent.

Roots

Porembski and Barthlott (1988) surveyed roots of 344 species in 262 genera. They described 12 types of roots using number of velamen layers, stratification of layers, presence of secondary wall thickenings and pores, thickness of exodermal cell walls, and presence of *Stabkörper* (= tilosomes) and also 'tracheoidal idioblasts' in the cortex. Ten of these types are named after representative genera: *Calanthe, Bulbophyllum, Pleurothallis, Malaxis, Spiranthes, Coelogyne, Dendrobium, Epidendrum, Cymbidium,* and *Vanda*. The other two types are 'unspecified' and 'velamen absent' (denoting a simple rhizodermis). They suggested that the *Calanthe* type is primitive because it lacks wall thickenings and is most common in ancestral taxa and that the *Cymbidium* and *Vanda* types are most derived. Pridgeon (1987) monographed the velamen and exodermis, summarizing observations in the literature and, with William L. Stern, adding new data for hundreds of species, including number of velamen layers, type of exodermal thickening, and presence and type of tilosomes. Figure 7A,B depicts typical tissues of roots of epiphytic Orchidaceae.

Hairs

Root hairs may arise not only from the simple rhizodermis but also from the outermost velamen layer, especially on the surface attached to a substrate (e.g. Vandeae). Hairs are usually unicellular (2- or 3-celled in isolated species of Spiranthinae; Stern, Morris et al. 1993c) but may branch as in *Bifrenaria* (Maxillariinae; Leitgeb 1864b),

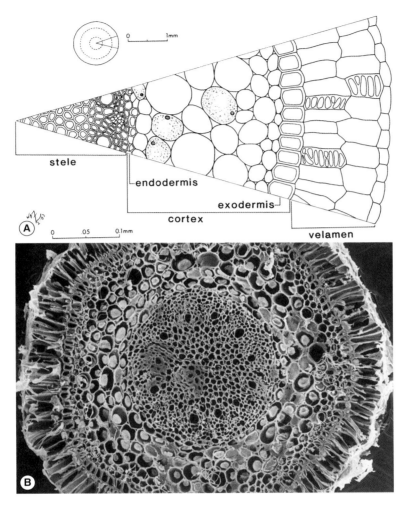

Fig. 7. Root TS. (A) Drawing from *Restrepiella ophiocephala* by Wendy Zomlefer showing typical root tissue zones of epiphytic Orchidaceae. (B) *Bulbophyllum micholitzianum*, with 1-layered, anticlinally elongated velamen, SEM photograph, ×135. Reprinted from Pridgeon (1987).

Chysis (Epidendreae; Meinecke 1894), and *Cyclopogon* (Spiranthinae; Stern et al. 1993c). They may also have spiral or reticulate thickenings (Leitgeb 1864b; Meinecke 1894; Porsch 1906; Stern and Judd 2001) or unroll into spiral bands (Leitgeb 1864b; Janczewski 1885). Hairs are bulbous-based in some taxa such as Stanhopeinae (Stern and Whitten 1999).

In addition to unicellular root hairs, multicellular trichomes in Diurideae (see **Stems**) occur on roots, droppers, and stolonoid roots of many Caladeniinae, all Pterostylidinae, and most Acianthinae. They are unbranched, non-glandular, and either uniseriate or multiseriate. The multiseriate type comprises up to five tiers of cells, up to 20 cells in circumference, and up to 10 central cells (Pridgeon 1994a; Pridgeon and Chase 1995). They are almost always associated with hyphae, and the collapsing apical cells are the likely ingress sites for mycorrhizal infection of underlying epidermal and cortical cells.

Simple rhizodermis/velamen

A single-layered rhizodermis is present in many terrestrial taxa and exclusively in Orchidinae, Pterostylidinae, Gastrodieae, and Nervilieae (Porembski and Barthlott 1988; Pridgeon and Chase 1995; Stern 1997a). It is difficult to generalize in other groups, for there are one to many species with velamen in Habenariinae (Stern 1997b) and Cranichideae (Stern, Morris et al. 1993c). Among Cypripedioideae, all genera have velamen except *Cypripedium*, and even one species of that genus (*C. irapeanum* Lex.) is reported to have velamen (Rosso 1966). Other terrestrials with velamen include Apostasioideae, some Vanilloideae, Tropidieae, Sobralieae, Calypsoeae, Collabieae, Diseae, most Diurideae, Malaxideae, some Arethuseae, and *Diceratostele* of Triphoreae (Engard 1944; Pridgeon 1987; Porembski and Barthlott 1988; Stern et al. 1993c; Pridgeon and Chase 1995). Too little is known about the anatomy of Chloraeeae and Codonorchideae of southern South America to draw any conclusions.

It is often difficult to distinguish a simple epidermis from a uniseriate velamen, so it is appropriate to define velamen here as that tissue arising from root dermatogen, consisting of dead cells at maturity, and subtended by an exodermis with passage cells capable of transporting water and solutes. A simple epidermis is never bordered internally by an exodermis. The number of cell layers constituting velamen varies from one (e.g. *Bulbophyllum*, Fig. 7B) to 24 in *Cyrtopodium*; the most common numbers are two, three, four, and five (Fig. 8A; Pridgeon 1987). Sanford and Adanlawo (1973) found that many African species growing in seasonally dry habitats possess more velamen layers, thicker cell walls, and more lignification than those growing in uniform moisture conditions. Number of layers also varies not only from species to species but also throughout a given root, among roots of the same plant, and among different plants of the same species and so it must be used carefully as a taxonomic character (Rüter and Stern 1994).

The outermost layer(s) in a multiseriate velamen was termed 'epi-velamen' [*sic*] by Sanford and Adanlawo (1973) 'to distinguish it from the velamen proper'. That layer differs only from internal layers by occasionally having unique wall ornamentations and perhaps more radially compressed cells, and 'epivelamen' seems to have

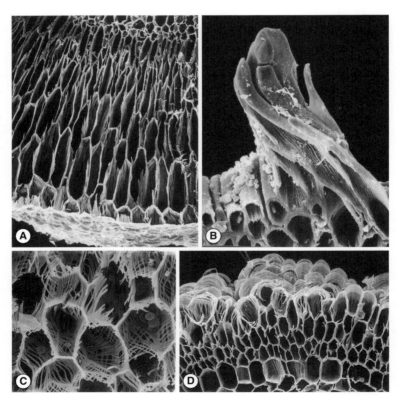

Fig. 8. Root TS, SEM photographs. (A) *Renanthera imschootiana*, showing multiseriate velamen, ×125; (B) *Tridactyle anthomaniaca*, with multicellular velamen tuft ridged internally, ×500; (C) *Lycaste xytriophora*, velamen showing secondary wall thickenings and linear files of small perforations, ×600; (D) *Trichocentrum tigrinum*, illustrating different patterns of secondary wall thickenings between outermost layer and subjacent layers, ×400. Reprinted from Pridgeon (1987).

been coined only as a term of convenience and adopted in the literature. However, because *epi-* is the Greek prefix meaning 'upon', the term epivelamen (= 'upon velamen' or 'on top of velamen') therefore refers to something *other than* velamen. A linguistically and anatomically more correct way to distinguish the outer layer would have been ectovelamen (Greek *ekto-*, outer, to contrast with endovelamen). Another approach would have been to refer to it simply as 'outermost velamen layer(s)', which is consonant with usage by Solereder and Meyer (1930) and leaves no doubt about its definition or derivation.

Palla (1889) reported multicellular conical tufts from the outermost velamen layer in *Campylocentrum* (Angraecinae), and Pridgeon (1987) and Carlsward et al. (2006a) illustrated tufts in *Tridactyle* from the same subtribe in which they are common (Fig. 8B). They also occur in *Bulbophyllum dearei* (Rchb.f.) Rchb.f. (Pridgeon 1987). These and similar configurations are essentially localized elongations of one or a few contiguous cells, often bearing conspicuous ridges of secondary wall thickenings and pores. Sanford and Adanlawo (1973) described hair-like projections from one or two cells of 'epi-velamen' in 15 of 76 West African taxa (all but one not specified) and found that the number of cells constituting the projections is consistent within a species.

Velamen cells often have characteristic secondary wall thickenings, first reported by Meyen (1837), that run parallel to one another, loop, spiral, and/or anastomose (Fig. 8C,D). The most plausible explanation of their function is that they provide support and prevent cellular collapse (Link 1849–50; Kraft 1949; Noel 1974; Pridgeon 1987; Porembski and Barthlott 1988). They occur especially in epiphytes but also in such terrestrials as *Drakaea* (Diurideae; Pridgeon and Chase 1995), *Stenorrhynchos* (Cranichideae; Stern, Morris et al. 1993c), and *Sobralia* (Sobralieae; Oudemans 1861; Leitgeb 1864b). Cells may also have slits or pores (Fig. 8C), the number and size of which can be specific for some taxa according to Porembski and Barthlott (1988), who identified 12 types of roots using six characters, among them four related to velamen: number of cell layers, stratification of layers and shape and size of cells, cell-wall characters (thickenings and pores), and *Stäbkorper* (tilosomes, discussed below). The other two characters related to exodermal cell-wall thickenings and spirally thickened idioblasts in the cortex. All of these have been used in combination to describe roots of many taxa throughout Orchidaceae (e.g. Solereder and Meyer 1930; Porembski and Barthlott 1988; collected works of W.L. Stern and colleagues in this volume).

Pridgeon, Stern et al. (1983) examined roots of 350 species in 175 genera for tilosomes, defined as the lignified/ suberized excrescences from cell walls of the innermost velamen layer and adjacent to thin-walled passage cells of the exodermis. They identified seven morphological types: spongy (e.g. Sobralieae, Figs 9A,B, 10A,D), lamellate (e.g. Coeliinae, Fig. 10B; Laeliinae, Fig. 10C; and *Bulbophyllum*), discoid (Fig. 11A,B; *Zootrophion*; Pleurothallidinae), webbed (Fig. 11C; Polystachyinae), meshed (Maxillariinae), baculate (Maxillariinae, some Spiranthinae), and plaited (*Trigonidium*; Maxillariinae). Tilosomes are most common in Polystachyinae, Sobralieae, Coelogyninae, Laeliinae, Pleurothallidinae, *Bulbophyllum* (Dendrobieae), and Maxillariinae; they are generally absent from Orchidoideae except for *Cryptostylis* (Diurideae; Stern, Morris et al. 1993c), Vandeae other than Polystachyinae, and most Cranichideae (but see Figueroa et al. 2008). Tilosomes may retard transpiration in exposed or xeric situations (Leitgeb 1864b; Haberlandt 1914; Benzing et al. 1982) or promote movement of water into the cortex when velamen is saturated (Leitgeb 1864b; Dietz 1930; Benzing et al. 1982). Much more study is needed to determine the function and ontogeny of tilosomes and whether or not the environment plays a role in the extent of their development. Preliminary transmission electron microscopy (TEM) of young roots of *Bulbophyllum micholitzianum* Kränzl. (Pridgeon, unpublished data) indicates that tilosomes in that species and others are probably laid down by protoplasts of those velamen cells abutting exodermal passage cells (Fig. 11D) before expiring rather than by the passage cells themselves; however, plasmodesmata between adjoining velamen and passage cells at an early stage of development may also facilitate transport of cell-wall building-blocks such as polysaccharides, phenylpropanoids, and enzymes from passage cells.

Leitgeb (1864b) was the first to describe *Deckzellen* or cover cells, which are two or three small velamen cells abutting passage cells of the exodermis. Their function is unknown, but Leitgeb speculated that they offer protection and promote condensation of water vapour and other gases. They are prominent in Vandeae (Fig. 12A; Leitgeb 1864b; Carlsward et al. 2006a) but also occur in *Epidendrum* (Mulay et al. 1958) and are probably more widespread than reported.

The last structural feature of velamen to be discussed here is pneumathodes (Fig. 12C), first observed by Leitgeb (1864b), named by Jost (1887), and ascribed a correct function by Schimper (1888)—gas exchange between velamen and photosynthetic cortex. They were also called *réservoirs aériens* by Janczewski (1885). Haberlandt (1914) described them as longitudinally oriented 'white spots' arranged in series, appearing when roots are wetted, and consisting of three parts: a wedge-shaped

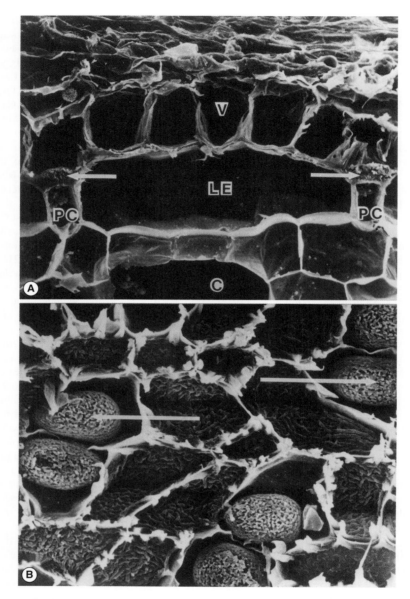

Fig. 9. Root and tilosomes, SEM photographs. (A) *Sobralia decora*, LS showing velamen (V), long exodermal cell (LE), passage cells (PC), and cortical parenchyma (C). Arrows above passage cells indicate spongy tilosomes, ×250; (B) *Sobralia* sp., tangential section through velamen, the arrows indicating spongy tilosomes, ×800. Reprinted from Pridgeon et al. (1983).

mass of velamen cells with numerous wall thickenings, 1–3 air-filled exodermal cells lacking inner tangential walls ('aeration cells'; Benzing et al. 1983), and, below those, one or more rows of rounded cortical parenchyma cells termed *cellules aquifères* by Janczewski (1885). In dorsiventral roots of leafless species of Angraecinae, they are present only on the ventral margins (Schimper 1888; Müller 1900; Benzing et al. 1983). Pneumathodes have been reported in Angraecinae (including Aerangidinae), Aeridinae, Laeliinae, Oncidiinae, and Dendrobieae (Leitgeb 1864b; Prillieux 1879; Janczewski 1885; Schimper 1888; Palla 1889; Porsch 1906; Haberlandt 1914; Dycus

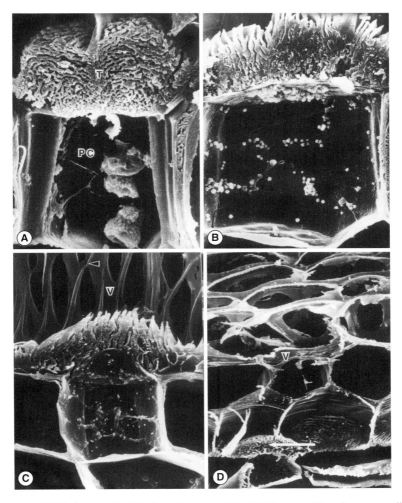

Fig. 10. Root and tilosomes, SEM photographs. (A) *Sobralia decora*, TS with tilosome (T) and passage cell (PC) visible, ×1360; (B) *Coelia triptera*, TS showing lamellate tilosome above passage cell, ×1600; (C) *Prosthechea radiata*, LS showing velamen (V) with secondary wall thickenings, lamellate-tufted tilosome, and passage cell, ×900; (D) *Stelis superbiens*, LS with multiseriate velamen (V) and spongy tilosome at arrow, ×450. Reprinted from Pridgeon et al. (1983).

and Knudson 1957; Benzing et al. 1983; Carlsward et al. 2006a).

Cortex

Exodermis—This outermost layer of the cortex (Engard 1944) consists of long and short (passage) cells (Fig. 9A). They may alternate with one another regularly or there can be as few as one passage cell for every 10 long cells (Sanford and Adanlawo 1973). Long cells are generally empty, dead at maturity, and possess a thin suberin lamella, on which are deposited layers of various apportionment and thickness (Guttenberg 1968). Proportions of lignin and suberin in secondary walls of long cells vary among taxa. Walls are rarely unthickened as in *Malaxis* (Malaxideae; Meinecke 1894) or else thickened in one of four patterns: (1) outer tangential wall thickened only; (2) outer tangential wall thickened with thickenings continued onto radial walls; (3) inner tangential wall thickened with thickenings continued onto radial walls; and (4) all walls uniformly thickened. More than one thickening pattern may occur in the same root, especially patterns 2 and 4 or 3 and 4. The most common pattern is a thickening on outer tangential and

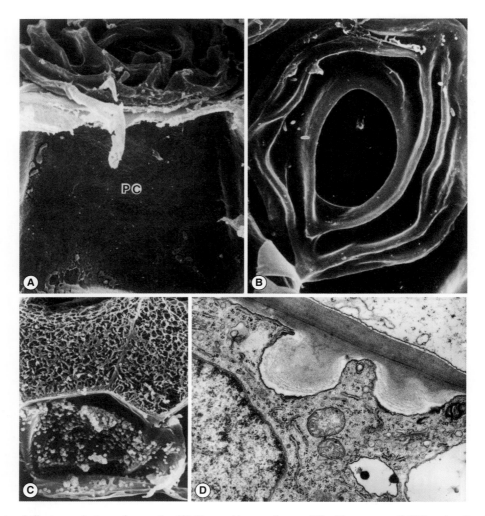

Fig. 11. Root and tilosomes, electron micrographs. (A) *Zootrophion gracilentum*, TS with passage cell (PC) and a discoid tilosome, SEM photograph, ×2050; (B) *Zootrophion gracilentum*, tangential section through velamen to show discoid tilosome with concentric ridges, SEM photograph, ×2700; (C) *Polystachya concreta*, LS illustrating passage cell and associated webbed tilosome, SEM photograph, ×1300; (D) *Bulbophyllum micholitzianum*, LS of young root (up to 2 mm long) showing two tilosome ridges forming on inner wall of innermost velamen cell. Note extensive rough endoplasmic reticulum, Golgi apparati, and nucleus to the left, TEM photograph, ×29,000. (A–C) Reprinted from Pridgeon et al. (1983); (D) photograph by A.M. Pridgeon.

radial walls (Pridgeon 1987). There are often tenuous scalariform secondary thickenings on the walls of long cells (e.g. Morris et al. 1996; Holtzmeier et al. 1998; Stern and Whitten 1999). In contrast, passage cells are living, densely cytoplasmic, and almost always have thin, cellulosic walls. Exceptions with thickened tangential and/or radial walls are rare (Palla 1889; Meinecke 1894). Plasmodesmata between passage cells and adjoining cortical parenchyma cells were reported by Benzing et al. (1982), enabling symplastic transport of water and solutes to internal tissues.

Generally, the exodermis is a single layer of cells. Chatin (1857) reported that there may be two layers in *Bulbophyllum* (Dendrobieae) and *Vanilla* (Vanilleae), and Häfliger (1901) observed three layers in *V. phalaenopsis* Rchb.f. ex Van Houtte, but Stern and Judd (1999) found only one layer in all species of *Vanilla* they examined. However, velamen is sloughed off dorsally in dorsiventral roots

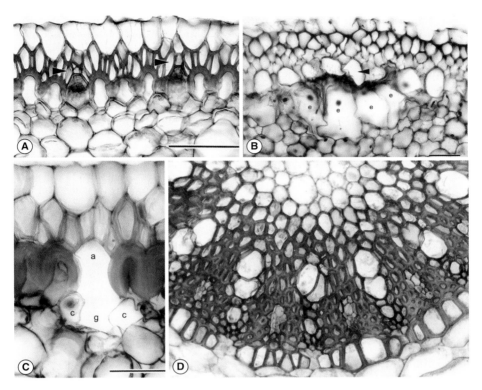

Fig. 12. Root TS. (A) *Solenangis clavata*, showing radially elongate outer velamen cells and cover cells (arrows) above passage cells of exodermis. Scale bar = 100 µm. (B) *Cyrtorchis praetermissa*, with exodermal proliferation (e) below the exodermis. Scale bar = 100 µm. (C) *Chamaeangis sarcophylla*, pneumathode showing aeration cell (a), modified cortical cells (c), and cortical region of gas exchange (g). Scale bar = 50 µm. (D) *Myoxanthus affinis*, showing endodermis with passage cells, vascular tissues, and parenchymatous pith, ×660. (A–C) Reprinted from Carlsward et al. (2006a); (D) reprinted from Pridgeon (1982a).

of many Vandeae, so that the exposed exodermis with thickened outer and radial walls is responsible for protection from injury and desiccation, whereas on the ventral, absorptive surface in contact with the substrate the velamen is 2-layered and exodermis with thinner walls is 2- or 3-layered (Porsch 1906). Passage cells are also more numerous ventrally in these situations (e.g. Müller 1900; Benzing et al. 1983). Carlsward et al. (2006a) noted and illustrated 'exodermal proliferations'—localized, internally directed hyperdevelopment of exodermal cells—in several genera of Vandeae (Fig. 12B).

Parenchyma—Cortical parenchyma cells adjoining the exodermis and endodermis are almost always smaller with fewer intercellular spaces than cells in the middle. Chloroplasts are most common in outer layers, whereas more internal layers store starch, either as cruciate starch granules or spiranthosomes (see below). As in the leaf and stem, water-storage idioblasts are common in the cortex, primarily in epiphytic taxa but also according to Porembski and Barthlott (1988) in some terrestrials such as *Govenia* (Calypsoeae), *Elleanthus* and *Sobralia* (Sobralieae), and *Liparis* and *Malaxis* (Malaxideae; referred to as 'pseudovelamen'). Holm (1904) also mentioned spirally thickened cortical cells in *Liparis*. They may be bifringent with spiral or reticulate thickenings and/or pores ('tracheoidal idioblasts') or non-birefringent with or without pores.

Large intercellular spaces or lacunae have been observed in *Goodyera* (Cranichideae; Holm 1904), *Aplectrum* (Calypsoeae; MacDougal 1899a,b), *Vanilla* (Vanilleae; Stern and Judd 1999), and *Habenaria* (Orchideae; Holm 1904; Stern 1997b). These lacunae may in some cases be mucilage idioblasts in epiphytes such as *Cycnoches* (Catasetinae; Stern and Judd 2001) or even canals as in *Epipactis* and *Cephalanthera* (Neottieae; Noack 1892).

Other cell inclusions in cortical parenchyma are discussed below.

Endodermis—The innermost layer of the cortex, the endodermis, has either thin walls with casparian strips or, more commonly, suberized and/or lignified wall thickenings. As a rule, the former characterizes terrestrial orchids, and the latter epiphytic orchids. Notable exceptions are thickened endodermal walls in Apostasioideae (Stern, Cheadle et al. 1993b) and thin ones in some Laeliinae (Stern and Carlsward 2009). Wall thickenings can be either uniform (O-shaped) or more extensive on the radial and inner or outer tangential walls (∩-shaped), but both occur opposite protophloem. Thin-walled passage cells are often visible opposite protoxylem (Fig. 12D).

Stele

Pericycle—The pericycle is generally 1-layered but may extend to three layers as in some species of Vanilleae (Solereder and Meyer 1930; Stern and Judd 2000), *Stanhopea* (Stern and Morris 1992), Apostasioideae (Stern, Cheadle et al. 1993b), Habenariinae (Stern 1997b), and Cranichideae (Stern, Morris et al. 1993c). Cell walls are usually thin opposite xylem and thicker (often sclerified) abutting phloem.

Vascular cylinder—Radially arranged, alternating strands of xylem and phloem embedded in either parenchyma or sclerenchyma constitute the vascular cylinder, ranging from diarch to 35-arch (Solereder and Meyer 1930). Protoxylem and protophloem are usually contiguous with the pericycle, whereas metaxylem and metaphloem are towards the centre. In some species, a few vascular strands may be medullary also as in *Apostasia* (Stern et al. 1993b) and some Cranichideae (Stern et al. 1993c). Meristeles have been observed in several taxa of Orchideae (Stern 1997a,b).

Xylem consists of tracheids with various wall thickenings and/or vessel elements with scalariform perforation plates and pitted lateral walls (Fig. 13A–D), xylem fibres, and xylem parenchyma. Phloem consists of sieve-tube elements, companion cells, phloem fibres, and phloem parenchyma.

Pith—Pith may be parenchymatous as in Apostasioideae (Stern et al. 1993b), Cypripedioideae (Rosso 1966), and Orchidoideae (Stern et al. 1993c; Kurzweil et al. 1995; Pridgeon and Chase 1995; Stern 1997a,b), but parenchymatous or sclerotic in Vanilloideae (Stern and Judd 1999, 2000) and Epidendroideae (Fig. 12D). Chloroplasts and starch granules may be present in pith cells (Stern and Judd 1999).

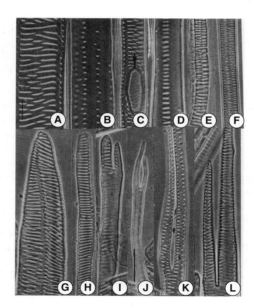

Fig. 13. **Tracheary elements in roots, shoots, inflorescence axes, and leaves of** *Apostasia* **and** *Neuwiedia*. (A–D) Pitting variations in vessel elements from roots. (E,F) Lateral wall pitting in tracheids from inflorescence axis and leaf, respectively. (G–L) Form and branching of tracheids in stem (G,H,J,K), root (I), and leaf (L). Scale bar = 20 μm except 100 μm in (J). Taxa: (A,G) *Neuwiedia veratrifolia*; (B,E,F) *Apostasia wallichii*; (C,J) *Neuwiedia singapureana*; (D) *Apostasia* sp.; (H,I,K) *Apostasia nuda*; (L) *Neuwiedia elongata*. Reprinted from Stern et al. (1993b).

Cell inclusions

Starch—As part of a study of vegetative anatomy of subfamily Spiranthoideae (now considered tribe Cranichideae except for *Cryptostylis* [Diurideae], *Diceratostele* [Triphoreae], and *Corymborkis* and *Tropidia* [Tropidieae]), Stern, Aldrich et al. (1993a) rediscovered specialized, globular amyloplasts with double membranes in root cortical cells and named them spiranthosomes. Such organelles had been reported earlier by Dietz (1930), among others. Stern et al. (1993a) surveyed 52 species, finding them only in 37 species of Cranichideae; they added a few more species with similar distribution in a companion paper (Stern, Morris et al. 1993c). However, Stern (1999) later observed spiranthosomes in the root cortex of *Wullschlaegelia* (Fig. 14A,B; Calypsoeae) and rhizome cortex of *Uleiorchis* (Gastrodieae), and Stern and Judd (1999) found them in the stem of *Vanilla* (Vanilloideae).

Silica—Stegmata with silica bodies may be associated with pericycle cells opposite phloem as in a few species of *Cymbidium* (Yukawa and Stern 2002).

Crystals—Raphides are present either free or bundled in cortical parenchyma and/or pith of many taxa and may also occur in idioblasts therein. Crystal sand was also reported in the cortex of *Cymbidium* (Yukawa and Stern 2002). Oliveira Pires et al. (2003) observed flavonoid crystals in the cortex and stele of *Prosthechea* (Laeliinae).

Mycorrhiza—Fungal hyphae are often seen in velamen cells, especially terrestrials such as Cranichideae (Fig. 14C; Stern et al. 1993c), *Vanilla* (Stern and Judd 1999), Diurideae (Pridgeon and Chase 1995), but are also well documented in epiphytes such as *Cymbidium* (Yukawa and Stern 2002). Pelotons are common in outer and middle layers of cortical parenchyma. Fungal coils are also present in the root tubercles of *Apostasia* and *Neuwiedia* (Stern et al. 1993b).

Others—Along with fungal hyphae, algae and other protists have been reported in velamen cells (e.g. Morris et al. 1996; Carlsward et al. 2006a). Malte (1902) reported elaiospheres (oil droplets) in aerial roots of *Acampe* (Vandeae) and tannin globules in roots of many Epidendroideae.

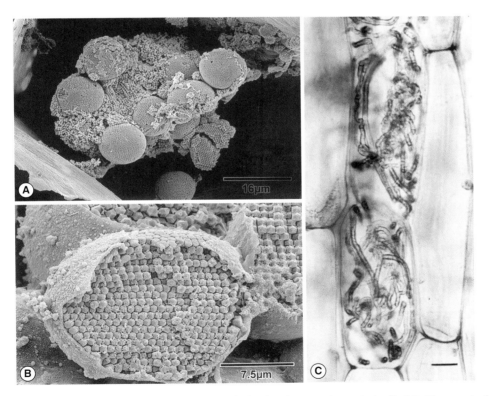

Fig. 14. Cell inclusions. (A,B) *Wullschlaegelia calcarata*, spherical spiranthosomes in cortical cell of fusiform roots. Scale bar as marked. (C) *Lepidogyne longifolia*, LS through root cortex showing cell with mycorrhizal pelotons, ×800. (A,B) Reprinted from Stern (1999); (C) reprinted from Stern et al. (1993c).

Root tubers

The following data refer to tribes Diseae (Kurzweil et al. 1995), Diurideae (Pridgeon 1994a; Pridgeon and Chase 1995), Orchideae (Stern 1997a,b), and Pterostylidinae (Pridgeon, this volume) of Orchidoideae and also to *Triphora* (Triphoreae, Epidendroideae; Carlsward and Stern 2009). Figures 15 and 16 depict underground axes of several taxa of Diurideae and Diseae, respectively. Figure 17A shows a longitudinal section through the past and developing tubers of a member of Diurideae.

Hairs

Unicellular root hairs are rare to common on tubers of many species of Diurideae, Orchideae, Pterostylidinae, and Diseae. Multicellular trichomes are rare on root tubers but do occur in some species of *Arthrochilus*, *Caladenia*, and *Microtis* (Diurideae).

Simple rhizodermis/velamen

Velamen of 1–3 layers is present on tubers of all Diurideae (Fig. 17B; Pridgeon and Chase 1995), Diseae, Pterostylidinae (Cranichideae), and subtribe Orchidinae (Orchideae) except for *Dactylorhiza fuchsii* (Druce) Soó, which has a simple rhizodermis. Subtribe Habenariinae (Orchideae) is characterized by a velamen of 1–5 layers (mostly two or three layers) of cells with pits on anticlinal walls. *Triphora* (Triphoreae) has a 2-layered velamen. Tilosomes are absent in all taxa.

Cortex

Exodermis—Subtending velamen is a uniseriate exodermis, the walls of which are uniformly thin or differentially thickened depending on the species. Passage cells are sparse, and Stern (1997a,b) reported scalariform bars in the walls of long cells.

Parenchyma—Assimilatory cells are filled with cruciate starch granules, especially in new or developing tubers. Larger cells containing mucilage (the major component of salep along with starch; Lawler 1984) are common in tubers of Orchideae. Other inclusions are mentioned below.

Endodermis—Endodermal cells, whether surrounding an undissected stele or individual meristeles, are thin-walled and provided with casparian strips.

Stele

In those species with an undissected siphonostele ('monostele'), the pericycle is 1-layered and vascular cylinder up to 100-arch (*Thelymitra*: Diurideae) with alternating xylem and phloem. In those species with a dissected siphonostele or meristeles ('polystelic'), the pericycle is again 1-layered, and archy varies from 1 to 7, even within the same tuber. Pith is parenchymatous in all cases and often contains mucilage idioblasts. Both types of vascular configurations can be found in Orchideae, Diurideae, and Diseae. The vascular cylinder of the root tuber of *Triphora* is monostelic and 33-arch. Xylem: tracheids have scalariform to pitted thickenings.

Cell inclusions

In addition to starch and mucilage in cortical parenchyma, raphide bundles are commonly seen, often within mucilage cells themselves. Fungal hyphae and/or pelotons occur in the ground tissue of some species, mainly Diurideae. Finally, Kurzweil et al. (1995) described crystal-like pellets composed of glucomannins in tubers of Diseae.

Droppers/stolonoid roots

Hairs

Multicellular, multiseriate trichomes are present on droppers and/or stolonoid roots of many species of Diurideae and Pterostylidinae (Fig. 17C). Root hairs occur on both organs in Diseae and Diurideae.

Simple rhizodermis/velamen

A simple rhizodermis is present in many Diurideae and Pterostylidinae but a velamen of 1–3 layers in others. Droppers of Diseae all have a velamen of 1–3 layers. Tilosomes are absent. Superficial stomata occur rarely in Diurideae, but guard cells appear non-functional.

Cortex

Exodermis—Walls of long cells of the 1-layered exodermis in species with velamen are often differentially thickened, either the outer wall only or both outer wall and radial walls.

Parenchyma—Outer and middle layers of cortical parenchyma are often infected with fungal hyphae and pelotons (Fig. 17C), especially in those taxa with multicellular trichomes. Hyphae are mostly absent from droppers of Diseae. Raphides are occasionally observed in the outer cortex. In *Caladenia* (Diurideae), lignified and suberized secondary wall thickenings develop and anastomose to form the common sheath through which subsequent droppers and new shoots grow. Tracheary remnants and multicellular trichomes of earlier droppers are conspicuous in the sheaths.

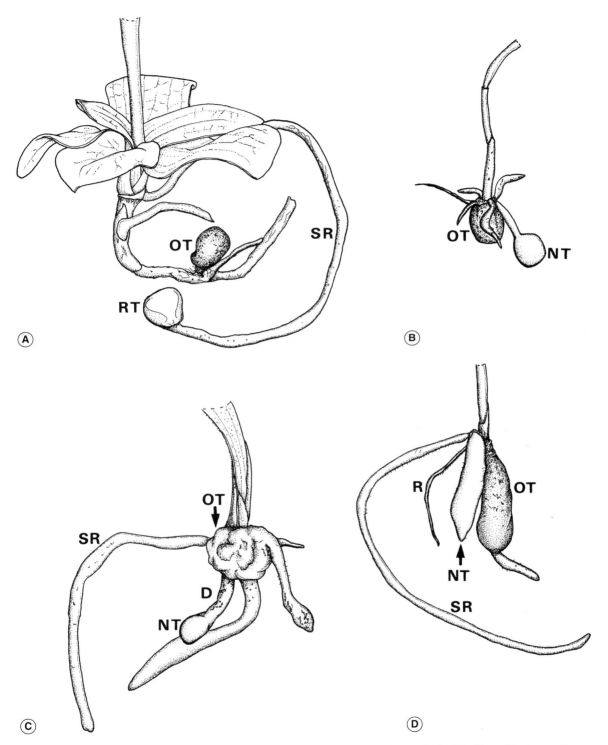

Fig. 15. Line drawings showing selected subterranean systems in subtribe Pterostylidinae and tribe Diurideae. (A) *Pterostylis curta*; (B) *Microtis unifolia*; (C) *Caladenia latifolia*; (D) *Diuris sulphurea*. D, dropper; NT, new (replacement) tuber; OT, old tuber; R, root; RT, reproductive tuber; SR, stolonoid root. All ×2. Drawn by Sarah Thomas. Reprinted from Pridgeon and Chase (1995).

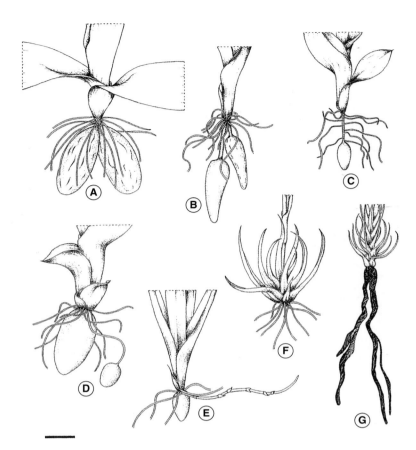

Fig. 16. Line drawings of stem base and underground organs of various species of Diseae. (A) *Satyrium carneum*; (B) *Satyrium stenopetalum*; (C) *Pterygodium caffrum*; (D) *Satyrium odorum*; (E) *Disa tripetaloides*; (F) *Ceratandra bicolor*; (G) *Ceratandra atrata*. Drawn from herbarium material. Scale bar = 1 cm. Reprinted from Kurzweil et al. (1995).

Endodermis—A 1-layered endodermis with casparian strips is present in both organs.

Stele

Among Diurideae, the stele of droppers and stolonoid roots is either a dictyostele with two or three rings of collateral vascular bundles enclosed by the pericycle (representing cauline tissue derived from the axillary bud on the parent axis) or an ectophloic siphonostele that may be dissected. In Diseae, the stele is dissected in some species of a few genera (*Disa*, *Monadenia*, *Satyrium*) but not others. In the droppers of Pterostylidinae, vascular bundles are collateral, arranged in 1–3 rings, the outer ring abutting pericycle. Xylem: tracheids have scalariform to pitted thickenings.

ACKNOWLEDGEMENTS

I wish to thank my long-time and respected colleague, Professor William Louis Stern, for his encouragement and collaboration spanning the last 30 years. His scholarly, methodical, and enthusiastic approach to matters anatomical has been a model for future generations to follow. Professor Shirley C. Tucker instilled in me (and many others) a love for plant anatomy and morphology that was to shape my professional career; her legacy lives on in this volume. I am also grateful to Mary Gregory and Dr David F. Cutler for their patience and bibliographic and editorial expertise on this long-running project. I thank The Linnean Society of London, the Botanical Society of America, the Royal Botanic Gardens, Kew, Cornell University Press,

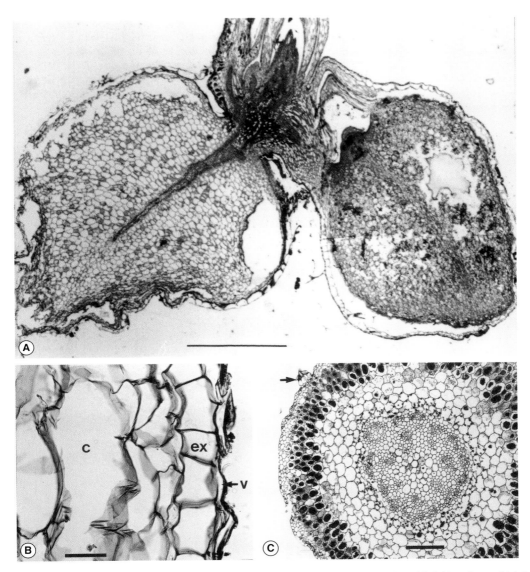

Fig. 17. Underground organs. (A) *Prasophyllum parvifolium*, LS of underground axes showing old (left) and new (right) tubers. (B) *Thelymitra venosa*, tuber TS showing velamen (v), exodermis (ex), and cortical parenchyma (c). (C) *Caladenia latifolia*, dropper TS with simple epidermis, multicellular trichome (arrow), pelotons in outer cortex, and stele with diverging trace. Scale bars: A = 500 μm; B = 50 μm; C = 300 μm. (A) Reprinted from Pridgeon (1999); (B,C) reprinted from Pridgeon and Chase (1995).

the American Orchid Society, Oxford University Press, Wendy Zomlefer, and Sarah Thomas for permission to reprint some figures. Lady Sainsbury, who has been a generous supporter of *Genera Orchidacearum* and orchid conservation research, helped to fund this contribution to the *Anatomy of the Monocotyledons* series as well.

Much of the laboratory research for this report was funded by the Australian Orchid Foundation, Asia and Pacific Science Foundation, Nell and Hermon Slade Foundation, Hermon Slade, and Trustees of Kew Foundation. Facilities were provided by Professor G.T. Prance, Professor Michael Bennett, and Dr David Cutler (Royal Botanic Gardens, Kew) and

Dr Roger Hnatiuk and Dr David Kay (Australian National Botanic Gardens). Dr Mark A. Clements and David Jones assisted in the collection of Diurideae in Australia and offered helpful discussions and logistical support along with Corinna Broers and Susan Winder (ANBG) and Professor Simon Owens and Chrissie Prychid (RBG, Kew).

Finally, I would like to acknowledge the shoulders of those giants on which we stand—the German and French anatomists of the nineteenth and twentieth centuries mentioned in this study. Their detailed observations using only basic light microscopes and their careful (and usually correct) interpretations continue to amaze their successors.

DESCRIPTIONS OF VEGETATIVE ANATOMY OF ORCHIDACEAE

SUBFAMILY APOSTASIOIDEAE GARAY

Apostasioideae consist of two genera: *Apostasia* and *Neuwiedia*. There are nine species of *Neuwiedia* and seven species of *Apostasia*. Plants are distributed from the Indian Himalayas and Sri Lanka in the west, through South East Asia, the Ryukyu Islands, the Malay and Philippine archipalagoes, and northern Queensland, Australia, to the Louisidade Archipelago in the east. *Neuwiedia* does not occur in the westernmost region of the group range. Interest in these genera arises from the fact that the end of the column is continued into a distinct style bearing an apical stigma and the anthers are joined to the style by separate filaments, whereas in all other orchids style and filaments are not separate but are intimately united to form the gynostemium.

Leaf surface

Hairs absent. **Epidermis**: cells 4–8-sided, elongated perpendicularly to veins in *Neuwiedia*, parallel to veins in *Apostasia*. **Stomata** ad- and abaxial, basically tetracytic (Fig. 18A,B), occasionally anomocytic.

Leaf TS

Cuticle thin to very thin. **Epidermis**: cells rectangular, outer walls mostly flat to rounded, domed to papillate near margins and larger veins. **Stomata** superficial, substomatal chambers small, outer ledges small, inner ledges inconspicuous. **Fibre bundles** absent. **Mesophyll** homogeneous, 5–8 cells wide; cells oval, thin-walled (Fig. 18D). **Vascular bundles** collateral, in one row, bundle sheath cells thin-walled. Sclerenchymatous bundle caps at xylem and phloem poles (Fig. 18D). **Stegmata** arranged linearly along vascular bundle fibres containing conical, rough-surfaced silica bodies (Fig. 18C). Raphide bundles in thin-walled saccate idioblasts, circular in TS, 3–6 times as long as chlorophyllous cells.

Stem TS

Hairs absent, except for thin-walled, unicellular hairs on raised multicellular cushions in *N. zollingeri*. **Cuticle** thin. **Epidermis**: cells mostly rectangular, outer walls smooth in *Apostasia*, corrugated in *Neuwiedia*. **Stomata** similar to those in leaves, present in *Neuwiedia*. **Cortex**: cells parenchymatous, oval, fairly thick-walled, broader in *Neuwiedia* than *Apostasia*. **Endodermoid layer**: cells thin-walled, but thick-walled in *A. wallichii* (Fig. 18E) and *N. veratrifolia*; casparian strips absent. **Ground tissue** heterogeneous in *Apostasia* and *N. veratrifolia* (Fig. 19B), consisting of outer, multiseriate, large-celled parenchymatous layer and central core of mostly thin-walled parenchyma cells. Ground tissue in *N. zollingeri* homogeneous and lacking the outer multiseriate parenchymatous layer (Fig. 19A,C). **Vascular bundles** collateral (Fig. 19A,C), scattered throughout ground tissue, smaller bundles aggregated towards periphery of ground tissue, larger bundles concentrated towards centre of stem. **Stegmata** occurring in peripheral bundles with conical rough-surfaced silica bodies (Fig. 19D). In *Neuwiedia veratrifolia*, conical silica bodies occurring in stegmata just internal to endodermoid layer adjacent to outer margin of sclerenchyma layer of heterogeneous ground tissue (Fig. 20A). Raphides occurring in cortical cells of *Apostasia* and in cortical and pith cells of *Neuwiedia*.

Aerial root TS

Presence of **velamen** questionable in root of *A. nuda*. Hairs unicellular, lageniform, harbouring hyphae and protists. **Exodermis** with U-thickened cell walls (Fig. 21A). **Cortex**: cells parenchymatous, thin-walled, rounded to somewhat angular (Fig. 21A). Pelotons occurring in some cortical cells in some specimens. **Endodermis** 1-layered, cell walls O-thickened in *Apostasia* (Fig. 21B), U-thickened in *Neuwiedia*. **Pericycle** 1–3-layered, cells with evenly thickened walls; thin-walled cells of endodermis and pericycle occurring opposite xylem strands (Fig. 21B). **Vascular cylinder** polyarch, xylem and phloem components alternating, vascular tissue embedded in very thick-walled sclerenchyma. Xylem occurring peripherally in radial rows or groups and centrally as isolated strands of randomly arranged tracheary elements in the pith. Phloem in *Apostasia* occurring like the xylem; *Neuwiedia* without medullary phloem. **Pith** comprising rounded to angular, somewhat thick-walled parenchyma cells. Raphides present in cortical cells in *Neuwiedia*; none seen in *Apostasia*.

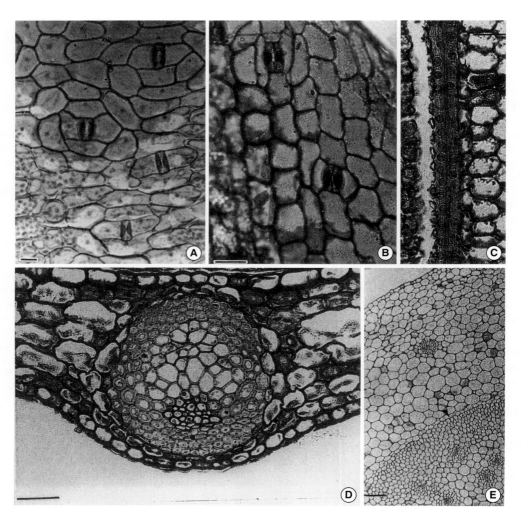

Fig. 18. Apostasioideae. *Neuwiedia* **and** *Apostasia* **leaf and stem**. (A,B) *N. borneensis*, epidermal leaf scrapings to show tetracytic stomata. (C) *Apostasia* species, leaf scraping exposing vascular bundle sclerenchyma with rows of stegmata bearing conical silica bodies. (D) *A. wallichii*, TS leaf midrib with ad- and abaxial sclerenchyma caps. (E) *A. wallichii*, TS stem showing heterogeneous ground tissue, sharply defined endodermoid layer (arrowhead), subtending multilayered sclerenchyma, and thick-walled pith cells. Scale bars: A,C = 20 µm, B = 50 µm, D,E = 100 µm. Reprinted from Stern et al. (1993b).

Terrestrial root TS

These resemble aerial roots. **Velamen** in *Apostasia wallichii* and *A. nuda* 1-layered (Fig. 21C), cell walls unsculptured. Collapsed velamen cells occurring adjacent to thick-walled exodermal cells in some roots of *A. nuda* (Fig. 20C). **Tilosomes** absent. **Exodermis** in *A. wallichii* poorly differentiated and cell walls only slightly thickened in same way as in aerial roots. In *Neuwiedia borneensis*, exodermis prominent, with thick-walled cells alternating with thin-walled passage cells. **Cortex**: cells large, very thin-walled, containing pelotons. **Endodermis** and **pericycle**: cells thin-walled; casparian strips occurring in endodermal cells. **Vascular cylinder** 6–9- or 10-arch with alternating strands of xylem and phloem in *N. veratrifolia* (Fig. 22A). Both *Apostasia* and *Neuwiedia* having isolated strands of metaxylem in pith (Figs 20D, 21C). **Pith** cells thick-walled in *Apostasia*, thin-walled in *Neuwiedia*. Raphides occurring in cortical cells of *Neuwiedia*.

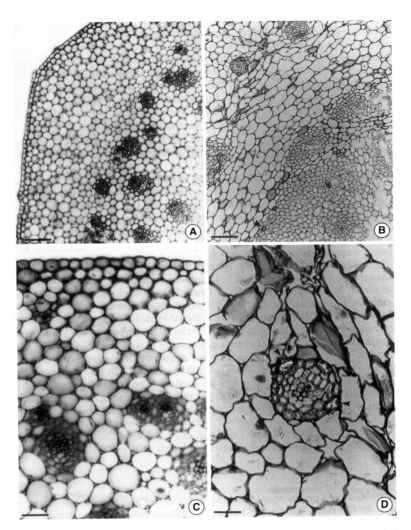

Fig. 19. Apostasioideae. *Neuwiedia* stem. (A) *N. zollingeri*. TS showing homogeneous ground tissue and absence of endodermoid and sclerenchyma layers. (B) *N. veratrifolia*, TS showing heterogeneous ground tissue, ill-defined endodermoid layer (arrowhead), and subtending multilayered sclerenchyma. (C) *N. zollingeri*, TS showing regularly shaped epidermal cells, homogeneous ground tissue, rounded parenchyma cells of cortex and ground tissue, and vascular bundles ensheathed in sclerenchyma. (D) *N. veratrifolia*, TS with cortical vascular bundle bearing conical silica bodies (arrowheads) in stegmata. Scale bars: A = 200 μm, B,C = 100 μm, D = 20 μm. Reprinted from Stern et al. (1993b).

Tubercle TS

Short, swollen branch roots produced in terrestrial roots of both genera (Fig. 20B), bearing multicellular, irregularly shaped papillae (Fig. 22B). **Epidermis**: cells thin-walled; exodermis absent. **Cortex** stratified, outer layer of cells with small nuclei and no hyphae, inner layer of cells with large nuclei and hyphae (Fig. 22C); cortical cells thin-walled, oval, those with pelotons enlarged (Fig. 22C,D). **Endodermis** and **pericycle**: cells thin-walled. **Vascular cylinder** tetrarch; xylem and phloem embedded in sclerenchyma. **Pith** cells thin-walled. **Tracheary cells**: simple perforation plates occurring in vessel elements of *Apostasia* and *Neuwiedia* in transverse or elongated end walls (Fig. 23A,B). Both tracheids and vessel elements present in roots of these genera, but only tracheids in the shoot system. Lateral wall pitting

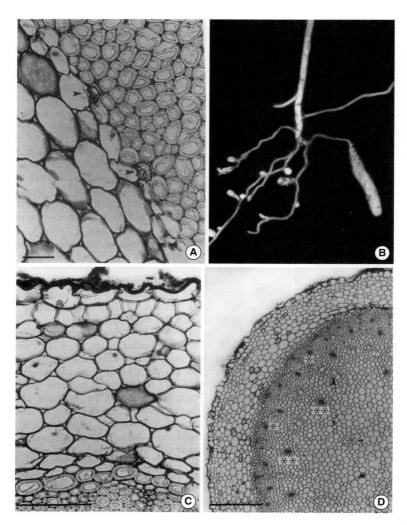

Fig. 20. Apostasioideae. *Neuwiedia* **and** *Apostasia*. (A) *N. veratrifolia*, TS stem with conical silica bodies (arrowheads) in stegmata internal to endodermoid layer at outer margin of sclerenchyma layer of heterogeneous ground tissue. (B) *A. wallichii*, terrestrial root system with tubercles. Natural size. (C) *A. nuda*, TS root showing collapsed cupulate, partially degenerated cells of 1-layered velamen, underlying exodermal layer, and thick-walled cortical cells. (D) *A. nuda*, TS root showing degenerated 1-layered velamen, polyarch vascular cylinder, peripheral xylem (arrowheads), medullary xylem (arrows), peripheral phloem (single asterisk), and medullary phloem (double asterisks). Scale bars: A = 20 μm, C,D = 100 μm. Reprinted from Stern et al. (1993b).

varying from none in some areas to regularly scalariform in others (Fig. 23B). Lateral wall pitting in tracheids similar to that in vessels.

Material examined

Apostasia (3), *Neuwiedia* (5). Full details appear in Stern, Cheadle, and Thorsch (1993b), Stern and Warcup (1994).

Reports from the literature

Brown (1833): relationships of *Apostasia*. Carlquist (2012): tracheary elements in apostasiads. Cheadle (1943): origin and trends of specialization of vessels in monocotyledons. Cheadle (1944): specialization of vessels in monocot organs. Cheadle and Kosaki (1982): occurrence and kinds of vessels in orchids. Dahlgren and Clifford (1982): comparative study of monocotyledons. Dahlgren,

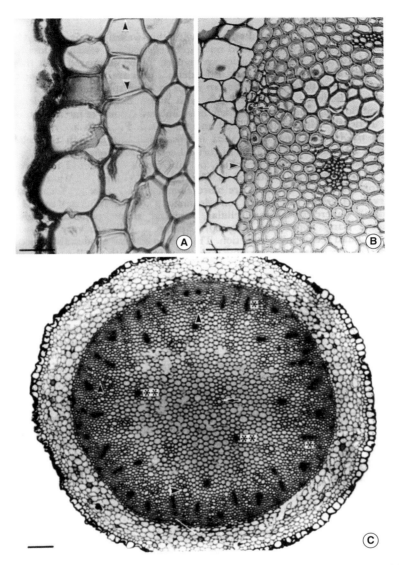

Fig. 21. Apostasioideae. *Apostasia nuda* root. (A) TS showing cells of velamen, exodermal cells with thickened anticlinal walls (arrowheads), and thin-walled cortical cells. (B) TS with O-thickened endodermal cells, 1-layered pericycle with O-thickened cell walls (arrowhead), thin-walled passage cells of endodermis and pericycle next to xylem cell clusters, and multilayered sclerenchymatous matrix in which vascular strands are embedded. (C) TS showing 1-layered velamen, peripheral (single asterisk) and medullary (double asterisks) phloem, multilayered sclerenchyma investing vascular strands, and thick-walled pith cells. Scale bars: A = 20 μm, B = 40 μm, C = 100 μm. Reprinted from Stern et al. (1993b).

Clifford, and Yeo (1985): families of monocotyledons. Dressler (1993): classification of orchids. Garay (1972): origin of Orchidaceae. Judd, Stern, and Cheadle (1993): phylogenetic position of *Apostasia* and *Neuwiedia*. Kocyan and Endress (2001): floral structure and relationships of Apostasioideae. Kocyan et al. (2004): phylogeny of Apostasioideae. Porembski and Barthlott (1988): simple rhizodermis in two species of *Apostasia*. Rao (1969 [1990], 1974): floral anatomy and relationships of the apostasias. Rolfe (1889): morphology and systematics of Apostasieae. Siebe (1903): anatomy of Apostasiinae. Solereder and Meyer (1930): comprehensive survey

48 *Descriptions of Vegetative Anatomy of Orchidaceae*

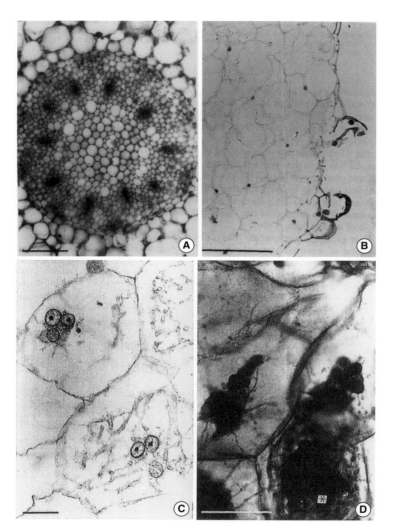

Fig. 22. Apostasioideae. *Neuwiedia* **and** *Apostasia* **root.** (A) *N. veratrifolia*, TS immature vascular cylinder to show casparian strips in endodermal cell walls. This thick section shows the still thin endodermal cell walls and unthickened walls of cells investing the vascular strands. (B–D) *A. nuda*. (B) LS root tubercle showing warty papillae, right. (C) Pelotons (hyphal coils) and hypertrophied cortical nuclei. (D) Thick section showing pelotons and dead hyphal masses, the hyphae having been digested by the host cells. Scale bars: A,B,D = 100 μm, C = 20 μm. Reprinted from Stern et al. (1993b).

of orchid anatomy. Stern, Cheadle, and Thorsch (1993b): apostasiads, systematic anatomy, and origins of Orchidaceae. Stern and Warcup (1994): root tubercles of apostasiads. Vogel (1969): monograph of Apostasieae.

Taxonomic notes

Because of the unique floral structure, i.e. a distinct style bearing an apical stigma and anthers joined to the style by separate filaments, the apostasiads have been proposed as progenitors of modern orchids. The appearance in these genera of predominantly simple perforation plates in the vessel elements of roots, an apomorphic or derived feature, precludes the idea that other orchids have evolved from the apostasiads (Judd, Stern, and Cheadle 1993). On the contrary, it is hypothesized that they diverged from other orchids in vessel development sometime in the past. The character of simple perforation plates in the

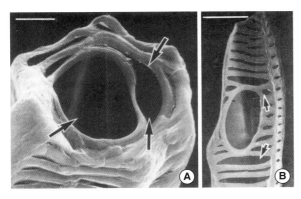

Fig. 23. Apostasioideae. *Neuwiedia* root. (A) *N. veratrifolia*, SEM depicting simple perforation in end wall of vessel element. (B) *N. borneensis*, simple perforation in end wall of vessel element. Scale bars: A = 40 μm, B = 100 μm. Reprinted from Stern et al. (1993b).

apostasiads does not lend support to the hypothesis that plants in the orchid subfamilies Cypripedioideae, Orchidoideae, Vanilloideae, and Epidendroideae were derived from the apostasiads. It would appear that the apostasiads constitute a sister group to the orchids in that, at least anatomically, they are the most closely related plants to them (see also Kocyan et al. 2004; Carlquist 2012).

SUBFAMILY CYPRIPEDIOIDEAE GARAY

Cypripedioideae consist of five genera: *Cypripedium*, *Mexipedium*, *Paphiopedilum*, *Phragmipedium*, and *Selenipedium*. These are terrestrial, lithophytic, or epiphytic rhizomatous herbs, ranging from tall terrestrial plants in *Selenipedium* to short, colony-forming herbs in *Mexipedium*. The several genera may be divided vegetatively on the vernation of their leaves: *Cypripedium* and *Selenipedium* have plicate leaves; *Mexipedium*, *Paphiopedilum*, and *Phragmipedium* have conduplicate leaves. Common names for the cypripedioids, slipper orchids and lady's slipper orchids, arise from the pouch-shaped or saccate lips. The slipper orchids, except for *Selenipedium*, are highly prized by hobbyists. *Cypripedium* is a terrestrial genus of about 50 species distributed across temperate regions from Europe east to Asia and the Aleutian Islands across North America and south to Guatemala and Honduras. *Selenipedium* has five terrestrial species in tropical America growing from Panama across northern South America. *Paphiopedilum* with 60 or more terrestrial, lithophytic, and epiphytic species is distributed from India and southern China to the Philippines, the Malay Archipelago, New Guinea, the Solomon Islands, and South East Asia. *Phragmipedium*, with about 15 terrestrial, lithophytic, or epiphytic species, grows from southern Mexico and Guatemala across Central America to Bolivia, Peru, and Brazil in South America. *Mexipedium* is a monospecific, lithophytic, narrow endemic only from Oaxaca in southern Mexico. Cypripedioids are absent from Africa and Australia. Species of *Cypripedium* are considered medicinal in North America, Mexico, Guatemala, Colombia, Russia, India, and Japan.

Leaf surface

Hairs: glandular and non-glandular hairs associated with both surfaces of plicate-leaved genera (Fig. 24A,B), absent from conduplicate-leaved genera. **Epidermis**: cells polygonal to rhomboidal; walls straight-sided in *Selenipedium* and conduplicate-leaved genera; undulating in most *Cypripedium* species (Fig. 24C), straight-sided in a few species, adaxial cells interdigitating in *C. debile* (Fig. 24D); adaxial cells of *Paphiopedilum* palisade-like. **Stomata** occurring on both leaf surfaces in *Selenipedium* and most *Cypripedium* species (Fig. 24D), but in some other *Cypripedium* species only on abaxial surface; stomata absent from adaxial surfaces in conduplicate-leaved genera. Stomata tetracytic in *Selenipedium*, *Phragmipedium*, and several species of *Paphiopedilum*; anomocytic in *Cypripedium* (Fig. 24D), surrounded by 4–7 epidermal cells.

Leaf TS

Cuticle thin in plicate-leaved genera (Fig. 25A), thick in conduplicate-leaved genera. **Epidermis** 1-layered (Figs 24B, 25A,B). **Mesophyll** homogeneous in plicate-leaved genera (Figs 24B, 25B); heterogeneous in many conduplicate-leaved genera with palisade and spongy layers: *Paphiopedilum bellatulum*, *P. niveum*, *P. rothchildianum*, *P. venustum*, and *Phragmipedium caudatum* having two layers of palisade cells. **Vascular bundles** in one row, each surrounded by several layers of fibres in conduplicate-leaved genera; sclerenchyma layer surrounding larger bundles but smaller veins lacking this or with it only abaxially in plicate-leaved genera. **Stegmata**: conical silica bodies occurring in leaves of *C. irapeanum*.

Leafy stem TS

Hairs non-glandular, characterized by an attenuated apical cell, and glandular with a bulbous apical cell. **Epidermis** 1-layered; cells spherical to rectangular; cells markedly elongated in *Cypripedium*, but only slightly in other genera. **Stomata** occurring only in *Cypripedium*

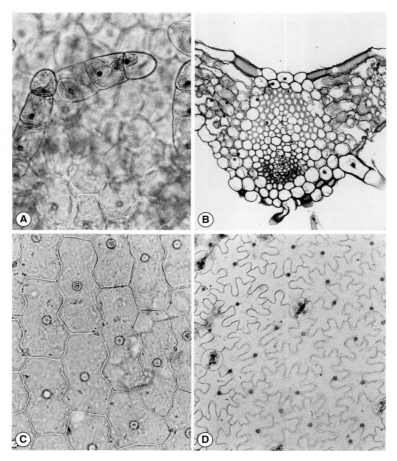

Fig. 24. Cypripedioideae. *Cypripedium* leaf. (A) *C. daliense*, adaxial scraping illustrating 4- and 5-celled non-glandular hairs, ×350. (B) *C. flavuum*, TS vein with thin-walled sheath cells, xylem-side arcuate sclerenchyma band, and mesophyll, ×380. (C) *C. henryi*, adaxial scraping with undulating epidermal cell walls, ×950. (D) *C. debile*, adaxial scraping showing interdigitating epidermal cell walls, ×350. Photographs by A.M. Pridgeon.

and *Selenipedium*; subsidiary cells absent in both genera. **Cortex** relatively narrow, bounded internally by sclerotic sheath of pericyclic fibres; fibres O- and U-thickened in conduplicate-leaved genera. Cortical cells generally cylindrical; cells in *Cypripedium irapeanum* have arm-like projections. Ground-tissue cells in *Cypripedium* parenchymatous and thick-walled in lower portions of stem; in *Paphiopedilum* and *Phragmipedium* cell walls usually sclerified; in *Selenipediium* parenchyma cells thick-walled adjacent to pericyclic sheath but thin-walled toward centre of stem. **Vascular bundles** amphivasal in basal parts of stems in plicate-leaved genera and throughout stems of conduplicate-leaved genera. Collateral bundles present in stems above basal internodes in plicate-leaved genera. Phloem sclerenchyma caps characteristic of bundles in *Cypripedium* and *Selenipedium* but may be present along the xylem as well. Bundles scattered in *Cypripedium* except relatively crowded in basal regions of stem; bundles peripheral in *Selenipedium*. Bundles scattered throughout ground tissue in conduplicate-leaved genera.

Rhizome TS

Two types of rhizome occurring in cypripedioids: highly condensed conical rhizomes characteristic of conduplicate-leaved genera and more or less elongated axes in plicate-leaved genera. Scale leaves and adventitious roots associated with all rhizomes.

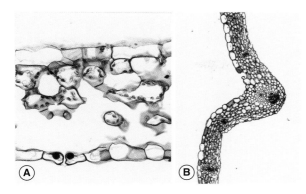

Fig. 25. Cypripedioideae. *Cypripedium* leaf. (A) *C. debile*, TS with abaxial stoma, large mesophyll intercellular spaces, and 1-layered epidermises, ×950. (B) *C. plectrochilum*, TS with 1-layered epidermises and homogeneous mesophyll, ×150. Photographs by A. M. Pridgeon.

Epidermis 1-layered, hairs and stomata absent. **Cortex** extensively developed in *Cypripedium*, much reduced in *Selenipedium* and conduplicate-leaved genera. **Endodermis** of *Cypripedium* and *Selenipedium* usually 1-layered; multilayered in conduplicate-leaved genera; cell wall thickenings varying by species: walls O-thickened, U-thickened, or thin. In endodermis of plicate-leaved genera thickenings various; in conduplicate-leaved genera and in four species, they are O-thickened; U-thickened in *Selenipedium palmifolium*. **Vascular cylinder** bounded by pericyclic zone surrounding core of scattered or fused vascular bundles immersed in parenchymatous ground tissue. *Cypripedium* noted for compact cylinders of vascular bundles; in *Selenipedium* vascular bundles relatively few, distinct, and scattered near periphery; in conduplicate-leaved genera vascular bundles numerous and distributed across ground tissue. Vascular bundles amphivasal in conduplicate-leaved genera and probably in *Cypripedium* as well; vascular bundles collateral in *Selenipedium* with sclerenchyma caps at phloem poles and sometimes at xylem poles. Starch occurring in ground-tissue cells of all genera.

Root TS

Fleshy roots typical of *Paphiopedilum* and *Phragmipedium*; fibrous roots occurring in *Selenipedium* and *Cypripedium*. **Hairs** unicellular, present in all genera except *Selenipedium*. **Velamen** multiseriate in *Paphiopedilum* and *Phragmipedium* varying from six to 10 layers; 1-layered in *Selenipedium* and *Cypripedium irapeanum*; other *Cypripedium* species having a 1-layered epidermis of living cells. **Tilosomes**: lamellate, present in one out of 19 *Paphiopedilum* species and two out of five *Phragmipedium* species examined (Pridgeon et al. 1983). **Exodermis** present between velamen and cortex in conduplicate-leaved genera, also in *Selenipedium* and *C. irapeanum*; long cells evenly thickened or O- and U-thickened in *Paphiopedilum*; U-thickened in *Phragmipedium* species, *Selenipedium*, and *C. irapeanum*. **Cortex** consisting of parenchyma in all genera. **Endodermis** 1-layered in *Paphiopedilum venustum*; in *Phragmipedium caudatum* two layers occurring in localized areas; cells unthickened or U-thickened in *Cypripedium acaule*, and *C. fasciculatum*, U-thickened in *C. arietinum*. *C. californicum*, and *C. reginae*, thin-walled or O-thickened in *C. calceolus* and *C. fasciculatum*, O-thickened in *C. candidum*; O-thickened in *Paphiopedilum insigne* and *P niveum*, U-thickened in *P. venustum*, U-thickened in *Phragmipedium caudatum*, and U-thickened in *Selenipedium palmifolium*. **Pericycle** multilayered in *Paphiopedilum*, *Phragmipedium*, and *Selenipedium*, composed of sclerenchyma cells; 1-layered in *Cypripedium*. **Vascular cylinder** polyarch with linear xylem strands and circular phloem strands alternating with each other. **Pith** present in *Cypripedium californicum* and *C. irapeanum*; present or absent in other species of *Cypripedium*.

Material examined

Cypripedium (11), *Paphiopedilum* (5), *Phragmipedium* (3), *Selenipedium* (1). Information mostly from Rosso (1966).

Reports from the literature

Atwood (1984): relationships of Cypripedioideae with summary of anatomical observations and additional data on leaf anatomy of *Paphiopedilum*. Atwood and Williams (1978, 1979): electron microscope examination of foliar surfaces in *Paphiopedilum* and *Phragmipedium* and use of epidermal features in identifying sterile plants. Dressler (1993): phylogeny and classification of orchids. Faber (1904): comparative study of vegetative anatomy of *Paphiopedilum*, *Phragmipedium*, and *Cypripedium*; apparently Faber had no material of *Selenipedium*. Karasawa and Aoyama (1981): leaf anatomy of 36 *Paphiopedilum* species. Pridgeon et al. (1999): comprehensive review of Orchidaceae including vegetative anatomy. Rosso (1966): the only modern all-around treatment of cypripedioid vegetative anatomy, though lacking *Mexipedium*, which was not described until 1992.

Additional literature

Barthlott and Capesius (1975 [1976], 1976): velamen of *Paphiopedilum* and *Phragmipedium*. Benzing and

Pridgeon (1983): trichomes of *Paphiopedilum*. Bernard (1904): mycorrhiza of *Cypripedium*. Borsos (1977, 1980, 1982b): leaf and stem anatomy of *Cypripedium*. Camus (1908, 1929): general anatomy of *Cypripedium*. Cyge (1930): leaf anatomy of *Cypripedium*. D'Amelio and Zeiger (1988): non-chlorophyllous guard cells in *Paphiopedilum*. Dietz (1930): underground organs of *Cypripedium* and *Paphiopedilum*. Duruz (1960): stomata of *Cypripedium calceolus* and *Paphiopedilum bellatulum*. Fricke (1926): bract and foliage leaves of *Cypripedium* × *sedenii*. Fuchs and Ziegenspeck (1925, 1926): root and stem anatomy of *Cypripedium*. Gao et al. (2009): mycorrhiza of four *Cypripedium* species. Guan et al. (2011): ecological leaf anatomy of *Cypripedium* and *Paphiopedilum*. Holm (1904): root anatomy of *Cypripedium*. Holm (1908): vegetative anatomy of *Cypripedium pubescens*. Holm (1927): leaf anatomy of *Cypripedium pubescens*. Kaushik (1983): anatomical study of two Indian species of *Paphiopedilum*. Krüger (1883): anatomy of aerial organs of '*Cypripedium insigne*, *C. barbatum*, and *C. venustum*' (these are probably species of *Paphiopedilum*, not *Cypripedium*!). Lawton et al. (1992): leaf anatomy of three species of *Paphiopedilum*. MacDougal (1899b): root anatomy of *Cypripedium*. Macfarlane (1892): vegetative anatomy of *Cypripedium* and hybrids. Meinecke (1894): anatomy of aerial roots of *Paphiopedilum* and *Cypripedium*. Möbius (1887): leaf anatomy of *Paphiopedilum* and *Cypripedium*. Møller and Rasmussen (1984): silica bodies in leaf of *Cypripedium irapeanum* only. Mulay and Panikkar (1956): velamen of *Cypripedium speciosum*. Müller (1908): root anatomy of *Cypripedium*. Namba et al. (1981b): drug plant anatomy of *Cypripedium*. Nelson and Mayo (1975): non-chlorophyllous guard cells in *Paphiopedilum* leaf. Nestler (1907): glandular hairs of *Cypripedium*. Neubauer (1978): inflorescence axis of *Paphiopedilum* and *Phragmipedium*. Porembski and Barthlott (1988): velamen in *Paphiopedilum* and *Phragmipedium*. Pridgeon et al. (1983): tilosome presence or absence in root of *Paphiopedilum* and *Phragmipedium*. Rao (1998): leaf anatomy of *Paphiopedilum*. Rūgina et al. (1987): leaf and root anatomy of *Paphiopedilum* and *Phragmipedium*. Rutter and Wilmer (1979): TEM of epidermis in *Paphiopedilum* showing lack of chloroplasts in guard cells of three species. Salokhin et al. (2005): leaf epidermis of *Cypripedium*. Shershevskaya (1963): anatomy of three *Cypripedium* species. Silva et al. (2006): leaf epidermis of *Cypripedium*. Solereder and Meyer (1930): leaf and axis anatomy of *Cypripedium* and *Paphiopedilum* and leaf of *Phragmipedium*. Sun (1995): leaf SEM of *Paphiopedilum*. Tatarenko (1995): mycorrhiza of *Cypripedium*. Zeiger et al. (1985): stomata of *Paphiopedilum*. Zhang et al. (2012): stomata of *Paphiopedilum*.

Taxonomic notes

Anatomical similarities and differences appear among the cypripedioid genera which may have taxonomic value/significance. Velamen is absent in all *Cypripedium* species, except *C. irapeanum*. It is present in all other genera. Velamen is multilayered in *Phragmipedium* and *Paphiopedilum* species. Exodermis occurs in *C. irapeanum* but is absent in other species of *Cypripedium*. It is present in *Selenipedium*, *Phragmipedium*, and *Paphiopedilum*. Pericycle consists of thick-walled cells in *C. irapeanum*; cells are thin-walled in all other *Cypripedium* species and in *Selenipedium*, *Phragmipedium*, and *Paphiopedilum*. Pericycle is multiseriate in *Selenipedium*, *Phragmipedium*, and *Paphiopedilum* but not in *Cypripedium*. Pith is present in roots of *C. californicum* and *C. irapeanum*; it is absent in other species of *Cypripedium*. Pith occurs in roots of *Selenipedium*, *Phragmipedium*, and *Paphiopedilum*. Endodermis is present in aerial stems of *C. irapeanum*; absent in other species of *Cypripedium*. It is present in aerial stems of *Selenipedium*, *Phragmipedium*, and *Paphiopedilum*. Vascular bundles in aerial stems of *Cypripedium* and *Selenipedium* are collateral; amphivasal in *Phragmipedium* and *Paphiopedilum*. Among the species of *Cypripedium*, *C. irapeanum* is distinct anatomically in the presence of velamen, exodermis in roots, pericycle of thick-walled cells, presence of pith in roots, and endodermis in aerial stems.

SUBFAMILY ORCHIDOIDEAE LINDL.

Tribe Diseae (Lindl. ex Benth.) Dressler

Diseae are glabrous, usually terrestrial herbs producing annual root tubers. Stems are mostly upright; leaves conduplicate, basal, or cauline, mostly narrowly lanceolate or linear, sometimes ovate, non-articulate. Diseae comprise 400 species and 11 genera assorted into five subtribes mostly of African and Madagascan distribution with two genera ranging into Asia. *Disa*, endemic to southern Africa and adjacent islands, is recorded as medicinal in this region. Some species of *Disa*, *Brownleea*, and *Satyrium* are amenable, but difficult to cultivate.

Leaf surface

Leaves mostly glabrous; **hairs** prominent in some species of *Disa*, unicellular non-glandular and multicellular glandular; unicellular in *Disperis villosa*. **Cuticle** striate to reticulate or punctate. **Epidermis**: cells irregularly isodiametric to regularly polygonal (Figs 26A–C, 27A–C); anticlinal walls ranging from straight to undulate. **Stomata**

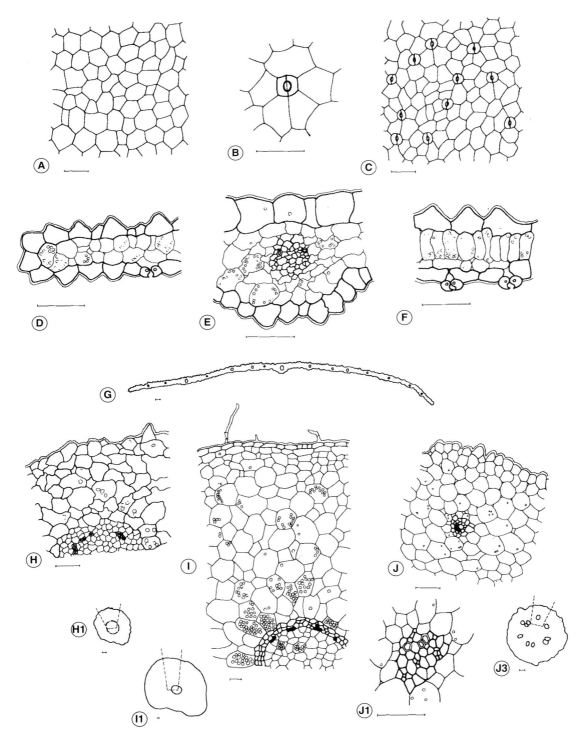

Fig. 26. Diseae. *Disperis johnstonii*. (A–G) Leaf. (A) Adaxial surface. (B) Anomocytic stoma. (C) Abaxial surface with anomocytic stomata. (D) TS margin with homogeneous mesophyll and conical epidermal cells. (E) TS midvein showing position of vascular bundle. (F) TS with heterogeneous mesophyll and abaxial stomata. (G) TS showing one row of vascular bundles. (H,I) TS root with 1-layered velamen, thin-walled cortical cells with starch granules, thin-walled endodermal and pith cells. (J) TS stem with ring of vascular bundles and thin-walled ground-tissue cells. Scale bars for line drawings = *c*.100 μm throughout the book.

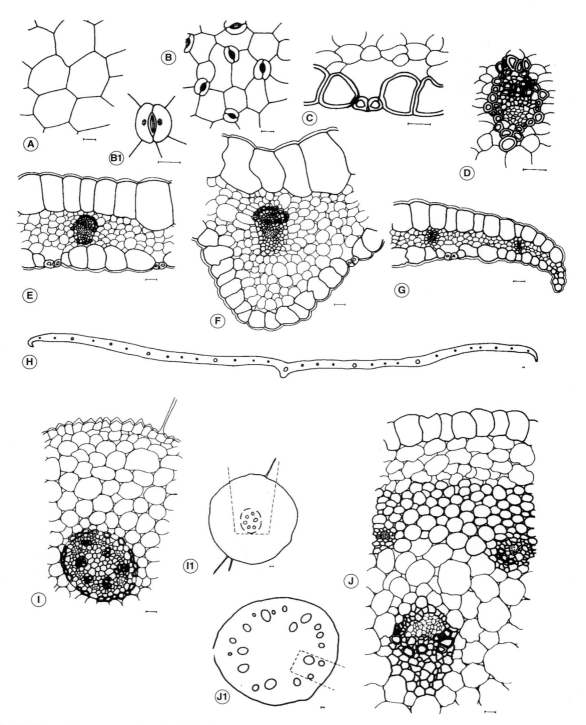

Fig. 27. Diseae. *Satyrium nepalense*. (A–H) Leaf. (A) Adaxial surface. (B) Abaxial surface with anomocytic stomata. (C) TS leaf epidermis with stoma and substomatal chamber. (D) Vascular bundle with adaxial sclerenchyma. (E,G) TS showing homogeneous mesophyll, thin-walled columnar adaxial epidermal cells, position of vascular bundles. (F) TS midvein showing adaxial sclerenchyma cap. (H) TS with one row of vascular bundles. (I) TS root with 1-layered velamen, thin-walled endodermal cells mostly opposite xylem, thick-walled cells opposite phloem. (J) TS stem to show thin-walled epidermal and cortical cells, subjacent layer of thin-walled sclerenchyma cells, and vascular bundles embedded in thin-walled ground tissue.

lacking subsidiary cells (Figs 26B,C, 27B); usually abaxial, but ad- and abaxial in some *Disa* species (including all species of sections *Herschelianthe* and *Monadenia*), all species of *Schizodium*, and a few species of *Ceratandra*, *Corycium*, *Disperis*, *Evotella*, and *Pachites*.

Leaf TS
Cuticle thin (Figs 26E, 27F) to moderately thick; up to 10 μm thick in *Disa* sections *Stenocarpa* and *Herschelianthe*. **Epidermis** 1-layered, cells rounded-square or rounded-rectangular (elongated); adaxial cells enlarged (Fig. 27E–G), some possibly for water storage, especially in *Satyrium* and some *Pterygodium* species. **Stomata** superficial (Fig. 27C,E) to slightly raised (Fig. 26F), ledges minute. **Fibre bundles** absent. **Mesophyll** generally homogeneous (Fig. 27E–G) or weakly heterogeneous. Mesophyll configuration labile with a distinctly heterogeneous mesophyll (Fig. 26F) occurring sporadically in all taxa but characteristic of *Schizodium* and *Disa* section *Micranthae*. Mesophyll cells of all species containing starch granules and raphide bundles. **Vascular bundles** in one row (Figs 26G, 27H) and generally collateral. Bundle sheath parenchymatous in all species; cells thin-walled and frequently containing raphide bundles. Sclerenchyma caps in all species of *Disa* sections *Emarginatae*, *Repanda*, and *Amphigena* and sporadically in *Ceratandra*. Sclerenchyma caps developed solely or more extensively at phloem poles of bundles (Fig. 27D–G) while largest bundles totally enclosed in sclerenchyma. **Stegmata** not seen in the species studied.

Aerial stem TS
Stems in Diseae usually unbranched and solitary. Aerial stems erect, leafy throughout or with basal rosettes. In some sections of *Disa* and in *Satyrium* species, leaves borne on separate stems (sterile shoots) differing from flowering stems that vary extensively in height from a few centimetres to over 1 metre in *Pterygodium magnum*. All adventitious roots arising from basal portions of stems. Aerial stems separated from tuber by comparatively thin underground stems and roots borne along entire length of these stems.
  Hairs unicellular only in *Disa glandulosa* and in some *Disperis* species, including *D. bolusiana*, *D. capensis*, *D. fanniniae*, and *D. villosa*. Papillae in some species of *Satyrium*. **Cuticle** thin to very thin. **Epidermis** 1-layered (Figs 26J, 27J), cells rounded-square to rounded-rectangular to tangentially elongate. **Stomata** occasionally present in most species. **Cortex**: cells generally thin-walled (Fig. 27J); walls strongly thickened in *Schizodium obliquum*; cells containing starch granules and raphides. **Fibre bundles** occurring in *Disa pulchra*. All species with a compact sclerenchymatous ring of cells between cortex and ground tissue (Fig. 27J). **Ground tissue** parenchymatous (Figs 26J, 27J) but central regions frequently sclerified; cell walls thin; cells sometimes containing raphide bundles and starch granules in some species of *Pterygodium*. **Vascular bundles** collateral (Fig. 27J), occurring in 1–5 rings. Bundles usually lacking a sheath. Sclerenchyma caps occurring at phloem pole in many species of *Satyrium* and a few species of *Disa*; a definite parenchyma cap of small cells developed at xylem pole in many species. **Stegmata** not seen in association with sclerenchyma bundle caps.

Underground stem TS
Stems arising from apex of tubers and growing towards soil surface. Underground stems commonly enveloped in a sheath derived from scale leaves or velamen and exodermis of dropper of previous season. **Epidermis** simple. **Cortex** comprising 60–80% of stem; cells may contain pelotons. Supportive tissue concentrated in central part of stem. No sclerenchyma ring as in aerial stems. **Vascular bundles** in two rings; xylem and phloem elements very irregularly arranged.

Root TS
Hairs sparse, uni- or multicellular, varying from shortly papilla-like to elongate (Fig. 26I). **Velamen** usually 1-layered (Figs 26H, 27I); cells elliptical to elongate with unthickened to slightly thickened walls. **Tilosomes** not seen. **Exodermis** 1-layered; cells isodiametric, radial walls slightly thickened. **Cortex** constituting greatest volume of root (Figs 26I, 27I), consisting of large isodiametric cells; most cells of outer cortex usually thin-walled with large pelotons; raphide bundles abundant, starch granules rare. **Pericycle** irregular, cells isodiametric, thin-walled (Fig. 27I). **Vascular cylinder**: xylem and phloem alternating around central pith (Fig. 27I). Vascular tissue embedded in thin-walled parenchyma except cells with thickened walls in *Pachites bodkinii*. **Pith** usually parenchymatous (Figs 26I, 27I), sometimes sclerenchymatous.

Root tuber TS
Underground storage tubers are swollen roots enveloped in a **velamen** of 1–3 layers of elliptical or periclinally compressed cells. Tubers hairy or smooth. **Exodermis** 1-layered, constituting outer cortical margin. Fungal hyphae generally absent. Cells containing starch granules. **Vascularization** 'monostelic' or 'polystelic' depending upon whether siphonostele dissected or not. 'Monosteles'

comprising a ring of alternating xylem and phloem elements surrounding parenchymatous pith; 'monosteles' occurring in most genera, including *Disa bracteata* (as *Monadenia bracteata*), *D. sagittalis*, and most *Disperis* species. 'Polystelic' tubers present in all species of *Brownleea*, *Satyrium*, *Disa* sections *Amphigena*, *Stenocarpa* (except *D. tenella*), and *Herschelianthe*, *Disperis lindleyana*, and *D. tysonii*; individual strands of 'polysteles' (i.e. meristeles) surrounded by an endodermis and pericycle.

An unusual underground 'organ' occurring in three species of *Ceratandra* section *Ceratandra*: fascicles constituting an underground stem surrounded by a sheath of roots, the ends of some of which produce tubers.

Dropper TS

New tubers are formed on vertical stalks called droppers that arise from the stem base and have root structure.

Velamen 1–3-layered. **Exodermis** large-celled. **Cortex** comprising large, thin-walled cells; pelotons usually absent. **Stele** dissected in some species and undissected in others.

Material reported

Descriptions above are based largely on the detailed monograph of Kurzweil et al. (1995), where full details are given for the genera listed here, and also other sources cited below.

Brownleea (3), *Ceratandra* (4), *Corycium* (5), *Disa* (34 plus 5 spp. previously in *Monadenia* and 13 spp. previously in *Herschelianthe*), *Disperis* (9), *Evotella* (1), *Huttonaea* (3), *Pachites* (1), *Pterygodium* (9), *Satyrium* (33 plus 1 sp. previously in *Satyridium*), *Schizodium* (3).

Reports from the literature

Chesselet (1989): report on systematic implications of leaf anatomy in subtribes Disinae and Coryciinae. Kurzweil et al. (1995): thorough review of vegetative anatomy of Diseae with cladistic analyses of the taxa and extensive discussion of vegetative characters in tribe Diseae and subtribes Satyriinae, Disinae, Brownleeinae, Huttonaeinae, and Coryciinae. Linder and Kurzweil (1990): phylogenetic study of subtribe Disinae based on floral morphology. Linder and Kurzweil (1994): study of phylogeny and classification of tribe Diseae based almost exclusively on floral morphology with a few references to vegetative morphology of tubers and leaves. Moreau (1913): root tuber anatomy of *Disa*, *Huttonaea*, and *Satyrium*. Porembski and Barthlott (1988): velamen of *Disa*. Pridgeon et al. (2001): comprehensive survey of tribe Diseae. Sasikumar (1975a): root and tuber anatomy of *Satyrium*. Singh (1981): stomata of *Satyrium*. Staudermann (1924): hairs of *Disa* and *Disperis*.

Taxonomic notes

Anatomical data do not materially change the phylogenetic hypothesis presented by Linder and Kurzweil (1994). The most informative characters are the presence of sclerenchyma in foliar vascular bundles and the occurrence of 'polysteles' in tubers. The amphistomatic condition characterizes *Evotella*, *Pachites*, *Schizodium*, and *Disa* sections *Monadenia* and *Herschelianthe*, but occurs sporadically in *Ceratandra*, *Corycium*, and *Disperis* and is therefore of little taxonomic significance. The heterogeneous mesophyll is characteristic of *Schizodium* and *Disa* section *Micranthae* but it also occurs among other genera where it has probably evolved independently. Sclerenchyma caps and sheaths in vascular bundles of stems are common in Satyriinae, occur sporadically in Disinae, vary within some of the species, and are of limited taxonomic value. *Schizodium* is treated as a separate genus in Pridgeon et al. (2001), but is now considered to be included in *Disa*.

Tribe Diurideae (Endl.) Lindl.

By Alec M. Pridgeon

Diurideae comprise nine subtribes, 37 genera and about 900 species, distributed in India and Nepal through South East Asia and Japan to Australia, New Zealand, Fiji, Samoa, and the Society Islands.

Plants are generally small, although inflorescences of some species of *Caladenia* may reach heights of 1 m, and their root systems depths of 1 m as well. Growth habit is mainly terrestrial, rarely epiphytic. Leaves are spiral, distichous, basal, or absent; convolute to weakly plicate, soft herbaceous to somewhat fleshy and coriaceous, dorsiventral to terete. Adventitious root systems are generally well developed, with some or all of the following depending on the taxon (Fig. 15): (1) absorbing storage roots; (2) droppers, often enclosed in fibrous sheaths, which arise endogenously from: (a) undifferentiated cortical parenchyma of stem base; (b) the previous dropper; (c) the distal pole of the previously formed dropper; or (d) coalescence of an axillary bud with root primordium on parent axis; droppers usually grow downwards either vertically or obliquely and terminate in replacement root tubers; (3) stolonoid roots, which originate in the same way as droppers but grow mainly horizontally and terminate in reproductive tubers, leading to formation of tufts, clumps, or colonies (e.g. *Corybas*); (4) root tubers (Fig. 17A), or swollen storage roots bearing root buds that produce new shoots and inflorescences; and (5) stem tubers, differentially swollen underground shoots with scale leaves subtending axillary buds (*Rhizanthella* only).

Leaf surface

Hairs: if present, of three main types: multicellular glandular (Fig. 1D), multicellular non-glandular, and bi- to triseriate glandular (only in *Aporostylis*, Caladeniinae); all hairs unbranched. Papillae present, one per cell, in some genera (Fig. 1E). **Cuticle** variously sculptured, most often striate (Fig. 1F). **Epidermis** with rectangular to polygonal cells, anticlinal walls sinuous or straight, thin- or thick-walled, with or without conspicuous pit-fields. **Stomata** on one or both surfaces (Fig. 2E); anomocytic (Fig. 1F).

Leaf TS

Dorsiventral to terete. **Stomata** superficial; outer and inner ledges small, more or less equally developed. **Hypodermis**: present only in *Praecoxanthus* (Caladeniinae). **Mesophyll** homogeneous (Fig. 2E) or heterogeneous depending on taxon. **Vascular bundles** collateral, generally in one row in centre of mesophyll. Sclerenchyma bundle caps mostly absent, and bundle-sheath cells mostly thin-walled. **Stegmata** absent. Crystals either rod-shaped or raphides, in assimilatory cells or mesophyll idioblasts.

Stem TS

Stem often differentiated into two regions: (1) an aerial portion sometimes with multicellular hairs and a ring or sheath of mechanical tissue in inner cortex, and (2) a basal, barely subterranean portion ('collar') with multicellular, mycorrhizal hairs (Fig. 6A), cortical infection zones, and an endodermis with casparian strips. **Cuticle** usually thin, striate. **Stomata** present or absent depending on taxon. **Vascular bundles** collateral, usually in one or two rings. **Stegmata** absent.

Tuber TS

Root tubers cylindrical to globose, often enclosed in a few- to multilayered fibrous sheath, extending up the dropper as well in some species. **Hairs** present or not; if present usually multicellular and associated with mycorrhiza. **Velamen** or simple epidermis depending on taxon (Fig. 17B). **Exodermis** 1-layered (Fig. 17B). **Tilosomes** absent. **Ground tissue** parenchymatous (Fig. 17B), filled with starch granules in new tubers. **Endodermis** 1-layered with casparian strips. **Pericycle** as for root. **Stele** dissected or not depending on taxon. Crystals in ground tissue either rod-shaped or raphides.

Absorbing root TS

Hairs unicellular, multicellular/uniseriate, or multicellular/multiseriate, usually associated with mycorrhiza; unbranched. **Velamen** 1–3-layered in most taxa. Simple epidermis in Acianthinae, some Prasophyllinae, and *Glossodia* (Caladeniinae). **Tilosomes** absent, except in *Cryptostylis* (Cryptostylidinae). **Exodermis** 1-layered. **Cortex** parenchymatous. **Endodermis** 1-layered with casparian strips. **Pericycle** 1-layered, cells elliptical to polygonal and thin-walled. **Stele** medullated. Xylem: tracheids with various thickening patterns. Phloem with sieve-tube elements, companion cells, and parenchyma.

Dropper/stolonoid root TS

Hairs often present, mycorrhizal (Fig. 17C). **Velamen** or simple epidermis depending on taxon (Fig. 17C). **Exodermis** 1-layered. **Tilosomes** absent. **Cortex**: cells often with secondary, banded, suberized wall thickenings, especially in Caladeniinae. **Endodermis** 1-layered with casparian strips. **Pericycle** as for root. **Stele** dissected or not.

Subtribe Acianthinae (Lindl.) Schltr.

Glabrous, tuberous, geophytic herbs. Tubers small, globose, naked. Leaf solitary, convolute, sessile or petiolate, as wide as long or nearly so, erect to prostrate, entire or lobed, thin to membranous. Five genera, c.160 species.

Leaf surface

Hairs absent. **Cuticle** smooth to striate at margins in *Acianthus*, minutely punctate with radiating ridges in *Cyrtostylis*, micropapillate to ridged adaxially in *Corybas*, minutely punctate in *Townsonia*. **Epidermis**: cells polygonal; walls thin (except moderately thick with conspicuous pit-fields in *Townsonia*), sinuous in *Acianthus*, and often sinuous in *Corybas*. **Stomata** on abaxial surface only; guard-cell pairs remarkably large in *Cyrtostylis* (92 μm long, 98 μm wide) and *Corybas macranthus* (88 μm long, 88 μm wide).

Leaf TS

Cuticle thin. **Epidermis**: cells isodiametric, elliptical to polygonal; outer walls thickened. Adaxial cells usually larger than abaxial cells; in *Cyrtostylis* adaxial cells large, occupying about ½ leaf volume, the abaxial cells about ¼. **Stomata** superficial (slightly elevated in *Cyrtostylis*), substomatal chambers shallow. **Mesophyll** homogeneous in all taxa except *Cyrtostylis* with two palisade layers. **Vascular bundles** as for tribe. Raphides in mesophyll idioblasts; druses in epidermis of *Cyrtostylis*.

Collar/stem TS

Hairs on collar multiseriate and associated with mycorrhiza in *Acianthus*, some species of *Corybas*, and *Townsonia*; uniseriate, buttress-type hairs in *Cyrtostylis* and some species of *Corybas*. **Cuticle** thin, smooth. **Epidermis**: cells elliptical to polygonal, anticlinally elliptical in *Townsonia*; outer wall thickened. **Stomata** present only in *Acianthus amplexicaulis* (as *A. sublestus*). **Ground tissue** parenchymatous, outer layers often infected with hyphae and pelotons. Outer 3–5 layers in *Townsonia* disrupted with large lacunae; inner one or two layers sclerotic, forming a sclerenchyma sheath. **Endodermis** of collar: cells elliptical to polygonal. **Vascular bundles** in one ring (with smaller bundles in flanges in *Corybas*). Stele of collar medullated, 4–13-arch. Xylem: tracheids mostly with helical or scalariform thickenings. Raphides present in ground tissue.

Tuber TS

Hairs unicellular on 1-layered sheath enclosing tuber; hairs absent in *Corybas* except uniseriate, buttressed type in *C. unguiculatus*. **Velamen** 1- or 2-layered, cells elliptical to polygonal, periclinally compressed. Walls suberized, of uniform thickness, but outer and/or radial walls thicker in *Acianthus*. **Exodermis**: cells isodiametric to polygonal; walls suberized, of uniform thickness or differentially thickened depending on taxon. **Ground tissue** parenchymatous, outer layers often dissociated, inner layers with starch granules. **Endodermis**: cells elliptical to polygonal. **Stele** usually dissected into 2–7 discrete vascular strands, each surrounded by endodermis and 3–6-arch. Stele medullated, undissected, 4- to 5-arch in some species of *Corybas*. Xylem: tracheids mostly with helical or scalariform thickenings. Raphides in ground tissue of *Acianthus* and *Corybas*; starch granules in all taxa.

Root TS

Hairs multiseriate, associated with mycorrhiza, in *Acianthus*, some species of *Corybas*, and *Townsonia*. **Epidermis**: cells elliptical to polygonal; walls of uniform thickness or outer walls thicker depending on taxon. **Cortex** parenchymatous, outer layers infected with hyphae and pelotons, inner layers often with starch granules. **Endodermis**: cells elliptical to polygonal. **Stele** medullated (sclerotic in *Cyrtostylis*), 3–10-arch. Xylem: tracheids mostly with scalariform thickenings. Raphides in cortex.

Dropper TS

Hairs uniseriate, associated with mycorrhiza, in *Cyrtostylis* and *Corybas unguiculatus*, otherwise multiseriate. **Epidermis**: cells isodiametric to polygonal; walls of uniform thickness or outer walls thicker, depending on species. **Cortex** parenchymatous, outer layers infected with hyphae and pelotons, inner layers infrequently with starch granules. **Endodermis**: cells elliptical to polygonal. **Stele** medullated, 3–6-arch. Xylem: tracheids mostly with scalariform or pitted thickenings. Raphides rare in cortex.

Material examined

Acianthus (3), *Corybas* (6), *Cyrtostylis* (1), *Townsonia* (1). Full details appear in Pridgeon (1994b).

Subtribe Caladeniinae Pfitzer

Hirsute, tuberous, geophytic herbs. Tubers partly or fully enclosed in a multilayered fibrous tunic. Leaf solitary, convolute, sessile, usually longer than wide, erect to prostrate. Nine genera, *c.*280 species.

Leaf surface

Hairs multicellular, glandular or non-glandular, in *Aporostylis*, *Caladenia*, *Elythranthera* (Fig. 1D), *Eriochilus*, and *Glossodia*. *Aporostylis* unique in having bi- to triseriate glandular trichomes. Papillae, one per cell, present in *Praecoxanthus* and some species of *Eriochilus*. **Cuticle** sculpturing variable, mainly at generic level, ranging from smooth, crinkled, punctate to striate. **Epidermis**: cells polygonal or rectangular to elliptical and often axially elongated. Anticlinal walls variable in thickness, but relatively thick with conspicuous pit-fields in all genera except *Adenochilus*; walls sinuous on one or both surfaces in *Adenochilus*, *Aporostylis*, some species of *Caladenia* and *Cyanicula*, and *Elythranthera*; walls straight in all other taxa. **Stomata** restricted to abaxial epidermis in *Adenochilus*, some species of *Caladenia*, *Cyanicula*, *Eriochilus*, *Glossodia*, and *Praecoxanthus*; in other genera stomata distributed over both surfaces. Size ranges, mean dimensions, and mean length: width ratios of guard-cell pairs for most taxa provided in Pridgeon (1993, 1994b).

Leaf TS

Cuticle thin (<3 μm). **Epidermis**: cells elliptical, isodiametric to polygonal, outer walls often differentially thickened, especially at leaf margins. Adaxial cells many times larger than abaxial cells in some species of *Caladenia*, *Cyanicula*, and especially in *Leptoceras*. **Stomata** superficial, substomatal chambers shallow. **Hypodermis**: present only in *Praecoxanthus*, adaxial, 2- or 3-layered. **Mesophyll** heterogeneous or homogeneous according to taxon, homogeneous in *Praecoxanthus*. Spongy mesophyll lacunar to stellate in *Elythranthera*, *Eriochilus*, and *Glossodia*. **Vascular bundles** as for tribe. Sclerenchyma bundle caps absent, but some isolated fibres occurring on abaxial pole of largest bundles in *Eriochilus dilatatus* subsp. *multiflorus* (as *E. multiflorus*). Bundle sheath 2-layered in *Aporostylis*. Raphides in unmodified chlorenchyma cells and/or mesophyll idioblasts.

Collar/stem TS

Hairs on collar unicellular or multicellular (Fig. 6A), the multicellular type uniseriate or multiseriate, unbranched, associated with mycorrhiza. Hairs on upper portion of stem in *Caladenia*, *Cyanicula*, and *Elythranthera* multicellular, uniseriate, unbranched, non-glandular or glandular depending on species. **Cuticle** thin, smooth to striate. **Epidermis**: cells rectangular or elliptical to polygonal, outer walls usually thickened. **Stomata** present or absent depending on taxon. **Ground tissue** parenchymatous, outer and middle layers of collar usually infected with hyphae and pelotons; inner layers of stem sclerotic, forming a multilayered sclerenchyma sheath. Cell walls often provided with secondary suberized, banded thickenings. **Endodermis** of collar: cells elliptical to polygonal with casparian strips. **Vascular bundles** often arranged in one or two rings with or without a few central bundles, outer ring frequently abutting sclerenchyma sheath. Xylem: tracheids primarily with helical or scalariform thickenings, although annular and pitted occurring in some species. Raphides and/or rod-shaped crystals in cortex and/or ground tissue of most species.

Tuber TS

Hairs absent or present depending on taxon; if present, multiseriate, unbranched, associated with mycorrhiza. **Velamen** 1–3-layered, cells elliptical to polygonal and often periclinally compressed; walls suberized, of uniform thickness. Simple epidermis in *Caladenia barbarossa*, *Eriochilus multiflorus*, and *E. scaber*. **Exodermis**: cells elliptical to polygonal, walls of uniform thickness or outer and/or radial walls thicker, depending on species. **Ground tissue** parenchymatous; outer layers often infected with hyphae and pelotons, unconnected with age of tuber. **Endodermis**: cells elliptical to polygonal. **Stele** dissected or not depending on taxon; if dissected, 2–10 discrete vascular strands, each 1–13-arch and enclosed by endodermis; if undissected, medullated. Xylem: tracheids mainly with scalariform or pitted thickenings. Starch granules and raphides and/or rod-shaped crystals in ground tissue.

Root TS

Hairs unicellular or multiseriate, associated with mycorrhiza; absent in *Glossodia*. **Velamen** 1–2(–4)-layered, cells elliptical to polygonal; walls of uniform thickness or outer and/or radial walls of outermost layer thickened, depending on taxon. Simple epidermis in *Glossodia*, cells anticlinally elliptical to irregular, outer walls very thick and suberized. **Exodermis**: cells elliptical to polygonal, walls of uniform thickness. **Cortex** parenchymatous, occasionally infected with hyphae and pelotons, occasionally with starch granules. **Endodermis**: cells elliptical to polygonal. **Stele** medullated, 2–12-arch. Xylem: tracheids with helical, scalariform, or pitted thickenings. Raphides or rod-shaped crystals in cortex.

Dropper TS

Hairs multiseriate, unbranched, associated with mycorrhiza (Fig. 17C); absent in *Aporostylis*. **Velamen** 1–3-layered, cells elliptical to polygonal, walls of uniform thickness or outer walls suberized-thickened. Simple epidermis in species of *Caladenia* (Fig. 17C) (except *C. flava*), *Elythranthera*, *Leptoceras*, and *Praecoxanthus*; cells isodiametric, dome-shaped, elliptical to polygonal. **Exodermis**: cells polygonal, walls of uniform thickness. **Cortex** parenchymatous, outer and/or middle layers often infected with hyphae and pelotons (Fig. 17C); starch granules infrequent; walls with secondary, banded, suberized thickenings in most species examined. **Endodermis**: cells elliptical to polygonal. **Stele** dissected or not depending on species (Fig. 17C). Older droppers in *Caladenia* and other genera serving as fibrous sheath with secondary wall thickenings, enclosing one or more vascular strands as well as tracheary remnants of older strands. Raphides or rod-shaped crystals in cortex or pith of medullated steles.

Material examined

Adenochilus (1), *Aporostylis* (1), *Caladenia* (23), *Cyanicula* (4), *Elythranthera* (2), *Eriochilus* (3), *Glossodia* (1),

Leptoceras (1), *Praecoxanthus* (1, formerly *Caladenia aphyllus* Benth.). Full details appear in Pridgeon (1993, 1994a,b), Pridgeon and Chase (1995).

Subtribe Cryptostylidinae Schltr.

Glabrous, geophytic herbs. Tubers absent, roots long and fleshy. Leaves absent or one or more per shoot, convolute, longer than wide, erect, coriaceous, petiolate. Two genera, *c.*25 species.

Leaf surface

Hairs absent. **Cuticle** papillate to striate. **Epidermis**: cells polygonal to axially elongated and rectangular or elliptical; walls thin, pit-fields not conspicuous. **Stomata** on abaxial surface only. Guard cells in *Coilochilus* with pigment or mucilage droplets.

Leaf TS

Cuticle thin (*Cryptostylis*) or thick (*Coilochilus*), depending on species. **Epidermis**: cells elliptical to rectangular, often anticlinally oriented abaxially; outer wall thickened. **Stomata** superficial, substomatal chambers shallow. **Mesophyll** homogeneous in *Coilochilus*, heterogeneous (1–3-layered palisade) in *Cryptostylis*. **Vascular bundles** as for tribe. Bundle sheath 1–3-layered, cells thin-walled. Raphides in mesophyll idioblasts.

Stem TS

Hairs absent. **Cuticle** thin, smooth to slightly papillose. **Epidermis**: cells elliptical, rectangular to polygonal, outer walls thickened. **Stomata** absent. **Cortex**: outer layers parenchymatous, inner 3–5 layers in *Cryptostylis subulata* forming a sclerenchyma sheath. **Ground tissue** parenchymatous. **Vascular bundles** in two rings. Crystals not observed.

Root TS

Hairs unicellular. **Velamen** 1-layered (*Coilochilus*) or 2–5-layered (*Cryptostylis*), cells elliptical to anticlinally rectangular, with minute pores in *Coilochilus*; walls of uniform thickness. **Tilosomes** present in *Cryptostylis subulata*. **Exodermis**: cells square to anticlinally rectangular, outer and radial walls thickened. **Cortex** parenchymatous, infected with hyphae and pelotons. **Endodermis**: cells elliptical. **Stele** medullated, 8–11-arch. Xylem: tracheids with helical thickenings, vessel elements with scalariform perforation plates. Cruciate starch granules and raphide bundles throughout cortex.

Material examined

Coilochilus (1), *Cryptostylis* (2). Full details appear in Stern et al. (1993b), Pridgeon and Chase (1995).

Subtribe Diuridinae Lindl.

Glabrous, tuberous, geophytic herbs. Tubers elongate, naked. Leaves one to several, convolute, sessile, much longer than wide, erect, channelled. Two genera, *c.*50 species.

Leaf surface

Hairs absent, except papillae in *Diuris sulphurea* abaxially at midrib and margins. **Cuticle**: sculpturing variable. **Epidermis**: cells rectangular, trapezoidal, rhomboidal, axially elongated; walls mostly thin without conspicuous pit-fields, somewhat sinuous in *D. semilunulata*. **Stomata** on both surfaces.

Leaf TS

Cuticle thin. **Epidermis**: cells isodiametric, elliptical to ovate, often anticlinally elliptical or S-shaped; outer walls thickened. **Stomata** superficial (slightly elevated in *D. longifolia*), substomatal chambers shallow, except connected to long, transverse lacunae in mesophyll of *D. lanceolata*. **Mesophyll** homogeneous or heterogeneous, depending on species. Palisade mesophyll in one or two layers, abaxial or bifacial. Transverse lacunae present in some species. **Vascular bundles** as for tribe. Raphides and rod-shaped crystals common in mesophyll idioblasts.

Stem TS

Hairs absent. **Cuticle** thin, smooth to rugose in *Diuris*, striate in *Orthoceras*. **Epidermis**: cells isodiametric, rectangular, elliptical to polygonal; outer walls thickened in all species except *D. sulphurea*, inner walls also thickened in *Orthoceras*. **Stomata** present, superficial in *D. aurea*, *D. semilunulata*, and *Orthoceras*. **Ground tissue** parenchymatous, innermost layers forming a sclerenchyma sheath. **Vascular bundles** as for tribe, in 1–4 rings, outer ring abutting sclerenchyma sheath if present. Xylem: tracheids mostly with helical and scalariform thickenings. Raphides or rod-shaped crystals in cortex and/or pith of many species.

Tuber TS

Hairs absent. **Velamen** 1–3-layered, cells elliptical to polygonal, often periclinally compressed; walls of uniform thickness or outer walls thicker. **Exodermis**: cells elliptical to polygonal, walls of uniform thickness or outer walls thicker. **Cortex** parenchymatous, outer layers often dissociated, infrequently infected with hyphae and pelotons. **Endodermis**: cells elliptical to polygonal. **Stele** medullated, 5–34-arch depending on species. Xylem: tracheids with helical, scalariform or pitted thickenings. Starch granules and raphides or rod-shaped crystals in cortex of some species.

Root TS

Hairs absent. **Velamen** 1- or 2-layered, cells elliptical to irregularly polygonal, often periclinally compressed; walls of uniform thickness or outer and/or radial walls of outermost layer thicker depending on species; protists present in some cells. **Exodermis**: cells elliptical to polygonal, walls of uniform thickness (except outer walls thicker in *D. longifolia*, *D. magnifica*, and *D. picta*). **Cortex** parenchymatous, outer and/or middle layers often infected with hyphae and/or pelotons. **Endodermis**: cells elliptical to polygonal. **Stele** medullated, 4–9-arch. Xylem: tracheids with helical, scalariform, or pitted thickenings. Raphides or rod-shaped crystals in cortex.

Material examined

Diuris (12), *Orthoceras* (1). Full details appear in Pridgeon (1994b).

Subtribe Drakaeinae Schltr.

Tuberous, mostly glabrous, geophytic herbs. Tubers globose or ovoid, naked. Leaves either absent, solitary, distichously paired or up to six spirally arranged in a rosette, encircling scape or on a lateral growth, convolute, sessile, erect to prostrate. Five genera, *c.*50 species.

Leaf surface

Cuticle: sculpturing striate in *Caleana*; striate-anastomosing in *Arthrochilus* (Fig. 1F); crinkled in *Chiloglottis*; smooth, punctate to striate in *Drakaea*; and smooth in *Spiculaea*. Papillae, one per cell, on both surfaces in *Caleana major* and adaxially in *Drakaea* (Fig. 1E; limited to margins in *D. livida* and *D. thynniphila*). **Epidermis**: cells elliptical to polygonal and axially elongated on abaxial surface of *Caleana*; anticlinal walls thick with conspicuous pit-fields, sinuous abaxially in *Arthrochilus*, *Drakaea*, and *Paracaleana nigrita* (as *Caleana nigrita*). **Stomata** on both surfaces (Figs 1F, 2E), except restricted to abaxial surface in *Drakaea*. Size ranges, mean dimensions, and mean length–width ratios of guard-cell pairs in Pridgeon (1994b).

Leaf TS

Cuticle thin. **Epidermis**: cells isodiametric, elliptical to polygonal, outer walls differentially thickened (inner walls as well in *Caleana*). Adaxial cells appreciably larger than abaxial cells in *Arthrochilus* and *Chiloglottis*. **Stomata** superficial, substomatal chambers shallow. **Mesophyll** heterogeneous (up to six palisade layers in *Spiculaea*) in all taxa except *Chiloglottis reflexa* (Fig. 2E; as *C. trilabra*) and *C. trapeziformis*. **Vascular bundles** as for tribe. Bundle sheath 2-layered in *Drakaea*. Raphide bundles in mesophyll idioblasts of all taxa except *Spiculaea*. Brownish aggregations surrounded by a transparent sheath, not birefringent in polarized light, in most epidermal cells of *Spiculaea*.

Collar/stem TS

Hairs on collar multiseriate, associated with mycorrhiza, in *Arthrochilus*, *Chiloglottis*, and *Spiculaea*; uniseriate, apical cell supported by buttress of 4–5 epidermal cells, in *Caleana* and *Drakaea*. **Cuticle** thin, smooth to minutely striate or rugose. **Epidermis**: cells elliptical, rectangular to polygonal, outer walls usually thickened. **Stomata** present or absent depending on taxon. **Ground tissue** of collar: parenchymatous, often infected with hyphae and pelotons. Chlorenchyma dense in the leafless species *Arthrochilus huntianus*. Inner layers of cortex of *Arthrochilus* and *Caleana* forming sclerenchyma sheath. **Endodermis** of collar: cells elliptical to polygonal. **Vascular bundles** usually arranged in 1–3 rings. Xylem: tracheids with annular, helical, scalariform or pitted thickenings. Raphide bundles or rod-shaped crystals in ground tissue observed in all taxa except *Spiculaea*.

Tuber TS

Hairs absent in most taxa; multiseriate, associated with mycorrhiza in *A. huntianus*; unicellular, filamentous, non-glandular in *A. sabulosus*. **Velamen** 1–3-layered, cells elliptical to polygonal, compressed periclinally; outer and/or radial walls infrequently thickened; cells in *Drakaea livida* pitted. **Exodermis**: cells rectangular, elliptical to polygonal; walls of uniform thickness or outer and/or radial walls thicker, depending on species. **Ground tissue** parenchymatous, filled with starch granules; outer layers in *A. huntianus* infected with hyphae and pelotons. **Endodermis**: cells elliptical to polygonal. **Stele** medullated, 4–14-arch, depending on species; dissected into nine discrete strands in *A. huntianus*. Xylem:

tracheids with scalariform or pitted thickenings. Starch granules and raphides in ground tissue.

Root TS

Hairs multiseriate, associated with mycorrhiza, in *A. huntianus* and *Spiculaea*; hairs in other taxa unicellular, from outermost velamen layer. **Velamen** 1–2(–3)-layered, cells rectangular to polygonal; walls of uniform thickness or outer walls thickened, depending on taxon. Walls with reticulate thickenings in *Drakaea* and with pores in a reticulate pattern in *Spiculaea*. **Exodermis**: cells rectangular to polygonal, walls of uniform thickness or outer walls thickened. **Cortex** parenchymatous, outer layers infected with hyphae and pelotons, innermost layers occasionally with starch granules; cells in *Drakaea* with reticulate wall thickenings. **Endodermis**: cells elliptical to polygonal. **Stele** medullated, 3–10-arch; pith sclerotic in *Chiloglottis reflexa* (as *C. trilabra*). Xylem: tracheids with helical, scalariform or pitted thickenings. Raphide bundles in cortex of some species.

Material examined

Arthrochilus (3), *Caleana* (1, and 1 sp. now in *Paracaleana*), *Chiloglottis* (3), *Drakaea* (3), *Spiculaea* (1). Full details appear in Pridgeon (1994a,b).

Subtribe Megastylidinae Schltr.

Glabrous, tuberous or rhizomatous, geophytic herbs. Tubers elongate or globose, naked, rarely absent. Leaves solitary or rarely clustered, rarely absent, sessile or petiolate, convolute, flat or channelled, erect. Seven genera, 15 species.

Leaf surface

Hairs absent, but papillae on both surfaces in *Pyrorchis*. **Cuticle** smooth to papillate. **Epidermis**: cells rectangular to polygonal; walls thick with conspicuous pit-fields; anticlinal walls sinuous in *Leporella*, *Pyrorchis*, and *Rimacola*. **Stomata** restricted to abaxial surface, except on both surfaces in *Leporella* and *Pyrorchis* and on adaxial surface only in *Lyperanthus*.

Leaf TS

Cuticle thin (<3 µm). **Epidermis**: cells elliptical, isodiametric to polygonal, outer walls often differentially thickened, especially at leaf margins. **Stomata** superficial, substomatal chambers shallow. **Hypodermis**: absent. **Mesophyll** heterogeneous or homogeneous according to taxon. Palisade mesophyll in *Lyperanthus* unusually abaxial with chloroplasts oriented in an adaxial arc in cells. Spongy mesophyll lacunar to stellate in *Leporella*. **Vascular bundles** as for tribe. Xylem: tracheary elements bibrachiate in *Megastylis glandulosa*; largest vascular bundles in *M. gigas* and *M. latilabris* hourglass-shaped. Sclerenchyma bundle caps present only as one row of fibres ad- and abaxially in *M. gigas*. Bundle-sheath cells thin-walled; 2- or 3-layered in *M. latilabris*. Bundle sheath 2-layered in *Rimacola*. Raphides in unmodified chlorenchyma cells and/or mesophyll idioblasts in all but *Rimacola*. Inclusions of unspecified nature, not birefringent in polarized light and often surrounded by a transparent sheath, in many epidermal cells of *Burnettia* (bract only).

Collar/stem TS

Hairs on collar unicellular or multicellular, the multicellular type uniseriate in *Burnettia*, multiseriate in *Lyperanthus*, *Pyrorchis*, and *Rimacola*, unbranched, associated with mycorrhiza; absent in *Leporella* and *Megastylis*. **Cuticle** thin, smooth to striate. **Epidermis**: cells rectangular, elliptical to polygonal, outer walls usually thickened; heavily sclerified in *Lyperanthus*. **Stomata** present or absent depending on taxon. **Ground tissue** parenchymatous, outer and middle layers of collar usually infected with hyphae and pelotons; inner layers of stem forming a multilayered sclerenchyma sheath. Cell walls often provided with secondary suberized, banded thickenings. Several subepidermal layers in *Lyperanthus serratus* heavily sclerified. **Endodermis** of collar: cells elliptical to polygonal with casparian strips. **Vascular bundles** often arranged in one or two rings with or without a few central bundles, outer ring frequently abutting sclerenchyma sheath. Xylem: tracheids primarily with helical or scalariform thickenings, although annular and pitted tracheids occurring in some species. Raphides and/or rod-shaped crystals in cortex and/or ground tissue of most species.

Tuber TS

Hairs absent or present depending on taxon; if present, multiseriate, unbranched, associated with mycorrhiza. **Velamen** 1–3-layered, cells elliptical to polygonal and often periclinally compressed; walls suberized, of uniform thickness. **Exodermis**: cells elliptical to polygonal, walls of uniform thickness or outer and/or radial walls thicker depending on species. **Ground tissue** parenchymatous; outer layers often infected with hyphae and pelotons, unconnected with age of tuber. **Endodermis**: cells elliptical to polygonal. **Stele** dissected or not depending on taxon; if dissected, 2–10 discrete vascular strands, each 1–13-arch and enclosed by endodermis; if undissected, medullated. Xylem: tracheids mainly with scalariform or pitted thickenings. Starch granules and raphides and/or rod-shaped crystals in ground tissue.

Root TS

Hairs unicellular, associated with mycorrhiza; absent in *Burnettia*. **Velamen** 1–2(–4)-layered, cells elliptical to polygonal; walls of uniform thickness or outer and/or radial walls of outermost layer thickened, depending on taxon. **Exodermis**: cells elliptical to polygonal, walls of uniform thickness. **Cortex** parenchymatous, occasionally infected with hyphae and pelotons, occasionally with starch granules. **Endodermis**: cells elliptical to polygonal. **Stele** medullated, 2–18-arch. Xylem: tracheids with helical, scalariform, or pitted thickenings. Raphides or rod-shaped crystals in cortex.

Material examined

Burnettia (1), *Leporella* (1), *Lyperanthus* (2), *Megastylis* (4), *Pyrorchis* (1), *Rimacola* (1). Full details appear in Pridgeon (1994a,b), Pridgeon and Chase (1995) (*Pyrorchis nigricans* (R.Br.) D.L.Jones & M.A.Clem. was described as *Lyperanthus nigricans* R.Br.).

Subtribe Prasophyllinae Schltr.

Tuberous, geophytic herbs. Tubers globose to elongate, naked. Leaf solitary, cylindrical, terete, sessile, erect. Three genera, *c.*140 species.

Leaf surface

Hairs adaxial in many species of *Prasophyllum*, 2–5-celled, filamentous, unbranched, glandular, clustered; hairs unicellular or bicellular on adaxial surface of bract of *Genoplesium*, absent in *Microtis*. **Cuticle** smooth to striate. **Epidermis**: adaxial epidermis often dissociated in *Microtis* and *Prasophyllum*. Abaxial epidermal cells elliptical, rectangular to polygonal, often axially elongated; walls thin to thick, depending on species. **Stomata** on abaxial surface only; absent in *Genoplesium*.

Leaf TS

Terete, sheathing, in *Prasophyllum* grading to ±dorsiventral, grooved on free portion at apex. **Cuticle** thin. **Epidermis**: adaxial cells (where present) isodiametric, ovate, peg-shaped, elliptical to rectangular; outer walls slightly thickened. Abaxial cells anticlinally elliptical or rectangular, S-shaped to polygonal; outer walls thickened, extending to radial walls in *M.* aff. *unifolia*. **Stomata** superficial to slightly elevated; substomatal chamber shallow but occasionally continuous with large central lacuna in *Microtis*. **Mesophyll** homogeneous in *Genoplesium* and *Microtis* (except *M.* aff. *unifolia*), heterogeneous in other species, one or two abaxial palisade layers with large interfascicular lacunae. **Vascular bundles** as for tribe, in one row in *Genoplesium* or in one ring in centre of mesophyll or near adaxial surface in *Microtis* and *Prasophyllum*. Sclerenchyma bundle caps absent, except isolated fibres present abaxially on some vascular bundles in some species and abaxial caps on largest bundles in *M. orbicularis*. Raphides or rod-shaped crystals in mesophyll.

Stem TS

Hairs absent, except unicellular and associated with mycorrhiza in *M. atrata*; occasional papillae in *M. atrata*. **Cuticle** thin, smooth to striate or ridged. **Epidermis**: cells square, anticlinally rectangular or elliptical to polygonal, often interrupted; outer walls thickened and striate, thickenings occasionally extending to radial walls. **Stomata** present in some species, superficial. **Cortex** parenchymatous, inner layers forming a sclerenchyma sheath. **Ground tissue** parenchymatous, often with large lacuna in centre. Outer or middle layers occasionally infected with hyphae and pelotons. Innermost layers of cortex in *M.* aff. *unifolia* forming a sclerenchyma sheath. **Vascular bundles** in Y-shaped configuration, in two arcs or in one or two rings; if two rings, one often outside sclerenchyma sheath and one inside, the outer ring homologous with leaf vasculature. Xylem: tracheids with helical, scalariform, or pitted thickenings. Raphides and/or rod-shaped crystals in cortex or pith.

Tuber TS

Hairs occasional, unicellular, associated with mycorrhiza in *M. atrata*. **Velamen** 1–3-layered, cells elliptical to polygonal, often periclinally compressed; walls with varying thickening patterns. Outer walls of *P. cyphochilum* with scalariform wall ornamentation. **Exodermis**: cells isodiametric, elliptical to polygonal, walls with varying thickening patterns. **Cortex** parenchymatous, new tubers filled with elliptical to oblong starch granules; outer layers in old tubers dissociated; outer layers in *M. atrata* infected with hyphae and pelotons. **Endodermis**: cells elliptical to polygonal. **Stele** medullated, 2–10-arch. Dissected into five tetrarch vascular strands in *M.* aff. *unifolia*. Steles of *P. fimbria* and *P. giganteum* apparently dissected, representing fusion of 11–13 separate tubers, each with individual cortex and highly compressed and fused velamen along common boundaries. Xylem: tracheids with helical, scalariform, or pitted thickenings. Raphides and/or rod-shaped crystals in cortex.

Root TS

Hairs absent. **Velamen** in *Prasophyllum* and *M.* aff. *unifolia* 1–3-layered, cells elliptical to polygonal with varying wall thickening patterns. Simple epidermis in *Genoplesium* and other species of *Microtis*; cells elliptical to polygonal; outer walls thickened. **Exodermis**: cells elliptical to polygonal, walls uniformly thickened or outer and radial walls thickened, depending on species. **Cortex** parenchymatous, occasionally infected with hyphae and pelotons. **Endodermis**: cells elliptical to polygonal. **Stele** medullated, 3–8-arch. Xylem: tracheids with annular, helical, scalariform, or pitted thickenings. Raphides and rod-shaped crystals in cortex.

Dropper/stolonoid root TS

Hairs multiseriate, associated with mycorrhiza, in some species of *Prasophyllum*; absent in *Microtis*. **Epidermis**: cells elliptical to polygonal, often anticlinally oriented; outer walls thickened; stomata occasionally present. **Cortex** parenchymatous, filled with elliptical to oblong starch granules, often infected with hyphae and pelotons. **Endodermis**: cells elliptical to polygonal. **Stele** medullated, 4–15-arch. Xylem: tracheids with helical, scalariform, or pitted thickenings. Raphides or rod-shaped crystals in cortex.

Material examined

Genoplesium (2), *Microtis* (4), *Prasophyllum* (13). Full details appear in Pridgeon (1994b).

Subtribe Rhizanthellinae R.S.Rogers

Glabrous, mostly subterranean, holomycotrophic, geophytic herbs. Rhizomes perennial, thickened, branching laterally, with or without tubers. Roots absent. Leaves reduced to colourless fleshy scale leaves. One genus, two species.

Stem tuber TS

Hairs absent. **Cuticle** thick, heavily suberized, smooth. **Epidermis**: cells elliptical to polygonal, outer walls slightly thickened. **Stomata** absent. **Ground tissue** parenchymatous, cells with relatively thick walls and large nuclei, some cells infected with pelotons. **Vascular bundles** collateral, in one ring. Xylem: tracheids with helical or scalariform thickenings. Crystals not observed.

Material examined

Rhizanthella slateri (Rupp) M.Clem. & Cribb; ANBG, Groeneveld 104.

Subtribe Thelymitrinae Lindl.

Glabrous or hirsute, tuberous, geophytic herbs. Tubers elongate, naked. Leaf solitary, rarely absent, sessile, either cylindrical and terete or convolute, flat or channelled, sessile, erect, sometimes spirally twisted. Three genera, *c.*70 species.

Leaf surface

Cuticle smooth to striate depending on species. Papillae present, one per cell, in *Calochilus* and various species of *Thelymitra*. **Epidermis**: cells elliptical, rectangular to polygonal, often axially elongated; walls thick with conspicuous pit-fields, except in *Epiblema*. **Stomata** on both surfaces of dorsiventral leaves or all surfaces of terete leaves.

Leaf TS

Dorsiventral or terete depending on species. **Cuticle** thin to moderately thick. **Epidermis**: cells isodiametric, elliptical, rectangular to papillate; outer walls very thick; cells in *Thelymitra benthamiana* rich in anthocyanins. **Stomata** superficial, substomatal chambers shallow. **Mesophyll** variable depending on species; if heterogeneous, 1–5 palisade layers on both surfaces (abaxial only in *Epiblema*, *Thelymitra cicumsepta* (as *T. retecta*) and *T. rubra*). Extensive lacunae present in some species. **Vascular bundles** as for tribe, in one ring in unifacial species immediately internal to palisade mesophyll, or in one or two rows with or without accessory bundles in midrib below midvein. Raphides or rod-shaped crystals in mesophyll idioblasts of most species and in epidermis of *Calochilus herbaceus*.

Stem TS

Enclosed by bract, 2–5 cells wide in *Epiblema*; vascularized by three collateral strands. **Hairs** absent. **Cuticle** thin, smooth to striate. **Epidermis**: cells isodiametric, elliptical, rectangular, polygonal to S-shaped; outer walls thickened. Pelotons present in epidermal cells of *Thelymitra benthamiana*. **Stomata** absent or present and superficial, depending on species. **Cortex** parenchymatous, innermost layers sclerotic, forming a sclerenchyma sheath in all species except *Epiblema*, *T. cyanea*, and *T. graminea*. **Vascular bundles** in 1–3 rings, outer ring abutting sclerenchyma sheath. Xylem: tracheids with annular, helical, scalariform, or pitted thickenings. Raphides or rod-shaped crystals infrequent in cortex and/or pith.

Tuber TS

Hairs absent. **Velamen** 1- or 2-layered, cells elliptical, dome-shaped to polygonal, often periclinally compressed (Fig. 17B); walls of uniform thickness or differentially thickened, depending on species. **Exodermis**: cells elliptical to polygonal; walls of uniform thickness or differentially thickened, depending on species (Fig. 17B). **Cortex** parenchymatous, filled with starch granules to varying degrees (Fig. 17B). Outer layers dissociated in some species. **Endodermis**: cells elliptical, rectangular to polygonal. **Stele** medullated, polyarch (9–100, depending on species). Xylem: tracheids with helical, scalariform, or pitted thickenings. Raphides in outer cortex of many species.

Root TS

Hairs absent. **Velamen** 1- or 2-layered, cells elliptical to irregularly polygonal, often periclinally compressed; walls uniformly thickened or outer and/or radial walls thicker, depending on species. Protists present in some velamen cells and in cupule-like excrescences in *Epiblema*. **Exodermis**: cells elliptical to polygonal, walls uniformly thickened or outer and radial walls thicker, depending on species. **Cortex** parenchymatous, variously infected with hyphae and pelotons. **Endodermis**: cells elliptical to polygonal. **Stele** medullated, 3–9-arch. Xylem: tracheids with annular, helical, scalariform, or pitted thickenings. Raphides and/or rod-shaped crystals in cortex.

Material examined

Calochilus (3), *Epiblema* (1), *Thelymitra* (14). Full details appear in Pridgeon (1994b).

Reports from the literature for the tribe

Burgeff (1909, 1932): mycorrhiza and vegetative anatomy of *Corysanthes* (= *Corybas*). Clements (1988): mycorrhiza of *Caladenia* and *Thelymitra*. Dietz (1930): anatomy of underground organs of *Thelymitra*. Dixon et al. (1990): tuber anatomy of *Rhizanthella gardneri*. Groom (1895–7): vegetative anatomy of *Corysanthes* (= *Corybas*). Hadley and Williamson (1972): mycorrhiza of *Cryptostylis*. Hall (1976): mycorrhiza of *Corybas macranthus*. Huynh et al. (2004): mycorrhiza of *Caladenia formosa*. Kores (1995): systematics and anatomy of *Acianthus*. Pittman (1929): mycorrhiza of several Diurideae. Porembski and Barthlott (1988): velamen of *Cryptostylis*. Pridgeon (1993): leaf anatomy of *Caladenia*. Pridgeon (1994a): leaf anatomy of Caladeniinae. Pridgeon (1994b): trichomes in Diurideae. Pridgeon and Chase (1995): subterranean axes in Diurideae. Pridgeon et al. (2001): comprehensive taxonomic treatment of genera in Orchidaceae. Staudermann (1924): hairs of *Caladenia*. Stern et al. (1993c): vegetative anatomy of Spiranthoideae, including *Cryptostylis*. Warcup (1981): mycorrhiza of Australian orchids, including many Diurideae. Warcup (1985): mycorrhiza of *Rhizanthella gardneri*. Wright and Guest (2005): mycorrhiza of *Caladenia tentaculata*.

Taxonomic notes for the tribe

With some exceptions, characters from vegetative anatomy of underground organs of Diurideae are of limited systematic utility above the genus level (often the species level), so high is the degree of homoplasy (Pridgeon 1994a; Pridgeon and Chase 1995). However, some leaf characters such as presence and types of trichomes, cuticular sculpturing, anticlinal walls of epidermal cells, and heterogeneity of chlorenchyma support the segregation of *Lyperanthus nigricans* R.Br. and *L. forrestii* F.Muell. as the genus *Pyrorchis* D.L.Jones & M.A.Clem. (Pridgeon 1994b), *Caladenia menziesii* R.Br. as *Leptoceras* (R.Br.) Lindl. (Pridgeon 1993, 1994b), and *Caladenia sericea* Lindl. and relatives as the genus *Cyanicula* Hopper & A.P.Brown, ms. (Pridgeon 1993, 1994b). Such characters, however, do not support the segregation of *Caladenia barbarossa* Rchb.f. and relatives as *Drakonorchis* Hopper & A.P.Brown (which supports molecular evidence; M.A. Clements, personal communication).

Tribe Orchideae Dressler & Dodson

Orchideae are mostly terrestrial plants with spirally arranged, basal or caulescent, non-articulate, soft, convolute leaves. Plants have variously-shaped storage root tubers and slender absorbing roots. Orchideae comprise a single subtribe, Orchidinae, which subsumes subtribe Habenariinae. There are about 53 genera and 1800 species distributed in the Old World, except *Habenaria*, *Galearis*, and *Pseudorchis* which also occur in the New World. Only *Habenaria* species occur in tropical America. Several endemic genera are found in Madagascar. Some species are adaptable to cultivation, e.g. *Stenoglottis fimbriata*, *Ophrys apifera*, *Orchis mascula*, and *Habenaria rhodocheila*.

Medicinal uses have been reported as follows: *Dactylorhiza* (including *Coeloglossum*) in Iran, Afghanistan, and western Himalayas; *Gymnadenia* in China; *Habenaria* in North America, eastern Asia, India, China, and Indonesia; *Ophrys* species in Brazil and North Africa; species of *Orchis* in Nepal, India, Iran, parts of Europe, New York State, and North Africa.

Leaf surface
Hairs absent, except in *Holothrix* (Fig. 28A,B) with elongate unicellular hairs on both surfaces and in *Platanthera dilatata* where blunt-tipped unicellular hairs with brown deposits present. **Cuticle** smooth to rugulose. **Epidermis**: cells polygonal, anticlinal walls straight-sided or curvilinear (Fig. 28D–F), but lobed in some cells of several species. **Stomata** abaxial (Fig. 29A,B), except with few adaxial in some *Platanthera* species and ab- and adaxial in *Platanthera nivea* and *Serapias vomeracea*; anomocytic (Fig. 29B), except anomocytic and tetracytic in a few species (Fig. 29A).

Leaf TS
Cuticle generally <2.5 μm thick, except in *Ophrys tenthredinifera* where adaxial cuticle ranging between 7.5 and 12.5 μm thick and abaxial cuticle up to 25 μm thick. **Epidermis**: adaxial cells anticlinal; modified for water storage in *Holothrix villosa* and other species (Figs 28C, 29C,D, 30A); abaxial cells variously anticlinal to isodiametric and periclinal (Fig. 29C,D). **Stomata** superficial to slightly raised, substomatal chambers small to moderate. Outer and inner guard-cell ledges small to moderately sized, inner ledges may be minute or absent. **Hypodermis** and **fibre bundles** absent throughout. **Mesophyll** homogeneous (Figs 29C,D, 30A), cells thin-walled. In *Habenaria repens* mesophyll highly lacunose around midrib and major veins (Fig. 30B) and in *Platanthera blephariglottis* and *P. cristata* lacunae present on one or both sides of midvein. Crystalliferous idioblasts longitudinally saccate and up to 18 times length of chlorenchyma cells. **Vascular bundles** collateral (Fig. 30C), in a single series across leaf blade. In some species of *Platanthera* vascular bundles reaching from epidermis to epidermis. Bundle-sheath cells thin-walled (Fig. 30C) and chlorophyllous. Vascular bundle sclerenchyma and **stegmata** absent.

Stem TS
Hairs absent. **Cuticle** smooth, <2.5 μm thick. **Epidermis**: cells isodiametric (Fig. 30D) to periclinal. **Stomata** occasional in *Habenaria floribunda* (as *H. odontopetala*) and frequent in *Platanthera cristata*; lacking in other species studied. **Cortex** ranging from three or four cells wide in *Habenaria monorrhiza* to 20 cells wide in *Holothrix villosa*; cells thick-walled in *H. villosa*; in other species outer cells thin-walled, inner cells thick-walled (Fig. 30D). Cortex in *Habenaria repens* consisting of a reticulum marked by large circular to elliptical lacunar interstices (Fig. 30E) separated by uniseriate rows of thin-walled cells. In *Holothrix villosa* thin-walled endodermal cells with casparian strips separating cortex from vascular cylinder. **Ground tissue**: cells of outer layers thick-walled in *Brachycorythis macrantha*, *Habenaria floribunda*, *H. monorrhiza*, *H. plantaginea*, and *H. rhodocheila* (Fig. 30F); all cells thick-walled in *Holothrix villosa*; inner cells thin-walled in *Brachycorythis macrantha*, *Habenaria monorrhiza*, *H. odontopetala*, and *H plantaginea*. Central ground tissue in *Habenaria repens* resembling reticular cortical configuration. **Vascular bundles** collateral, scattered throughout ground tissue (Fig. 30D), apparently in two series in *Holothrix villosa*, but absent from central ground tissue in this species. Xylem in *Habenaria floribunda* arcuate, partially enclosing phloem. In *Habenaria repens*, some inner bundles laterally fused and in *H. rhodocheila* xylem frequently binary. **Sclerenchyma** and **stegmata** absent. Raphide bundles occurring in unmodified cells of *Brachycorythis macrantha*.

Roots
Roots of two kinds in Orchideae: variously enlarged and shaped storage roots and slender absorbing roots.

Absorbing root TS
Velamen present in *Bonatea*, *Cynorkis*, *Stenoglottis*, and all species of *Habenaria*, except *H. distans* and *H. repens*; varying from 1-layered in *H. rhodocheila* to 3- or 4-layered in *Bonatea steudneri*; non-velamentous in *Anacamptis pyramidalis*, *Brachycorythis macrantha*, *Dactylorhiza fuchsii*, *Galearis spectabilis*, *Himantoglossum robertianum* (as *Barlia robertiana*), *Holothrix secunda*, *Neotinea maculata*, *Ophrys* species, *Orchis* species, *Platanthera* species, *Ponerorchis graminifolia* (as *Gymnadenia graminifolia*), and *Serapias* species. Unicellular hairs occurring throughout. **Tilosomes** absent. **Exodermis**: cells thin-walled, isodiametric, and periclinal (Fig. 32A); dead cells with tenuous scalariform bars; exodermis indiscernible in *Anacamptis pyramidalis*, *Galearis spectabilis*, *Orchis* species, and *Platanthera* species. **Cortex**: 4–15 cells wide, 3-layered in some species; cells thin-walled, rounded, polygonal. Inner cortical tissues of *Habenaria*

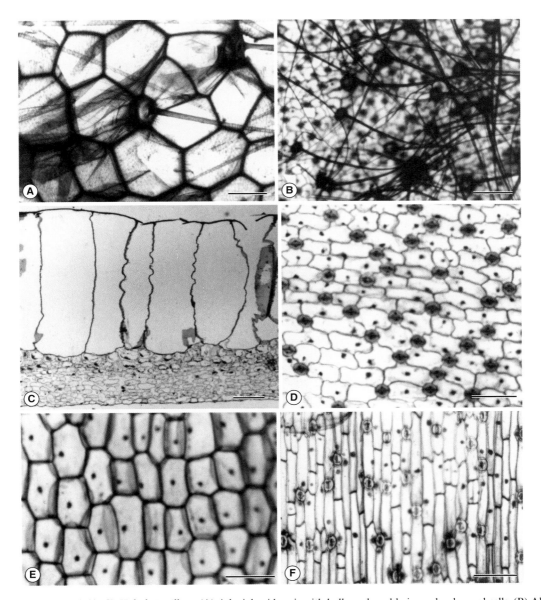

Fig. 28. Orchideae leaf. (A–C) *Holothrix villosa*. (A) Adaxial epidermis with bulbous-based hairs and polygonal cells. (B) Abaxial epidermis. (C) TS with greatly enlarged water-storage epidermal cells. (D) *Orchis anthropophora*, abaxial surface showing lobed cells, anomocytic and tetracytic stomata. (E,F) *Anacamptis pyramidalis*. (E) Adaxial epidermis with polygonal cells. (F) Abaxial surface showing elongate epidermal cells and anomocytic stomata. Scale bars = 10 μm. Reprinted from Stern (1997a).

repens featuring anticlinal lacunae separated from each other by uniseriate strands of rounded, thin-walled cells (Fig. 32B). **Endodermis** and **pericycle** 1-layered, cells thin-walled, isodiametric to periclinal; endodermal cells with casparian strips. **Vascular cylinder** 4–18-arch (Fig. 32B), xylem and phloem components embedded in parenchyma and alternating regularly. Twin vascular cylinders in *Habenaria distans*; three central coalescent vascular cylinders in *H. monorrhiza*, each with its own endodermis and pericycle. In *Stenoglottis longifolia* tracheary

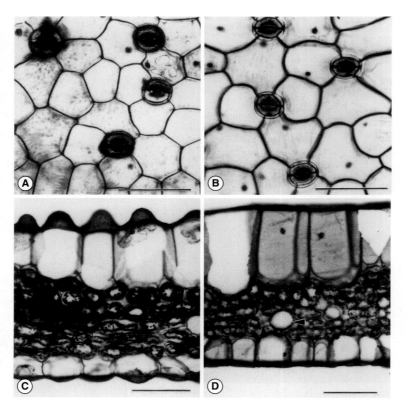

Fig. 29. Orchideae leaf. (A) *Habenaria odontopetala*, abaxial surface showing polygonal cells, curvilinear cell walls, tetracytic and anomocytic stomata. (B,C) *Stenoglottis fimbriata*. (B) Abaxial epidermis with polygonal cells and nearly circular stomata. (C) TS with papillose adaxial epidermal cells. (D) *H. rhodocheila*, TS with anticlinally enlarged epidermal cells, homogeneous mesophyll, and circular crystalliferous idioblasts (arrowhead). Scale bars = 100 μm. Reprinted from Stern (1997b).

elements scattered in pith. **Pith** parenchymatous, cells thin-walled, polygonal to rounded; intercellular spaces absent. Starch granules rare in pith cells.

Tuberous root TS

Root tubers in habenariads exhibit two distinct classes of vascularization: taxa with a single central vascular cylinder and those with a series of meristeles. Vascular tissue in root tubers of orchiads consists of dispersed meristeles only.

Velamen present in all root tubers, except *Dactylorhiza fuchsii* with a simple epidermis. Velamen cells thin-walled, periclinal and anticlinal to isodiametric in *Stenoglottis*, walls unmarked or with elliptical, circular, or oval pits; with widely spaced spiral thickenings in *Holothrix secunda*. Hairs unicellular, bulbous-based throughout, sometimes forming a dense felt. **Exodermis** with thin-walled cells, thickened in some *Habenaria* species; passage cells with thick outer walls, entirely thin-walled in *Stenoglottis* species; dead cells with scalariform bars. **Cortex** (habenariads): assimilatory cells thin-walled, with cruciate starch granules; homogeneous with or without mucilage cells in *Habenaria* species, depending on taxon. Mucilage cells thin-walled, irregular, rounded to sharply angular with tan to brown contents and a single raphide bundle (Fig. 32C). Cortex in *Stenoglottis* heterogeneous with many smaller, starch-storing assimilatory cells and fewer larger, dead, water-storage cells (Fig. 31A). Mucilage cells absent in *Stenoglottis*. Meristeles dispersed in ground tissue (Fig. 31D). Each meristele in *Habenaria* and *Stenoglottis* and in genera of orchiads surrounded by an **endodermis** and **pericycle** of thin-walled cells, endodermal cells having casparian strips. Meristeles embedded in parenchyma. Vascular arrangement varying from 1-arch to 8-arch in *Stenoglottis* species. In those habenariads with a vascular cylinder the surrounding

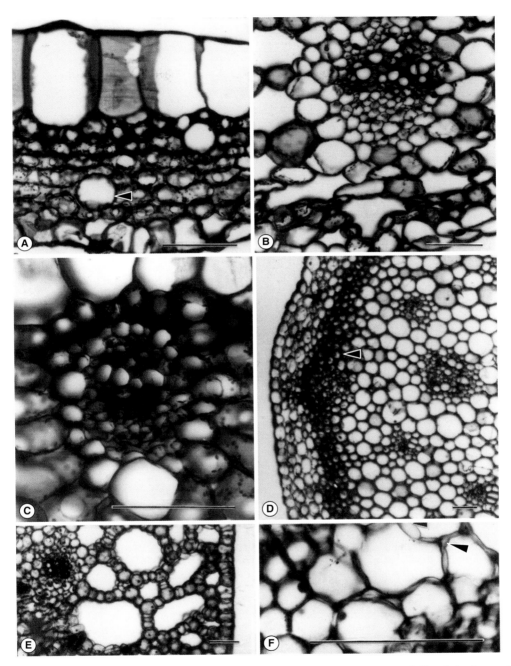

Fig. 30. Orchideae. *Cynorkis* and *Habenaria*. (A) *C. fastigiata*, TS leaf with homogeneous mesophyll and circular crystalliferous idioblasts (arrowhead). (B) *Habenaria repens*, TS leaf midvein showing surrounding lacunae. (C) *C. fastigiata*, TS leaf collateral vascular bundle showing bundle sheath cells. (D) *H. monorrhiza*, TS stem cortical cells subtended by band of thick-walled cells (arrowhead), and collateral vascular bundles scattered in ground tissue. (E) *H. repens*, TS stem with circular to elliptical cortical lacunae. (F) *H. rhodocheila*, TS stem ground tissue with collenchymatous wall thickenings (arrowhead). Scale bars = 100 μm. Reprinted from Stern (1997b).

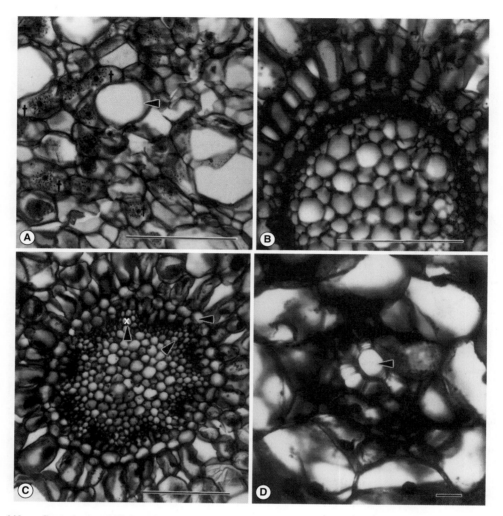

Fig. 31. Orchideae. *Stenoglottis* and *Habenaria* root tuber. (A) *S. woodii*, TS showing heterogeneous cellular construction and absence of mucilage cells. (B) *Habenaria arenaria*, TS vascular cylinder showing surrounding sheath of thick-walled cells. (C) *S. woodii*, TS vascular cylinder with surrounding sheath and conspicuous intercellular spaces, endodermis (arrowhead), 1-layered pericycle, and primary xylem (arrowhead x). (D) *H. rhodocheila*, TS monarch meristele of root tuber showing primary xylem (arrowhead). Scale bars = 100 μm. Reprinted from Stern (1997b).

endodermis consisting of thin-walled cells with obvious casparian strips. **Ground tissue** containing many smaller, living, starch-bearing assimilatory cells and fewer larger, dead cells that may store water or contain mucilage. **Vascular cylinder** embedded in parenchyma and ranging from 11- to 65-arch in *Habenaria* (Fig. 31B) but only 6–8-arch in *Stenoglottis* (Fig. 31C). **Pith** parenchymatous, cells sometimes containing cruciate starch granules; thin-walled raphide-bearing mucilage cells occurring in several *Habenaria* species.

Roots in Platanthera

Thin roots of some *Platanthera* species with a typical central vascular cylinder surrounded by a homogeneous cortex; thicker roots in other platantheras having a central vascular cylinder and a heterogeneous cortex with meristeles. Platantheras with globoid tubers characterized by a ground tissue interspersed with meristeles originating as branches from central vascular cylinder. Thin, thick, and globoid roots of platantheras covered by a simple epidermis; velamen absent. Vascular tissue surrounded by an

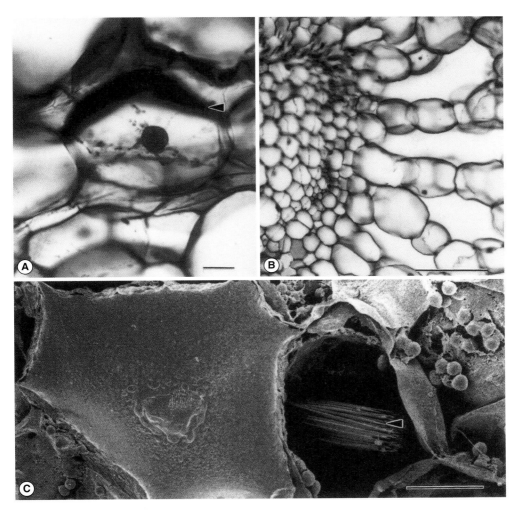

Fig. 32. Orchideae. *Habenaria* root. (A) *H. rhodocheila*, TS showing exodermal passage cell with thickened outer wall (arrowhead). (B) *H. repens*, TS cortex with radiating lacunae separated by uniseriate rows of cells. (C) *H. rhodocheila*. SEM showing mucilage cell (left) and raphide bundle (arrow) extruded from it. Scale bars: A = 10 μm, B,C = 100 μm. Reprinted from Stern (1997b).

endodermis and pericycle of thin-walled cells embedded in parenchyma; endodermal cells with casparian strips.

Material examined

Subtribe Orchidinae: *Anacamptis* (1), *Brachycorythis* (1), *Dactylorhiza* (1), *Galearis* (1), *Himantoglossum* (1, as *Barlia*), *Holothrix* (2), *Neotinea* (1), *Ophrys* (7), *Orchis* (6, including 1 sp. as *Aceras* and 2spp. now in *Anacamptis*), *Platanthera* (8), *Ponerorchis* (1, as *Gymnadenia*), *Serapias* (3). Subtribe Habenariinae: *Bonatea* (1), *Cynorkis* (1), *Habenaria* (15), *Stenoglottis* (4). Full details appear in Stern (1997a,b).

Reports from the literature

Aybeke et al. (2010): vegetative anatomy of *Dactylorhiza*, *Ophrys*, and *Orchis*. Borsos (1977, 1980, 1982, 1990): leaf, stem, and tuber anatomy of *Anacamptis*, *Dactylorhiza*, *Gymnadenia*, *Ophrys*, *Orchis*, and *Platanthera*. Camus (1908, 1929): general anatomy of many genera of Orchideae. Dressler (1993): phylogeny and classification of Orchidaceae. Fuchs and Ziegenspeck (1925, 1927b,c): root, leaf, and stem anatomy of many genera of European Orchideae. Jurcak (1999): vegetative anatomy of *Dactylorhiza*, *Gymnadenia*, *Orchis*, *Platanthera*, and

Traunsteinera. Möbius (1886, 1887): leaf and stem anatomy of *Anacamptis*, *Gymnadenia*, *Orchis*, and *Platanthera*. Pridgeon et al. (2001): comprehensive taxonomic treatment of genera in Orchideae. Rasmussen (1981a): stomata of subfamily Orchidoideae. Solereder and Meyer (1930): general survey of orchid anatomy including *Gymnadenia*, *Herminium*, and *Orchis*. Stern (1997a,b): anatomical descriptions of Orchidinae and Habenariinae.

Additional literature

Azevedo (1970–1): mycorrhiza of *Serapias*. Barroso and Pais (1985): mycorrhiza of *Ophrys lutea*. Baytop (1968): tuber anatomy of *Orchis*. Bernard (1902): tuber and seedling development of *Orchis*. Birger (1906–7): tuber anatomy of *Habenaria*, *Herminium*, and *Orchis*. Borriss et al. (1971): mycorrhiza of *Habenaria*. Burgeff (1909, 1959): mycorrhiza and vegetative anatomy of *Ophrys* and *Platanthera*. Chatin (1857): vegetative anatomy of *Gymnadenia* and *Orchis*. Cribb and Gasson (1982): bulbil and tuber anatomy of *Cynorkis*. Currah et al. (1990): mycorrhiza of *Coeloglossum* (= *Dactylorhiza*) and *Platanthera*. Cyge (1930): leaf anatomy of *Coeloglossum* (= *Dactylorhiza*), *Gymnadenia*, *Orchis*, and *Platanthera*. Dangeard and Armand (1898): mycorrhiza of *Ophrys aranifera*. Das and Paria (1992): stomata of *Habenaria*. Dietz (1930): anatomy of underground organs of *Cynorkis* and *Habenaria*. Fabre (1855): tuber anatomy of *Himantoglossum*. Filipello Marchisio et al. (1985): mycorrhiza of *Anacamptis*, *Dactylorhiza*, *Gymnadenia*, and *Platanthera*. Freytag (1956): mucilage in *Orchis* tubers. Fricke (1926): leaf and bract anatomy of *Cynorkis*. Gillot (1898): leaf and stem anatomy of *Orchis*. Gonuz (2001): ecological anatomy of leaf and stem of *Orchis*. Hadley et al. (1971): TEM mycorrhiza of *Dactylorhiza*. Hofsten (1973): mycorrhiza of *Ophrys insectifera*. Holm (1904): root anatomy of *Habenaria*, *Orchis*, and *Platanthera*. Holm (1927): leaf anatomy of *Orchis*. Inamdar (1968): stomatal development in *Habenaria marginata*. Irmisch (1853): developmental anatomy of *Orchis*. Kasapligil (1961): leaf anatomy of *Orchis*. Kaushik (1983): vegetative anatomy of *Habenaria* and *Herminium*. Kumazawa (1956, 1958): sinker anatomy of *Pecteilis* and *Platanthera* (including *Perularia*). Kutschera and Lichtenegger (1982): root anatomy of *Dactylorhiza*, *Gymnadenia* (including *Nigritella*), and *Orchis*. Lal et al. (1979, 1980, 1982): tuber anatomy of *Habenaria*. Latr et al. (2008): mycorrhiza of *Dactylorhiza majalis*. Lausi et al. (1989): leaf anatomy of *Habenaria*. Link (1849): tuber anatomy of *Orchis*. Meyer (1886): tuber anatomy of *Orchis*. Mohana Rao et al. (1989) vegetative anatomy of *Orchis*. Mollberg (1884): mycorrhiza of *Orchis* and *Platanthera*. Moreau (1913): root tuber anatomy of *Holothrix*. Mulay et al. (1956): root anatomy of *Habenaria*. Neubauer (1978b): axis idioblasts of *Stenoglottis*. Nobécourt (1920, 1921, 1922): tuber anatomy of *Ophrys* and relatives. Ogura (1953, 1964): studies of subterranean parts of several genera of Orchideae. Olatunji and Nengim (1980): tracheoids of *Habenaria*. Olivier (1881): root anatomy of *Himantoglossum*. Prillieux (1865): bulb anatomy of *Herminium*, *Loroglossum* (now in *Himantoglossum*), *Ophrys*, *Orchis* (including *Aceras*), and *Platanthera*. Raunkiaer (1895–9): vegetative anatomy of *Ophrys*, *Orchis*, and *Platanthera*. Ravnik and Susnik (1964): anatomy of *Gymnadenia* (including *Nigritella*). Samuel and Bhat (1994): stomatal development in *Stenoglottis fimbriata*. Sasikumar (1973a, 1975b): anatomy of subterranean organs of *Habenaria*. Sgarbi and Del Prete (2005): leaf anatomy of Mediterranean *Orchis* species. Sharman (1939): sinker anatomy of *Orchis*. Silva et al. (2006): leaf anatomy of *Habenaria*. Singh (1981): stomata of *Habenaria* and *Herminium*. Singh (1983): vessels in stem of *Habenaria* and *Herminium*. Staudermann (1924): hairs of *Habenaria* and *Holothrix*. Stojanow (1916–17): tuber development of *Platanthera*. Sun et al. (1999): leaf epidermis of Chinese *Neottianthe*. Tatarenko (1995): mycorrhiza of *Habenaria* and *Tulotis* (now in *Platanthera*). Vij et al. (1991): leaf epidermis of *Habenaria*. Wahrlich (1886): mycorrhiza of *Gymnadenia*, *Ophrys*, *Orchis*, *Platanthera*, and *Serapias*. Weiss (1880): tuber anatomy of *Gymnadenia* and *Orchis*. Weltz (1897): axis anatomy of *Platanthera*. White (1907): polystely in roots of *Habenaria* and *Orchis*. Williams (1979a): stomata of *Habenaria*, *Orchis*, and *Platanthera*. Williamson (1973): mycorrhiza of *Dactylorhiza*.

Taxonomic notes

Subtribes Orchidinae and Habenariinae have been considered more or less distinctive, although with some reservations (Dressler 1993). Pridgeon et al. (2001) have carried these reservations further, contending that 'the limited molecular data available for Habenariinae appear to confirm the blurring of the boundary between subtribe Orchidinae and Habenariinae.' They feel the data argue for incorporating Habenariinae into a broadly based Orchidinae. Yet, there are several anatomical features that could continue to support the maintenance of the two taxa, e.g. absorbing roots in Orchidinae have a simple epidermis, but in Habenariinae, except for *Habenaria distans* and *H. repens*, absorbing roots have a velamen. The tuberous roots of the orchiads are characterized by a series of meristeles, whereas the root tubers of the habenariads are of two kinds: those with a single vascular cylinder and those with a series of meristeles. Still the overwhelming similarities would tend to uphold the union of the two subtribes: in both taxa stomata are mostly abaxial, mesophyll is homogeneous, there is a single series of collateral vascular bundles across the lamina, there is no foliar sclerenchyma nor are there stegmata containing silica bodies, fibre bundles are absent in both taxa, hypodermis is absent in both groups, absorbing roots in both taxa have a thin-walled exodermis, vascular tissue in roots is embedded in parenchyma, endodermal cells have casparian strips, and pith is parenchymatous. Root tubers are largely velamentous in both taxa. Thus, anatomy would tend to support the hypothesis of Pridgeon et al. (2001) that subtribes Orchidinae and Habenariinae should be merged in tribe Orchideae.

Tribe Chloraeeae (Rchb.f.) Pfeiff.

Plants are perennial, terrestrial, sympodial herbs with fleshy, clustered roots and spiral, convoluted, non-articulate leaves. Chloraeeae are distributed in South America from Peru and Bolivia in the north to Tierra del Fuego in the south. They also occur on the Juan Fernandez Islands in the Pacific Ocean and the Falkland Islands in the southern Atlantic Ocean. There are four genera: *Chloraea*, the largest, has about 46 species, *Bipinnula* has about 10 species and *Gavilea* about 13; *Geoblasta* is monospecific. Most species of *Chloraea* are Andean. *Chloraea membranacea* appears to be the only species in continuous cultivation.

Leaf surface

Hairs absent and cuticle smooth in *Chloraea*. **Epidermis**: cells thin-walled; rectangular to polygonal in *Chloraea*. **Stomata** abaxial, anomocytic.

Leaf TS

Cuticle thin in *Chloraea*. **Epidermis**: cells rectilinear. **Stomata** superficial; substomatal chambers shallow but often continuous with lacunose mesophyll in *Chloraea*. **Mesophyll** homogeneous; cells globular, parenchymatous, crystalliferous; lacunae in abaxial layers in *Chloraea*. **Vascular bundles** lacking phloem. **Stegmata** absent in *Gavilea lutea*.

Stem TS

Aerial stems. **Cortical** parenchyma free of bundles. **Central cylinder** of fibres with vascular bundles embedded in cylinder and scattered in stem centre.

Subterranean rhizomes. Central sclerenchymatous cylinder absent. **Vascular bundles** scattered, organized in short horizontal groups. Raphides present in some parenchyma cells.

Root TS

Root hairs clothing entire length of roots. **Velamen** 1-layered in *Chloraea*, with spirally sculptured cell walls. **Tilosomes** absent in *C. piquichen* (as *C. virescens*). **Exodermis**: cells elliptical, rectangular to polygonal, with variable wall thickening patterns. **Cortical** parenchyma extensive; cortical cells containing starch and endotrophic mycorrhizae. **Endodermis**: cells with lignified casparian strips present along radial walls. **Vascular cylinder** polyarch, of alternating xylem and phloem sectors. **Stegmata** absent in *Gavilea lutea*.

Material reported

The only published work on anatomy of Chloraeeae, which forms the basis of the observations above, is that of Reiche (1910), in which he briefly discusses the structure of three species of *Chloraea* and an unnamed *Bipinnula*. Additional data from an unpublished description of leaf and root of *Chloraea piquichen* (as *C. virescens*) (ANBG 861123) by A.M. Pridgeon.

Reports from the literature

Cisternas et al. (2012): phylogenetic analysis based on DNA sequences. Møller and Rasmussen (1984): stegmata absent in *Gavilea lutea*. Reiche (1910): external and internal morphology of vegetative organs of Chloraeeae. The reports are incomplete and provide little detail.

Taxonomic notes

The phylogenetic analysis by Cisternas et al. (2012) showed a lack of resolution among species of this group.

Tribe Codonorchideae P.J.Cribb

Codonorchideae are terrestrial, sympodial, stoloniferous herbs with filamentous roots and globose tubers produced on the stolons. Stems bear 2–4 small, non-articulate leaves carried on the lower portion of the scape, which emerges apically from the stem tip. There are only two temperate species of eastern Brazil, Chile, Argentina, and the Falkland Islands. The only published record of anatomy is a brief description by Reiche (1910) for *Codonorchis lessonii* (as *Pogonia lessonii*). Pridgeon et al. (2003) provided a taxonomic treatment of the tribe.

Leaf

Venation: markedly reticulate. **Epidermis**: cell walls undulate; cells elongated, lacking chlorophyll but containing crystals of calcium oxalate. **Mesophyll** with scattered, slightly enlarged, translucent cells containing raphide bundles.

Stem

Vascular bundles in upper part of stem situated within a weak ring of sclerenchyma; sclerenchyma ring absent from lower part of stem; vascular bundles scattered throughout.

Root

Cortical parenchyma with endotrophic mycorrhizae.

Taxonomic notes

Anatomy is too incomplete to draw any conclusions.

Tribe Cranichideae (Lindl.) Endl.

Cranichideae are terrestrial herbs, rarely epiphytic, with fleshy roots, clustered or scattered along a thick rhizome, or with tubers in subtribe Pterostylidinae. Leaves are non-articulate, convolute, spirally arranged along the stem, or in a basal rosette, and often differentiated into lamina, with a prominent midrib, and petiole. Cranichideae are widespread in both Old and New Worlds, mostly tropical, some temperate, on all continents except Antarctica. They are absent on some Pacific islands. There are 95 genera and about 1500 species (Stern, Morris et al. 1993c; Pridgeon et al. 2003) aggregated into six subtribes, the largest being Spiranthinae with 40 genera, followed by Goodyerinae with 35 genera and Cranichidinae with 17 genera; Galeottiellinae, Manniellinae, and Pterostylidinae each comprise a single genus. Figueroa et al. (2008) based their work on Dressler's 1993 classification in which subtribe Prescottiinae is separate from Cranichidinae. Several species of Cranichideae are grown by hobbyists, principally in the genera *Goodyera*, *Ludisia*, and *Spiranthes*.

Anoectochilus has been used medicinally in China, Taiwan, and Japan; species of *Goodyera* have been used medicinally in western and central USA, and in India, Indonesia, China, and Taiwan; *Spiranthes* species are used medicinally in Nepal, China, Taiwan, and Trinidad and Tobago.

Leaf surface

Hairs mostly absent; thin-walled, uniseriate hairs up to 10 cells long on both surfaces in *Ponthieva maculata* (Fig. 33A). **Cuticle** smooth; rugose to rugulose in several species; striate in some *Pterostylis* species; multipapillate over each cell in *Goodyera oblongifolia*. **Epidermal cells**: polygonal (Fig. 33C), roughly isodiametric; anticlinal walls straight-sided, except sinuous in most *Pterostylis* species. Adaxial epidermal cells of *Cyclopogon argyrifolius* elongated parallel to midrib in contrast to other species. **Stomata** abaxial (Figs 33B, 34A), except amphistomatal in *Spiranthes* and some *Pterostylis* species; organizational patterns varying: tetracytic, diacytic, anisocytic, and anomocytic (Fig. 34A).

Leaf TS

Cuticle thin to very thin. **Epidermis**: outer tangential cell walls generally flat to slightly convex or domed on both surfaces; sharply conical adaxially in *Anoectochilus* species (Fig. 34B) and *Ludisia discolor*; similarly modified cells occurring in other species. Adaxial cells square or rectangular, ovate or elliptical; abaxial cells flattened tangentially or radially in different species. **Stomata** superficial, outer ledges small to moderately sized, inner ledges minute to obscure. Substomatal chambers small to moderately sized in most species, moderately large to large in *Altensteinia citrina*, *A.* aff. *virescens*, *Cranichis ciliata*, *Cyclopogon elatus*, *Platythelys vaginata*, and *Ponthieva tuerckheimii*. **Hypodermis** and **fibre bundles** absent. **Mesophyll** homogeneous, but heterogeneous in *Anoectochilus* (Fig. 34B) and some species of *Pterostylis*; cells thin-walled. Cells of palisade varying from tall- to short-columnar. Spongy cells nearly circular or irregularly shaped, with frequent intercellular spaces. **Vascular bundles** collateral, in one row; tracheary elements of midrib paired side by side (Fig. 34C) separated by parenchyma, or bilobed. **Stegmata** absent. Bundle sheath cells thin-walled, chlorophyllous. Raphide idioblasts elongated parallel to midrib, larger in all dimensions than adjacent assimilatory cells. Plumose crystals in epidermal and/or mesophyll cells of some *Pterostylis* species.

Stem TS

Hairs absent. **Cuticle** thin, smooth to rugose and rugulose. **Epidermal cells**: square to rectangular, elliptical to ovate, spherical to domed. **Stomata** occasional, superficial in most species with small substomatal chambers; sunken in *Ludisia discolor*. **Cortex** up to 23 cells wide in *Ligeophila clavigera* and *Microchilus plantagineus* (as *Erythrodes plantaginea*); cortex in other species wider than 12 cells. Cells rounded to polygonal, mostly thin-walled, parenchymatous, chlorophyllous, especially in outer layers; thick-walled in *Prescottia stachyodes*;

Subfamily Orchidoideae 75

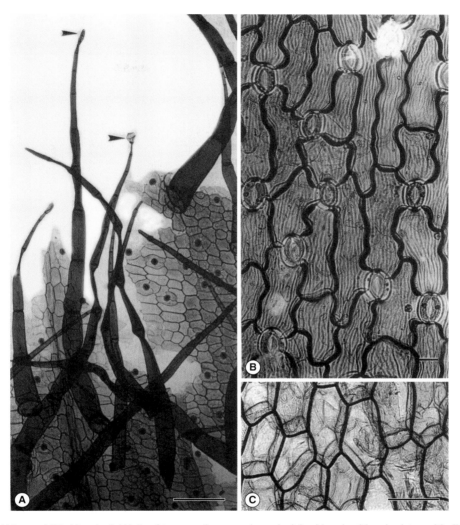

Fig. 33. Cranichideae and Diurideae leaf. (A) *Ponthieva maculata*, scraping, adaxial epidermis with uniseriate multicellular hairs with collapsed bulbous tips (arrowheads) and straight-sided anticlinal cell walls. (B) *Cryptostylis subulata*, scraping, abaxial epidermis with stomata lacking subsidiary cells. (C) *Goodyera macrophylla*, scraping, adaxial epidermis showing straight to curvilinear anticlinal cell walls. Scale bars = 10 μm. Reprinted from Stern, Morris et al. (1993c).

outer layers collenchymatous in *Goodyera rubicunda* (as *G. grandis*), *Hylophila gracilis*, and *Vrydagzynea elongata* (as *V. pachyceras*). Some cells with cruciate starch granules; spiranthosomes (modified starch granules) occurring in *Pelexia laxa*, *Prescottia stachyodes*, and *Vrydagzynea elongata*. Rod-shaped crystals in many *Pterostylis* species. Thick-walled sclereids with ramiform pits usually occurring among cortical parenchyma cells in *Goodyera pubescens*. Cortical parenchyma cells merging with ground tissue surrounding vascular elements.

No identifiable **endodermis** or endodermoid layer in most species, but endodermoid layer of tangentially flattened cells present between cortex and ground tissue in *Goodyera oblongifolia*, *G. pubescens*, and *Ligeophila clavigera* (Fig. 34D); a true endodermis separating cortex and ground tissue in *Microchilus hirtellus* (as *Erythrodes hirtella*), *M. plantagineus*, *Pelexia laxa*, *Platythelys vaginata* (Fig. 35A), *Prescottia stachyodes*, and *Vrydagzynea elongata*. Aerial stem in *Microchilus hirtellus* having an endodermoid layer and rhizome a true endodermis

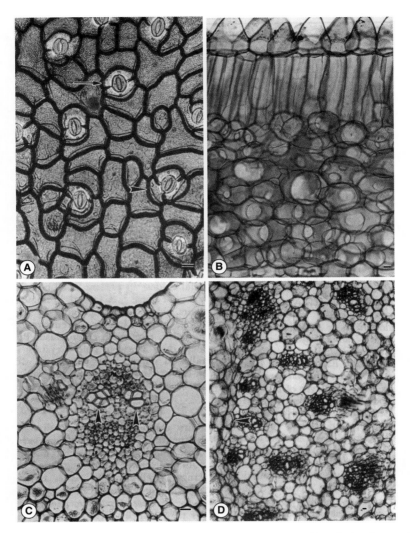

Fig. 34. **Cranichideae leaf and stem.** (A) *Prescottia stachyodes*, leaf scraping, abaxial epidermis showing curvilinear anticlinal cell walls, diacytic (arrow) and anisocytic (arrowhead) stomata. Arrowhead points to a T-shaped cell. (B) *Anoectochilus* species, TS leaf with heterogeneous chlorenchyma and conical adaxial epidermal cells. (C,D) *Ligeophila clavigera*. (C) TS foliar midrib showing paired tracheal clusters (arrowheads) and complete absence of vascular bundle sclerenchyma. (D) TS stem with scattered vascular bundles lacking associated sclerenchyma and presence of endodermoid layer of tangentially flattened cells. Scale bars = 10 μm. Reprinted from Stern, Morris et al. (1993c).

with casparian strips. **Ground tissue** cells parenchymatous, thin-walled, rounded to polygonal, lacking starch, except cruciate starch granules in ground-tissue cells of *Goodyera oblongifolia* and *Vrydagzynea elongata* and spiranthosomes in thick-walled ground-tissue cells of *Prescottia stachyodes*; chlorophyllous in *Ligeophila clavigera* and *Ludisia discolor*. **Vascular bundles** collateral (Figs 34D, 35A) 15–48, scattered, or in one or more rings; sclerenchyma and **stegmata** absent. Raphides occurring in unmodified cortical cells in *Microchilus hirtellus* and *Goodyera pubescens*; in enlarged thin-walled idioblasts in *Ligeophila clavigera*, *Ludisia discolor*, and other species. Ground-tissue raphides present in *Goodyera rubicunda* and *Ligeophila clavigera*. Rhomboids occupying entire cell volume in ground-tissue cells of *Platythelys vaginata*.

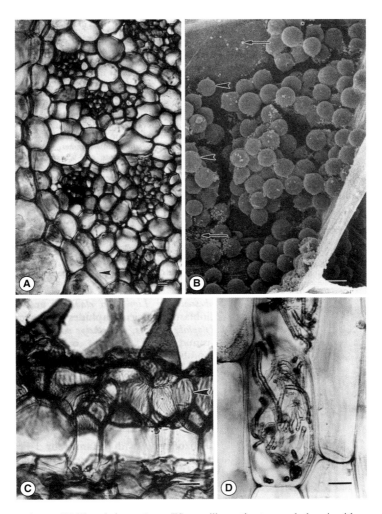

Fig. 35. Cranichideae stem and root. (A) *Platythelys vaginata*, TS stem illustrating true endodermis with casparian strips (arrowhead) and vascular bundles lacking sclerenchyma. (B) *Cranichis ciliata*, SEM of cortical root cell showing globular spiranthosomes (amyloplasts, arrowheads) and starch granules (arrows) released from ruptured spiranthosomes. (C) *Stenorrhynchos speciosum*, TS root with velamen cells showing finely anastomosing strands of cell wall material (arrowhead) and thin-walled exodermal cell. (D) *Lepidogyne longifolia* (as *L. minor*), LS root cortex showing cell with mycorrhizal pelotons (hyphal coils). Scale bars = 10 μm. Reprinted from Stern, Morris et al. (1993c).

Root TS

Rhizodermis: simple, 1-layered in many species. Root hairs usually single, unbranched cells; both unicellular and 2- to several-celled uniseriate hairs in *Sarcoglottis acaulis* and *Stenorrhynchos speciosum*; multiseriate or uniseriate in *Pterostylis*. Stomata sporadic in rhizodermis of *Pristiglottis montana*. **Velamen** present in some species, usually 1-layered (Fig. 35C), but 2–5-layered in *Prescottia plantaginea*, *P. stachyodes*, and *Stenorrhynchos speciosum*. Walls of velamen cells with radially anastomosing fine strands of cell wall material (Fig. 35C). **Tilosomes** absent in subtribes Goodyerinae, Cranichidinae, and Manniellinae; present, lamellate in subtribe Spiranthinae and *Galeoglossum tabulosum* (as *Prescottia tubulosa*); baculate in *Prescottia stachyodes*. **Exodermis**: dead cells, thin-walled (Fig. 35C); longitudinal walls with tenuous scalariform thickenings. **Cortex** 8–25 cells wide, homogeneous except in *Goodyera pubescens* where three distinct layers. Cells parenchymatous, thin-walled, variably shaped, containing spiranthosomes (Fig. 35B). Hyphae and pelotons (Pl. 1B) occurring widely in cortical cells (Fig. 35D). **Endodermis** 1-layered, all walls thin (Fig. 36), casparian strips present. **Pericycle** 1-layered, cells evenly thin-walled. **Vascular cylinder** paired in specimens of *Hetaeria oblongifolia* and *Stenorrhynchos speciosum*, otherwise solitary; 4-arch in *Vrydagzynea elongata*, 6–7-arch in *Goodyera pubescens* (Pl. 1A) to 23- or 24-arch in *Sarcoglottis acaulis* and *Spiranthes vernalis*. Specimens of *Eurystyles standleyi*, *Goodyera* species, *Ligeophila clavigera*, *Microchilus hirtellus*, and *Zeuxine strateumatica* having one or two strands of vascular tissue embedded in pith in addition to usual peripheral strands. Medullary conductive strands usually consisting of one or a few cells. Conductive tissue of *Microchilus hirtellus* dispersed throughout ground tissue of vascular cylinder; in *M. plantagineus* xylem alternating regularly with phloem as in most specimens studied (Fig. 36). **Pith** cells thin-walled, parenchymatous (Fig. 36); somewhat thick-walled in *Altensteinia citrina*, *Platythelys vaginata*, and *Sarcoglottis acaulis*. Starch granules mostly absent; spiranthosomes present in cells of most Australian taxa. Raphides usually in unmodified cortical and medullary cells, including cortex of *Pterostylis* species; idioblasts large-celled in cortex of *Cyclopogon elatus*, *Manniella gustavi*, and *Ponthieva tuerckheimii*. Distinguishing subtribal characters presented in tables 3–6 in Stern, Morris et al. (1993c).

Root tuber TS (Pterostylis only)

Hairs rare, unicellular, from outermost velamen layer. **Velamen** 1–3-layered, cells elliptical to polygonal, often periclinally compressed; outer walls of outermost layer usually suberized-thickened. **Exodermis**: cells elliptical to polygonal, walls of uniform thickness and/or radial walls suberized-thickened depending on species. **Cortex** parenchymatous, in new tubers usually filled with starch granules; outer layers of older tubers often dissociated. **Endodermis**: cells elliptical to polygonal. **Stele** medullated, undissected in most species, 8–40-arch depending on species, rarely dissected into two (*P. uliginosa*) or three (*P. daintreana*) discrete vascular strands. Xylem: tracheids with scalariform to pitted thickenings. Raphides in outer cortex, often subepidermal idioblasts, occasionally in pith; rod-shaped crystals in cortex of some species.

Material examined

Altensteinia (3), *Anoectochilus* (1), *Cheirostylis* (3), *Cranichis* (2), *Cyclopogon* (4), *Eurystyles* (1), *Galeoglossum* (1, in *Prescottia*), *Gonatostylis* (1), *Goodyera* (5), *Hetaeria* (1), *Hylophila* (1), *Lepidogyne* (1), *Ligeophila* (1), *Ludisia* (1), *Manniella* (1), *Microchilus* (2, as *Erythrodes*), *Pachyplectron* (1), *Pelexia* (2), *Platythelys* (1), *Ponthieva* (3), *Prescottia* (1), *Pristiglottis* (1), *Pterostylis* (28), *Sarcoglottis* (1), *Spiranthes* (5), *Stenorrhynchos* (1), *Vrydagzynea* (1), *Zeuxine* (2). Full details appear in Stern, Morris et al. (1993c) and for *Pterostylis* in Pridgeon (1994a), Pridgeon and Chase (1995).

Reports from the literature

Ackerman and Williams (1981): pollen morphology of Chloraeinae including Pterostylidinae placed in Cranichideae by Pridgeon et al. (2003). Ames (1922): mycorrhiza of *Goodyera pubescens*. Aybeke (2012): vegetative anatomy of *Spiranthes spiralis*. Burgeff (1909): mycorrhiza of *Goodyera*. Burgeff (1932): vegetative anatomy of *Cystorchis*, *Myrmechis*, and *Zeuxine*. Burgeff (1959): mycorrhiza of *Cystorchis*. Cameron, Chase, Hills, and Bridges (1992): a phylogenetic analysis of Orchidaceae. Campbell (1970): rhizome and mycorrhiza of *Yoania* (now = *Danhatchia*). Camus (1908, 1929): anatomy of *Spiranthes* and *Goodyera*. Clements (1988): mycorrhiza and seedling anatomy of *Pterostylis*. Das and Paria

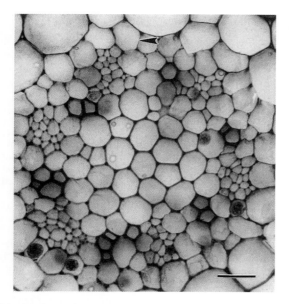

Fig. 36. Cranichideae. *Pachyplectron neocaledonicum.* TS vascular cylinder of root showing unthickened endodermal cell walls (arrowhead) and vascular tissue totally surrounded by thin-walled parenchyma cells. Scale bar = 10 μm. Reprinted from Stern, Morris et al. (1993c).

(1992): stomata of *Zeuxine*. Dietz (1930): anatomy of underground organs of *Anoectochilus*, *Haemaria* (now in *Ludisia*), *Myrmechis*, *Pterostylis*, and *Stenorrhynchos*. Diskus and Kiermayer (1954): crystals in *Haemaria* (now in *Ludisia*). Dressler (1990): taxonomic position of Spiranthoideae. Dressler (1993): orchid classification and phylogeny. Figueroa et al. (2008): root anatomy of many Cranichideae. Fuchs and Ziegenspeck (1925, 1926, 1927b): root, stem, and leaf anatomy of *Goodyera* and *Spiranthes*. Garay and Christenson (1995): *Yoania* renamed *Danhatchia*. Groom (1895–7): anatomy of *Spiranthes australis*. Guttenberg (1968): root anatomy of *Hetaeria*. Holm (1904): root anatomy of *Goodyera* and *Spiranthes*. Holm (1927): leaf anatomy of *Goodyera*. Irmisch (1853): root tuber anatomy of *Spiranthes*. Janse (1897): mycorrhiza of *Myrmechis*. Kaushik (1983): anatomy of *Zeuxine*. Liang and Zheng (1984): stem and root anatomy of *Anoectochilus* and *Goodyera*. Lin and Namba (1981a,b): drug plant anatomy of *Anoectochilus* and *Goodyera*. Lin and Namba (1982a,b): drug plant anatomy of *Spiranthes sinensis* and vars. MacDougal (1899b): root anatomy of *Gyrostachys* (now in *Spiranthes*), and *Peramium* (now in *Goodyera*). Masuhara and Katsuya (1992): mycorrhiza of *Spiranthes sinensis*. Meindl and Kiermayer (1987): crystals in *Haemaria* (now in *Ludisia*). Meinecke (1894): aerial root anatomy of *Goodyera*. Möbius (1887): leaf anatomy of *Haemaria* (now in *Ludisia*) and *Macodes*. Mollberg (1884): mycorrhiza of *Goodyera*. Narayanaswamy (1950): root anatomy and mycorrhiza of *Spiranthes australis*. Nieuwdorp (1972): mycorrhiza of *Goodyera*. Olatunji and Nengim (1980): 'tracheoids' in *Cystorchis* and *Hetaeria*. Pirwitz (1931): 'storage tracheids' in *Spiranthes*. Pittman (1929): mycorrhiza in *Pterostylis*. Porembski and Barthlott (1988): velamen of *Altensteinia*. Pridgeon (1994a): trichomes in tribe Diurideae including *Pterostylis*. Pridgeon and Chase (1995): morphology and anatomy of subterranean axes in Diurideae inclusive of *Pterostylis* placed in Cranichideae by Pridgeon et al. (2003). Rasmussen and Whigham (2002): mycorrhiza of *Goodyera*. Raunkiaer (1895–9): leaf anatomy of *Goodyera*. Salazar et al. (2009): phylogenetic relations of Cranichideae. Schmucker (1927): vegetative anatomy of *Haemaria* (now in *Ludisia*). Silva et al. (2006): leaf anatomy of *Prescottia*. Singh (1981): stomata of *Zeuxine*. Solereder and Meyer (1930): comprehensive report on orchid anatomy including original observations on *Spiranthes* root. Staudermann (1924): trichomes of *Spiranthes* and *Zeuxine*. Stern, Aldrich et al. (1993a): study of spiranthosomes, a newly reported specialized amyloplast in spiranthoid orchids. Stern, Morris et al. (1993c): comprehensive analysis of spiranthoid vegetative anatomy. Tatarenko (1995): mycorrhiza of *Goodyera* and *Spiranthes*. Vij et al. (1991): leaf epidermis of *Goodyera*, *Spiranthes*, and *Zeuxine*. Vij et al. (1985): mycorrhiza of *Spiranthes lancea*. Wahrlich (1886): mycorrhiza of *Goodyera*. Warcup (1981): mycorrhiza of *Pterostylis* and *Spiranthes*. Williams (1975, 1979): specialized stomatal subsidiary cells in *Ludisia discolor*. Zhang et al. (1992): vegetative anatomy of *Anoectochilus*.

Taxonomic notes

Dressler's (1993) subfamily Spiranthoideae is nearly congruent with Pridgeon et al.'s (2003) tribe Cranichideae, except for the addition by them of subtribes Pterostylidinae and Galeottiellinae, the inclusion of Prescottiinae in Cranichidinae, and the removal from Spiranthoideae of tribes Tropidieae and Diceratosteleae. They incorporated tribe Diceratosteleae as subtribe Diceratostelinae in tribe Triphoreae leaving tribe Tropidieae as possibly related to *Palmorchis* in subfamily Epidendroideae. Pterostylidinae may be anomalous in Cranichideae, all members of which have fleshy roots scattered along a rhizome, whereas *Pterostylis* has root tubers. In addition, root hairs in Cranichideae are usually unicellular, although 2- to several-celled uniseriate hairs occur in some taxa, but *Pterostylis* has multiseriate or uniseriate hairs, depending upon the species.

The study of Figueroa et al. (2008) on roots of Cranichideae established the presence of tilosomes in roots of several taxa of Cranichideae in contrast to previous work on this group by Stern, Morris et al. (1993c). Lamellate tilosomes are a synapomorphy grouping all members of subtribe Spiranthinae distinguishing it from subtribes Goodyerinae, Cranichidinae, and Manniellinae. Garay and Christenson (1995) have transferred the New Zealand endemic *Yoania australis* from Calypsoeae to *Danhatchia australis* and incorporated it into Cranichideae. Phylogenetic relations within the tribe have been investigated by Salazar et al. (2009) using nuclear and plastid DNA studies.

SUBFAMILY VANILLOIDEAE SZLACH.

Tribe Pogonieae Pfitzer ex Garay & Dunsterv.

Terrestrial, perennial, mostly sympodial, mycorrhizal herbs, foliated except for *Pogoniopsis*, which is achlorophyllous and mycoheterotrophic; roots elongate, fibrous, or tuberous. Leaves non-articulate. There are six genera: *Cleistes* has one species native to south-eastern USA and about 30 species distributed throughout tropical South America, Panama, and Costa Rica, with three species separated as *Cleistesiopsis*; *Duckeella* has three species in Venezuela, Colombia, and Brazil; *Isotria* with two species of eastern North America; *Pogonia* has four species divided between eastern North America and eastern Asia; and *Pogoniopsis* has two species localized in southeastern Brazil. It does not appear that species of any Pogonieae are in cultivation.

The information on this tribe has been compiled entirely from the literature but the names now accepted have been used where possible. Where no data are given for any organ or tissue, it is because none was available.

Leaf surface
Epidermis: cells thin-walled, papillose on upper surface of *Isotria verticillata*; cells elongate in *Cleistes rosea*, *Cleistesiopsis divaricata* (Fig. 37A,B) (as *Cleistes divaricata*), *Duckeella adolphii*, and *Pogonia minor*; walls sinuous in *Isotria verticillata* (Fig. 37C,D), *Pogonia japonica*, and *P. ophioglossoides*. **Stomata** on both surfaces (*Cleistes* sp., *Cleistesiopsis divaricata*, *Duckeella*), or abaxial (*Cleistes rodriguesia* (as *C. gracilis*), *Isotria*, *Pogonia*); subsidiary cells absent in *Cleistes* sp., *Cleistesiopsis divaricata*, *Isotria verticillata*, *Pogonia japonica*, and *P. ophioglossoides*.

Leaf TS
Cuticle thin, smooth, except striated in *Cleistes rodriguesia*. **Mesophyll** homogeneous; starch and hyphae present in *Cleistes rodriguesia*; raphides present in *Cleistesiopsis divaricata*, *Isotria verticillata*, *Pogonia minor*, and *P. ophioglossoides*. **Vascular bundles** lacking sclerenchyma; embedded in mesophyll in *Isotria verticillata*.

Root TS
Velamen 1-layered, lacking helical thickenings, with small pits, in *Cleistesiopsis divaricata*, *Duckeella pauciflora*, and *Pogonia ophioglossoides*. Hairs numerous in *Cleistesiopsis divaricata* and *Pogonia ophioglossoides*, fewer in *Isotria verticillata*; borne on epidermal papillae in *Pogoniopsis*; persistent in *Pogoniopsis* but replaced by an exodermis in other genera. **Tilosomes** absent. **Hypodermis**: cells thin-walled in *Pogonia ophioglossoides*. **Cortex** comprising 6–9 layers of thin-walled cells with frequent large intercellular spaces in *Pogonia ophioglossoides*, but compact in other species; three or four layers in *Pogoniopsis*. Hyphae occurring in cortex of all species. **Endodermis**: cells thin-walled throughout in

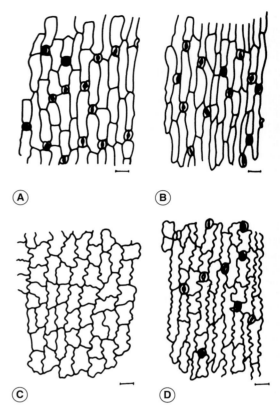

Fig. 37. Pogonieae. (A,B) *Cleistesiopsis divaricata* leaf, stomata amphiepidermal, anomocytic, epidermal cells elongate. (A) Adaxial surface. (B) Abaxial surface. (C,D) *Isotria verticillata* leaf. (C) Adaxial surface with broad epidermal cells with sinuous walls. (D) Abaxial surface with anomocytic stomata and narrow epidermal cells with sinuous walls.

Cleistesiopsis divaricata and *Pogonia ophioglossoides* but in *Isotria verticillata* with a single thick-walled endodermal cell abutting each phloem strand. **Pericycle** cells thin-walled in *P. ophioglossoides*; pericycle continuous in *P. ophioglossoides* but interrupted in *Cleistesiopsis divaricata* and *Isotria verticillata*. Roundish phloem strands alternating with xylem rays; 5-arch in *Pogonia ophioglossoides*, 6-arch in *Cleistesiopsis divaricata*, and 8-arch in *Isotria verticillata*; with few xylem elements in centre of root in *Pogoniopsis*.

Material reported

Cleistes (2), *Cleistesiopsis* (1, as *Cleistes* or *Pogonia*), *Duckeella* (2), *Isotria* (1), *Pogonia* (3), *Pogoniopsis* (1).

Reports from the literature

Cameron (2009): phylogeny of Vanilleae. Cameron and Chase (1999): relationships of pogoniad orchids. Cameron and Dickison (1998): architecture of leaves in *Cleistes*, *Duckeella*, *Isotria*, and *Pogonia*. Carlson (1938): origin of shoots from roots of *Pogonia ophioglossoides*. Dressler (1993): phylogeny and classification of orchids. Holm (1900): vegetative anatomy of *Pogonia ophioglossoides*. Holm (1904): root anatomy of *Pogonia divaricata* (now = *Cleistesiopsis*), *P. verticillata* (now = *Isotria*), and *P. ophioglossoides*. Holm (1927): leaf anatomy of *Isotria verticillata*. Johow (1889): root morphology and anatomy of *Pogoniopsis* aff. *nidus-avis*. MacDougal (1899b): root anatomy of *Pogonia ophioglossoides*. Moreau (1912): anatomical study of five purported Madagascan species of *Pogonia* that have since been referred to *Nervilia*. Porembski and Barthlott (1988): velamen of *Cleistes* (now = *Cleistesiopsis*), *Duckeella*, and *Pogonia*. Pridgeon et al. (2003): generic treatment of pogoniad orchids. Silva et al. (2006): leaf anatomy of *Cleistes gracilis* (now = *C. rodriguesii*). Tatarenko (1995): mycorrhiza of *Pogonia*. Williams (1979): stomatal subsidiary cells in *Cleistes*, *Isotria*, and *Pogonia*.

Taxonomic notes

Published material on pogoniad orchids is so sparse and incomplete it would be difficult to discern how at this stage it could contribute to any meaningful taxonomic interpretations. This tribe requires a complete anatomical analysis. Its occurrence as a vanilloid taxon is arguable (F.N. Rasmussen 1982) but has recently been demonstrated by Cameron (2009), and given the rather thorough anatomical treatments of other vanilloid taxa (Cameron and Chase 1998, 1999; Cameron and Dickison 1998; Cameron, Chase et al. 1999; Stern and Judd 1999, 2000), a comprehensive study of the anatomy of Pogonieae is desirable.

Tribe Vanilleae Blume

Vanilleae comprise terrestrial, epiphytic, hemiepiphytic, and saprophytic, monopodial or sympodial, autotrophic or achlorophyllous, mycotrophic, foliate or non-foliate vines, and upright herbs. Vanilleae are represented by a group of geographically disjunct genera: *Vanilla* is pantropical, *Epistephium* is South American, *Dictyophyllaria* is a Brazilian endemic, *Eriaxis* and *Clematepistephium* are New Caledonian endemics, *Cyrtosia*, *Erythrorchis*, *Galeola*, *Lecanorchis*, and *Pseudovanilla* are distributed across Oceania and South East Asia. *Erythrorchis cassythoides* and *Pseudovanilla* are eastern Australian endemics, and at least two species of *Pseudovanilla* are endemic to Fiji and Ponape in the Pacific (Pridgeon et al. 2003). *Lecanorchis* comprises only about 15 species, most of which are native to the island of Honshu, Japan (Pridgeon et al. 2003). Most genera have few species; three are monospecific; *Vanilla* with over 100 species is by far the largest and probably most diverse of the genera. Dressler (1993) divided tribe Vanilleae into three subtribes: Galeolinae, Vanillinae, and Lecanorchidinae with a single genus.

Galeola and *Vanilla* have been used medicinally but the main use of *Vanilla* is as a flavouring agent.

Leaf surface (bract of *Pseudovanilla*; *Vanilla* described separately, see below)

Hairs absent. **Epidermis**: cells polygonal (Fig. 48E), straight-sided in *Eriaxis*, with undulate margins in some *Clematepistephium* and *Epistephium* species (Fig. 48A), irregularly shaped in *Pseudovanilla*. **Stomata** abaxial, except ad- and abaxial in *Pseudovanilla* and some species of *Epistephium*; tetracytic (Fig. 48E), some anomocytic in *Clematepistephium* (Fig. 47D) and *Eriaxis*; anomocytic in *Epistephium* (Fig. 48A), some tetracytic in *Epistephium*; anomocytic, probably non-functional in *Pseudovanilla* (Fig. 46F).

Leaf TS

Cuticle smooth (Fig. 47A,E), adaxial 5.0–22.0 µm thick, abaxial 5.0–12.5 µm thick; 2.5 µm thick in *Pseudovanilla*. **Epidermis**: cells periclinal to isodiametric (Fig. 47A–C). **Stomata** somewhat sunken in *Clematepistephium*, superficial in *Epistephium* and *Eriaxis*, adaxial superficial, abaxial somewhat raised in *Pseudovanilla*; outer and inner ledges large in *Clematepistephium*, small to medium in *Epistephium*; outer ledges large, inner ledges small to large in *Eriaxis*; outer and inner ledges small in *Pseudovanilla*; substomatal cavities small to large. **Hypodermis**: 1-layered (Fig. 47B,E), ad- and abaxial in *Clematepistephium* and *Pseudovanilla*, adaxial in *Eriaxis*, absent in *Epistephium*. **Fibre bundles** absent. **Mesophyll** homogeneous (Fig. 47E), cells thin-walled in *Clematepistephium*, *Eriaxis*, and *Pseudovanilla*, unevenly thick-walled in *Epistephium*; cells circular to oval in *Clematepistephium* and *Epistephium*; lobulate to polygonal in *Eriaxis*, rounded to polygonal in *Pseudovanilla*; idioblasts with raphide bundles frequent (Fig. 48B). **Vascular bundles** collateral, in one row; each surrounded by a sclerenchymatous sheath in *Clematepistephium*, sclerenchyma occurring at both xylem and phloem poles and laterally in *Epistephium* and *Eriaxis*; xylem composed of discontinuous arcs of tracheary cells parallel to epidermis, separated by groups of parenchyma cells, each subtended by arcuate cluster of phloem in *Pseudovanilla* (Fig. 47C). Bundle sheath cells thin-walled, bundle sheath sometimes incomplete or indistinct. **Stegmata** absent.

Stem TS

Hairs and stomata absent in *Clematepistephium*, *Cyrtosia*, *Erythrorchis*, and *Galeola*; hairs absent in *Epistephium*, *Eriaxis*, *Lecanorchis*, and *Pseudovanilla*. **Stomata** superficial in *Epistephium* and *Eriaxis* with prominent outer ledges and small inner ledges; superficial in *Lecanorchis*, superficial to raised in *Pseudovanilla*, with moderately sized outer and inner ledges; small to moderate substomatal chambers. **Cuticle** smooth, ±2.5 µm thick, except ±25 µm thick in *Clematepistephium smilacifolium*, 2.5–10.0 µm thick in *Epistephium* species, and 10.0–12.5 µm thick in *Eriaxis*; rough-surfaced, undulate, and grooved between cells in *Eriaxis*. **Epidermis**: cells isodiametric in *Clematepistephium*, *Epistephium*, and *Erythrorchis*, isodiametric to anticlinal in *Galeola* and *Lecanorchis*, anticlinal in *Cyrtosia*, *Eriaxis*, and *Pseudovanilla*; chlorophyllous in *Clematepistephium*. **Hypodermis**: 1-layered in *Cyrtosia* (Fig. 46C), *Eriaxis*, and *Galeola* (Fig. 46E) only, of isodiametric cells. **Cortex**: cells variable, thin-walled to moderately thick-walled, oval, circular, and polygonal (Fig. 46C,E); chlorophyllous in *Clematepistephium* and *Eriaxis*; small, triangular intercellular spaces to large irregularly shaped spaces resembling those in hydrophytes in *Epistephium subrepens* (Fig. 48C); cruciate starch granules and crystalliferous idioblasts present in some species; banded, branched, cell wall thickenings in *Erythrorchis*; water-storage cells in *Galeola*; an innermost band 1–6 cells wide (Fig. 46D) (5–7 in *Lecanorchis*) surrounding ground tissue and consisting of thick-walled polygonal cells lacking intercellular spaces. **Ground tissue**: cells circular to oval and polygonal, outer cells thick-walled, walls anisotropic; inner cells thin-walled, walls isotropic, except all cells thin-walled in *Galeola*; cruciate starch granules present in some cells. **Vascular bundles** collateral, scattered; bundles surrounded by 1–4 layers of polygonal sclerenchyma cells in *Cyrtosia*, *Erythrorchis*, *Galeola*, and *Pseudovanilla*; by one or two layers of flattened cells in *Clematepistephiium*; polygonal thin-walled sclerenchyma cells associated with phloem or surrounding bundles in *Epistephium*; small clusters of polygonal sclerenchyma cells associated with phloem in *Eriaxis*; sclerenchyma absent in *Lecanorchis*. **Stegmata** absent.

Rhizome TS (Cyrtosia septentrionalis only)

Hairs and stomata absent. **Cuticle** smooth, <2.5 µm thick. **Epidermis**: cells anticlinal and isodiametric (Fig. 46A); outer and radial walls thickened, inner walls thin (Fig. 46A). **Hypodermis**: cells rounded to anticlinal; outer and radial walls thickened, inner walls thin. **Cortex**: cells thick-walled, almost circular, parenchymatous, with cruciate starch granules; raphide bundles in enlarged cells; inner band of thick-walled cells surrounding ground tissue. **Ground tissue** cells circular to oval, thick-walled, parenchymatous, with cruciate starch granules. **Vascular bundles**: a central complex of four clusters of scattered masses of xylem cells intermingled with thin-walled phloem cells and collateral bundles randomly dispersed across ground tissue (Fig. 46B); outer series of bundles smaller than inner bundles; smaller bundles encircled by polygonal sclerenchyma cells.

Root TS

Velamen 1-layered (Fig. 49B), cells anticlinal in *Galeola*, *Lecanorchis*, and *Pseudovanilla*, anticlinal, polygonal, thick-walled in *Clematepistephium*, periclinal in *Erythrorchis*; simple epidermis in *Epistephium* and *Eriaxis*; hairs unicellular, present only in *Erythrorchis* and *Lecanorchis*, with hyphae in *Lecanorchis*. **Tilosomes** absent. **Exodermis**: cells anticlinal in *Erythrorchis*, *Lecanorchis*, and *Pseudovanilla*, outer and radial walls heavily thickened in *Lecanorchis*; narrow, polygonal, anticlinal in *Clematepistephium*; absent in *Epistephium*, *Eriaxis*, and *Galeola*. **Cortex**: in *Clematepistephium* outer layer with anticlinal cells, remainder 2-layered, outer layer with thin-walled rounded chlorophyllous cells, inner layer with thick-walled rounded cells lacking chloroplasts; in *Epistephium*, *Erythrorchis*, *Galeola*, and *Pseudovanilla*, cells thin-walled, somewhat circular to polygonal, some cells with cruciate starch granules; in *Galeola* with water-storage cells; in *Eriaxis* cells slightly thick-walled, outer cells polygonal, inner cells rounded to oval; in *Lecanorchis* cells circular to oval with dead fungal masses. **Endodermis** 1-layered, with O-thickened walls opposite phloem, thin-walled opposite xylem in *Clematepistephium*, *Erythrorchis*, and *Pseudovanilla*; in *Epistephium* cells walls thin, casparian strips present; in *Eriaxis* and *Lecanorchis* cells O- or U-thickened opposite phloem (Fig. 49C), thin-walled opposite xylem; in *Galeola* all walls thin. **Pericycle** 1-layered, cells O-thickened opposite phloem, thin-walled opposite xylem in *Erythrorchis*, *Lecanorchis*, and *Pseudovanilla*, one to several layers in *Clematepistephium*, 1–3 layers in *Epistephium*, one or two layers in *Eriaxis*, entirely thin-walled in *Galeola*. **Vascular cylinder** 5- or 6-arch in *Lecanorchis*, 13–18-arch in most genera, 32-arch in *Clematepistephium*; xylem alternating with phloem strands, all embedded in thick-walled sclerenchyma (Fig. 49D); in *Epistephium* xylem alternating with phloem strands; in *Eriaxis* xylem and phloem alternating in a matrix of outer sclerenchyma cells and inner thin-walled parenchyma cells; vascular tissue embedded in sclerenchyma in *Clematepistephium* (Fig. 47F), *Erythrorchis*, and *Lecanorchis* (Fig. 49C), in parenchyma in *Galeola* and *Pseudovanilla*. **Pith**: cells circular to oval, sclerotic or parenchymatous (Fig. 49A); some cells completely filled with a single starch granule in *Clematepistephium*, cells circular, thin-walled in *Epistephium*; cells rounded to polygonal, somewhat thick-walled, with a central medullary lacuna in *Eriaxis* (Fig. 49A), chlorophyllous in *Pseudovanilla*; intercellular spaces absent in *Lecanorchis*.

Vanilla Plum. ex Mill.

Vanilla is a pantropical genus of green-stemmed vines with aerial clasping roots and terrestrial absorbing roots, foliated and non-foliated with scale leaves. Cured fruits of several species of *Vanilla* yield the flavouring agent vanilla.

Foliage leaf surface

Hairs absent. **Epidermis**: cells polygonal, walls straight or curvilinear (Fig. 38A,B). **Stomata** abaxial, tetracytic (Fig. 38A), but ad- and abaxial in *Vanilla insignis* and *V. siamensis*. Crystals in epidermal cells (Fig. 38C).

Foliage leaf TS

Cuticle smooth, 2.5–7.5 µm thick adaxially, 2.5 µm or less abaxially. **Epidermis**: cells periclinal to isodiametric (Fig. 38D,E). **Stomata** superficial, substomatal chambers small to moderate, outer ledges small to moderate, inner ledges small to minute, but large in *V. polylepis*. **Hypodermis**: 1-layered (Fig. 38D), abaxial; ad- and abaxial in *V. polylepis*, absent from five species examined. **Mesophyll** homogeneous (Fig. 38E); crystalliferous idioblasts in series at upper and lower parts of mesophyll; raphide cells present (Fig. 38F). **Vascular bundles** in one row (Fig. 38E), collateral; sclerenchyma present or not; where present in larger bundles it surrounds entire bundle. Bundle sheath cells thin-walled, chlorophyllous; absent from *V. polylepis*.

Scale leaf TS

Cuticle smooth, 2.5 µm thick or less in all species. **Epidermis**: adaxial cells periclinal and isodiametric; anticlinal in *V. dilloniana* (Fig. 39D). **Stomata** abaxial, except amphistomatal in *V. poitaei*, superficial; substomatal chambers very small, outer and inner ledges minute to absent. **Hypodermis**: 1-layered ad- and abaxially in *V. barbellata*, *V. dilloniana*, and *V. poitaei*, absent from *V. madagascariensis*. **Mesophyll** homogeneous (Fig. 39A–D); raphide bundles in very thin-walled idioblasts. **Vascular bundles** collateral, in one row (Figs 39A, 40A); sclerenchyma absent in *V. barbellata* and *V. madagascariensis*.

Stem TS

Stems of leafless species bilobate with opposite grooves (Fig. 40B); stems of leafy species terete. **Hairs** absent. **Cuticle** smooth (Fig. 40C), 2.5–7.5 µm thick, but 10 µm thick in *V. barbellata* and *V. madagascariensis* and 25 µm thick in *V. claviculata*. **Epidermis**: cells isodiametric (Fig. 40D) to somewhat periclinal, anticlinal in *V. africana*; all walls thickened. **Stomata** superficial, mostly with small substomatal chambers, large in several species. **Hypodermis**: 1-layered (Fig. 40C,D), cells chlorophyllous. **Cortex**: cells circular to oval, walls thin, chlorophyllous; raphide bundles present. Cortex and ground tissue separated by a sclerenchyma band in leafy species only (Fig. 41A,B). In *V. imperialis*, sclerenchyma band replaced by one layer of thin-walled cells containing aggregations of spiranthosomes (Fig. 45C). **Ground tissue** consisting of thin-walled cells; in some species with only assimilatory cells; in other species with assimilatory cells in stem centre surrounded by water-storage cells, and in still other species with mixed assimilatory and water-storage cells (Figs 41C, 43A). Spiranthosomes at cortex/ground tissue transition containing starch granules in *V. imperialis* (Fig. 45C). **Vascular bundles** collateral; smaller bundles in peripheral ring, larger bundles centrally scattered (Fig. 43B,C). Sclerenchyma completely surrounding vascular bundles (Fig. 41C,D), occurring only at phloem pole of smaller bundles (Fig. 41B), or lacking completely.

Root TS

Velamen 1-layered (Figs 42A, B, 43D), cells anticlinal in aerial and terrestrial roots. Cell wall sculpturing varying from absent to minute pits, to marked stripings (Fig. 42A); hyphae and protists infesting cells. Hairs absent in aerial roots, present in terrestrial roots (Fig. 42C). **Exodermis** 1-layered (Figs 42C, 43D); cells polygonal and anticlinal; ladder-like thickenings occurring along longitudinal walls of dead cells; cells thin-walled in *V. insignis* (Fig. 42C), thickened on outer and lateral walls in in *V. pompona* (Fig. 42D): passage cells intermittent. **Cortex**: cells thin-walled, except thick in *V. africana* (Fig. 44A), mostly circular and oval (Figs 42D, 44A); cells in aerial roots

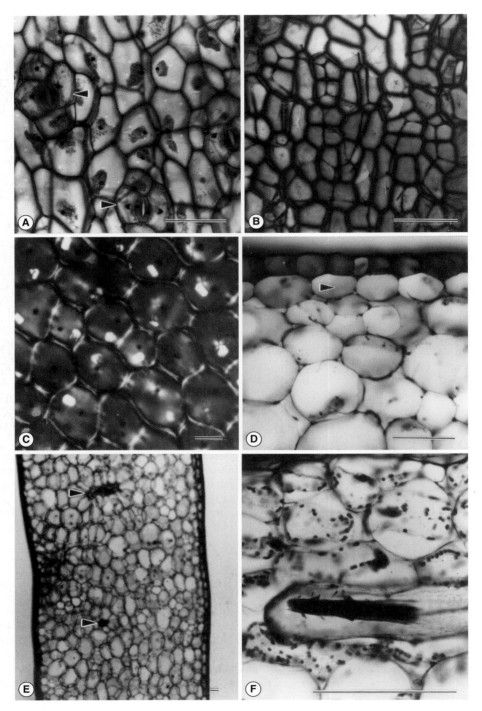

Fig. 38. Vanilleae. *Vanilla* **leaf.** (A,B) *V. pompona*. (A) Abaxial epidermis with tetracytic stomata (arrowheads). (B) Adaxial epidermis showing elongated cells. (C) *V. planifolia*, polarized light, adaxial epidermis showing crystals in cells. (D) *V. pompona*, TS showing 1-layered adaxial hypodermis (arrowhead). (E) *V. planifolia*, TS, homogeneous mesophyll with large and small vascular bundles (arrowhead). (F) *V. africana*, TS, raphide bundle in mesophyll idioblast. Scale bars: A,B,D,E,F = 100 μm, C = 10 μm. Reprinted from Stern and Judd (1999).

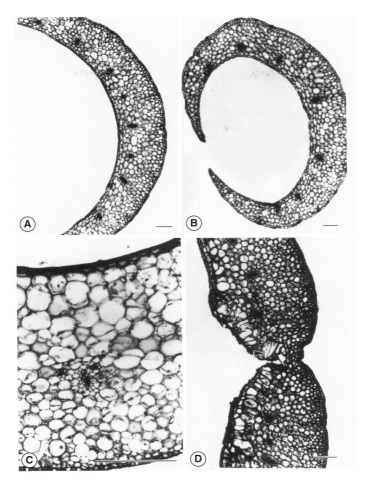

Fig. 39. Vanilleae. *Vanilla* **scale leaf**. (A) *V. barbellata*, TS crescentiform scale leaf with one adaxial row of vascular bundles. (B) *V. madagascariensis*, TS C-shaped scale leaf and adaxial row of vascular bundles. (C) *V. poitaei*, TS with homogeneous mesophyll. (D) *V. dilloniana*, TS scale leaf; adaxial groove lined with thick-walled anticlinal cells (arrowhead). Scale bars = 100 μm. Reprinted from Stern and Judd (1999).

chlorophyllous; raphide bundles in thin-walled idioblasts; hyphae and pelotons infesting cells (Fig. 44D). **Endodermis**: cell walls variously thickened opposite phloem (Fig. 44E), thin-walled opposite xylem (Fig. 44B,C). **Pericycle** 1-layered, cells O-thickened opposite phloem, thin-walled opposite xylem. **Vascular cylinder**: vascular tissue embedded in thin- or thick-walled sclerenchyma (Fig. 44F), except parenchyma in aerial roots of *V. insignis* and aerial and terrestrial roots of *V. madagascariensis*; xylem and phloem strands alternating around circumference of vascular cylinder; 8–16-arch in aerial roots and 6–18-arch in terrestrial roots. Metaxylem elements always wider in terrestrial roots (Fig. 45B) than in aerial roots of same species (Fig. 45A). **Pith** cells parenchymatous in aerial and terrestrial roots, except sclerotic in aerial roots of *V. dilloniana* and terrestrial roots of *V. siamensis*. Chloroplasts occurring in pith cells of aerial roots of several species and in aerial and terrestrial roots of *V. africana*.

Material examined

Clematepistephium (1), *Cyrtosia* (1), *Epistephium* (2), *Eriaxis* (1), *Erythrorchis* (1), *Galeola* (1), *Lecanorchis* (2), *Pseudovanilla* (1), *Vanilla* (17). Full details appear in Stern and Judd (1999, 2000).

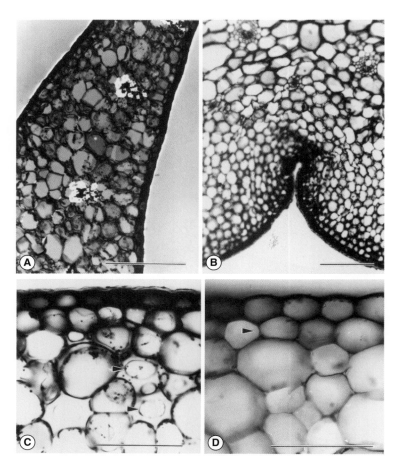

Fig. 40. Vanilleae. ***Vanilla* leaf and stem**. (A) *Vanilla* sp., TS polarized light; scale leaf tissue showing discontinuous series of sclerenchyma cells surrounding vascular bundle. (B) *V. claviculata*, TS stem showing one of two opposite invaginations. (C) *V. barbellata*, TS stem showing thick cuticle, isodiametric epidermal cells, 1-layered hypodermis, chloroplasts, and primary pit fields (arrowheads) in cortical cells. (D) *V. phaeantha*, TS stem with one layer of more or less isodiametric hypodermal cells (arrowhead). Scale bars = 100 μm. Reprinted from Stern and Judd (1999).

Reports from the literature

Alconcero (1968a,b, 1969a,b): anatomy of *Vanilla* roots and mycorrhiza. Baruah (1998): vegetative anatomy of *Vanilla pilifera*. Baruah and Saikia (2002): vegetative anatomy of *V. planifolia*. Bouriquet (1954): taxonomy of vanillas of the world. Burgeff (1909, 1932, 1959): mycorrhiza and root anatomy of *Galeola* and *Lecanorchis*. Cameron (2009): phylogeny of Vanilloideae. Cameron and Dickison (1998): foliar structure of vanilloid orchids. Chatin (1856, 1857): vegetative anatomy of *Vanilla*. Cordemoy (1904): mycorrhiza of *Vanilla*. Costantin (1885): root anatomy of *Vanilla*. Dressler (1993): description of subtribes in Vanilleae. Dycus and Knudson (1957): velamen of *Vanilla* roots. Ecott (2004): history, economics, and sociology of the vanilla industry. Engard (1944): root anatomy of *Vanilla*. Garay (1986): support for the establishment of family Vanillaceae. Groom (1895–7): anatomy of *Galeola* and *Lecanorchis*. Guttenberg (1968): root anatomy of *Galeola*. Häfliger (1901): anatomy of *Vanilla* species. Hamada (1939): anatomy and mycorrhiza of *Galeola*. Hashimoto (1990): taxonomic review of Japanese species of *Lecanorchis*. Heckel (1899): anatomy of leafless vanillas. Holm (1915): vegetative anatomy of *V. planifolia*. Janse (1897): mycorrhiza of *Lecanorchis*. Kausch and Horner (1983): crystals in roots of

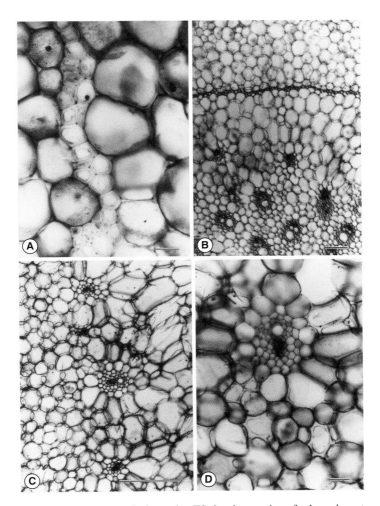

Fig. 41. Vanilleae. **Vanilla** stem. (A) *V. planifolia*, a leafy species; TS showing portion of sclerenchymatous band separating cortex (right) and ground tissue (left). (B) *Vanilla* sp., a leafy plant; TS with intact sclerenchymatous band separating cortex (top) and ground tissue (bottom). Cells immediately below sclerenchymatous band represent turgid water-storage tissue surrounding cortical assimilatory cells in which vascular bundles are embedded. (C) *V. barbellata*, TS; this leafless species lacks a sclerenchymatous band separating cortex (left) and ground tissue (right). (D) *V. hartii*, TS ground tissue composed entirely of assimilatory cells. Scale bars: A = 10 μm, B–D = 100 μm. Reprinted from Stern and Judd (1999).

V. planifolia. Krüger (1883): aerial organs of *Vanilla*. Kuttelwascher (1964, 1965): aerial root development of *Vanilla*. Leitgeb (1864b): aerial root anatomy of *Vanilla*. Liu and Zhou (1987): mycorrhiza of *Galeola faberi*. Mangin (1882): stem of *V. planifolia*. Milanez et al. (1966): leaf epidermis of *Vanilla*. Möbius (1887): leaf anatomy of *Vanilla*. Mollenhauer and Larson (1966): raphides of *V. planifolia*. Napp-Zinn (1953): aerial root anatomy of *Vanilla*. Nayar et al. (1976): epidermis of *Vanilla*. Neubauer (1961a,b): structure and development of aerial root and leaf of *V. planifolia*. Olivier (1881): root anatomy of *Vanilla*. Oudemans (1861): aerial roots of *Vanilla*. Pompilian (1883): anatomy of *Vanilla* stems. Pridgeon et al. (2003): discussion of and classification of genera in tribe Vanilleae. Queva (1894, 1895): stem anatomy of *Vanilla*. Roux (1954): morphology and histology of *Vanilla*. Segonzac (1958): phloem of *V. planifolia*. Shirley and Lambert (1918): anatomy of stems of *Galeola*. Stern and Judd

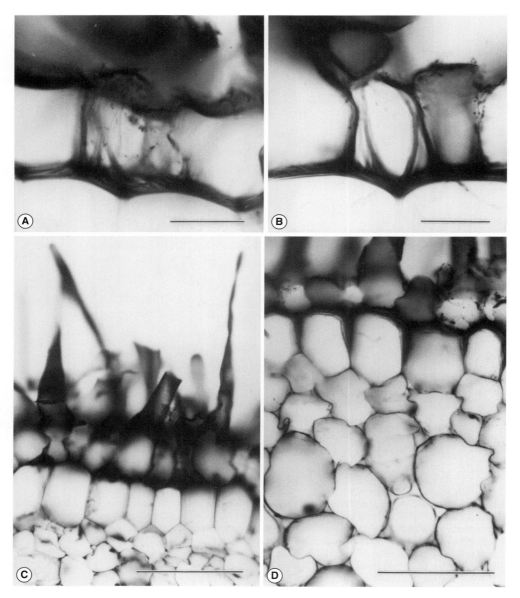

Fig. 42. Vanilleae. *Vanilla* **root.** (A,B) *V. madagascarieneis*, TS terrestrial root showing sculptured velaminal cells. (C) *V. insignis*, TS terrestrial root with hairs and thin-walled exodermal cells. (D) *V. pompona*, TS terrestrial root showing hair bases and thick-walled exodermal cells. Scale bars: A–C = 10 μm, D = 100 μm. Reprinted from Stern and Judd (1999).

(1999): comprehensive anatomical study of foliated and non-foliated taxa of *Vanilla* including a cladistic analysis. Stern and Judd (2000): anatomical analysis of the subtribes in tribe Vanilleae with a cladistic interpretation of their relationships. Suryanarayana Raju (1996): anatomy of *V. wightiana*. Tischler (1910): root anatomy of *Vanilla*. Tominski (1905): leaf anatomy of *Vanilla*. Wallach (1938–9): aerial roots of *Vanilla*. Went (1895): root anatomy of *Vanilla*. Williams (1979): stomata of *Vanilla*. Zhao and Wei (1999): stem anatomy of *Vanilla*.

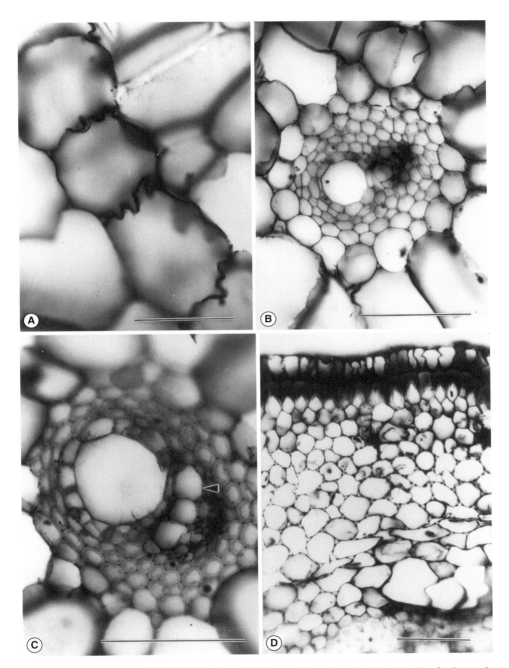

Fig. 43. Vanilleae. *Vanilla* **stem and root**. (A,B) *V. barbellata*. (A) TS stem illustrating pleated cell walls of exhausted water-storage cells. (B) TS stem showing vascular bundle embedded in parenchyma ground tissue, wide, thin-walled metaxylem cell, and thin-walled bundle sheath cells. (C) *V. claviculata*, TS stem showing vascular bundle embedded in sclerenchyma sheath and wide, thin-walled metaxylem cell associated with thick-walled sieve cells of the phloem (arrowhead). (D) *V. dilloniana*, TS aerial root with anticlinal velamen cells and very thick-walled exodermal cells. Scale bars = 100 µm. Reprinted from Stern and Judd (1999).

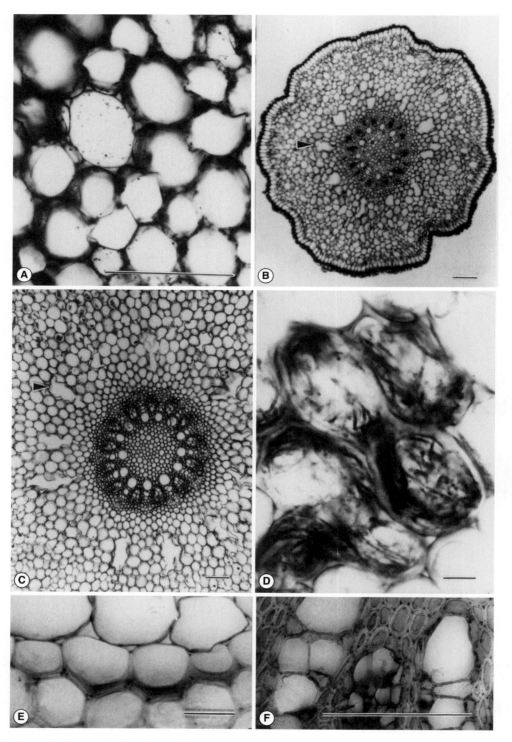

Fig. 44. Vanilleae. *Vanilla* root. (A) *V. africana*, TS terrestrial root showing thick-walled cortical cells. (B,C) *Vanilla* sp. (B) Aerial root. (C) Terrestrial root, both showing lacunae radiating from vascular cylinder (arrowheads). (D) *V. hartii*, TS terrestrial root cortex demonstrating cells infested with pelotons. (E) *Vanilla* sp., terrestrial root illustrating U-thickened endodermal cell walls. (F) *V. pompona*, TS root and portion of vascular cylinder with xylem and phloem embedded in thick-walled sclerenchyma cells. Scale bars: A–C,F = 100 µm, D,E = 10 µm. Reprinted from Stern and Judd (1999).

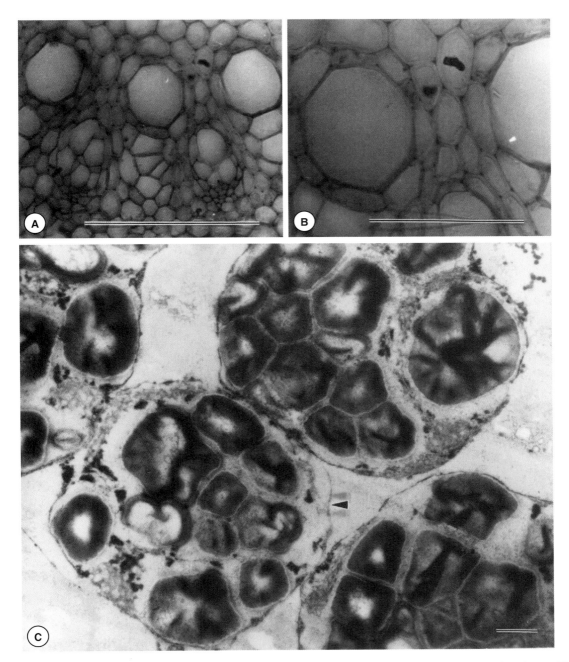

Fig. 45. Vanilleae. (A,B) *Vanilla* sp. (A) TS aerial root metaxylem. (B) Terrestrial root metaxylem. (C) *V. imperialis*, TS stem, TEM with spiranthosomes containing starch granules (arrowhead) at cortex/ground tissue transition. Scale bars: A,B = 100 μm, C = 1 μm. Reprinted from Stern and Judd (1999).

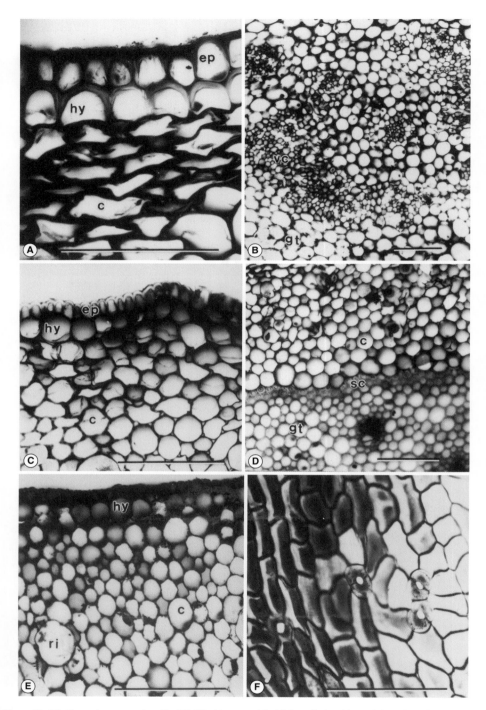

Fig. 46. Vanilleae. (A–D) *Cyrtosia septentrionalis.* (A) TS rhizome with thick-walled epidermal (ep), hypodermal (hy), and cortical (c) cells. (B) TS rhizome showing portion of vascular complex ground tissue (gt). (C) TS stem with thick-walled epidermal cells (ep), thin-walled hypodermal (hy), and cortical cells (c). (D) TS stem with sclerenchyma band (sc) separating cortex (c, top) and ground tissue (gt, bottom). (E) *Galeola nudifolia,* TS stem illustrating hypodermis (hy) and raphide idioblast (ri) embedded among cortical cells (c). (F) *Pseudovanilla foliata,* leaf scraping to show elongated epidermal cells and non-functional stomata. Scale bars = 100 μm. Reprinted from Stern and Judd (2000).

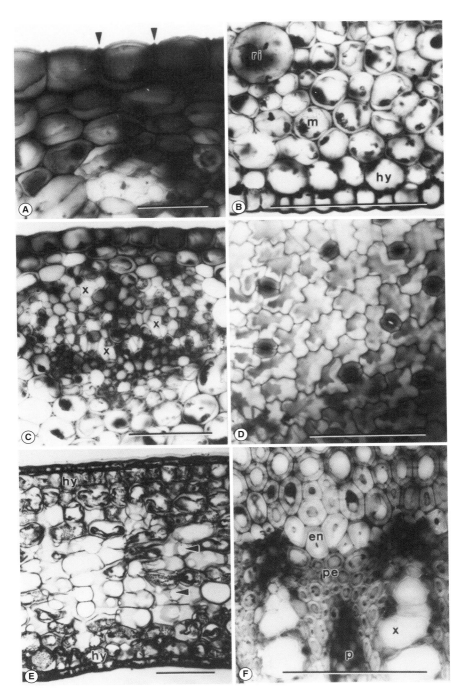

Fig. 47. Vanilleae. *Pseudovanilla* and *Clematepistephium*. (A–C) *P. foliata*. (A) TS leaf showing protuberances of cell wall material (arrowheads), which appear as ridges protruding from the grooves between adjacent epidermal cells. (B) TS leaf with homogeneous mesophyll (m) raphide idioblasts (ri), and abaxial hypodermis (hy). (C) TS leaf showing vascular bundles with discontinuous arcs of xylem (x) and lack of well-defined bundle sheaths. (D–F) *C. smilacifolium*. (D) leaf scraping to show undulate margins of abaxial epidermal cell walls, tetracytic and anomocytic stomata. (E) TS leaf with homogeneous mesophyll, ad- and abaxial hypodermises (hy); note circular primary pit fields (arrowheads). (F) TS root with very thick-walled endodermal cells (en) and pericycle cells (pe), vascular tissue (x = xylem, p = phloem) embedded in sclerenchyma. Scale bars: A = 25 μm, B,D–F = 100 μm, C = 50 μm. Reprinted from Stern and Judd (2000).

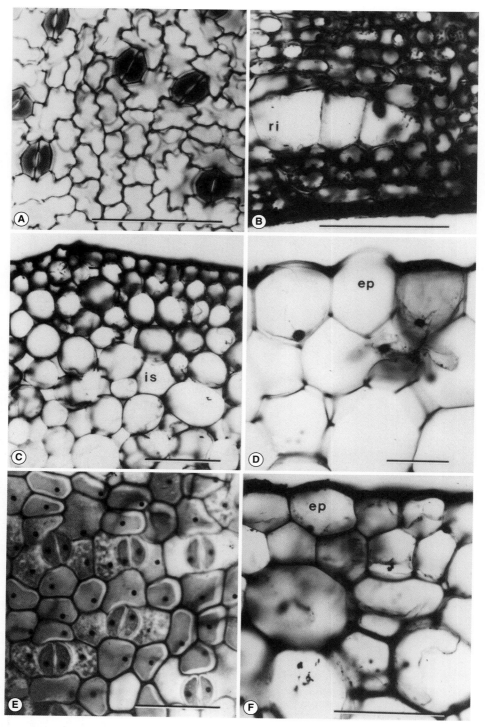

Fig. 48. Vanilleae. *Epistephium* **and** *Eriaxis*. (A,B) *Epistephium lucidulum*. (A) abaxial leaf scraping with anomocytic stomata. (B) TS leaf with chambered raphide idioblast (ri) and homogeneous mesophyll. (C,D) *E. subrepens*. (C) TS stem to show enlarged intercellular spaces (is) among cortical cells. (D) TS root illustrating unmodified epidermal cells (ep). (E,F) *Eriaxis rigida*. (E) Abaxial leaf scraping with tetracytic stomata and straight-sided to curvilinear cell walls. (F) TS root with unmodified epidermal cells (ep). Scale bars = 100 μm. Reprinted from Stern and Judd (2000).

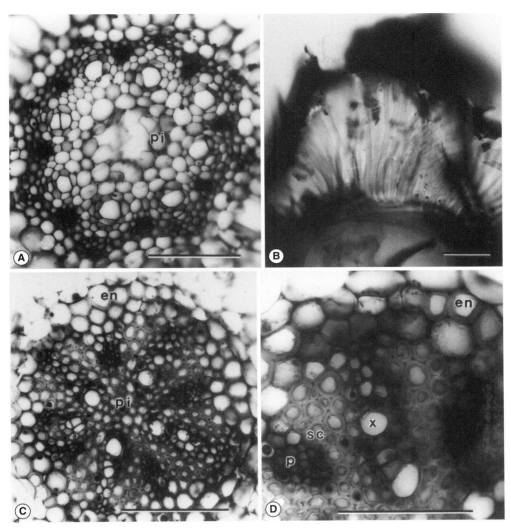

Fig. 49. Vanilleae. *Eriaxis* **and** *Lecanorchis*. (A) *E. rigida*, TS root vascular cylinder with thin-walled parenchymatous pith cells (pi), and medullary lacunae. (B) *Lecanorchis multiflora*, velamen cell with anticlinal thickening strips of cell wall material. (C,D) *L. nigrica*. (C) TS root showing vascular cylinder with thick-walled pith cells (pi) and U-thickened endodermal cells (en). (D) TS root vascular cylinder with U-thickened endodermal cells (en) and vascular tissue (x = xylem, p = phloem) embedded in thick-walled sclerenchyma (sc). Scale bars: A,C = 100 µm, B = 12 µm, D = 50 µm. Reprinted from Stern and Judd (2000).

Taxonomic notes

The study by Cameron et al. (1999) found strong support for the monophyly of *Vanilla* and *Epistephium*, a conclusion borne out by anatomy and in another study by Cameron (2009). It is suggested, based on anatomical grounds, that more detailed analyses of Vanilleae are called for using additional species and both morphological/anatomical and molecular features to strengthen phylogenetic hypotheses.

SUBFAMILY EPIDENDROIDEAE LINDL.

Tribe Arethuseae Lindl.

Subtribe Arethusinae Benth.

Arethusinae are terrestrial plants with reed-like or cormous stems and linear to lanceolate distichous conduplicate leaves. They are temperate to tropical herbs of eastern Asia, eastern North America, and the northern

Caribbean. There are five genera: *Anthogonium* with one species of tropical and subtropical mountainous regions of Asia; *Arundina* from South East Asia has one polymorphic species; *Arethusa* of eastern and central North America is monospecific; *Calopogon* has five species of eastern North America and the northern Caribbean; and *Eleorchis* has two species from northern Japan and south-eastern Russia. *Arundina*, with large attractive flowers, is easily cultivated. *Arundina* has been reported as used medicinally in Asia south to Java.

Leaf surface

Hair bases seen ab- and adaxially in *Arundina*. **Epidermis**: cells large in *Calopogon tuberosus* (as *C. pulchellus*) and adaxial cells very large in *Arundina* (Fig. 51A). **Stomata** abaxial (Fig. 51B,C), but ab- and adaxial in *Arethusa bulbosa* (Fig. 50A,B) and *C. tuberosus*, anomocytic, tending to tetracytic in *Arundina*.

Leaf TS

Cuticle thin to very thin in *Arundina* (Fig. 51F–H), thin in *C. tuberosus*. **Stomata** superficial in *Arundina* (Fig. 51D,F). **Mesophyll** homogeneous (Fig. 51F–H), tending toward heterogeneous in *Arundina* and with abaxial air lacunae in *C. tuberosus*. **Fibre bundles** present between vascular bundles in *C. tuberosus*. **Vascular bundles** in *Arundina* and *C. tuberosus* collateral, in one row (Fig. 51E,I), each surrounded by a sclerenchymatous sheath, smaller bundles having only phloem. Raphides and druses in mesophyll of *Arundina*.

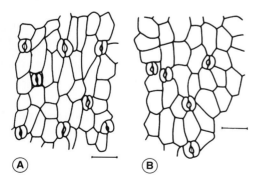

Fig. 50. Arethusinae. *Arethusa bulbosa* leaf epidermis. Stomata amphiepidermal. (A) Adaxial. (B) Abaxial. Stomata anomocytic, epidermal cells polygonal, cell walls straight-sided and curvilinear.

Root TS

Velamen 1-layered in *Arethusa bulbosa* and *C. tuberosus*, 1- or 2-layered in *Arundina* (Fig. 51J), cell walls without spiral thickenings but with small pits; cells with hyphae. **Exodermis**: cells thin-walled throughout in *Arethusa bulbosa*, *Arundina* (Fig. 51J), and *C. tuberosus*. **Cortex**: cells thin-walled in these genera and containing pelotons in *C. tuberosus* and *C. multiflorus*. **Endodermis**: cell walls O-thickened in *Arundina*, thin-walled in *Arethusa bulbosa* and *C. tuberosus*; endodermal cells of *C. tuberosus* with casparian strips. **Pericycle** cells thin-walled opposite xylem and thick-walled opposite phloem in *C. tuberosus*; entirely thin-walled in *Arethusa bulbosa*. **Vascular cylinder**: xylem and phloem embedded in sclerenchyma in *C. tuberosus* and alternating regularly around vascular cylinder. **Pith** parenchymatous in *Arundina* (Fig. 51J); parenchyma cells slightly thick-walled in *Calopogon* species.

Material reported

Information above has been assembled from several published sources noted below. Where data are lacking, none was available in the literature. There are no reports for *Anthogonium* and *Eleorchis*.

Reports from the literature

Carlson (1943): ontogeny of *Calopogon tuberosus* (as *C. pulchellus*) with skimpy descriptions of mature roots and leaves. Chiang and Chou (1971): leaf anatomy of *Arundina*. Engard (1944): velamen and exodermis of *Arundina*. Goh (1975): leaf anatomy of *Arundina*. Goh et al. (1992): mycorrhiza of *Arundina*. Hadley and Williamson (1972): mycorrhiza of *Arundina*. Holm (1904): root structure in *Calopogon* species and *Arethusa bulbosa*. Möbius (1887): leaf anatomy of *Calopogon tuberosus* (as *C. pulchellus*). Porembski and Barthlott (1988): velamen morphology in *Arethusa bulbosa*, *Arundina*, and *Calopogon tuberosus* (as *C. pulchellus*). Pridgeon (1987): velamen and exodermis of *Arundina*. Pridgeon et al. (2005): systematic review of Arethusinae. Singh (1981): stomata and stomatal development in *Arundina*. Singh (1986): root anatomy of *Arundina*. Solereder and Meyer (1930): summary of published studies of orchid anatomy. Tham et al. (2009): mycorrhizal infection in *Arundina*. Tominski (1905): leaf anatomy of *Arundina*. Williams (1979): subsidiary cells in stomata of *Calopogon tuberosus* and *Arundina*.

Taxonomic notes

Anatomical studies are too diffuse to be relevant towards defining any taxonomy.

Subfamily Epidendroideae 97

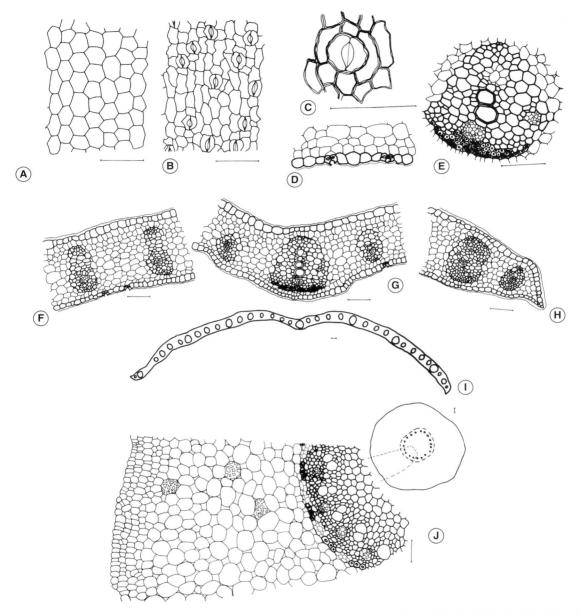

Fig. 51. Arethusinae. *Arundina graminifolia*. (A) Adaxial epidermis; cells polygonal, walls straight-sided. (B) Abaxial epidermis, cells elongated, polygonal, walls curvilinear, stomata anomocytic. (C) Single stoma. (D) TS abaxial portion of leaf with superficial stomata. (E) TS foliar midvein with abaxial sclerenchyma cap. (F) TS leaf with vascular bundles lacking bundle sheaths. (G) TS foliar mid-section with large midrib vascular bundle and smaller flanking vascular bundles, all lacking bundle sheaths. (H) TS foliar marginal vascular bundles. (I) TS lamina to show one row of vascular bundles. (J) TS root with parenchymatous cortex and pith, 2-layered velamen, thin-walled exodermal cells, O-thickened endodermal cells opposite phloem and thin-walled passage cells opposite xylem.

Subtribe Coelogyninae Benth.

Coelogyninae are epiphytic, lithophytic, or uncommonly terrestrial herbs with distinct rhizomes, corms, and pseudobulbs. Leaves are convolute or conduplicate, sometimes plicate; deciduous or perennial, articulate; plants one- to many-leaved. There are 21 genera and about 680 species, widely distributed in Australasia, tropical Asia, Japan, and the Pacific. They are absent from the New World and Africa. Many species of *Coelogyne* are cultivated by hobbyists as are a few species of *Dendrochilum* and *Panisea*. Species of *Coelogyne* are held to be medicinal in Nepal and China, as are *Bletilla* species in Nepal, China, Taiwan, and Japan, and *Pholidota* in India and China.

The information on this subtribe has been compiled entirely from the literature and the line drawings but the names now accepted have been used where possible. Where no data are given for any organ or tissue, it is because none was available.

Leaf surface

Hairs usually few-celled, uniseriate, sunken, on both surfaces, in *Bletilla*, *Chelonistele*, *Coelogyne*, *Dendrochilum*, *Geesinkorchis*, *Nabaluia*, *Neogyna*, and some *Otochilus* and *Pholidota* species; not sunken in *Pleione*; absent in *Panisea*, some *Pholidota* species, and *Thunia*. Hairs surrounded by three or four cells in *Bletilla*; cuticle forming thickened rings around hair bases in *Coelogyne punctulata* (as *C. ocellata*) and in species of several other genera; hair cells with oil droplets in *Pleione*. **Epidermis**: cells polygonal in *Bletilla* (Fig. 52A,B), *Coelogyne*, *Dendrochilum*, *Neogyna*, *Otochilus* (Fig. 53A,B), *Pholidota* (mostly hexagonal; Fig. 54A,B), *Pleione* (Fig. 55A,B), and *Thunia* (Fig. 56A); rectangular to polygonal in *Panisea*; cells thick-walled in *Dendrochilum*; adaxial cells very large (Fig. 55A), thin-walled, and abaxial cells variable in *Pleione* (Fig. 55B); with small oil droplets in *Neogyna*. **Stomata** abaxial (Figs 52B,H, 53B, 54B, 55B, 56B,C), tetracytic in *Bletilla*, *Chelonistele*, *Geesinkorchis*, *Nabaluia*, and *Thunia*, tetracytic with occasionally five or six subsidiary cells in *Coelogyne*, *Dendrochilum*, *Neogyna*, *Otochilus* (Fig. 53B), *Panisea*, and *Pholidota* (Fig. 54B); abaxial, lacking subsidiary cells in *Pleione* (Fig. 55B); with a group of strongly thickened sclerenchyma cells overlying each stoma in *Coelogyne punctulata*; oil droplets occurring in guard and subsidiary cells in *Coelogyne* and in guard cells in *Pholidota* and *Pleione*.

Leaf TS

Cuticle thin in *Neogyna*, *Pleione* (Fig. 55C–E), and *Thunia* (Fig. 56D–F); thin to thick in *Dendrochilum*; thick in *Chelonistele*, *Coelogyne*, *Geesinkorchis*, *Nabaluia*, *Otochilus* (Fig. 53E–G), and *Panisea*; rugose, warty, moderately well developed on both surfaces in *Pholidota*. **Epidermis**: oil droplets in cells in *Coelogyne*, crystalliferous in *Dendrochilum*, with water-storage cells in *Otochilus* and adaxial water-storage cells in *Pholidota*. **Stomata** superficial to sunken (Figs 52H, 53C,F,G, 54C, 55C–E, 56E,F). **Hypodermis**: 1-layered adaxially in *Dendrochilum* and *Geesinkorchis* (with water-storage cells), ad-and abaxially in *Chelonistele* (with water-storage cells), *Nabaluia* (with water-storage cells and some chloroplasts in abaxial cells) and *Pholidota* (some cells with spiral thickenings); 1- or 2-layered in *Coelogyne*, with ad- and abaxial water-storage cells (only adaxial in *C. punctulata*), and cells with wide spiral thickenings; 1- or 2-layered in some species of *Otochilus* (Fig. 53G); absent in *Neogyna*, *Panisea*, and *Pleione*. **Fibre bundles** occurring between vascular bundles in *Thunia* and between smaller adaxial bundles in *Pholidota imbricata*. **Mesophyll** homogeneous in *Bletilla* (Fig. 52D–E), *Neogyna*, *Panisea*, and *Pleione* (Fig. 55C–E), in some species of *Coelogyne*, and more or less so in *Geesinkorchis*; chlorenchyma cells anticlinal and without spiral thickenings in *Bletilla* and *Thunia*; 7- or 8-layered in *Neogyna* (cells without spirals); homo- to heterogeneous in *Dendrochilum* (cells with a columnar crystal or druse), *Otochilus* (cells with raphide bundles and without spirals), and *Pholidota* (some cells with spirals). Heterogeneous in *Chelonistele* with one or two layers of palisade and about seven layers of spongy chlorenchyma, in some *Coelogyne* species with two layers of palisade and about six layers of spongy cells, in *Nabaluia* with one layer of nearly palisade-like cells and four or five layers of spongy cells. Raphide bundles in chlorenchyma cells of *Coelogyne* and *Pholidota*; druses and single crystals of calcium oxalate in *Coelogyne punctulata*. **Vascular bundles** in one row (Figs 53H, 54F, 55F, 56G), except in two rows in *Coelogyne viscosa* and one or two rows in *Pholidota*; capped with ad- and abaxial thin-walled sclerenchyma cells, except in *Coelogyne punctulata* and *Pleione yunnanensis* (Fig. 55D), where sclerenchyma heavily thickened; surrounded by sclerenchyma in *Bletilla* (Fig. 52C) and *Thunia*; alternating larger and smaller bundles in *Bletilla* (Fig. 52F), *Dendrochilum*, *Neogyna*, and *Otochilus*, with ad- and abaxial sclerenchyma caps (Fig. 53D), these thin-walled and narrow-lumened in *Neogyna*; bundles of varying sizes in *Panisea*. Lacunae between bundles of adaxial row in *Pholidota*. **Stegmata** with conical silica bodies present adjacent to vascular bundle sclerenchyma in *Bletilla*, *Coelogyne*, *Dendrochilum*, *Glomera*, *Neogyna*, *Otochilus*, *Panisea*, *Pholidota*, and *Thunia*; absent in *Pleione*.

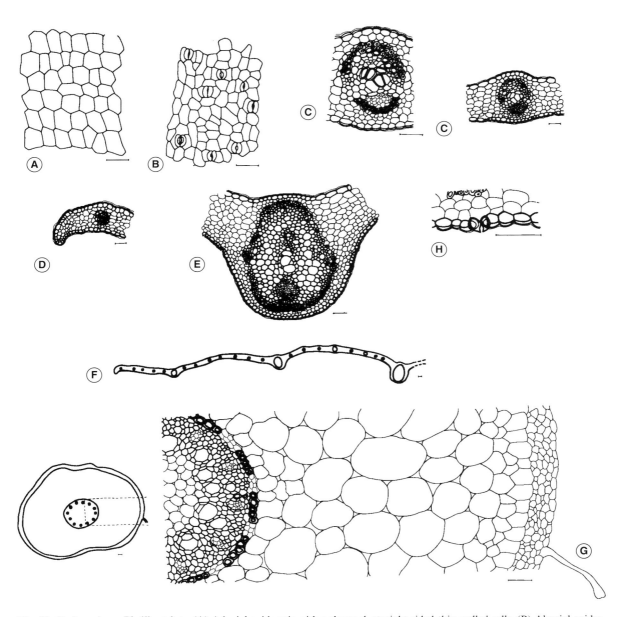

Fig. 52. Coelogyninae. *Bletilla striata*. (A) Adaxial epidermis with polygonal, straight-sided thin-walled cells. (B) Abaxial epidermis with tetracytic stomata, polygonal cells, with mostly curvilinear thin walls. (C) Collateral vascular bundle with ad- and abaxial arcs of sclerenchyma. (D,E) TS leaves to show position of vascular bundles and structure of major vascular bundle with ad- and abaxial arcs of sclerenchyma and homogeneous mesophyll. (F) TS lamina to show one row of alternating larger and smaller vascular bundles. (G) TS root with several-layered velamen, unicellular root hair, thin-walled exodermal and cortical cells, O-thickened endodermal cell walls opposite phloem, thin-walled endodermal cells opposite xylem, and thin-walled pith cells. (H) TS leaf, stoma.

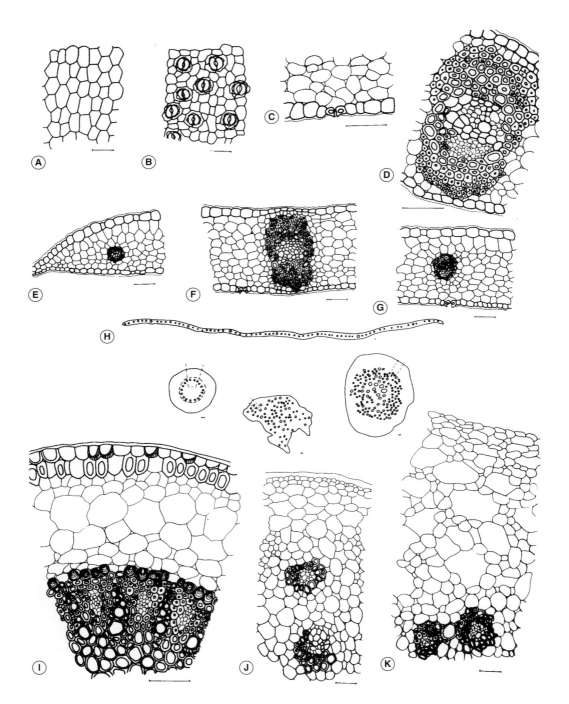

Fig. 53. Coelogyninae. *Otochilus albus*. (A) Adaxial epidermis with polygonal, thin-walled cells. (B) Abaxial epidermis with polygonal cells and anomocytic stomata. (C) TS leaf to show superficial stomata. (D,F) Midvein vascular bundle with ad- and abaxial sclerenchyma layers. (E) TS leaf margin with vascular bundle. (G) TS leaf with vascular bundle midway between epidermises and homogeneous mesophyll. (H) TS lamina showing one row of vascular bundles. (I) TS root with 1-layered velamen. Some velamen cells thick-walled internally, exodermal and endodermal cell walls O-thickened except for thin-walled endodermal passage cells, vascular tissue embedded in thick-walled sclerenchyma cells, pith cells thick-walled. (J,K) TS pseudobulb with scattered vascular bundles having sclerotic sheaths at phloem poles and parenchymatous ground tissue. (K) Large lacunae interrupting homogeneous ground tissue.

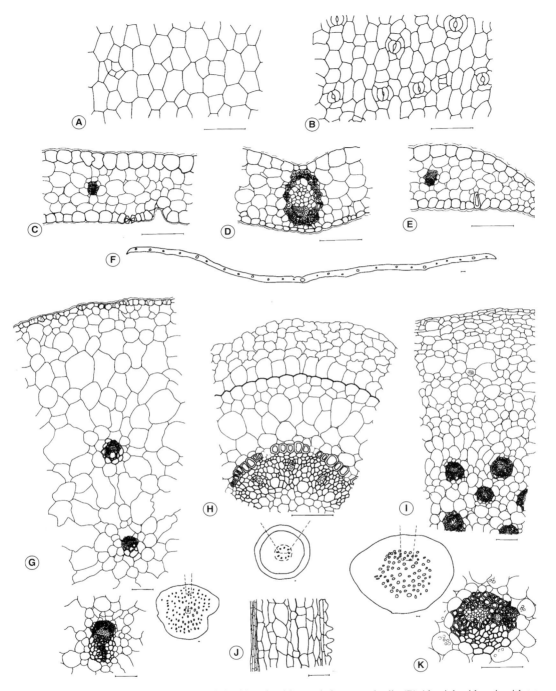

Fig. 54. Coelogyninae. *Pholidota chinensis*. (A) Adaxial epidermis with mostly hexagonal cells. (B) Abaxial epidermis with tetracytic stomata, but sometimes with five subsidiary cells. (C) TS leaf with superficial stomata, small vascular bundle, and homogeneous mesophyll. (D) TS leaf midvein vascular bundle with thick-walled bundle sheath cells. (E) TS leaf margin with small vascular bundle. (F) TS lamina with one row of widely spaced vascular bundles. (G,I) TS pseudobulb with scattered vascular bundles and lacunae in parenchymatous ground tissue. (H) TS root with 5-layered velamen, thin-walled exodermal and cortical cells, endodermal cell walls O-thickened, passage cells thin-walled, vascular tissue surrounded by parenchyma cells. (J) LS pseudobulb with papillate-pointed epidermal cells. (K) TS pseudobulb, detail of vascular bundle.

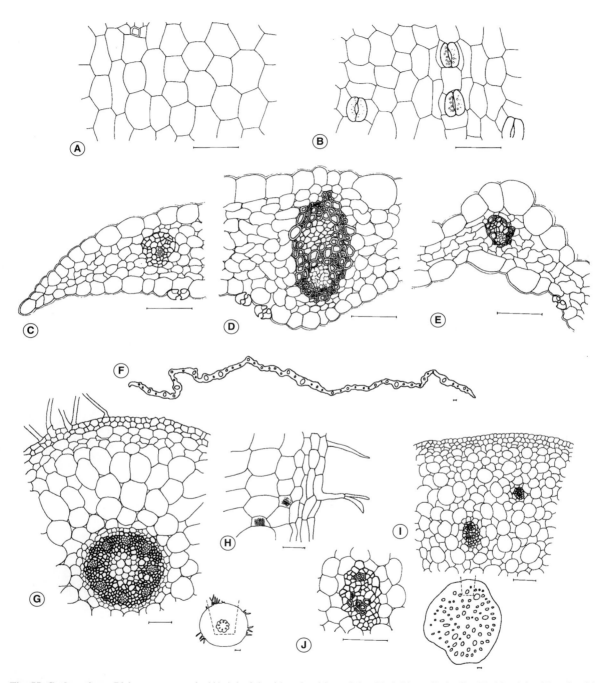

Fig. 55. Coelogyninae. *Pleione yunnanensis*. (A) Adaxial epidermis with straight-sided thin-walled cells. (B) Abaxial epidermis with polygonal cells and anomocytic stomata, oil droplets in guard cells. (C–E) TS leaf showing superficial stomata, homogeneous mesophyll, and vascular bundles. Midvein vascular bundle with sclerenchymatous bundle sheath cells and ad- and abaxial sclerenchyma caps. (F) TS lamina illustrating one row of vascular bundles. (G) TS root with 2-layered velamen and simple hairs, thin-walled exodermal and endodermal cells, vascular tissue embedded in thin-walled sclerenchyma and parenchymatous pith with thin-walled cells. (H) LS root with simple and bifurcated hairs and raphides in cortical cells. (I,J) TS pseudobulb with mucilage cells and lacunae in ground tissue and scattered vascular bundles.

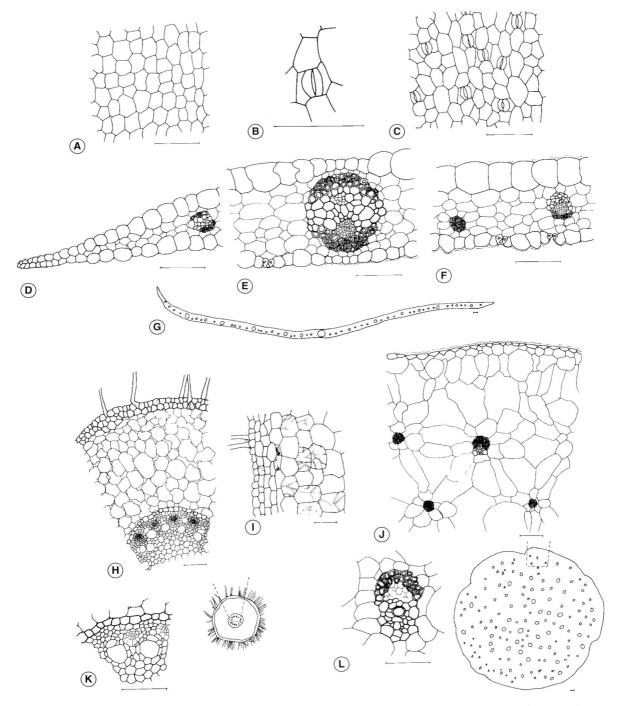

Fig. 56. Coelogyninae. *Thunia alba*. (A) Adaxial epidermis with polygonal cells and curvilinear walls. (B,C) Abaxial tetracytic stomata. (D) TS leaf margin with vascular bundle. (E) TS foliar midvein vascular bundle having ad- and abaxial sclerenchyma caps, homogeneous mesophyll, and superficial stomata. (F) TS leaf with homogeneous mesophyll and superficial stomata. (G) TS lamina showing one row of vascular bundles. (J,L) TS stem with scattered vascular bundles, these with only adaxial sclerenchyma cells, parenchymatous ground tissue with prominent lacunae. (H,K) TS and (I) LS root with 2-layered velamen, thin-walled exodermal and endodermal cells, cortical cells with hyphae.

Pseudobulb surface (*Coelogyne, Dendrochilum, Panisea, Pholidota,* and *Pleione*)

Epidermis: cells squarish with thick wavy anticlinal walls in *Coelogyne*, cells elongated, sometimes sclerotic, with very thick wavy anticlinal walls in *Dendrochilum*, cells elongated with wavy anticlinal walls in *Pholidota*, cells polygonal with straight, slightly thickened anticlinal walls in *Pleione*. **Stomata** scattered in *Pleione*, lacking subsidiary cells.

Pseudobulb TS

Cuticle thin in *Pleione*, thin or thick in *Pholidota*, thick in *Coelogyne* and *Panisea*, smooth and well developed in *Dendrochilum*. **Epidermis**: cells isodiametric in *Dendrochilum*, isodiametric with sinuous, strongly thickened anticlinal walls and numerous pit canals in *Coelogyne*. **Hypodermis**: 2–4 layers of sclerified cells in *Coelogyne*, 1–5 layers of thickened cells lacking spiral thickenings in *Dendrochilum*, composed of large cells in *Pholidota* (Fig. 54J); absent from *Pleione*. **Ground tissue** parenchymatous (Figs 53J,K, 54G, 55I); with large mucilage-containing cells, smaller chlorophyllous cells, and lacunae (Figs 53K, 54I), sometimes with spiral thickenings and starch granules, in *Coelogyne*; mucilage cells and chlorophyllous cells, lacking spirals, in *Dendrochilum*; large water-storage cells, mucilaginous cells and smaller chlorophyllous and starch-containing cells in *Panisea*; with spirally thickened mucilage cells and chlorenchyma cells containing starch granules in *Pholidota*; with chlorophyllous, mucilaginous, and raphide-containing cells in *Pleione*. **Vascular bundles** scattered (Figs 54I, 55I); smaller bundles having sclerenchyma only at phloem pole (Figs 54K, 55I), larger bundles with sclerenchyma at both phloem and xylem poles (Fig. 53K). **Stegmata** with conical silica bodies adjacent to vascular bundle sclerenchyma in *Dendrochilum* and *Pholidota*, adjacent only to phloem sclerenchyma in *Coelogyne*.

Stem TS (*Thunia* only)

Epidermis: cells with undulating anticlinal walls in surface view. **Ground tissue** parenchymatous; cells of outer layers chlorophyllous; other cells mucilaginous; some cells with spirally thickened walls. **Vascular bundles** and fibre bundles scattered (Fig. 56J). Phloem accompanied by sclerotic sheath (Fig. 56L); xylem lacking sheath. **Stegmata** with conical silica bodies associated with phloic sclerenchyma.

Root TS

Velamen layers: one in *Glomera*, two in *Pleione* (Fig. 55G,H), two or three in *Thunia* (Fig. 56H), 2–8 in *Pholidota* (Fig. 54H), three or four in *Bletilla* (Fig. 52G), 3–6 in *Dendrochilum*, one or five in *Otochilus* (Fig. 53I) and *Panisea*; up to 8–10 in *Coelogyne*; walls with fine spirals and small pits in *Coelogyne, Dendrochilum,* and *Pholidota*, with wide bands of parallel thickenings in *Panisea*, with helical thickenings in *Thunia*, walls without thickenings in *Pleione*, walls with small pits in *Bletilla* and *Glomera*. Hairs unicellular (Figs 55G, 56H), except bicellular/bifurcate in *Pleione* (Fig. 55H). **Tilosomes** spongy, present in *Dendrochilum*, present or absent in *Coelogyne* and *Pholidota*; absent in *Bletilla, Glomera, Pleione,* and *Thunia*. **Exodermis** 1-layered (Figs 52G, 53I, 54H, 55G, 56H), cells O- or ∩-thickened in *Coelogyne*, O-, ∩-, or U-thickened in *Dendrochilum* and *Pholidota*, O-thickened in *Panisea*, outer walls thickened in *Bletilla*, long cells slightly thickened in *Pleione*, cells unthickened in *Thunia* (Fig. 56H,I); cells alternating with thin-walled passage cells in *Coelogyne*. **Cortex** with spirally thickened cell walls in *Coelogyne, Dendrochilum, Pholidota,* and *Thunia*; lacking such cell walls in *Pleione*. Raphides present in *Pleione* (Fig. 55H). **Endodermis** 1-layered (Figs 52G, 53I, 54H, 55G, 56H); cells O- or U-thickened, except unthickened in *Pleione* (Fig. 55G) and *Thunia* (Fig. 56K). **Pericycle** 1-layered in *Coelogyne, Dendrochilum, Otochilus, Pholidota,* and *Thunia*. **Vascular cylinder** 6–12-arch in *Dendrochilum*, 7–12-arch in *Bletilla* (Fig. 52G) and *Pleione* (Fig. 55G), 7–12-arch in *Pholidota* (Fig. 54H), 12–26-arch in *Coelogyne*, 18-arch in *Thunia* (Fig. 56H), and 22- or 23-arch in *Otochilus* (Fig. 53I). **Pith** sclerotic in *Pholidota* and some species of *Coelogyne*, sclerotic or parenchymatous in *Dendrochilum* and *Pleione* (Fig. 55G), parenchymatous in *Bletilla* (Fig. 52G), *Panisea,* and *Thunia* (Fig. 56H).

Material reported

The treatment of Coelogyninae was prepared from multiple published sources varying in numbers of taxa, and line drawings, yielding the composite report above. The genera are: *Bletilla, Chelonistele* (leaf only), *Coelogyne, Dendrochilum, Geesinkorchis* (leaf only), *Glomera* (root only), *Nabaluia* (leaf only), *Neogyna* (leaf only), *Otochilus, Panisea, Pholidota, Pleione,* and *Thunia*.

Reports from the literature

Dressler (1993): systematic analysis of the phylogeny and classification of Orchidaceae. Gravendeel (2000): taxonomic reorganization of *Coelogyne* with a brief anatomical account. Kaushik (1983): anatomy of *Coelogyne, Otochilus,* and *Pholidota*. Meineke (1894): anatomy of aerial roots of *Coelogyne, Dendrochilum* (as *Platyclinis*),

Pholidota, and *Thunia*. Möbius (1887): anatomy of leaves of *Bletilla*, *Coelogyne*, *Dendrochilum* (as *Platyclinis*), *Otochilus*, *Pholidota*, *Pleione*, and *Thunia*. Mohana Rao and Khasim (1987b): anatomy of *Coelogyne*, *Otochilus*, *Pholidota*, and *Pleione*. Mohana Rao et al. (1988): vegetative anatomy of *Panisea*. Moreau (1913): anatomy of pseudobulbs of *Coelogyne* and *Pholidota*. Pridgeon et al. (1983): tilosome presence or absence in roots of *Bletilla*, *Coelogyne*, *Dendrochilum*, *Pholidota*, and *Thunia*. Pridgeon et al. (2005): comprehensive analysis of genera in subtribe Coelogyninae. Rosinski (1992): comprehensive study of functional leaf anatomy in major genera of epiphytic Coelogyninae. Rŭgina et al. (1987): root and leaf anatomy of *Coelogyne*, *Dendrochilum*, and *Pholidota*. Singh (1986): root anatomy of *Coelogyne cristata*, *Pholidota articulata*, and *P. imbricata*. Solereder and Meyer (1930): comprehensive review of orchid anatomy and notes on leaf of *Coelogyne* and leaf and root of *Pholidota*. Tominski (1905): leaf anatomy of *Coelogyne odoratissima* and *Pholidota imbricata*. Williams (1979): subsidiary cells in some *Coelogyne*, *Dendrochilum*, and *Pholidota* species. Zörnig (1903): thorough report on vegetative anatomy of Coelogyninae.

Additional literature

Banerjee and Rao (1978): epidermis of *Coelogyne*. Burr and Barthlott (1991): velamen-like tissue in root cortex of *Thunia*. Chen (1970): root anatomy of *Pleione formosana*. Chiang and Chen (1968): aerial organs of *Pleione formosana*. Das and Paria (1992): stomata of *Pholidota*. Dietz (1930): root structure of *Thunia alba*. Duruz (1960): anatomy of leaf of *Coelogyne pulverula* (as *P. dayana*) and *C. pandurata*. Guttenberg (1968): root anatomy of *Pholidota*. Hadley and Williamson (1972): mycorrhiza of *Chelonistele*, *Coelogyne*, *Dendrochilum*, and *Dilochia*. Katiyar et al. (1986): mycorrhiza of *Coelogyne* and *Pleione*. Khasim and Mohana Rao (1984): velamen–exodermis complex of *Coelogyne*. Khasim and Mohana Rao (1986): leaf and root anatomy of *Pholidota imbricata*. Kroemer (1903): root anatomy of *Coelogyne*. Krüger (1883): structure of leaf of *Coelogyne cristata*. Löv (1926): leaf anatomy of *Coelogyne*, *Dendrochilum*, and *Pholidota*. Metzler (1924): leaf anatomy of *Coelogyne*. Mohana Rao et al. (1989): vegetative anatomy of *Pholidota*. Mohana Rao et al. (1991): anatomy of *Coelogyne odoratissima*. Møller and Rasmussen (1984): stegmata in *Bletilla striata*. Mulay et al. (1958): velamen and exodermis in *Coelogyne*. Pedersen (1997): taxonomy and anatomy of *Dendrochilum* in the Philippines. Porembski and Barthlott (1988): velamen structure in *Bletilla*. Pridgeon (1987): velamen and exodermis in *Bletilla*. Sasikumar (1973a, 1974): root anatomy of *Coelogyne nitida* (as *C. ochracea*). Schindler and Toth (1950): leaf anatomy of *Coelogyne flaccida*. Singh (1981): so-called floating stomata of *Pholidota articulata*. Singh and Singh (1974): stomata of *Coelogyne* and *Pholidota*. Vij et al. (1991): leaf epidermis of *Coelogyne*, *Pholidota*, and *Thunia*. Weltz (1897): axis anatomy of *Coelogyne* and *Dendrochilum* (as *Platyclinis*). Went (1895): root anatomy of *Coelogyne*.

Taxonomic notes

The major genera of Coelogyninae (i.e. *Coelogyne* with about 200 species, *Dendrochilum* with 263 species, and *Pholidota* with 31 species) seem to hold together rather well: each has a well-defined foliar hypodermis, stegmata with conical silica bodies next to foliar vascular bundle sclerenchyma, tetracytic stomata, heterogeneous mesophyll in each genus (although some species of *Coelogyne* and *Dendrochilum* may have a homogeneous mesophyll), pith sclerotic in each genus, cells of root cortex have spiral thickenings, and spongy tilosomes are present in roots. There are a few major differences, however: *Pleione* has deciduous leaves and lacks subsidiary cells and stegmata. *Panisea*, *Pleione*, and *Thunia* lack the sunken foliar hairs present in other genera.

Tribe Calypsoeae (Camus) Dressler

Calypsoeae (Corallorhizinae in Dressler (1981) except *Calypso*) are a small group consisting of only 11 genera and about 62 species (Freudenstein 2005). Plants are terrestrial, mostly cormous and/or rhizomatous, leafless in *Wullschlaegelia* and *Corallorhiza* and rootless in *Corallorhiza*. Plants are usually mycorrhizal. Several genera are monospecific or oligospecific, *Govenia* being the richest with about 30 species. Plants are distributed widely from Europe, northern Asia, and North America to tropical Central America, the Caribbean, Brazil, and Argentina. Calypsoeae are absent from Africa, Australia, islands of the East Indies and Pacific Ocean. *Aplectrum hyemale* has been employed medicinally by North American Indians and *Corallorhiza* species have been used medicinally in central and eastern North America and in parts of eastern Europe; *Cremastra appendiculata* has been used in China and Japan.

Several species of *Govenia* are cultivated, as is *Calypso bulbosa*.

Leaf surface

Hairs absent in *Aplectrum*, *Cremastra*, and *Govenia tingens*. Hairs 2-celled on both surfaces in *Calypso* (Fig. 57F),

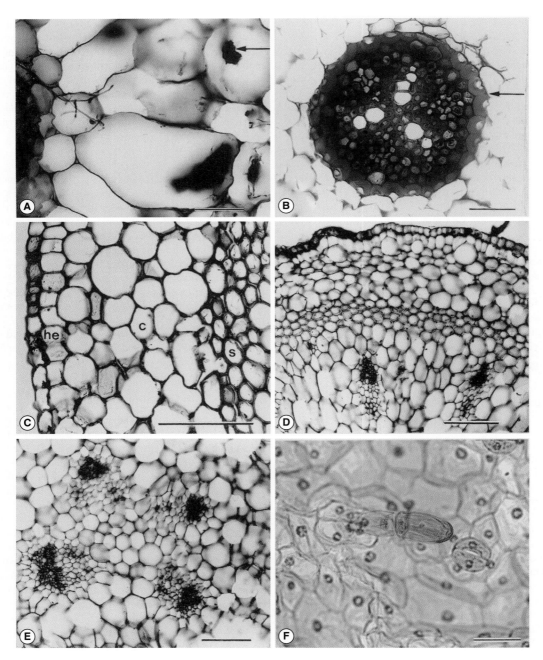

Fig. 57. Calypsoeae. *Wullschlaegelia* **and** *Calypso*. (A–E) *W. calcarata*. (A) TS filiform root illustrating anticlinal cells of third cortical layer with decayed fungal masses (arrow). (B) TS filiform root with Y-shaped vascular tissue and U-thickened endodermal cells opposite phloem (arrow). (C) TS stem showing 1-layered epidermis, hypodermis/exodermis (he), cortex (c), and sclerenchyma band (s). (D) TS stem with two of several elliptical peripheral vascular bundles. (E) TS rhizome with four ±circular vascular bundles in ground tissue. (F) *C. bulbosa*, abaxial leaf scraping; hairs 2-celled, basal cell clear, shorter than dark-staining apical cell. Scale bars: A–E = 100 μm, F = 50 μm. (A–E) reprinted from Stern (1999); (F) reprinted from Stern and Carlsward (2008).

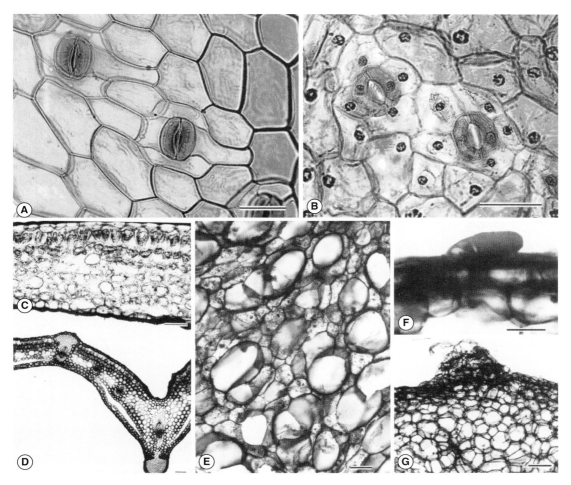

Fig. 58. Calypsoeae. (A) *Govenia tingens*, abaxial foliar epidermis with tetracytic stomata, lateral cells intruding among adjacent epidermal cells. (B) *Calypso bulbosa*, abaxial epidermis with anomocytic stomata. (C) *Tipularia discolor*, TS leaf with chlorophyllous, upright hypodermal cells and spongy mesophyll. (D,E) *Aplectrum hyemale*. (D) TS leaf showing pronounced abaxial midvein sclerenchyma, xylem sclerenchyma clusters resulting in abaxial ribbing. (E) TS underground stem (rhizome) with starch granules in smaller nucleated assimilatory cells and larger enucleate water-storage cells. (F) *Calypso bulbosa*, corm bearing 2-celled hairs. (G) *Cremastra appendiculata*, rhizome with excrescences bearing tufts of unicellular hairs. Scale bars: A,B = 50 μm, C–G = 100 μm. Reprinted from Stern and Carlsward (2008).

Govenia superba, and probably *Tipularia*, base embedded among a cluster of small epidermal cells; in *Calypso* basal cell of adaxial hairs clear, apical cell bulbous, darkly staining; basal cell of abaxial hairs much shorter than blunt-tipped, darkly staining, elongated apical cells (Fig. 57F). **Epidermis**: cells polygonal (Fig. 58A,B) on both surfaces, except adaxial cells in *Calypso* may be elongated; walls straight-sided or curvilinear. **Stomata** abaxial, except ad- and abaxial in *Aplectrum*; occasionally stomata noted adaxially in other taxa. Stomata basically tetracytic in *Aplectrum*, *Cremastra*, *Govenia*, and *Tipularia*, with a few anomocytic. Lateral cells of stomata in *G. tingens* often elongated serpent-like, intruding between adjoining epidermal cells (Fig. 58A); in *Tipularia* lateral cells and sometimes apical cells protruding among other epidermal cells; in *Calypso* stomata entirely anomocytic (Fig. 58B). Guard cells typically reniform and stomata parallel to long axis of guard-cell pair.

Leaf TS

Cuticle smooth, somewhat granulose in *Calypso*; 2.5 μm or less thick. **Epidermis**: cells isodiametric to somewhat periclinal in *Govenia* and to a certain extent in *Tipularia*. **Stomata** superficial; substomatal chambers large in *Calypso*, *Govenia*, and *Tipularia*; moderate in *Aplectrum* and *Cremastra*. Stomatal ledges usually poorly defined in TS; outer ledges apparent in *Aplectrum*, *Calypso*, *Cremastra*, *Govenia*, and *Tipularia*, but most pronounced in *Aplectrum*; inner ledges obscure but apiculate in *G. tingens* and *Tipularia*. **Hypodermis**: 1-layered abaxially, cells globose or inflated, sparsely provided with chloroplasts in *Aplectrum*; 2-layered in *Tipularia*, outer layer of more or less inflated upright cells (Fig. 58C), inner layer cells isodiametric; cells of both layers rich in chloroplasts. Hypodermis absent in *Calypso*, *Cremastra*, and *Govenia*. **Fibre bundles** absent throughout, except in *G. tingens*. **Mesophyll** homogeneous, 4–7 cells wide; cells thin-walled, mostly oval and circular with small triangular and polyhedral intercellular spaces; in *Tipularia* (Fig. 58C), cells associated with intercellular spaces organized as in a eudicotyledon, i.e. with highly lobed cells. Raphide-bearing idioblasts circular in TS, saccate, blunt-ended in LS. **Vascular bundles** collateral in a single series. In larger bundles of *Aplectrum*, *Cremastra*, and *Govenia* both xylem and phloem subtended by patches of thin-walled sclerenchyma; on xylem side in *Aplectrum* and *Govenia*, but not phloem side, these producing bulges (TS) resulting in ridges on adaxial leaf surface. Midvein in *Aplectrum*, *Cremastra*, and *Govenia* subtended opposite phloem by a massive cluster of sclerenchyma creating a pronounced keel (Fig. 58D). In *Tipularia* sclerenchyma associated only with xylem. Vascular bundles in *Calypso* lacking sclerenchyma. **Stegmata** absent from *Aplectrum*, *Calypso*, and *Tipularia*, but present in *Cremastra* and *Govenia*. Conical, rough-surfaced silica bodies in stegmata occurring along sclerenchyma opposite xylem and phloem in *Cremastra*, only along phloem sclerenchyma in *Govenia*, and associated with fibre bundles in *G. tingens*. Bundle sheath cells circular, thin-walled, and chloroplast-bearing in all taxa except chloroplasts absent in some sheath cells of *Calypso*.

Scale leaves occurring along rhizome in *Corallorhiza*. **Epidermis**: cells periclinal and isodiametric, subtended by an abaxial, 1-layered hypodermis. **Stomata** abaxial. **Vascular bundles** in one layer, phloic only, embedded in an undifferentiated matrix of thin-walled, rounded to angular cells.

With respect to the statement in Pridgeon et al. (2005) that hairs on leaves of *Aplectrum* appear in a single row, Solereder and Meyer (1930) indicate that each hair comprises a single row of cells, i.e. they are *'einzellreihige'*, not that hairs occur in a single row on leaves. Our observations conclude that leaves of *Aplectrum* are glabrous.

Stem TS

Subterranean storage, perennating, and connective organs (rhizomes) several in Calypsoeae (Fig. 57C–E), as noted by Pridgeon et al. (2005). Vascular bundles consisting of adnate strands of xylem and phloem.

Corm (Aplectrum, Calypso, Cremastra, Govenia, Tipularia)

Hairs present only in *Calypso*, 2-celled, thick-walled; apical cell clavate, darkly staining (Fig. 58F). **Cuticle** only in *Calypso*, smooth, ±5 μm thick. **Epidermis**: cells isodiametric, except periclinal in *Cremastra*. **Stomata** absent except in *Govenia tingens*. **Hypodermis**: present only in *Calypso*, 1-layered, cells tending to be periclinal. **Cortex** absent, except two or three cells wide in *Aplectrum*. **Endodermis** in *Cremastra* only; cells thin-walled, angular, surrounding each vascular bundle. **Pericycle** absent. **Ground tissue**: numerous large, water-storage cells surrounded by smaller, variously shaped assimilatory cells with cruciate starch granules in *Aplectrum*, *Cremastra*, and *Tipularia*. **Vascular bundles** many, collateral, scattered. **Sclerenchyma** and **stegmata** absent.

Rhizome TS (Aplectrum, Corallorhiza, Cremastra, Wullschlaegelia)

Hairs tufted, unicellular, borne on raised pyramidal cushions in *Corallorhiza* and *Cremastra* (Fig. 58G); at least 2-celled, thin-walled in *Wullschlaegelia*; absent in *Aplectrum*. **Cuticle** noted in *Aplectrum* only, smooth to rugose, ±2.5 μm thick. **Epidermis**: cells mostly isodiametric in *Aplectrum* and *Cremastra*, somewhat anticlinal in *Corallorhiza*, larger than hypodermal cells, periclinal and isodiametric in *Wullschlaegelia*. **Stomata** noted in *Aplectrum* only, superficial; substomatal chambers large. **Hypodermis**: 1-layered in *Corallorhiza* and *Wullschlaegelia* and in some areas in *Aplectrum*; anticlinal walls with ladder-like thickenings in *Wullschlaegelia*. **Cortex** 4–5 cells wide in *Wullschlaegelia*, subtended by sclerenchyma band 1–4 cells wide; outer layer of cells in *Corallorhiza* infested with hyphae and pelotons, overlying five or six layers of larger cells containing opaque masses of dead hyphae and 5–7 layers of fungus-free cells with starch granules; many cells wide in *Cremastra*, cells crowded, variously shaped, walls moderately thick, water-storage cells circular, empty. Cortex absent in *Aplectrum*. **Endodermis** discontinuous in *Aplectrum*, cells isodiametric, thin-walled, with suggestion of casparian strips; 1- or 2-layered in *Corallorhiza*, cells periclinal, thin-walled;

in *Cremastra* cells thin-walled, rectangular to rounded, and isodiametric with casparian strips. **Pericycle** noted in *Cremastra* only, cells like endodermal cells but without casparian strips. **Ground tissue** cells thin-walled, parenchymatous, with cruciate starch granules in *Aplectrum* and *Wullschlaegelia*; enucleate water-storage cells scattered in *Aplectrum* (Fig. 58E). **Vascular bundles** 25–30 in *Aplectrum*, each surrounded by parenchymatous matrix of small, very thin-walled, angular cells. Bundles collateral, poorly formed in *Corallorhiza*, where vascular cylinder demarcated by two or three rows of mostly thin-walled periclinal cells lacking any special features and vascular bundles embedded in a parenchyma matrix of small, thin-walled rounded and angular cells lacking intercellular spaces; one or two xylem strands subtending adjacent strands of phloem cells; rhizome may have one or more clusters of vascular bundles that could be interpreted as meristelic. Vascular bundles collateral, elliptical in *Wullschlaegelia*, forming ring at periphery of ground tissue with four bundles in a central square (Fig. 57E). In *Cremastra* irregular series of discontinuous arcs of vascular tissue interspersed with collateral bundles. **Sclerenchyma** and **stegmata** absent.

Root TS

Velamen one cell layer wide in *Aplectrum* (Fig. 59A) and *Govenia*, one or two layers wide in *Cremastra*, two layers in *Calypso* (Fig. 59B), three to four layers in *Tipularia* (Fig. 59C). Cells thin-walled without secondary thickenings, isodiametric in *Aplectrum*, *Cremastra*, *Govenia*, and *Tipularia*; epivelamen cells periclinal, endovelamen cells isodiametric in *Calypso*. Unicellular hairs present in all taxa, borne on a raised tuft in *Corallorhiza maculata*. Hyphae, dead

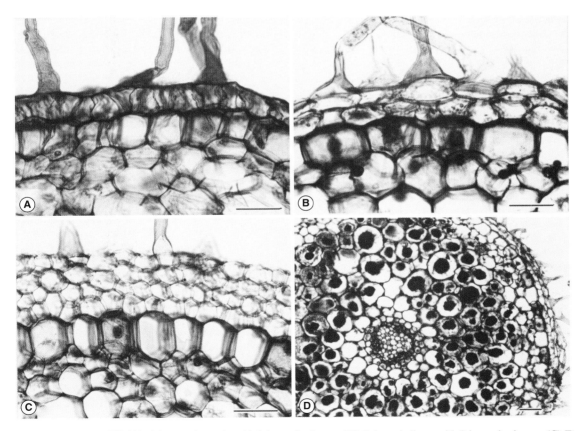

Fig. 59. Calypsoeae root TS. (A) *Aplectrum hyemale*, with 1-layered velamen. (B) *Calypso bulbosa*, with 2-layered velamen. (C) *Tipularia discolor*, with 4-layered velamen. (D) *Calypso bulbosa*, cortical cells with hyphal pelotons, dead fungal masses, and triarch vascular cylinder. Scale bars: A–C = 50 μm, D = 100 μm. Reprinted from Stern and Carlsward (2008).

cell masses or clots, and pelotons occurring in all taxa except *Cremastra* and *Govenia tingens*. **Tilosomes** absent. **Exodermis**: cells square in *Aplectrum* and polygonal to anticlinal, thin-walled throughout; passage cells intermittent. **Cortex** up to 10 cells wide in *Aplectrum*, 8–10 cells wide in *Calypso* (Fig. 59D), seven cells wide in *Cremastra*, 7–9 cells wide in *Govenia* and seven or eight cells wide in *Tipularia*; cells thin-walled; inner cells mostly circular, sometimes distorted. **Endodermis**: cells isodiametric (Fig. 59D), except rectangular in *Cremastra* and periclinal in *Govenia tingens*, entirely thin-walled; casparian strips in *Aplectrum* and *Tipularia*. **Pericycle** cells thin-walled throughout, mostly isodiametric, smaller than endodermal cells. **Vascular cylinder** 9-arch in *Aplectrum*, 3-arch in *Calypso*, 5-arch in *Cremastra*, 6-arch in *Govenia*, 4-arch in *Tipularia*. Xylem in short arms, cells clustered in *Govenia*; several intramedullary xylem clusters in *Aplectrum*; xylem alternating with strands of phloem cells. Vascular tissue embedded in very thin-walled sclerenchyma or thick-walled parenchyma. **Pith** parenchymatous, sometimes with thick walls in *Aplectrum*; cells polygonal, intercellular spaces absent. **Stegmata** absent.

Wullschlaegelia has two kinds of roots: fusiform and filiform (Fig. 60A).

Fusiform root

Velamen at least 1-layered, cells anticlinal or polygonal. **Exodermal** cells basically pentagonal, scalariform bars occurring along longitudinal walls of dead cells. **Cortex** eight cells wide, cells thin-walled, variably shaped (Fig. 60C), with banded wall thickenings in some cells; spiranthosomes present (Fig. 60E). **Endodermis** 1-layered; cell walls facing phloem thick, outer walls thin (Fig. 60D); cells facing xylem thin-walled throughout. **Vascular cylinder** with four- (Fig. 60B) or five-armed (Fig. 60D) xylem elements embedded in thick-walled sclerenchyma cells; phloem occurring in embayments between xylem arms. **Pith** cells thick-walled. Raphide bundles in unmodified cortical cells.

Filiform root

Velamen absent. **Epidermal** cells isodiametric, cell walls thin. **Cortex** four or five cells wide, cells thin-walled, outermost cells rounded to polygonal, cells of adjacent inner layer anticlinal, cells of innermost layer isodiametric (Fig. 57A). Cortical and epidermal cells with hyphae. **Endodermal** cells U-thickened opposite phloem; thin-walled opposite xylem. **Vascular cylinder** Y-shaped, arms containing xylem elements, phloem in embayments (Fig. 57B); vascular tissue surrounded by thick-walled sclerenchyma. **Pith** absent.

Material examined

Aplectrum (1), *Calypso* (1), *Corallorhiza* (3), *Cremastra* (1), *Govenia* (2), *Tipularia* (1), *Wullschlaegelia* (3). Full details are given in Stern and Carlsward (2008) and Stern (2009).

Reports from the literature

Campbell (1970): brief anatomical description of *Yoania*, a New Zealand endemic recently transferred from Calypsoeae to Cranichideae. Camus (1908, 1929): general anatomy of *Calypso* and *Corallorhiza*. Carlsward and Stern (2008): anatomy of *Corallorhiza*. Currah et al. (1988): mycorrhiza of *Calypso bulbosa*. Freudenstein (1997): monograph of *Corallorhiza*. Freudenstein and Rasmussen (1999): coded anatomical characters for many genera including *Aplectrum*. Fuchs and Ziegenspeck (1925, 1927a): root and stem anatomy of *Corallorhiza*. Holm (1904): root anatomy of *Aplectrum*, *Calypso*, and *Tipularia*. Holm (1913): vegetative anatomy of *Corallorhiza odontorhiza*. Holm (1927): leaf anatomy of *Aplectrum*, *Calypso*, *Corallorhiza*, and *Tipularia*. Jennings and Hanna (1899): mycorrhiza of *Corallorhiza*. Johow (1885, 1889): studies of stem and root of *Wullschlaegelia aphylla*. Kozhevnikova and Vinogradova (1999): pseudobulb of *Calypso*. MacDougal (1898, 1899a,b): mycorrhiza and anatomy of *Aplectrum*, *Calypso*, and *Corallorhiza*. Martos et al. (2009): mycorrhiza of *Wullschlaegelia*. Mollberg (1884): mycorrhiza of *Corallorhiza*. Møller and Rasmussen (1984): stegmata absent in *Calypso*, *Corallorhiza*, and *Wullschlaegelia*. Nieuwdorp (1972): mycorrhiza of *Corallorhiza*. Paschkis (1880): tuber anatomy of *Aplectrum hyemale* drug. Porembski and Barthlott(1988): root velamen of *Calypso*, *Corallorhiza*, *Govenia*, *Oreorchis*, and *Tipularia*. Pridgeon et al. (1983): root anatomy of *Govenia*. Rasmussen and Whigham (2002): root and mycorrhiza of *Tipularia*. Raunkiaer (1895–9): anatomy of *Corallorhiza*. Reinke (1873): rhizome anatomy of *Corallorhiza*. Singh (1981): stomata of *Oreorchis*. Solereder and Meyer (1930): comprehensive review of orchid anatomy and original data for *Aplectrum* and *Corallorhiza*. Stern (1999): anatomy of *Wullschlaegelia*. Stern and Carlsward (2008): anatomical analysis of Calypsoeae. Stern and Judd (2002): anatomy of Cymbidieae and also *Govenia*. Tatarenko (1995): mycorrhiza of *Oreorchis*. Thomas (1893): vegetative anatomy of *Corallorhiza*. Vij et al. (1991): leaf epidermis of *Oreorchis*. Wahrlich (1886): mycorrhiza of *Corallorhiza*. Weber (1981): rhizome anatomy and mycorrhiza of *Corallorhiza trifida*. Zelmer and Currah (1995): mycorrhiza of *Corallorhiza trifida*.

There appears to be a considerable discrepancy in reporting velamen cell layers in *Govenia* roots: Porembski

Subfamily Epidendroideae 111

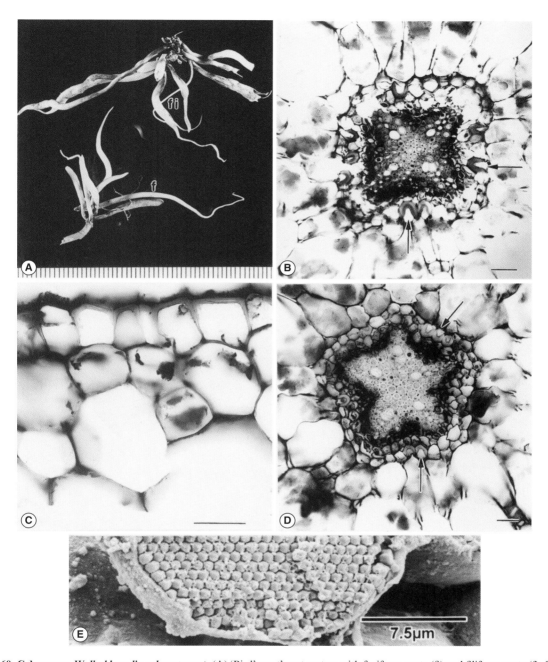

Fig. 60. Calypsoeae. *Wullschlaegelia calcarata* **root**. (A) 'Bird's nest' root system with fusiform roots (fi) and filiform roots (f). Scale lines = 1 mm. (B) TS tetrarch vascular cylinder in fusiform root with U-thickened endodermal cell walls opposite phloem (arrows). (C) TS cortex of fusiform root. (D) TS pentarch vascular cylinder in fusiform root with U-thickened endodermal cell walls opposite phloem (arrows). (E) SEM showing regular arrangement in cortical cell of polygonal starch granules in spiranthosomes of fusiform root. Scale bars: A = 1 mm, B–E = 7.5 μm. Reprinted from Stern (1999).

and Barthlott (1988) note 9–13 layers for *G. pauciflora* (a possible synonym of *G. purpusii*) while Stern and Carlsward show only a single layer in velamina of *G. superba* and *G. tingens*. Perhaps *G. pauciflora* has been misidentified.

Taxonomic notes

The hypothesis of Freudenstein (2005) establishing two groups of genera in Calypsoeae based on the origin of pollinial stipes and the presence of spurred lips cannot be substantiated by anatomical study alone. Cladistic analyses based on anatomy and morphology could not reliably support the grouping of *Aplectrum* and *Tipularia*, nor the grouping based on pollinarium features. More likely is the pairing of *Wullschlaegelia* and *Corallorhiza*, two leafless genera, as sister to each other. *Yoania australis*, a New Zealand endemic, has been renamed *Danhatchia australis* by Garay and Christenson and placed in Cranichideae (Freudenstein 2005).

Tribe Collabieae Pfitzer

The genera in Collabieae may be found under different tribal and subtribal names: Calanthinae, Bletiinae, Phajeae, Chrysoglossinae, Collabiinae, and Phajinae. As currently constituted under Collabieae there are 18 genera, the largest of which is *Calanthe* with 260 species, mostly confined to the Old World with a single species of *Calanthe* in tropical America. Plants range from central Africa, Madagascar, and the Indian Ocean islands northeastward to the Himalayas and south to southern Asia, northern Australia, New Guinea, and the Indonesian islands, Japan, China and Taiwan, the Philippines, and western Pacific islands. Collabieae are mostly terrestrial herbs with cylindrical stems, corms, and pseudobulbs. Leaves are distichous, convolute, and articulate. Several species of *Calanthe*, *Acanthephippium*, and *Phaius* appear in the horticultural trade.

Calanthe species are deemed medicinal in Nepal, Malaysia, China, Korea, Indonesia, Fiji, and other Pacific islands; *Cephalantheropsis* is noted as medicinal in China and Taiwan, and *Phaius* in Indonesia and Taiwan.

The information on this subtribe has been compiled entirely from the literature and the line drawings, but the names now accepted have been used where possible. Where no data are given for any organ or tissue, it is because none was available.

Leaf surface

Hairs present, sunken, in *Acanthephippium*, *Calanthe*, *Phaius*, and *Tainia* (abaxial); also thick-walled unicellular in *Calanthe*; absent in *Cephalantheropsis*. **Epidermis**: cells polygonal (Fig. 61A,B). **Stomata** abaxial, mostly tetracytic, in *Acanthephippium*, *Calanthe* (except on both surfaces in *C. brevicornu* and *C. discolor* and with two subsidiary cells in *C. calanthoides* (as *C. mexicana*)), *Cephalantheropsis*, *Gastrorchis*, *Nephelaphyllum* (with some anomocytic; Fig. 61B), *Phaius*, *Spathoglottis* (Fig. 62B), and *Tainia*; on both surfaces in *Ipsea speciosa*.

Leaf TS

Cuticle thin in *Acanthephippium*, thin to moderately thick in *Phaius* and *Spathoglottis plicata*, very thick and warty in *Ipsea*, but thin in *I. speciosa*. **Epidermis**: outer cell walls arched in *Ipsea*, walls thin in *Phaius*. **Fibre bundles** absent in *Calanthe*. **Mesophyll** homogeneous in *Acanthephippium*, *Calanthe*, *Cephalantheropsis*, *Ipsea*, *Nephelaphyllum* (Fig. 61C–E), *Phaius*, *Spathoglottis* (Fig. 62C–E), and *Tainia*; cells thin-walled, some with raphides in *Acanthephippium* and *Ipsea*; water-storage cells and raphides present in *Phaius* and *Tainia*. **Vascular bundles** in one row in *Acanthephippium* (few), *Calanthe*, *Cephalantheropsis*, *Nephelaphyllum* (Fig. 61F), *Phaius*, *Spathoglottis* (Fig. 62F), and *Tainia*; three bundles in midrib of *Ipsea*. Vascular sclerenchyma well developed in *Calanthe*, with small fibre caps at phloem poles only in *Acanthephippium* and in *Cephalantheropsis*, at both poles in *Ipsea speciosa*, and surrounding bundles in *Spathoglottis* (Fig. 62C–E) and *Nephelaphyllum* (Fig. 61D,E). **Stegmata** with conical silica bodies present in some specimens of *Chrysoglossum ornatum* (but absent in those named *C. erraticum*), *Eriodes barbata* (as *Tainiopsis barbata*), *Spathoglottis*, *Tainia viridifusca*, and in bracts of *Ipsea*; absent in *Acanthephippium*, *Calanthe*, and *Phaius*.

Stem TS

Epidermis: cells small, barrel-shaped, some with mycorrhiza. **Stomata** present in *Cephalantheropsis*. **Cortex**: cells thin-walled, globular to polygonal in *Cephalantheropsis*. **Vascular bundles** at periphery and intervening sclerenchyma forming a continuous cylinder in *Cephalantheropsis*, also large and small vascular bundles scattered in ground tissue with sclerenchyma caps at phloem poles only; scattered, with no sclerenchyma ring in *Nephelaphyllum* (Fig. 61G). **Stegmata** with conical silica bodies present in *Eriodes barbata* (as *Tainiopsis barbata*) and in some specimens of *Chrysoglossum ornatum* (but absent in specimens named *C. erraticum*), absent in *Nephelaphyllum pulchrum* and *Tainia viridifusca*.

Pseudobulb TS (Phaius and Tainia)

Cuticle smooth, usually thick. **Epidermis** 1-layered, of elongated, rectangular, or pentagonal cells with slightly

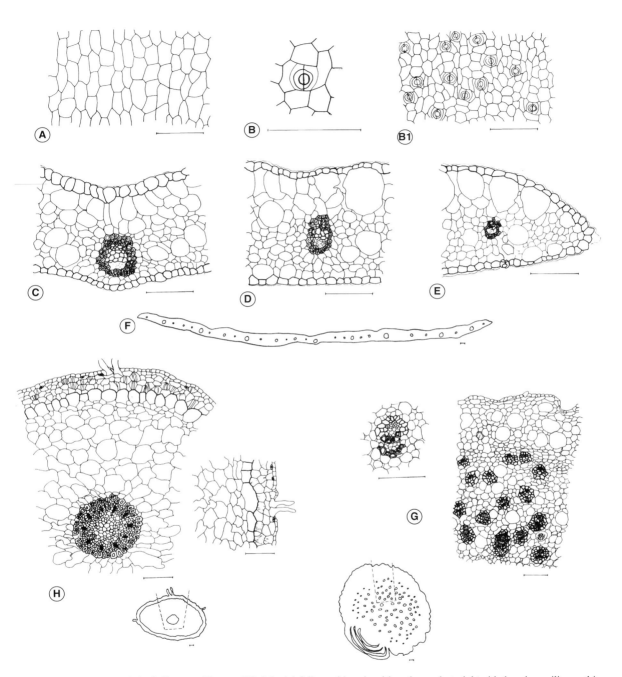

Fig. 61. Collabieae. *Nephelaphyllum tenuiflorum*. (A) Adaxial foliar epidermis with polygonal straight-sided and curvilinear thin-walled cells. (B) Abaxial foliar epidermis with polygonal, mostly curvilinear thin-walled cells and anomocytic stomata. (C–E) TS leaves with collateral vascular bundles and sclerenchyma sheaths, homogeneous mesophyll with large mostly adaxial lacunae. (F) TS lamina showing one row of vascular bundles. (G) TS stem with scattered vascular bundles having adaxial groups of sclerenchyma cells all surrounded by small-celled, lacunose ground tissue and a multilayered lacunose cortex, epidermal cells periclinal. (H) TS root with several-layered velamen and banded cell walls, exodermal cells moderately large, thin-walled, and isodiametric, ground-tissue cells thin-walled and variably shaped, endodermal cells having O-thickened walls opposite phloem strands and thin walls opposite xylem rays; pith comprising thin-walled cells.

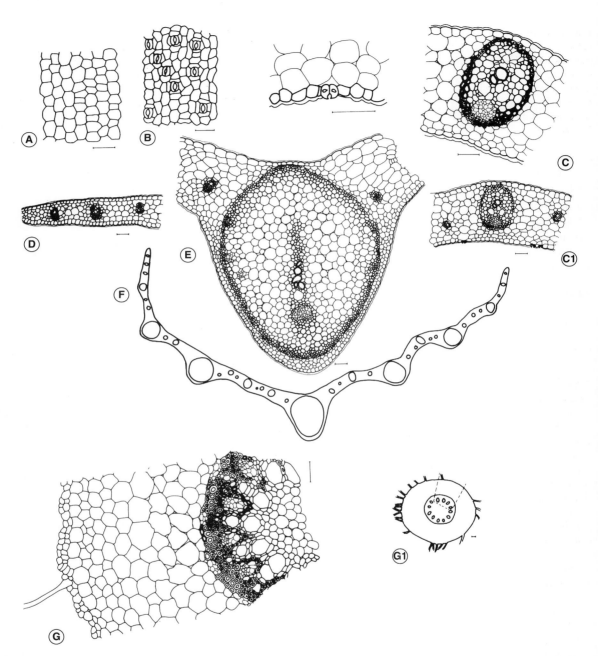

Fig. 62. Collabieae. *Spathoglottis plicata*. (A) Adaxial foliar epidermis with thin-walled polygonal cells. (B) Abaxial foliar epidermis with thin-walled polygonal cells and tetracytic stomata. (C) TS foliar vascular bundle encircled by ring of sclerenchyma cells or (C1) ad- and abaxial arcs of sclerenchyma. (D) TS leaf to show row of smaller vascular bundles. (E) Large vascular bundle with two lateral supernumerary bundles and homogeneous mesophyll. (F) TS lamina to show one row of alternating larger and smaller vascular bundles. (G) TS root with 2-layered velamen, unicellular hair, thin-walled exodermal and cortical cells, O-thickened endodermal cells opposite phloem and thin-walled endodermal cells opposite xylem, and thin-walled pith cells.

thickened outer walls in *Phaius*, of small, barrel-shaped cells in *Tainia*. **Stomata** few in *Phaius*. **Cortex** parenchymatous, cells thin-walled, starch granules present; some cells with mucilage, raphide bundles, or solitary crystals; large water-storage cells in *Tainia*; thick-walled sclerenchyma cells scattered in cortex in some *Phaius* species. **Vascular bundles** dispersed in ground tissue; those at periphery with crescentiform fibre sheaths linked by sclerotic parenchyma cells to form ring in *Phaius*. **Ground tissue** cells packed with starch granules; some cells mucilaginous and with raphide bundles or solitary crystals. **Stegmata** absent.

Rhizome TS (Calanthe and Phaius)

Cuticle thin or thick. **Epidermis** 1-layered. **Stomata** few. **Cortex** of thin-walled cells containing starch granules, some cells mucilaginous, tanniniferous, or with raphide bundles; scattered sclereids present. **Ground tissue** cells similar to those of cortex. **Vascular bundles**: peripheral bundles with circular or crescent-shaped fibre sheaths linked by sclerenchyma cells to form ring. Phloem sheath well developed, xylem sheath often lacking in *Phaius*. **Stegmata** absent in *Phaius*, present in *Chrysoglossum ornatum*.

Root TS

Velamen 1-layered in *Ancistrochilus*, cells lignified but without wall thickenings, hairs present in *A. thomsonianus* and *Phaius flavus*; 2–5(–13)-layered in *Acanthephippium*, *Calanthe*, *Cephalantheropsis*, *Nephelaphyllum* (Fig. 61H), *Phaius*, *Plocoglottis*, *Spathoglottis* (Fig. 62G), and *Tainia*; cells with helical thickenings and small pits in *Acanthephippium* and some *Calanthe* species, cells without helical thickenings but with small pits in some *Calanthe* species, *Cephalantheropsis longipes* (as *Phaius longipes*), *Plocoglottis*, and *Spathoglottis* and with fissured pits in *Tainia*; cells thick-walled with helical bands in *Phaius*, cells thin-walled with helical bands in *Nephelaphyllum* (Fig. 61H). **Tilosomes** absent in *Acanthephippium*, *Calanthe*, *Phaius*, *Plocoglottis*, *Spathoglottis*, and *Tainia*. **Exodermis**: cell walls thin in *Nephelaphyllum* (Fig. 61H), evenly thickened in *Acanthephippium*, *Ancistrochilus*, and *Calanthe*, or ∩-thickened in *Calanthe* and *Tainia*. **Cortex**: parenchymatous in *Nephelaphyllum* (Fig. 61H) and *Spathoglottis* (Fig. 62G), with scattered sclerenchyma cells in most *Calanthe* species; cells globular to polygonal, some with raphide bundles, in *Cephalantheropsis*; cells with starch granules, some with raphide bundles and mucilage, in *Calanthe* and *Phaius*. **Endodermis**: 1-layered, of oblong to pentagonal thick- or thin-walled cells in *Calanthe*; cells thin-walled in *Cephalantheropsis* and *Phaius*, O-thickened in *Nephelaphyllum* (Fig. 61H), *Spathoglottis* (Fig. 62G), and *Tainia*. **Pericycle** 1-layered; of rectangular or polygonal cells in *Calanthe*; of lignified fibres in *Cephalantheropsis* with passage cells opposite protoxylem strands; of slightly lignified fibres in *Tainia*; of thin-walled cells in *Phaius*. **Central ground tissue** with raphide bundles in *Phaius*. **Vascular cylinder** with alternating radii of xylem and phloem in *Calanthe*, *Cephalantheropsis*, *Nephelaphyllum* (Fig. 61H), *Phaius*, *Spathoglottis* (Fig. 62G), and *Tainia*. **Pith**: cells thin-walled, isodiametric (Fig. 61H), with starch granules in *Calanthe* (walls lignified in some *Calanthe* species), *Cephalantheropsis*, and *Phaius*, also some cells with mucilage and raphide bundles.

Material reported

The treatment of Collabieae was prepared from multiple published sources varying in numbers of taxa and parts described, and the line drawings, yielding the composite report above. A number of genera in Collabieae have not been studied anatomically.

Reports from the literature

Clements (1988): mycorrhiza of *Spathoglottis*. Das and Paria (1992): stomata of *Spathoglottis*. Dietz (1930): underground organs of *Phaius*. Engard (1944): root anatomy of *Spathoglottis*. Goh (1975): leaf anatomy of *Spathoglottis*. Goh et al. (1992): mycorrhiza of *Spathoglottis*. Guttenberg (1968): root anatomy of *Phaius*. Hadley and Williamson (1972): mycorrhiza of *Calanthe* and *Spathoglottis*. Janse (1897): mycorrhiza of *Phaius*. Kaushik (1983): anatomy of Himalayan Collabieae, especially *Cephalantheropsis*, for which this is the only complete description. Leitgeb (1864b): velamen in *Phaius grandifolius* with spiral 'fibres'. Lin and Namba (1981c): drug plants from Taiwan, including anatomy of *Calanthe* and *Phaius* species. Meinecke (1894): anatomy of aerial roots of *Tainia* and *Phaius*. Möbius (1887): leaf structure of *Calanthe* and *Phaius*. Mohana Rao et al. (1989): anatomy of *Calanthe*. Møller and H. Rasmussen (1984): presence or absence of silica bodies in *Calanthe*, *Chrysoglossum*, *Nephelaphyllum*, *Tainia viridifusca*, and *Tainiopsis* (*Eriodes*). Mulay et al. (1958): velamen of *Calanthe*. Mulay and Panikkar (1956): velamen of *Phaius*. Neubauer (1978): inflorescence axis idioblasts of *Phaius*. Porembski and Barthlott (1988): velamen in *Acanthephippium* and the *Calanthe* type of velamen in *Ancistrochilus*, *Calanthe*, *Phaius*, *Plocoglottis*, and *Spathoglottis*. Pridgeon (1987): velamen and exodermis of *Acanthephippium*, *Ancistrochilus*, *Calanthe*, *Phaius*, and *Tainia*. Pridgeon et al. (1983): lack of tilosomes in *Acanthephippium*, *Calanthe*, *Spathoglottis*, and *Tainia*. Pridgeon et al. (2005): comprehensive account of genera in tribe Collabieae. Sakharam Rao (1953): velamen of *Phaius*. Sanford and Adanlawo (1973): root anatomy of *Ancistrochilus*. Sasikumar (1973a, 1974): root anatomy of *Calanthe*. Singh (1981): stomata of *Calanthe* and *Phaius*. Singh and Singh (1974): stomata of *Calanthe brevicornu*. Solereder and Meyer (1930): review of orchid anatomy and original data on anatomy of *Acanthephippium*, *Ipsea*, and *Spathoglottis*. Staudermann (1924): trichomes of *Ipsea*. Tominski (1905): leaf anatomy of *Acanthephippium*, *Calanthe*, *Ipsea speciosa*, and *Phaius*. Vij et al. (1991): leaf epidermis of *Calanthe* and *Phaius*. Weltz (1897): stem anatomy of *Phaius*. Williams (1979): stomata of *Calanthe*, *Gastrorchis*, *Phaius*, and *Spathoglottis*.

Taxonomic notes

There is overall consistency in many features with but few exceptions: lack of hypodermis in leaves, homogeneous mesophyll, tetracytic, abaxial stomata (but two subsidiary cells in *Calanthe calanthoides* and ad- and abaxial stomata in *Calanthe brevicornu*, *C. discolor*, and *Ipsea*), absence of stegmata and silica bodies except present in leaves of *Spathoglottis*, *Tainia viridifusca*, bracts of *Ipsea*, and in leaves and stems of *Chrysoglossum ornatum* (except specimens described as *C. erraticum*) and *Eriodes barbata*, and lack of tilosomes. Overall, Collabieae are anatomically consistent, pending anatomical studies of neglected genera, in accord with conclusions of Pridgeon et al. (2005).

Tribe Epidendreae Kunth

Plants in tribe Epidendreae are exclusively New World. They are terrestrial epiphytes, lithophytes, and achlorophyllous saprophytes. Plants are velamentous. Leaves may be absent, or, when present, are articulate, narrowly conduplicate, plicate, membranous, or fleshy, planar or sometimes terete. Stems are reed-like, slender, or expanded into pseudobulbs. Epidendreae had included as many as 47 subtribes (Schlechter 1926) but have been narrowed markedly in recent years (Dressler 1993; Pridgeon et al. 1999–2014).

Subtribe Bletiinae Benth.

Bletiinae are terrestrial herbs, some of which lack roots. Stems are corms, tubers, or rhizomes bearing articulate inconspicuous or basal leaves, linear conduplicate in *Basiphyllaea*, plicate in *Bletia*, and nearly absent and achlorophyllous in *Hexalectris*. There are three neotropical genera and about 50 species distributed in the southern, eastern, and Midwestern USA, Caribbean, northern Mexico, Central America, and into South America as far as north-western Argentina. Several showy species of *Bletia* are grown ornamentally.

Bletia species are used medicinally in Mexico, Caribbean islands, China, Mongolia, Tibet, Japan, Korea, and the Philippines.

Leaf surface
Hairs occasional, adaxial in *Basiphyllaea*; uniseriate, 4-celled in *Bletia*. **Cuticle** granular. **Epidermis**: cells irregularly polygonal, isodiametric to elongated. **Stomata** ad- and abaxial in *Basiphyllaea*, abaxial in *Bletia*; subsidiary cells two in *Bletia*.

Leaf TS (mainly Basiphyllaea)
Cuticle moderately thick, convex above epidermal cells. **Stomata** superficial. **Hypodermis**: adaxial, 3- or 4-layered, of cuboidal water-storage cells; abaxial 1-layered, cells transversely elongated. **Fibre bundles** in abaxial layers of mesophyll. **Mesophyll** with mixed larger and smaller cells, some faintly reticulately thickened. **Vascular bundles** in one row. Raphides numerous in hypodermal and mesophyll cells. Water-storage cells with radial crystalline masses. Tannins copious in all tissues. **Stegmata** with conical silica bodies in *Bletia*.

Root TS (Bletia only)
Velamen 4- or 5-layered in *Bletia catenulata*; cells with fine spiral thickenings in *B. purpurea* (as *B. verecunda*). **Tilosomes** absent. **Exodermis**: outer cell walls thickened. **Pericycle** 2-layered with moderately thickened cell walls. Eight xylem rays alternating with circular strands of phloem cells. **Pith** wide, with thin-walled cells and large intercellular spaces.

Material reported

The information above was assembled from only a few published works, the major one being Baker's (1972) on the leaves of *Basiphyllaea*.

Reports from the literature

Baker (1972): leaf structure in Laeliinae including *Basiphyllaea sarcophylla*. Bernard (1904): mycorrhiza of *Bletia*. Holm (1904): anatomy of *Bletia purpurea* (as *B. verecunda*) roots. Pridgeon (1987): velamen and exodermis of *Bletia catenulata*. Solereder and Meyer (1930): review of orchid anatomy with original observations on leaf anatomy of *Bletia purpurea* (as *B verecunda*). Williams (1979): subsidiary cells in *Bletia*.

Taxonomic notes

Anatomical information is too sparse to interpret any taxonomy.

Subtribe Chysinae Schltr.

Chysinae are a neotropical, monogeneric group of herbs. *Chysis* has 8–10 species ranging from Mexico to Peru; most species occur in Mesoamerica. *Chysis* is characterized by clustered, pendulous pseudobulbs from the basal nodes of which the inflorescences are produced. Leaves are articulate, distichous, plicate, convolute. Several species are grown as ornamentals by hobbyists.

Leaf surface

Hairs ad- and abaxial or adaxial only; multicellular, branched in *Chysis bractescens*; lacking completely in *C. aurea*. **Stomata** abaxial, subsidiary cells two or rarely four, paracytic or tetracytic.

Leaf TS

Cuticle thin. **Epidermis**: cells filled with clear watery fluid in *C. limminghei*. **Hypodermis**: absent. **Fibre bundles** absent. **Mesophyll** homogeneous. **Vascular bundles** in a single series. **Stegmata** with conical silica bodies.

Pseudobulb TS

Cuticle thin, even. **Epidermis**: cell walls thickened externally, thickness decreasing along radial and inner walls. **Hypodermis**: two or three layers of weakly thickened, irregular cells of differing sizes. **Vascular bundles** scattered, lacking sclerenchyma bridges, with incomplete phloem sheath.

Root TS

Velamen 3-layered in *C. aurea*, first layer with plain walls, second and third layers with spiral thickenings; 2-layered in *C. bractescens*. **Tilosomes** lamellate in *C. aurea*. **Exodermis**: cells ∩-thickened along outer and radial walls in *C. aurea*. **Cortex**: parenchyma broad, cell walls occasionally slightly thickened. **Endodermis**: walls of 4–7 cells somewhat thickened with 2–4 thin-walled intervening passage cells. **Central cylinder** 16-arch. **Pith** cells parenchymatous, thin-walled.

Material reported

The information above has been assembled from several published sources noted below. Where data are lacking, none was available in the literature.

Reports from the literature

Krüger (1883): vegetative structure of aerial parts. Leitgeb (1864b): anatomy of aerial roots. Löv (1926): leaf anatomy. Meinecke (1894): anatomy of aerial roots. Möbius (1887): leaf anatomy. Moreau (1913): pseudobulb anatomy. Porembski and Barthlott (1988): velamen in *Chysis bractescens*. Pridgeon (1987): velamen and exodermis. Pridgeon et al. (1983): tilosomes in roots of *C. aurea*. Pridgeon et al. (2005): comprehensive review of systematics of Chysinae. Solereder and Meyer (1930): review of orchid anatomy with some original observations on leaf of *C. bractescens*. Weltz (1897): stem anatomy. Williams (1979): stomata and subsidiary cells.

Taxonomic notes

Anatomy is too sparse to develop any systematic analysis.

Subtribe Coeliinae Dressler

Coeliinae consist of only a single genus, *Coelia*, with five species growing in Cuba, Jamaica, and Mexico south in Central America into Panama. *Coelia* species are terrestrial, lithophytic, or epiphytic herbs with obpyriform pseudobulbs bearing linear-lanceolate, subplicate, subcoriaceous articulate leaves. Roots are thick, fleshy, rhizomes short, creeping. *Coelia triptera* is sometimes seen in cultivation.

Leaf surface

Hairs unicellular, papillose, ab- and adaxial. **Epidermis**: cells polygonal, often raphide-containing. **Stomata** abaxial, paracytic.

Leaf TS

Cuticle moderately thick. **Epidermis** 1-layered; adaxial cells elliptic to isodiametric. **Stomata** superficial. **Hypodermis**: adaxial 1-layered, abaxial absent in *C. bella*, *C. macrostachya*, and *C. triptera* (Pridgeon 1978), but ad- and abaxial in *C. triptera* (as *C. baueriana*) according to Möbius (1887). **Fibre bundles** absent. **Mesophyll** heterogeneous, palisade cells 2- or 3-layered. **Vascular bundles** in one row, larger bundles with well-developed abaxial sclerenchyma caps, adaxial sclerenchyma less well developed; smaller bundles with little sclerenchyma. **Stegmata** with silica bodies associated with vascular bundle sclerenchyma.

Stem TS

Cuticle thin in *C. triptera*. **Epidermis**: cells thin-walled, tangentially elongated. **Endodermis**: cells thin-walled, isodiametric. **Vascular bundles** scattered. Sclerenchyma bridge between xylem and phloem lacking. Silica cells present adjacent to sclerenchyma surrounding vascular bundles. Cruciate starch granules and raphides present in parenchyma cells.

Root TS

Velamen 4-layered with fine spiral 'fibres' in *C. triptera*. **Tilosomes**: lamellate, present with fine wall thickenings. **Exodermis**: cells with ∩-thickenings. **Endodermis** with five or six thick-walled cells alternating with two or three passage cells. **Vascular cylinder** with 20 xylem and phloem strands and slightly lignified central cells.

Material reported

The only anatomical report exclusively of *Coelia* (incorporating *Bothriochilus*) is that of Pridgeon (1978) where he reports on leaf anatomy of three species of *Coelia*. Additional information on anatomy has been extracted from Möbius (1887): leaf of two species, Weltz (1897): stem of *C. triptera* (as *C. baueriana*), and Meinecke (1894), root of *C. triptera* (as *C. baueriana*).

Reports from the literature

Dressler (1993): brief description of Coeliinae. Meinecke (1894): aerial root anatomy. Möbius (1887): leaf anatomy. Pridgeon (1978): attempts to show congeneric nature of *Coelia* and *Bothriochilus* using evidence from leaf anatomy. Pridgeon (1987): velamen of *Coelia triptera*. Pridgeon et al. (2005): comprehensive review of the systematics of subtribe Coeliinae. Solereder and Meyer (1930): review of orchid anatomy. Weltz (1897): stem anatomy. Williams (1979): stomata.

Taxonomic notes

Coelia was first described by Lindley in 1830 and subsequently *Bothriochilus* by Lemaire in 1856. A controversy arose over the distinctiveness of these two genera based on floral characteristics that apparently involved intergradations of features rather than sharp divisions among them. Using anatomical characteristics, Pridgeon (1978) was able to align structures from both taxa to demonstrate that segregation into two genera was unjustified.

Subtribe Laeliinae Benth.

Laeliinae are tropical and subtropical pseudobulbous plants ranging throughout the warmer regions of the Americas. They are usually epiphytic and as *Epidendrum magnoliae* they occur as far north as North Carolina along the eastern coast of North America. Species occur in South America as far south as Uruguay and northern Argentina. Owing to their beauty the larger-flowered members, e.g. *Broughtonia, Cattleya, Encyclia, Laelia*, and some species of *Epidendrum*, are highly prized and much desired by the public and hobbyists. *Cattleya* flowers are often perceived to represent the essence of the orchid family. There are 40 genera and almost 5000 species in Laeliinae and scores of artificial hybrids made to satisfy the desires of hobbyists.

Epidendrum species are recorded as medicinal in North and South America, Mexico, Antilles, Korea, and China, and *Laelia* species in Mexico, Central America, and Caribbean islands, *Nidema* has been used in Malaysia and *Scaphyglottis* in Mexico.

Leaf surface

Hairs absent throughout, except rare, glandular, sunken in *Meiracyllium*. **Cuticle** smooth in most taxa, rough in *Broughtonia, Cattleya forbesii, C. intermedia, Guarianthe skinneri, Leptotes, Psychilis, Rhyncholaelia, Scaphyglottis coriacea*, and *S. reflexa*, mammilate in *Domingoa*, ridged and vesiculate in *Barkeria*. **Epidermis**: cells typically polygonal on both surfaces with straight or curvilinear cell walls (Figs 65A,B, 66A,B, 67A,B–71A,B). **Stomata** abaxial, except amphistomatal in *Barkeria, Broughtonia, Domingoa, Meiracyllium* (Fig. 69A,B), and *Pseudolaelia*. In taxa with isobilateral leaves, exposed surface abaxial and surrounded by stomata. Stomata basically tetracytic (Figs 63A, 65B), but sometimes other configurations arising by additional divisions of subsidiary cells. In a few taxa stomata mostly anomocytic (Figs 66B, 68B, 71B) and in some both tetracytic and anomocytic on same leaf (Figs 68B, 69B, 70B). Guard cells C-shaped in *Cattleya sincorana*, differing from predominant reniform condition of most taxa.

Leaf TS

Cuticle mostly between 2.5 and 5.0 μm thick but up to 50 or 60 μm thick in a few species. **Epidermis**: cells variable, periclinal, square, isodiametric, regularly and consistently formed within taxa; abaxial cells dentiform in *Arpophyllum* (Fig. 63C). **Stomata** usually superficial, slightly sunken in some species (Fig. 66C), decidedly sunken in *Arpophyllum* (Fig. 65C), *Laelia anceps, Leptotes*, and *Scaphyglottis reflexa*. Substomatal chambers usually tiny, small to moderate in few species, large in *Pseudolaelia* and *Rhyncholaelia*. Stomatal ledges usually undeveloped or obscure, inner and outer ledges pronounced in *Pseudolaelia*, outer ledges prominent in *Meiracyllium*. **Hypodermis**: typical hypodermis usually absent, when present variable: either 1-layered throughout (Fig. 68D–F), 1-layered and 2-layered in part (Fig. 67D), or 2-layered overall. In *Cattleya forbesii*, hypodermis 2-layered along both surfaces

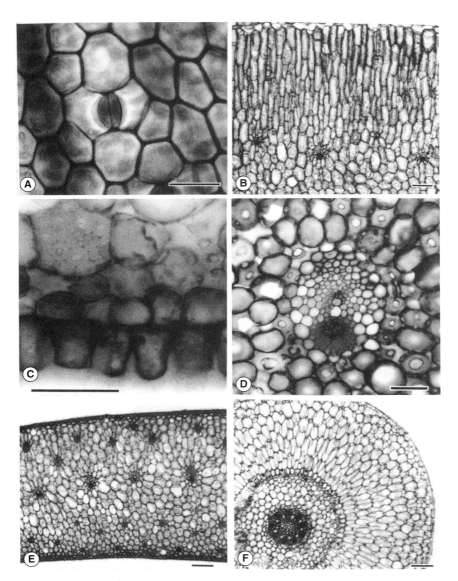

Fig. 63. Laeliinae. (A–E) Leaf. (A) *Caularthron bicornutum*, epidermal scraping, abaxial surface showing a tetracytic stoma. (B) *Guarianthe skinneri*, TS with heterogeneous mesophyll and vascular bundles with robust phloem sclerenchyma and weaker xylem sclerenchyma. (C,D) *Arpophyllum giganteum*. (C) TS showing dentiform abaxial epidermal cells. (D) TS vascular bundle showing robust phloem sclerenchyma. (E) *Broughtonia negrilensis*, TS with several adaxial rows of fibre bundles and one abaxial row. (F) *Prosthechea boothiana*, TS root with velamen cells of different configurations, O-thickened endodermal cell walls, parenchymatous cortex, and vascular tissue embedded in sclerenchyma. Scale bars: A,C = 50 μm, B,E,F = 200 μm, D = 100 μm. Reprinted from Stern and Carlsward (2009).

and cells with birefringent outer walls alternating with cells having non-birefringent outer walls; in *C. intermedia* adaxial hypodermis continuous, abaxial discontinuous, cells having fine birefringent banding parallel to width of cells; in *Guarianthe skinneri* hypodermal cells with birefringent walls, but no banding. Water-storage cells with birefringent walls forming second adaxial hypodermal layer in *Cattleya sincorana*. In *Leptotes* and *Scaphyglottis*

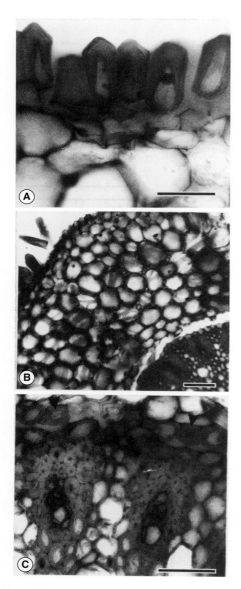

Fig. 64. Laeliinae. *Arpophyllum giganteum* root TS. (A) Exodermis with heavily thickened cell walls. (B) Banded cortical cell walls. (C) Vascular cylinder showing conductive elements embedded in thick-walled sclerenchyma cells. Scale bars: A,C = 50 μm, B = 100 μm. Reprinted from Stern and Carlsward (2009).

reflexa 1-layered hypodermis encircling leaf; in *Rhyncholaelia* scalariform thickenings in cell walls of second hypodermal layer. **Fibre bundles** present in most taxa, variable in three positions (Figs 63E, 70D–F), lacking in *Domingoa*, some *Encyclia* species, *Homalopetalum*, and *Meiracyllium*. In *Constantia* two large bundles, one at each leaf margin. In *Scaphyglottis imbricata* an abaxial row of fibre bundles alternating with rows of stomata. Fibre bundles in isobilateral leaves encircling rounded abaxial surface and absent from restricted adaxial surface. **Mesophyll** mostly homogeneous (Figs 66F, 67D–F, 69D–F), with some heterogeneity in a few species. Cells thin-walled, circular, oval, and irregular. Adaxial layers of palisade cells in heterogeneous mesophyll (Figs 63B, 65E, 70E, 71D), but abaxial cells rounded, lacking intercellular spaces. Heterogeneous in *Arpophyllum* with 7–9 layers of palisade cells (Fig. 65F,G), circular to oval abaxial cells and irregularly shaped adaxial cells forming tripartite mesophyll in places. Crystalliferous idioblasts ubiquitous. Water-storage cells occurring in mesophyll of *Encyclia*, *Jacquiniella*, and *Leptotes*. **Vascular bundles** collateral in a single series in most taxa (Figs 66G–70G), in several rows in *Arpophyllum* (Fig. 65H). Sclerenchyma caps at both xylem and phloem poles in all taxa but predominantly at phloem pole (Figs 65D, 66F, 67C,D, 68C); only at phloem pole in *Arpophyllum* (Fig. 63D). In *Cattleya forbesii*, *Leptotes*, and *Scaphyglottis imbricata* sclerenchyma surrounding vascular bundles (Figs 69C, 70C–F). **Stegmata** with conical rough-surfaced silica bodies typically occurring along vascular sclerenchyma and fibre bundles. Bundle sheath cells circular or oval, thin-walled and often chlorophyllous, but thick-walled in *Cattleya cernua* (Fig. 71C) and *Scaphyglottis imbricata* (Fig. 70F).

Stem TS

Hairs and **stomata** absent. **Cuticle** smooth, thickness ranging from 5.0 μm in *Jacquiniella* up to 50 μm in *Prosthechea radiata*; rough in *Leptotes*, *Pseudolaelia*, and *Scaphyglottis reflexa*. **Epidermis**: cells mostly thin-walled; thick-walled in *Epidendrum nocturnum*, *Orleanesia*, and *Pseudolaelia*; largely periclinal. **Hypodermis**: 1- or 2-layered (Figs 66H, 67H). **Cortex** lacking in most taxa; present in *Broughtonia*, *Cattleya forbesii*, *C. intermedia*, *Jacquiniella*, *Pseudolaelia*, and *Tetramicra*. **Fibre bundles** encircling ground tissue in *Broughtonia*. **Ground tissue**: few large, rounded, thin-walled empty water-storage cells surrounded by many smaller assimilatory cells with cruciate starch granules (Figs 65I, 66H, 70H, 71H); water-storage cells with banded cell wall thickenings in *Homalopetalum* and lacunae in *Cattleya cernua* (Fig. 71H). **Vascular bundles** many, collateral, mostly scattered in ground tissue (Figs 65I, 66H, 67I,J, 68H, 70H, 71H); bundles encircling stem and randomly scattered in *Jacquiniella*, *Nidema*, *Pseudolaelia*, *Scaphyglottis*

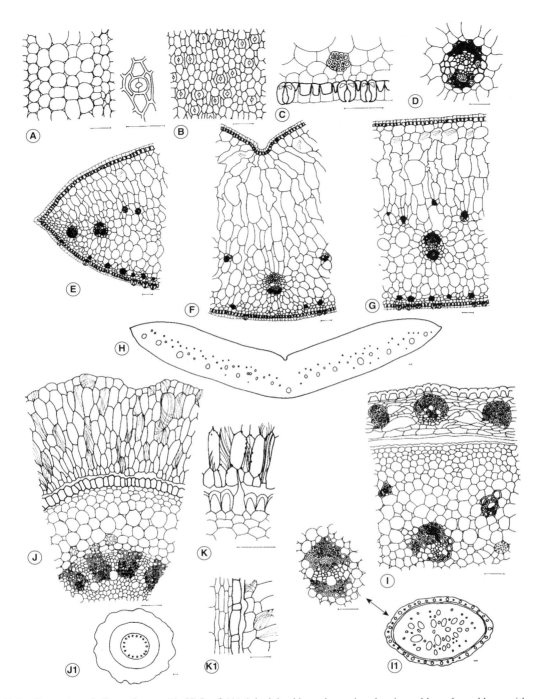

Fig. 65. Laeliinae. *Arpophyllum spicatum*. (A–H) Leaf. (A) Adaxial epidermal scraping showing subhypodermal layer with rounded cells. (B) Abaxial epidermis with variably shaped cells and curvilinear walls, and tetracytic stomata. (C) TS abaxial epidermal cells with U-thickened walls and falcate guard cell walls. (D) TS vascular bundle with ad- and abaxial sclerenchyma caps and oval thin-walled bundle sheath cells. (E–G) TS with heterogeneous mesophyll and several rows of vascular bundles, the larger mid-leaf, the smaller in ad- and abaxial rows. (H) TS with several rows of vascular bundles. (I) TS stem with sheathing leaf base, scattered vascular bundles in parenchymatous ground tissue and fibre bundles. (J,K) TS and (K1) LS root with multilayered velamen, squarish epivelamen cells, and anticlinal endovelamen cells all with spiral thickenings; exodermal cells with ∩-thickened cell walls, cortex with fibre bundles, endodermal cells with O-thickened walls opposite phloem, and parenchymatous pith.

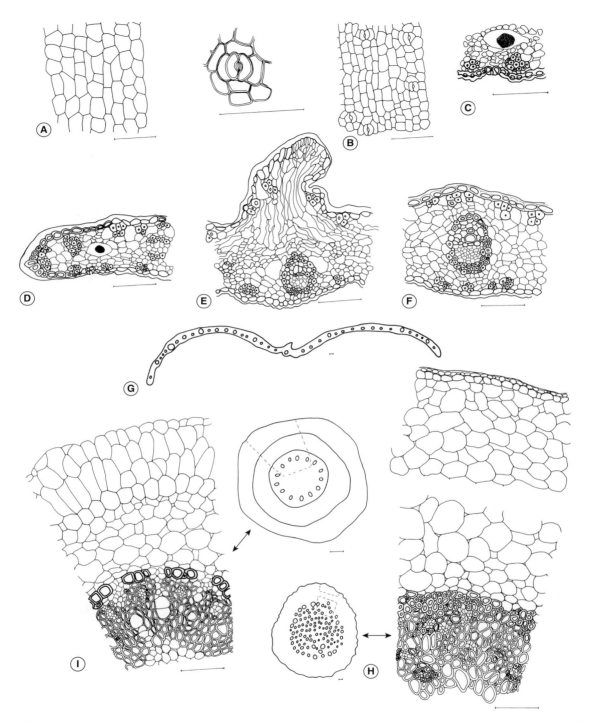

Fig. 66. Laeliinae. *Prosthechea boothiana*. (A–G) Leaf. (A) Adaxial surface with polygonal, curvilinear-walled cells. (B) Abaxial surface with polygonal cells and anomocytic stomata. (C) TS depressed stoma in abaxial epidermis. (D–F) TS with collateral vascular bundles having ad- and abaxial sclerenchyma caps, lacunae adjacent to midrib vascular bundle, and homogeneous mesophyll. (G) TS with one row of vascular bundles. (H) TS stem with parenchymatous ground tissue, scattered vascular bundles embedded in sclerenchyma, and small-celled hypodermis. (I) TS root with 3-layered velamen, ill-defined exodermis, endodermis with O-thickened cell walls, vascular tissue embedded in sclerenchyma, and parenchymatous pith.

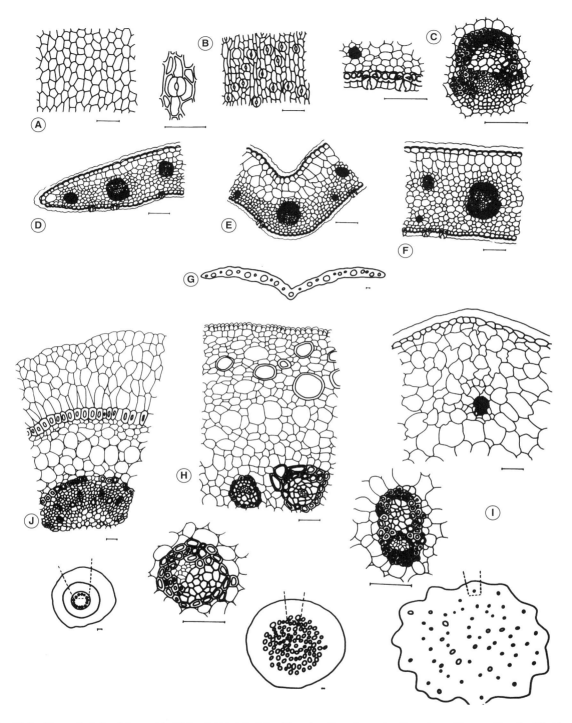

Fig. 67. Laeliinae. *Isabelia violacea*. (A–G) Leaf. (A) Adaxial epidermis with polygonal cells and curvilinear cell walls. (B) Abaxial surface with mostly anomocytic stomata. (C) TS vascular bundle with adaxial sclerenchyma cap. (D–F) TS with one or several adaxial hypodermal cell layers, homogeneous mesophyll, vascular bundles with ad- and abaxial sclerenchyma caps. (G) TS with one row of vascular bundles. (H) TS rhizome with 1-layered hypodermis, parenchymatous ground tissue, scattered vascular bundles encircled by sclerenchyma. (I) TS pseudobulb with scattered vascular bundles. (J) TS root with several-layered velamen, O-thickened exodermal cell walls, parenchymatous cortex, O-thickened endodermal cell walls, vascular tissue embedded in parenchyma, and parenchymatous pith.

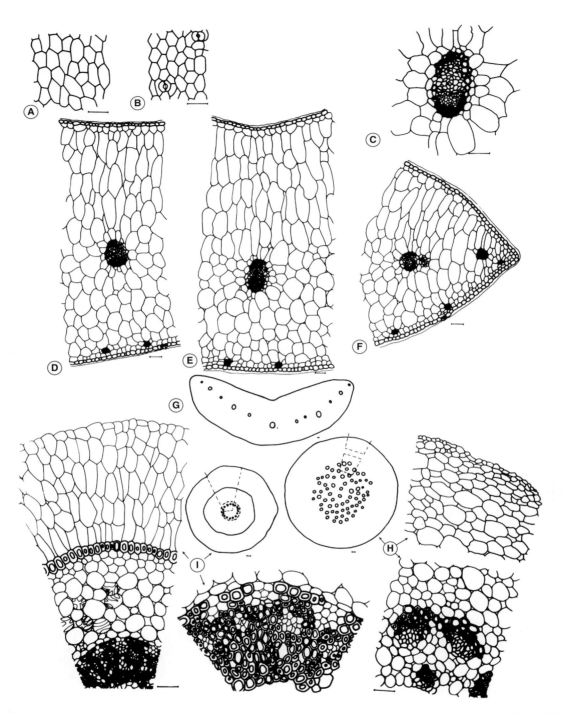

Fig. 68. Laeliinae. *Cattleya lundii*. (A–G) Leaf. (A) Adaxial surface showing polygonal cells and curvilinear cell walls. (B) Abaxial surface with polygonal cells with curvilinear walls, tetracytic and anomocytic stomata. (C) TS vascular bundle surrounded by thin-walled bundle sheath cells and ad- and abaxial sclerenchyma caps. (D–F) TS showing thick cuticle, 1-layered ad- and abaxial hypodermises, ad- and abaxial strands of sclerenchyma, abaxial fibre bundles, homo- to heterogeneous mesophyll, and vascular bundles with thin-walled bundle sheath cells. (G) TS with one row of vascular bundles. (H) TS stem with scattered vascular bundles and thin-walled ground-tissue cells. (I) TS root with multilayered velamen, O-thickened exodermal cell walls, banded-walled parenchymatous cortical cells, O-thickened endodermal cell walls, thin-walled pericycle cells, and vascular tissue embedded in sclerenchyma.

Subfamily Epidendroideae 125

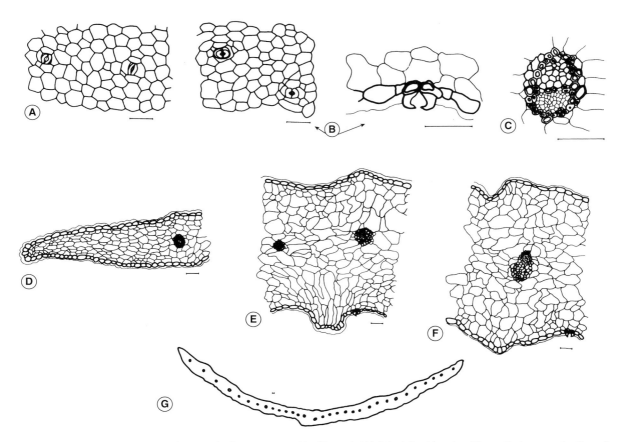

Fig. 69. Laeliinae. *Meiracyllium trinasutum* leaf; stomata amphiepidermal. (A) Adaxial epidermis with mostly hexagonal cells and tetracytic stomata. (B) Surface and TS abaxial epidermis with polygonal cells and tetracytic to anomocytic stomata. (C) TS collateral vascular bundle with thick- and thin-walled bundle sheath cells and sclerenchymatous ring surrounding vascular tissue. (D–F) TS with homogeneous mesophyll, one row of vascular bundles, and superficial stomata. (G) TS with one row of vascular bundles.

reflexa, and *Tetramicra*. Xylem and phloem capped by crescentiform mass of sclerenchyma (Figs 65I, 70H, 71H); phloem sclerenchyma always present, xylem sclerenchyma absent in some taxa (Fig. 70H); sclerenchyma completely surrounding vascular bundles in *Jacquiniella* and *Scaphyglottis reflexa*. **Stegmata** with conical, rough-surfaced silica bodies occurring in most taxa, associated only with phloem sclerenchyma, except in *Prosthechea radiata* where also next to xylem sclerenchyma.

Root TS
Velamen mostly 3–6 cells wide, ranging from three cells in *Cattleya forbesii*, *Leptotes*, and *Pseudolaelia* to 5–9 cells in *Arpophyllum* (Fig. 65J), 8–10 cells in *Laelia anceps*, and 3–10 cells in *Prosthechea boothiana* (Fig. 66I). Cell walls thin, except moderately thick to thick in *Barkeria*, *Brassavola*, *Caularthron*, *Laelia lyonsii*, *Nidema*, *Pseudolaelia*, and *Psychilis*. Cell wall thickenings various, ranging from branched anastomosing tenuous strips of cell wall material (Figs 65J, 71I) to broad bands in *Arpophyllum* and *Nidema* and ladder-like in *Orleanesia*. **Tilosomes**: spongy in *Arpophyllum*, lamellate in some *Encyclia*, *Epidendrum*, and *Prosthechea* species, and in *Laelia anceps*; also noted in *Quisqueya* and *Cattleya sincorana*. **Exodermis**: cells anticlinal (Figs 63F, 65J) or ±isodiametric (Fig. 71I); cell wall thickenings various (Figs 64A, 65J, 67J, 68I, 70I). Passage cells intermittent, thin-walled with large nuclei. **Cortex** stratified, from 4–6 cells wide in *Caularthron* up to 16–19 cells in *Epidendrum nocturnum*. Cells thin-walled, banded (Figs 64B, 71I); outer and inner layers of cells smaller, isodiametric or polygonal, outer layer cells

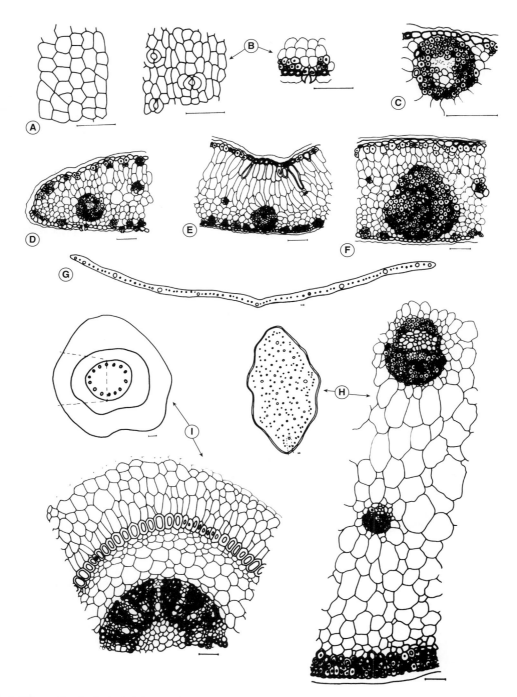

Fig. 70. Laeliinae. *Scaphyglottis imbricata*. (A–G) Leaf. (A) Adaxial surface with polygonal cells with mostly straight walls. (B) Surface and TS abaxial epidermis with polygonal cells, curvilinear walls, tetracytic and anomocytic stomata. (C) TS vascular bundle with ring of sclerenchyma surrounding vascular tissue. (D–F) TS with thick-walled epidermal cells in part, ad-, abaxial, and mid-mesophyll fibre bundles, heterogeneous and homogeneous mesophyll portions, vascular bundles surrounded by several layers of sclerenchyma. (G) TS lamina with one row of vascular bundles. (H) TS stem showing thick cuticle, several-layered subepidermal sclerenchyma band surrounding culm, scattered vascular bundles with thin-walled bundle sheath cells, thin-walled parenchymatous ground-tissue cells. (I) TS root with several-layered velamen, exodermal cells with O-thickened walls, parenchymatous cortex, endodermal cells with O-thickened walls, vascular tissue embedded in sclerenchyma, thin-walled parenchymatous pith cells.

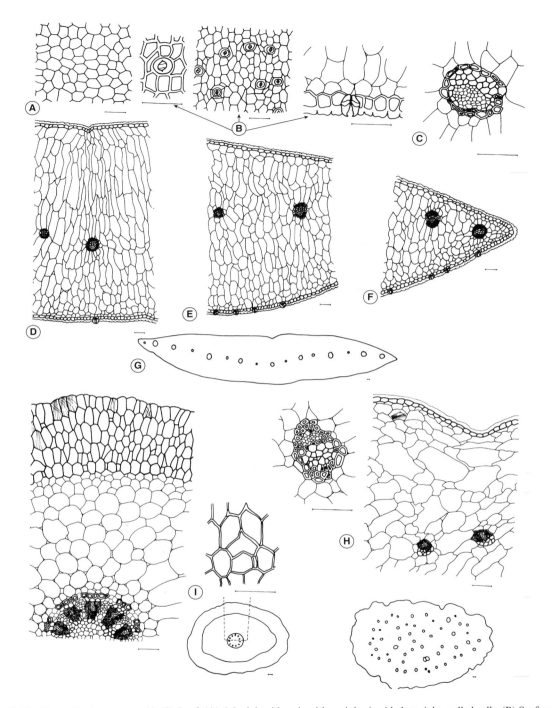

Fig. 71. Laeliinae. *Cattleya cernua*. (A–G) Leaf. (A) Adaxial epidermis with mainly six-sided straight-walled cells. (B) Surface and TS abaxial epidermis with anomocytic stomata. (C) TS vascular bundle with thick-walled bundle sheath cells. (D–F) TS with thick cuticle, superficial stomata, heterogeneous/homogeneous mesophyll, and one row of vascular bundles having ad- and abaxial sclerenchyma. (G) TS with one row of vascular bundles. (H) TS stem with thick cuticle, scattered vascular bundles with ad- and abaxial sclerenchyma embedded in parenchymatous ground tissue, and irregularly shaped lacunae in ground tissue. (I) TS root with many-layered velamen cells with slender spiral bands, thin-walled exodermal cell walls, O-thickened endodermal cell walls, thin-walled pericycle cells, parenchymatous cortex, vascular tissue surrounded by sclerenchyma, and parenchymatous pith.

lacking intercellular spaces; middle layer cells larger, oval and circular with numerous, triangular intercellular spaces; masses of dead hyphae common and with living pelotons in *Leptotes* and *Domingoa*. Middle layer cells with broad, birefringent bands of cell wall material in *Arpophyllum* (Fig. 64B), *Guarianthe skinneri*, *Laelia*, *Meiracyllium*, *Rhyncholaelia*, and *Scaphyglottis reflexa*; narrow bands in *Caularthron*, ladder-like bands in *Psychilis*; empty water-storage cells with birefringent cell walls occurring in *Epidendrum nocturnum*, *Nidema*, *Orleanesia*, *Psychilis*, *Quisqueya*, *Rhyncholaelia*, *Scaphyglottis coriacea*, and *Sophronitis* cf. *cernua*. **Endodermal** and **pericyclic** cells variously oriented and cell walls of both tissues O-thickened opposite phloem and thin opposite xylem (Figs 68I, 70I, 71I); endodermal cells usually isodiametric, anticlinal in *Barkeria* and *Nidema*; O-thickened; thin-walled throughout in *Barkeria*, *Guarianthe skinneri*, *Laelia anceps*, *Pseudolaelia*, and *Cattleya sincorana*; ∩-thickened in *Arpophyllum* (Fig. 65J,K). Pericycle cells usually resembling endodermal cells. **Vascular cylinder**: xylem poles ranging from seven in *Encyclia* to 24 in *Arpophyllum* (Fig. 65J) and *Laelia lyonsii*. Vascular tissue embedded in thin- or thick-walled (Figs 64C, 65J, 68I, 70I, 71I) sclerenchyma in all taxa except surrounded by parenchyma in *Barkeria*, *Epidendrum anceps*, and *Pseudolaelia*. **Pith** cells parenchymatous, thin-walled (Figs 65J, 67J, 70I, 71I), circular to oval; cell walls moderately thick in *Quisqueya*, polygonal and thick-walled in *Scaphyglottis reflexa* and *Rhyncholaelia*; sclerotic, angular in *Encyclia*. Starch granules occurring in *Epidendrum nocturnum*; cells with 1–3 large, cruciate starch granules occupying entire lumen in *Leptotes*.

Material examined

Arpophyllum (1), *Barkeria* (1), *Brassavola* (1), *Broughtonia* (1), *Cattleya* (4, 1 sp. now *Guarianthe*), *Caularthron* (1), *Constantia* (1), *Dimerandra* (1), *Domingoa* (1, was *Nageliella*), *Encyclia* (1), *Epidendrum* (2), *Guarianthe* (1, was *Cattleya skinneri*), *Homalopetalum* (1), *Jacquiniella* (1), *Laelia* (2), *Leptotes* (1), *Meiracyllium* (1), *Nidema* (1), *Orleanesia* (1), *Prosthechea* (2), *Pseudolaelia* (1), *Psychilis* (1), *Quisqueya* (1), *Rhyncholaelia* (1), *Scaphyglottis* (4), *Sophronitis* (2), *Tetramicra* (1). Full details appear in Stern and Carlsward (2009).

Additional information from Baker (1972)

Names of several taxa representing genera described by Baker (1972) have been changed by recent studies, e.g. *Dilomilis* to *Epidendrum*, *Dithonea* to *Epidendrum*, *Epidanthus* to *Epidendrum*, *Hexisea* to *Scaphyglottis*, *Hoffmannseggella* to *Laelia*, *Nidema* to *Epidendrum*, *Platyglottis* to *Scaphyglottis*, *Reichenbachanthus* to *Scaphyglottis*, *Schomburgkia* to *Laelia*, and *Sophronitella* to *Isabelia*.

Leaf material examined, in addition to genera described above: *Alamania* (1), *Artorima* (1), *Dinema* (1, as *Encyclia polybulbon*), *Hagsatera* (1, as *Encyclia brachycolumna*), *Isabelia* (2), *Microepidendrum* (1, as *Encyclia subulatifolia*), *Myrmecophila* (1, as *Schomburgkia wendlandii*).

Leaf surface

Hairs sparse or occasional, adaxial in *Alamania*, *Artorima*, *Dinema*, *Hagsatera*, adaxial or absent in *Isabelia*, on both surfaces in *Microepidendrum* and *Myrmecophila*; hairs usually 3-celled, stalk sunken in epidermis. **Epidermis**: cells irregularly polygonal in most taxa. **Stomata** abaxial, except on both surfaces in *Microepidendrum*; with 4–6 surrounding cells, little differentiated from other epidermal cells.

Leaf TS

Cuticle thin to moderately thick, granular and warty in *Dinema*, papillose to tuberculate in *Microepidendrum*; moderately thick, ridged in *Alamania*, *Artorima* (abaxial), *Hagsatera*, beaded in *Isabelia*, thick and smooth in *Artorima* (adaxial), thick and papillose in *Myrmecophila*. **Hypodermis**: adaxial: 1-layered in *Dinema* and *Hagsatera*, 1–3-layered, consisting of water-storage cells in *Alamania*, two or three layers of water-storage cells in *Artorima*, 2–4 layers of water-storage cells in *Microepidendrum*, one to several layers in *Isabelia*, one layer of reticulately thickened cells in *Myrmecophila*; abaxial: one layer of reticulately thickened cells in *Alamania*, one layer of unthickened cells in *Artorima*, one layer in *Dinema*, *Hagsatera*, *Isabelia*, and *Microepidendrum*, one or two layers of reticulately thickened cells in *Myrmecophila*. **Mesophyll** homogeneous in all taxa, with scattered spirally thickened cells in *Alamania* and often in *Microepidendrum*. **Fibre bundles** in one row (occasionally two) in abaxial mesophyll in *Artorima*, one abaxial subhypodermal row in *Hagsatera* (plus occasional adaxial bundles), *Dinema*, and *Microepidendrum*, one abaxial arc in *Isabelia pulchella*, one adaxial and one abaxial row in *Myrmecophila*, absent in *Isabelia virginalis* and *Alamania*. Raphides present in all taxa, mostly hypodermal or subhypodermal, in mesophyll in *Isabelia*, few in *Microepidendrum*. Spindle-shaped crystals and/or druses recorded in all taxa except *Myrmecophila*. **Stegmata** with conical silica bodies in all taxa.

Reports from the literature

Baker (1972): leaf anatomy of Laeliinae and its bearing on relationships of the group, including *Alamania, Artorima, Dinema* (as *Encyclia polybulbon*), *Isabelia, Microepidendrum* (as *Encyclia subulatifolium*), *Myrmecophila* (as *Schomburgkia wendlandii*), in addition to the genera listed above. Chatin (1856, 1857): vegetative anatomy of *Brassavola, Cattleya, Epidendrum,* and *Laelia* (also as *Schomburgkia*). Dressler (1993): list of taxa within Laeliinae, description of subtribe, discussion of related taxa, biology of the taxa. Meinecke (1894): aerial root anatomy of *Arpophyllum, Barkeria, Brassavola, Broughtonia, Cattleya, Epidendrum, Guarianthe* (as *Cattleya skinneri*), *Laelia,* and *Leptotes*. Möbius (1887): leaf anatomy of *Arpophyllum, Cattleya, Epidendrum* (and as *Hormidium*), *Laelia* (and as *Schomburgkia*), *Psychilus* (as *Encyclia atropurpureum*), *Rhyncholaelia* (as *Brassavola*), *Scaphyglottis,* and *Sophronitis*. Noguera-Savelli and Jauregui (2011): leaf anatomy of *Brassavola* and other species of Laeliinae. Oliveira and Sajo (1999, 2001): leaf and stem anatomy of *Encyclia* and *Epidendrum*. Porembski and Barthlott (1988): velamen of many genera of Laeliinae. Pridgeon (1987): velamen and exodermis of many genera of Laeliinae. Pridgeon et al. (1983): tilosomes of many genera of Laeliinae. Pridgeon et al. (2005): detailed discussion of Laeliinae: history, biology, systematics, taxonomy, phylogenetics. Solereder and Meyer (1930): review of systematic anatomy of orchids, including original data on *Brassavola, Cattleya, Epidendrum,* and *Laelia*. Stern and Carlsward (2009): vegetative anatomy of Laeliinae. Van den Berg et al. (2005): molecular analysis of phylogenetic relationships within Epidendroideae. Van den Berg et al. (2009): phylogenetic study of Laeliinae involving DNA sequences. Weltz (1897): anatomy of axis of *Arpophyllum, Brassavola, Cattleya, Epidendrum, Laelia* (and as *Schomburgkia*), *Scaphyglottis,* and *Sophronitis*. Williams (1979): stomata of many genera of Laeliinae.

Additional literature

Barthlott and Capesius (1975) [1976]: velamen of *Cattleya, Epidendrum,* and *Laelia*. Benzing (1982): mycorrhiza of *Encyclia* and *Epidendrum*. Benzing and Pridgeon (1983): leaf trichomes of *Encyclia*. Benzing et al. (1983): root anatomy of *Encyclia* and *Epidendrum*. Bernard (1904): mycorrhiza of *Brassavola, Cattleya,* and *Laelia*. Bloch (1935): root anatomy of *Cattleya* and *Epidendrum*. Bonates (1993): leaf anatomy of *Brassavola, Cattleya, Encyclia,* and *Epidendrum*. Bonnier (1903a,b): aerial root anatomy of *Cattleya* and *Laelia*. Burgeff (1909): mycorrhiza of *Epidendrum* and *Laeliocattleya*. Campos Leite and Oliveira (1987): leaf anatomy of *Cattleya*. Capesius and Barthlott (1975): velamen of *Cattleya* and *Epidendrum*. Dannecker (1898): anatomy of *Caularthron* (as *Diacrium bicornutum*) and *Myrmecophila* (as *Schomburgkia humboldtii*). Dexheimer and Serrigny (1983): mycorrhiza of *Epidendrum ibaguense*. Duruz (1960): stomata of *Cattleya*. Dycus and Knudson (1957): velamen of *Cattleya, Epidendrum,* and *Laelia*. Engard (1944): root anatomy of *Brassavola, Cattleya,* and *Epidendrum*. Garcia Cruz and Hágsater (1998): leaf anatomy of *Epidendrum anisatum* group. Janczewski (1885): root anatomy of *Epidendrum*. Khasim and Mohana Rao (1986): leaf and root anatomy of *Epidendrum*. Kohl (1889): silica bodies in *Epidendrum*. Kraft (1949): velamen of *Laelia*. Kroemer (1903): root anatomy of *Epidendrum*. Krüger (1883): aerial organs of *Brassavola, Cattleya, Epidendrum,* and *Laelia* (and as *Schomburgkia*). Kuttelwascher (1964, 1965): root anatomy of *Brassavola* and *Laeliocattleya*. Leitgeb (1864b): aerial root anatomy of *Arpophyllum, Cattleya,* and *Epidendrum*. Metzler (1924): leaf anatomy of *Brassavola, Cattleya, Epidendrum, Laelia,* and *Sophronitis*. Moreau (1913): pseudobulb anatomy of *Cattleya, Epidendrum,* and *Laelia*. Moreira et al. (2009): leaf and root anatomy of *Epidendrum secundum*. Mulay et al. (1958): velamen of *Epidendrum*. Napp-Zinn (1953): aerial root anatomy of *Cattleya, Epidendrum, Laelia,* and *Laeliocattleya*. Neubauer (1978): idioblasts in inflorescence axes of *Arpophyllum, Cattleya,* and *Epidendrum*. Oliveira Pires et al. (2003): leaf and root anatomy of *Prosthechea* separated from *Encyclia*. Olivier (1881): root anatomy of *Epidendrum*. Oudemans (1861): aerial root anatomy of *Cattleya, Encyclia,* and *Epidendrum*. Parrilla Diaz and Ackerman (1990): root anatomy of *Domingoa, Encyclia, Epidendrum, Jacquiniella, Scaphyglottis,* and *Sophronitis*. Prasad (1960): velamen of *Epidendrum*. Rŭgina et al. (1987): leaf and root anatomy of *Cattleya* and *Epidendrum*. Rüter and Stern (1994): root anatomy of *Brassavola, Cattleya,* and *Laelia* (as *Schomburgkia*). Schimper (1884): root anatomy of *Brassavola* and *Epidendrum*. Silva et al. (2006): leaf anatomy of *Epidendrum*. Stancato et al. (1999): stomata of *Epidendrum* species. Staudermann (1924): trichomes of *Laelia*. Wallach (1938–9): aerial root anatomy of *Laelia*. Widholzer and Oliveira (1994): leaf anatomy of *Sophronitis coccinea*. Zanenga Godoy and Costa (2003): leaf anatomy of four *Cattleya* species. Zankowski et al. (1987): aerial root of *Epidendrum*.

Taxonomic notes

Laeliinae have been considered differently by different authors since originally described by Bentham in 1881.

An extensive review of the various proposals is included in Pridgeon et al. (2005) and as a result several genera included by Dressler (1993) in this subtribe have been excluded in order to treat Laeliinae as a monophyletic entity. Accordingly, those excluded genera have been eliminated from the anatomical treatment above. Stern and Carlsward (2009) have considered Laeliinae to contain an anatomical medley of features, the subtribe lacking overall anatomical uniformity: (1) fibre bundles occur in leaves of almost all taxa but are lacking in leaves of *Domingoa*, *Encyclia*, *Homalopetalum*, and *Meiracyllium*; (2) hypodermis occurs in leaves of several taxa: *Arpophyllum*, *Brassavola*, *Broughtonia*, *Cattleya*, *Guarianthe*, *Jacquiniella*, *Laelia anceps*, *Leptotes*, *Meiracyllium*, *Rhyncholaelia*, *Scaphyglottis*, and *Sophronitis* cf. *cernua* but is absent in *Epidendrum*, *Nidema*, *Orleanesia*, *Prosthechea*, *Pseudolaelia*, *Psychilis*, *Quisqueya*, and *Tetramicra*; (3) mesophyll is mostly homogeneous, but heterogeneous in *Arpophyllum*, *Brassavola*, *Cattleya*, *Dimerandra*, *Laelia anceps*, *Rhyncholaelia*, *Scaphyglottis coriacea*, *S. imbricata*, *S. reflexa*, and *Tetramicra*; (4) stegmata are associated with foliar vascular bundle sclerenchyma in all taxa except *Broughtonia*, *Dimerandra*, *Homalopetalum*, *Psychilis*, *Quisqueya*, and *Scaphyglottis coriacea*; (5) stegmata are associated with fibre bundle sclerenchyma except in *Homalopetalum* and *Scaphyglottis imbricata*; (6) stegmata occur with cauline vascular bundle sclerenchyma except in *Broughtonia*, *Homalopetalum*, *Jacquiniella*, and *Tetramicra*; (7) exodermal cells are thin-walled in *Barkeria*, *Broughtonia*, *Domingoa*, *Guarianthe skinneri*, *Laelia anceps*, *Prosthechea radiata*, and *Pseudolaelia*; in all other taxa exodermal cells are variously thick-walled; (8) stomata occur only on abaxial surfaces of leaves except on both surfaces in *Barkeria*, *Broughtonia*, *Domingoa*, *Meiracyllium*, and *Pseudolaelia*. Baker (1972) could not locate clearly differentiated subsidiary cells, but they are distinct in many taxa. He indicated hairs were frequent on leaves of Laeliinae, but we failed to find these in most taxa. Perhaps hairs are detached during processing or are ephemeral in mature leaves.

In line with these inconsistencies, Pridgeon et al. (2005) stated that 'relationships among the main groups in Laeliinae are still unresolved.' Thus, it may be necessary to reconsider Laeliinae further and to exclude or add genera from/to Laeliinae based on anatomical studies as has been done by Pridgeon et al. (2005) using molecular studies. Indeed, further molecular and anatomical research has been carried out by van den Berg et al. (2009): who discriminated among generic alliances within Laeliinae to segregate *Isabelia*, *Domingoa*, *Encyclia*, *Scaphyglottis*, *Broughtonia*, and *Cattleya* alliances. Because of the great variations in morphological features, the large numbers of species, and infrageneric groupings, further research involving pollination mechanisms, habitat preferences, and biogeographic patterns needs to be undertaken (van den Berg et al. 2009).

Subtribe Pleurothallidinae Lind. ex G. Don

Pleurothallids are mostly diminutive New World epiphytes and lithophytes with single conduplicate leaves borne on non-pseudobulbous aerial stems (ramicauls) produced from short or elongate rhizomes. These plants are mostly denizens of deeply wooded, moist, mossy, shadowy forests usually at cooler elevations in the tropics and subtropics. Some, however, inhabit seasonally dry regions. There are 37 genera and more than 4000 species distributed from southern Florida and Caribbean islands through Mexico to southern Brazil and Argentina. The Andes of Colombia, Ecuador, and Peru and the cloud forests of Costa Rica and Panama are the most species-rich regions. The most spectacular of the pleurothallids for culture are the masdevallias, specifically *Masdevallia coccinea*, *M. limax*, *M. racemosa*, *M. uniflora*, and *M. veitchiana*. Some other pleurothallids are also desirable for culture owing to the jewel-like quality of the tiny flowers.

Leaf surface

Hairs variable; present or absent; glandular, sunken, and ephemeral; mucilage hairs (Fig. 72A) 1-, 2-, 3-, or 4-celled, uniseriate; ad- and/or abaxial in different species; laminar papillae only in *Dresslerella* (Fig. 72C). **Epidermis**: cells polygonal, tetragonal, pentagonal, hexagonal, wall thickness variable; radial walls extremely thick in *Octomeria*, moderately thick in *Stelis* (as *Physothallis*) and some species of *Pleurothallis*; thin in *Brachionidium*. **Stomata** abaxial; 2–7 subsidiary cells in most taxa (Fig. 72B), 4–8 subsidiary cells in *Dresslerella* and *Echinosepala*; usually tetracytic or anisocytic; cyclocytic in *Dresslerella* and *Echinosepala*. Coralloid raphide clusters in epidermal cells of some *Pleurothallis* species and oil droplets in most *Platystele* species.

Leaf TS

Cuticle <3 μm thick in *Brachionidium*, 6–14 μm thick in *Echinosepala* and *Stelis*, up to 22 μm thick in *Octomeria costaricensis* (as *O. valerioi*). **Epidermis**: cells elliptical, rectangular, isodiametric; domed in *Echinosepala* and *Myoxanthus exasperatus* (as *Pleurothallis peduncularis*);

Subfamily Epidendroideae 131

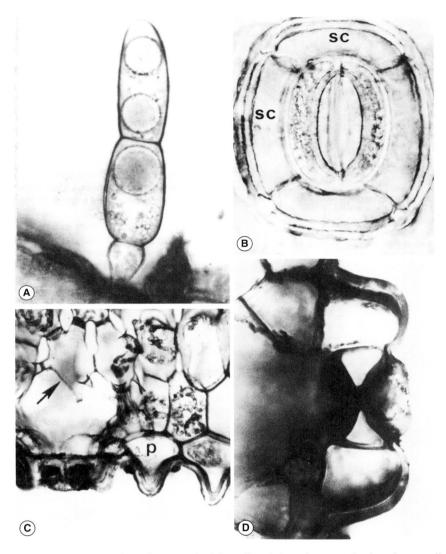

Fig. 72. **Pleurothallidinae leaf.** (A) *Acianthera glumacea*, abaxial mucilage hair on immature lamina; three mucilage drops visible, ×1000. (B) *Echinosepala pan*, abaxial stoma in surface view, four subsidiary cells (sc), ×875. (C) *Dresslerella hirsutissima*, TS abaxial epidermis and hypodermis; note papilla (p) with thickened outer tangential wall, elevated stoma (left), and stellately lobed mesophyll cell (arrow) bordering substomatal chamber, ×490. (D) *D. powellii*, TS abaxial stoma with subsidiary cells and elevated stomatal apparatus, ×1,250. Reprinted from Pridgeon (1982a).

marginal cells in *Octomeria* and *Restrepia* anticlinally rectangular, domed, or triangular. **Stomata** superficial, except raised 28 μm in *Dresslerella caesariata* and raised in other species of *Dresslerella* (Fig. 72C,D) and in *Echinosepala*. Adaxial **hypodermis** ranging up to 11 layers; at least two layers in most genera; 1-layered in *Neocogniauxia*; absent in *Brachionidium* and *Dracula*. Abaxial hypodermis 1-layered in most taxa (Fig. 73C),

except absent in *Brachionidium* and *Dracula*. Birefringent annular helical thickenings occurring in hypodermal cells of most taxa and in mesophyll idioblasts. **Fibre bundles** absent, except at leaf margins in *Barbosella australis*. **Mesophyll**: heterogeneous (Figs 72C, 73C), clearly differentiated in many genera; in other genera mesophyll homogeneous (Fig. 73A), abaxial chlorenchyma lacunose with stellately lobed (Fig. 72C)

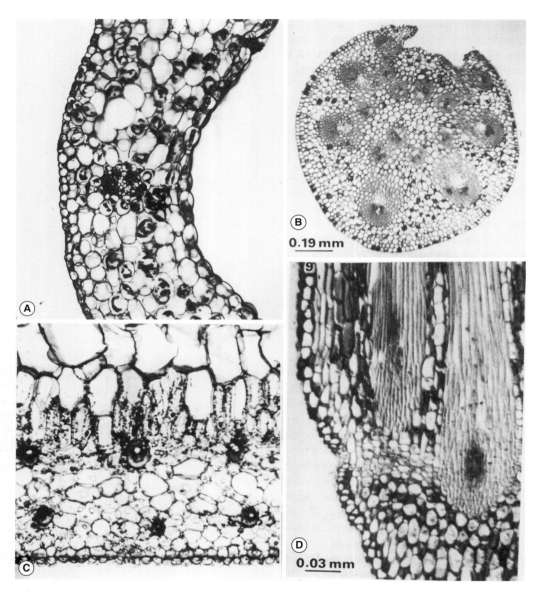

Fig. 73. Pleurothallidinae. (A) *Brachionidium* species, TS lamina through midrib, homogeneous mesophyll, and lack of distinct hypodermis, ×125. (B) *Pabstiella tripterantha*, TS stem with two series of vascular bundles each one surrounded by sclerenchyma, medullary and cortical parenchyma conterminous, smaller bundles migrated into medullary region of ramicaul, and adaxial invagination is evident. Scale bar = 0.19 mm. (C) *Octomeria costaricensis*, TS leaf, showing adaxial hypodermis, palisade mesophyll, and two rows of vascular bundles, ×140. (D) *Zootrophion hypodiscus*, LS stem with axially oriented rows of meristematic cells and intact vascular tissue passing through meristematic region. Scale bar = 0.03 mm. (A,C) Reprinted from Pridgeon (1982a); (B,D) reprinted from Stern et al. (1985).

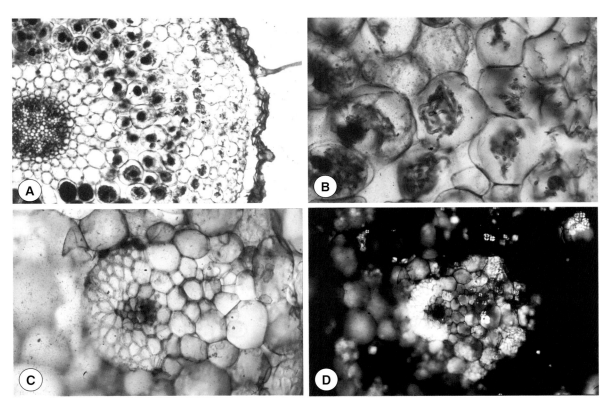

Plate 1. (A,B) Cranichideae. *Goodyera pubescens*, TS root. (A) One-layered velamen bearing root hairs, dead mycorrhizal masses in cortical cells, and 7-arch vascular cylinder. (B) Thin-walled exodermal cells and hyphal pelotons in cortical cells. **(C,D) Catasetinae**. *Clowesia glaucoglossa*, TS pseudobulb. (C) Starch granules in assimilatory cells and collateral vascular bundle. (D) Starch granules in assimilatory cells under polarized light. A × c.120, B,C × c.360, D × c.240. Photographs by W.L. Stern.

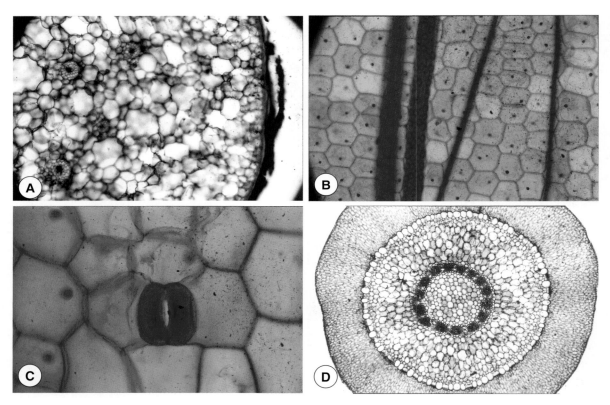

Plate 2. (A–C) Catasetinae. (A) *Clowesia glaucoglossa*, TS pseudobulb with assimilatory and water-storage cells. (B) *Catasetum planiceps*, adaxial leaf epidermis with polygonal cells and fibres with conical silica bodies. (C) *Catasetum pileatum*, abaxial leaf epidermis with tetracytic stomatal apparatus. **(D) Eulophiinae.** *Eulophia callichroma*, TS root with wide velamen, thin-walled exodermal cells, broad cortex, thin-walled endodermal cells, 16-arch vascular cylinder, and thick-walled pith cells. A × c.100, B × c.160, C × c.370, D × c.80. Photographs by W.L. Stern.

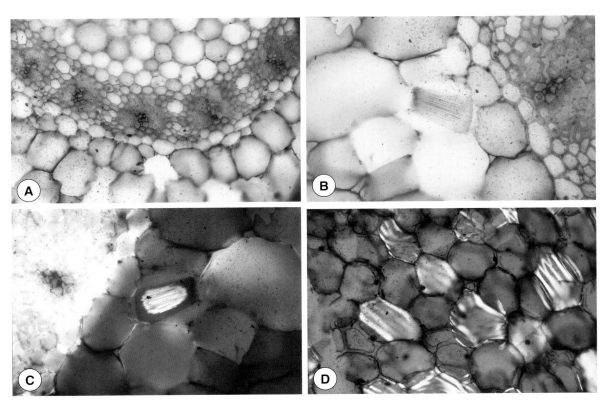

Plate 3. (A–C) Eulophiinae. *Eulophia callichroma*, TS root. (A) O-thickened endodermal cell walls and vascular tissue surrounded by sclerenchyma. (B) Raphide bundle in cortical cell and thick-walled endodermal cells opposite phloem. (C) Raphide bundle under polarized light. **(D) Angraecinae.** *Dendrophylax* species, TS leaf with banded mesophyll cell walls under polarized light. A × c.200, B–D × c.300. Photographs by W.L. Stern.

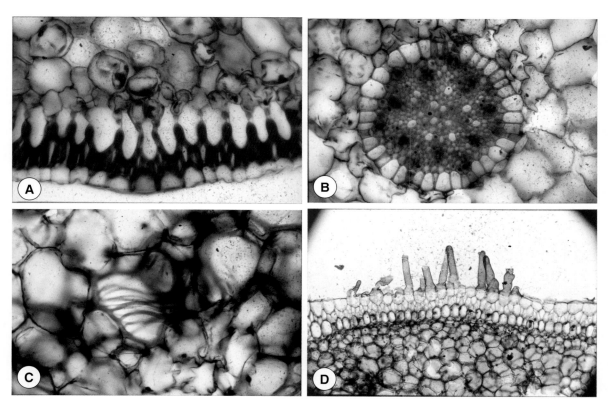

Plate 4. Angraecinae. (A) *Microcoelia macrantha*, TS root with 1-layered velamen and ∩-thickened exodermal cell walls. (B) *Campylocentrum pachyrrhizum*, TS root with thin-walled cortical cells and 12-arch vascular system surrounded by endodermis. (C,D) *Dendrophylax lindenii*. (C) TS leaf with banded mesophyll cell walls. (D) TS root with 2-layered velamen bearing root hairs. A,B × *c*.400, C × *c*.600, D × *c*.100. Photographs by W.L. Stern.

and/or isodiametric cells; water-storage cells in *Neocogniauxia*. **Vascular bundles** of several size classes, with one row at juncture of palisade and spongy mesophyll (Fig. 73C) or aligned in uppermost layers of chlorenchyma in homogeneous mesophyll; larger bundles typically with a sclerenchyma sheath, many with a very narrow layer of sclerenchyma or none at all at adaxial pole. One-layered bundle sheath of chlorenchyma cells may surround sclerenchyma. **Stegmata** with conical silica bodies associated with vascular sclerenchyma noted in *Myoxanthus* only. As in most Epidendreae, conical silica bodies most likely occur throughout the subtribe in association with foliar bundle sheaths and fibres in the stem (A. Pridgeon 2012, personal communication).

Stem TS

Stem single, unbranched, slender, called a ramicaul (formerly called a secondary stem); inflorescence produced toward its apex. Conspicuous ring or annulus encircling stem just proximal to emergence of inflorescence in many species; ring associated internally with a disc of tissue.

Cuticle variable, thick to very thick in some species, thin to moderately thin in others. **Epidermis** 1-layered (Fig. 73B); a distinctive hypodermis present or not; hypodermis 1- or 2-layered in *Pabstiella tripterantha* (as *Pleurothallis tripterantha*). **Cortex** parenchymatous, outermost cells sometimes sclerified. Number of **vascular bundles** in ramicauls a function of diameter of ramicaul; slender stems, as in *Dryadella elata*, having three or four large bundles and four or five smaller bundles; relatively stout stems, as in *Myoxanthus exasperatus* (as *Pleurothallis peduncularis*), having up to 75 bundles in internode, dispersed in two or more rings. Outer ring of nine bundles surrounding central cluster of nine bundles in *Neocogniauxia*. Vascular bundles always associated with sclerenchyma (Fig. 73B), as a concentric ring of tissue separating cortex and pith in which individual bundles embedded, or as discrete sheaths surrounding each individual bundle with cortex and pith merging between them (Fig. 73D). Tissues associated with the several-layered annular plate cutting through ramicaul as a somewhat lignified disc of cells; vascular bundles passing through disc unaltered and uninterrupted (Fig. 73D). **Pith** parenchymatous. **Stegmata** with conical silica bodies occurring along vascular fibres of *Myoxanthus*.

Root TS

Velamen 1-layered in *Phloeophila* and *Zootrophion* (as *Cryptophoranthus*), *Dryadella*, *Stelis* (as *Physosiphon*), *Platystele*, and *Trisetella*; 2-layered in *Dresslerella*, *Octomeria*, and *Stelis* species; 2–4(–5)-layered in other taxa (Fig. 74A). Epivelamen cells relatively small, elliptical to polygonal; endovelamen cells elliptical to rectangular or polygonal, elongated anticlinally (Fig. 74A). **Tilosomes**: discoid in *Zootrophion* species (as *Cryptophoranthus*), lamellate in a few *Anathallis* (as *Pleurothallis*), *Scaphosepalum*, and *Stelis* species, meshed in *Stelis batillacea* (as *Pleurothallis batillacea*), spongy in some *Pleurothallis* and *Stelis* species. **Exodermis** and **endodermis** 1-layered; exodermal cells mostly U-thickened; endodermal cells mostly O-thickened (Fig. 74C) opposite phloem and unthickened opposite xylem. Number of protoxylem poles varying from four in *Trisetella* and 4–6 in *Dresslerella* to 12–16 in *Echinosepala* and 9–20 in *Myoxanthus* (Fig. 74B) and *Restrepiella* (Fig. 74A). **Pith** parenchymatous in *Echinosepala*, *Neocogniauxia*, *Octomeria*, *Myoxanthus exasperatus*, and *Restrepiella*, and sclerotic in most other genera.

Material examined

Two figures are given, the first refers to species in Pridgeon (1982a,b), the second to species in Stern, Pridgeon et al. (1985, stem), where full details appear.

Acostaea (1,1, included in *Specklinia* in Pridgeon et al. 2005), *Barbosella* (5,2), *Brachionidium* (1,3), *Condylago* (1,1, now = *Stelis*), *Cryptophoranthus* (6,0, now = *Phloeophila* and *Zootrophion*), *Diodonopsis* (1, in *Masdevallia*), *Dracula* (6,1), *Dresslerella* (8,3), *Dryadella* (3,4), *Lepanthes* (1,9), *Lepanthopsis* (1,4), *Masdevallia* (20,5 including 1 sp. now in *Diodonopsis* and 1 in *Phloeophila*), *Myoxanthus* (13, see Pridgeon and Stern 1982), *Neocogniauxia* (1, see Stern and Carlsward 2009), *Octomeria* (3,3), *Ophidion* (0,1 + 3 now in *Phloeophila*), *Phloeophila* (species described as *Cryptophoranthus*, *Masdevallia*, *Ophidion*, some *Pleurothallis* spp.), *Physosiphon* (3,1, now = *Acianthera* or *Stelis*), *Physothallis* (2,1, now = *Stelis*), *Platystele* (6,2), *Pleurothallis* (99,28, including species now in *Acianthera*, *Anathallis*, *Echinosepala*, *Pabstiella*, *Trichosalpinx*, and some *Myoxanthus*, *Phloeophila*, *Specklinia*, and *Stelis* spp.), *Porroglossum* (4,3), *Restrepia* (3,3), *Restrepiella* (1,1), *Restrepiopsis* (2,1, now = *Pleurothallopsis*), *Salpistele* (1,1, now = *Stelis*), *Sarracenella* (0,1, now = *Acianthera*), *Scaphosepalum* (9,3), *Stelis* (4,8, also *Condylago*, *Physosiphon*, *Physothallis*, and *Salpistele* spp.), *Trichosalpinx* (0,5), *Trisetella* (2,1), *Zootrophion* (0,2).

Reports from the literature

Brieger, Butzin, and Senghas (1975): classification of Pleurothallidinae. Dressler (1993): phylogeny and classification of orchids. Hünecke (1904): early analysis of pleurothallid

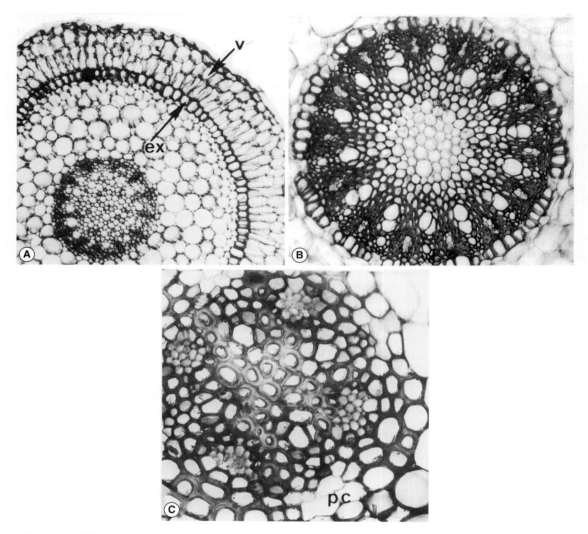

Fig. 74. Pleurothallidinae root. (A) *Restrepiella ophiocephala*, TS with multilayered velamen (v), exodermal cells (ex) with U-thickened and O-thickened walls, 13-arch vascular cylinder, ×140. (B) *Myoxanthus affinis*, TS with 14-arch vascular cylinder and parenchymatous pith, ×225. (C) *Pleurothallis* species, TS with O-thickened endodermal cell walls (pc = pericycle) and thick-walled pith cells, ×840. Reprinted from Pridgeon (1982a).

anatomy. Luer (1986): systematics of Pleurothallidinae and key to the genera. Neyland, Urbatsch, and Pridgeon (1995): phylogenetic analysis of Pleurothallidinae. Pridgeon (1981b): shoot anatomy of two species of *Dresslerella*. Pridgeon (1982a): diagnostic features among pleurothallids. Pridgeon (1982b): numerical review of pleurothallids and its significance in classification. Pridgeon, Solano, and Chase (2001): phylogenetic relationships in Pleurothallidinae from DNA sequences. Pridgeon et al. (2005): comprehensive source of information on all genera of Pleurothallidinae. Pridgeon and Stern (1982): vegetative anatomy of *Myoxanthus*. Pridgeon and Williams (1979): anatomical study of *Dresslerella*. Solereder and Meyer (1930): comprehensive overview of orchid anatomy and notes on leaf and axis anatomy of *Pleurothallis*. Stern and Carlsward (2009): vegetative anatomy of *Neocogniauxia*. Stern and Pridgeon (1984): refutation of term 'secondary stem' for branches of rhizome in pleurothallids. Stern, Pridgeon, and Luer (1985): relationship between stem structure and systematics of Pleurothallidinae.

Additional literature

Baker (1972): leaf anatomy of *Dilomilis*, *Neocogniauxia*, and *Zootrophion*. Benzing and Pridgeon (1983): trichomes of *Octomeria*, *Pleurothallis*, *Restrepia*, *Restrepiella*, *Scaphosepalum*, and *Stelis*. Chatin (1856, 1857): vegetative anatomy of *Lepanthes*, *Pleurothallis*, and *Physosiphon* (now = *Stelis*). Feldman and Alquini (1997, 1999): vegetative anatomy of *Pleurothallis*. Ferreira et al. (1994): leaf anatomy of *Octomeria*. Howard (1969): leaf epidermis of *Brachionidium*. Krüger (1883): anatomy of aerial organs of *Octomeria* and *Pleurothallis*. Leitgeb (1864b): aerial root anatomy of *Pleurothallis*. Löv (1926): leaf anatomy of *Masdevallia*. Macfarlane (1892): vegetative anatomy of *Masdevallia* and hybrids. Meinecke (1894): aerial root anatomy of *Cryptophoranthus* (now = *Phloeophila*), *Masdevallia*, *Octomeria*, *Physosiphon* (now = *Stelis*), *Pleurothallis*, *Restrepia*, and *Scaphosepalum*. Metzler (1924): leaf anatomy of *Masdevallia*, *Pleurothallis*, *Restrepia*, and *Stelis*. Möbius (1887): leaf anatomy of *Brachionidium*, *Lepanthes*, *Masdevallia*, *Octomeria*, *Pleurothallis*, *Restrepia*, and *Stelis* (including *Physosiphon*). Oliveira and Sajo (1999, 2001): leaf and stem anatomy of *Pleurothallis*. Oudemans (1861): aerial root of *Pleurothallis*. Parrilla Diaz and Ackerman (1990): root of *Brachionidium*, *Dilomilis*, *Lepanthes*, *Pleurothallis*, and *Stelis*. Pirwitz (1931): 'storage tracheids' in *Pleurothallis* and *Physosiphon*. Porembski and Barthlott (1988): velamen of *Barbosella*, *Dracula*, *Lepanthes*, *Masdevallia*, *Neocogniauxia*, *Octomeria*, *Pleurothallis*, *Restrepiella*, and *Stelis* (including *Physosiphon*). Pridgeon (1981a): trichomes of many pleurothallids. Pridgeon (1984): anatomy of *Restrepiella* compared with *Restrepiopsis* (now = *Pleurothallopsis*). Pridgeon (1987): velamen and exodermis of many genera of pleurothallids. Scatena and Nunes (1996): anatomy of *Pleurothallis rupestris* (now = *Acianthera teres*). Silva et al. (2006): leaf anatomy of *Pleurothallis*. Suárez et al. (2009): mycorrhiza of *Stelis*. Weltz (1897): axis anatomy of *Cryptophoranthus* (now = *Phloeophila*), *Masdevallia*, *Octomeria*, *Pleurothallis*, *Restrepia*, *Scaphosepalum*, and *Stelis* (including *Physosiphon*). Williams (1979): stomata of several pleurothallids.

Taxonomic notes

There is significant variation in character states among and within the genera of Pleurothallidinae (Pridgeon et al. 2005). *Pleurothallis*, for example, is far too diverse an assemblage to characterize anatomically as a single group with any consistency or meaning (Pridgeon 1982a,b). Pridgeon et al. (2005) note that there are only two non-molecular synapomorphies in Pleurothallidinae:

the stem annulus (Stern, Pridgeon, and Luer 1985) and the raised cyclocytic stomata of *Dresslerella* (Pridgeon and Williams 1979) and *Echinosepala*. Neyland et al. (1995), using morphological and anatomical characters, showed substantial homoplasy among the pleurothallid taxa they studied. Anatomy is not rewarding as a taxonomic guide to Pleurothallidinae.

Subtribe Ponerinae Pfitzer

Ponerinae are sympodial epiphytic or lithophytic herbs with bamboo- or grass-like foliage, slender, reed-like stems, and often fat roots. There are only three genera and about 22 species, widely distributed in tropical America but centred in Mexico and Guatemala where most species occur. *Nemaconia* is now included in *Ponera*. Anatomy of the three genera was described from original research (Stern and Carlsward 2009). Stem anatomy of *Isochilus* follows Weltz (1897). Leaf anatomy of *Ponera striata* is taken from Baker (1972).

Leaf surface

Hairs none, except adaxial in *Ponera striata*. **Cuticle** smooth, except papillate in *Isochilus* and bordered and 'pegged' in *P. striata*. **Epidermis**: adaxial and abaxial cells polygonal; rectangularly elongated parallel to veins in *Ponera*, radial walls plicate in *P. striata*. **Stomata** abaxial only, tetracytic; lateral subsidiary cells greatly flattened against guard cells in *Ponera juncifolia*.

Leaf TS

Cuticle 5.0–10.0 μm thick, 10–15 μm thick in *Ponera juncifolia*. **Epidermis**: cells rectangular, square to isodiametric. **Stomata** superficial, somewhat sunken in *Helleriella*, sunken in *P. striata*. **Hypodermis**: absent in *Isochilus*; adaxial, 1-layered in *Helleriella* and *P. juncifolia*, 2-layered in *P. striata*; cells lacking chloroplasts in *Helleriella* and *P. juncifolia*, chlorophyllous in *P. striata*; cells squarish in *Helleriella*; cells in *P. juncifolia* thick-walled, with isodiametric and elongated cells alternating irregularly, thin-walled cells nucleated. **Fibre bundles** absent. **Mesophyll** heterogeneous in *Ponera*, palisade cells adaxial, oval, thin-walled, 'spongy' cells irregularly shaped; homogeneous in *Isochilus*, and stratified in *Helleriella* with three layers; thick-walled fibre-like cells with anisotropic walls scattered among mesophyll cells in *Helleriella*, *Isochilus*, and *Ponera*. Raphide cells abaxial, druses abundant in *Ponera*. **Vascular bundles** collateral, in one row, with sclerenchyma bundle caps. Sclerenchyma more strongly developed at phloem pole. **Stegmata** with conical silica bodies associated with both xylem and phloem sclerenchyma.

Stem TS

Cuticle thin in *Isochilus*. **Epidermis**: cells isodiametric. **Hypodermis**: substantial, consisting of two or three layers of massive thick-walled rounded cells with intercellular spaces. Inner sclerenchyma ring consisting of 6–8 layers of narrow-lumened fibres. **Vascular bundles** collateral, partly embedded in sclerenchyma ring and also scattered in ground tissue; obvious phloem sheath in ground-tissue bundles, but no distinct xylem sheath, whereas noticeable xylem sheath present in bundles of sclerenchyma ring. **Stegmata** apparently present, associated with phloem sclerenchyma.

Root TS

Velamen eight cells wide in *Helleriella*, three or four cells wide in *Isochilus* and *Ponera striata* and three cells wide in *P. juncifolia*. Epivelamen cells isodiametric with wide bands in *Isochilus*, periclinal in *Helleriella* and *P. juncifolia*; endovelamen cells anticlinal in *Helleriella* and *Isochilus* and isodiametric in *P. juncifolia*. **Tilosomes** absent. **Exodermis**: cells anticlinal, cell walls ∩-thickened in *Helleriella*; outer cell walls thickened in *P. striata*; cells isodiametric, all walls thin in *Isochilus* and *P. juncifolia*; passage cells intermittent. **Cortex** >26 cells wide in *Helleriella* and 15–18 cells in *P. juncifolia*; cells thin-walled in *Helleriella*, *Isochilus*, and *P. juncifolia*; assimilatory cells circular to polygonal through crowding, much smaller than water-storage cells; latter with birefringent walls. Variably-shaped cells with anisotropic, stratified walls scattered in *Helleriella*. **Endodermis**: cells square with thick laminated walls opposite phloem in *Helleriella* and *Isochilus*, circular with barely thickened walls in *P. juncifolia*. **Pericycle**: cells smaller than endodermal cells, with polygonal, thickened, stratified walls in *Helleriella* and *Isochilus*, polygonal in *P. juncifolia* with barely thickened walls. **Vascular cylinder** 8–15-arch; radial rows of xylem alternating with oval and circular phloem clusters. Vascular tissue embedded in thick- or thin-walled sclerenchyma. **Pith** cells circular and angular, parenchymatous; some with starch granules in *Helleriella*.

Material examined

Helleriella guerrerensis (1), *Isochilus linearis* (1), *Ponera juncifolia* (1). Full details appear in Stern and Carlsward (2009).

Reports from the literature

Baker (1972): foliar anatomy of Laeliinae including *Isochilus linearis* and *Ponera striata*; Baker included these genera under subtribe Laeliinae; he noted foliar hairs in *Isochilus* and *P. striata* in contrast with our findings and an ad- and abaxial hypodermis in *Isochilus* whereas we cite it as absent. Möbius (1887): leaf anatomy of *Isochilus* and *Ponera striata*. Parrilla Diaz and Ackerman (1990): root anatomy of *Helleriella*. Porembski and Barthlott (1988): velamen of *Isochilus* (their *Ponera prolifera* is now *Maxillariella prolifera* in Maxillariinae). Pridgeon (1987): velamen and exodermis of *P. striata*. Pridgeon et al. (2005): comprehensive review of Ponerinae. Stern and Carlsward (2009): comprehensive study of vegetative anatomy of Laeliinae. Weltz (1897): stem anatomy of *Isochilus linearis*. Williams (1979): stomata of *Helleriella*, *Isochilus*, and *P. glomerata*.

Taxonomic notes

The genera described here were included in the anatomical study of Laeliinae by Stern and Carlsward (2009) and the anatomical characters for the genera now in Ponerinae showed no significant differences from the genera of Laeliinae studied.

There have been no published anatomical descriptions of *Helleriella* or *Ponera* prior to those outlined here. There are some anatomical consistencies among the three genera constituting Ponerinae: stomatal apparatuses are all abaxial and tetracytic, stegmata all contain conical silica bodies, and tilosomes are lacking throughout. Foliar hairs are absent, except adaxial in *Nemaconia* (= *Ponera*) and mesophyll is heterogeneous, except homogeneous in *Isochilus* and stratified in *Helleriella*. Pridgeon et al. (2005) record the presence of an ad- and abaxial single-layered hypodermis in leaves of *Isochilus*. We did not see a hypodermis in our material of *Isochilus*, although it was present adaxially in *Helleriella* and *Ponera* and was recorded abaxially in *Nemaconia* (= *Ponera*) by Pridgeon et al. Overall, the three genera are anatomically consistent with the few exceptions noted above.

Tribe Gastrodieae Lindl.

Gastrodieae consist of six genera of terrestrial, saprophytic, achlorophyllous, mycorrhizal plants widely distributed across the Old and New Worlds, Indian Ocean, and Pacific island regions. One genus, *Uleiorchis*, is strictly confined to the New World tropics. These are bizarre plants rarely appearing above ground except when flowering. Nothing is known of the anatomy of *Auxopus* or *Neoclemensia* and little of *Didymoplexiella, Didymoplexis,* or *Gastrodia*.

Gastrodia species have been considered medicinal in China, Taiwan, Japan, and Korea.

Rhizome TS

Rhizome swollen and periodically constricted, sometimes with short roots lacking root hairs in *Gastrodia*; rhizome warty with non-vascularized scales and lacking mycorrhizae in *Uleiorchis* (Fig. 75A). **Epidermis** 1-layered; cells suberized and tangentially flattened in *Gastrodia*; cells periclinal, thick-walled, subtended by a 1-layered **hypodermis** of isodiametric to periclinal cells in *Uleiorchis*. (Fig. 75A). **Cortex** 4–11 cells wide, cells thin-walled, containing starch granules and occasional raphides and infected with fungal pelotons in *Gastrodia*; 3–5 cells wide, cells thin-walled, raphide bundles and periclinal water-storage cells present in *Uleiorchis* (Fig. 75B). **Endodermis** between cortex and ground tissue in *Uleiorchis* (Fig. 75A); cells with broad casparian strips. Ground-tissue cells parenchymatous, thin-walled, packed with starch granules in *Gastrodia*; cells thin-walled, polygonal to isodiametric, containing spiranthosomes in *Uleiorchis* (Figs 75C, 76). **Vascular bundles** collateral, scattered throughout ground tissue in *Didymoplexis, Gastrodia,* and *Uleiorchis*, lacking sclerenchyma in *Gastrodia* and *Uleiorchis*. **Stegmata** absent in *Didymoplexiella, Didymoplexis,* and *Gastrodia*, not seen in *Uleiorchis*.

Rhizome of *Didymoplexis* has a hairless epidermis of living cells, weakly suberized exodermal cells with raphides, thick-walled endodermal cells over phloem, thin-walled over xylem, triarch to tetrarch and pentarch vascular cylinder.

Stem TS (Uleiorchis only, Fig. 75D)

Cuticle thin. **Epidermis**: cells isodiametric, cell walls thin. **Cortex** three or four cells wide; cells thin-walled, rounded. Crystalliferous cells occurring intermittently around cortical circumference. Ground tissue separated from cortex by sclerenchymatous band of 1–3 cell layers, cells thick-walled, angular, living, without intercellular spaces. Ground-tissue cells thin-walled, rounded to angular. **Stegmata** absent (also absent in stem of *Didymoplexiella* and *Gastrodia exilis* (as *G. siamensis*)).

Root TS

Roots filiform in *Uleiorchis*. **Velamen** absent. **Epidermis**: cells isodiametric, tetragonal, polygonal, thin-walled in *Uleiorchis*, soon breaking down in *Didymoplexis*. **Hypodermis**: cells with suberized outer walls and passage cells in *G. abscondita*, subjacent layer with raphide cells. **Cortex** five cells wide, cells thin-walled, rounded to polygonal in *Uleiorchis*; with fungal hyphae in *Didymoplexis* and *Gastrodia*. Inner starch sheath in *Gastrodia*. **Endodermis** and **pericycle** 1-layered, cells thin-walled, isodiametric in *Uleiorchis* (Fig. 75E); cells thick-walled over phloem, thin-walled with casparian strips over xylem in *Didymoplexis*. **Vascular cylinder** 4-arch in *Uleiorchis*, xylem and phloem embedded in parenchyma, alternating around circumference (Fig. 75E); 3–5-arch in *Didymoplexis*, 5- or 6-arch in *Gastrodia*. **Pith** parenchymatous, cells thin-walled, polygonal (Fig. 75E). Cell inclusions and mycorrhizae absent in *Uleiorchis*. Nodules developed on roots of *Didymoplexis* infected by fungal hyphae.

Coralloid roots present in *Didymoplexis* species and *Gastrodia javanica*; aggregate starch granules and hyphae in cortex; central cylinder very reduced, ±diarch.

Material examined/reported

Didymoplexis (3), *Gastrodia* (5), *Uleiorchis* (1). Full details for *Uleiorchis* appear in Stern (1999). Other reports are from the literature listed below.

Reports from the literature

Burgeff (1932): vegetative anatomy of *Gastrodia* and *Didymoplexis*. Burgeff (1959): mycorrhiza of *Gastrodia*. Campbell (1962): anatomy of rhizome of *G. cunninghamii*. Dietz (1930): underground organs of *Gastrodia javanica*. Dong and Zhang (1986): mycorrhiza of *G. elata*. Kusano (1911): rhizome and mycorrhiza of *G. elata*. Martos et al. (2009): mycorrhiza of *Gastrodia*. McLennan (1959): anatomy of *G. sesamoides*. McLuckie (1923): structure of rhizome in *G. sesamoides*. Møller and Rasmussen (1984): stegmata in orchids including absence in *Didymoplexiella* and *Gastrodia*. Porembski and Barthlott (1988): simple rhizodermis in root of *Didymoplexis pallens*. Poulsen (1911): root anatomy of *Didymoplexis cornuta*. Pridgeon et al. (2005): a comprehensive systematic review of tribe Gastrodieae. Stern (1999): anatomy of saprophytic *Uleiorchis* compared with *Wullschlaegelia*. Wang and Wang (1992): mycorrhiza of tuber of *G. elata*. Xu and Fan (2001): mycorrhiza of corms of *Gastrodia*.

Taxonomic notes

In Gastrodieae there are few molecular tools to evaluate the affinities of these highly specialized, non-photosynthetic orchids owing to the lack of gene-bearing plastids. It is of interest that, except for the study of *Uleiorchis* by Stern

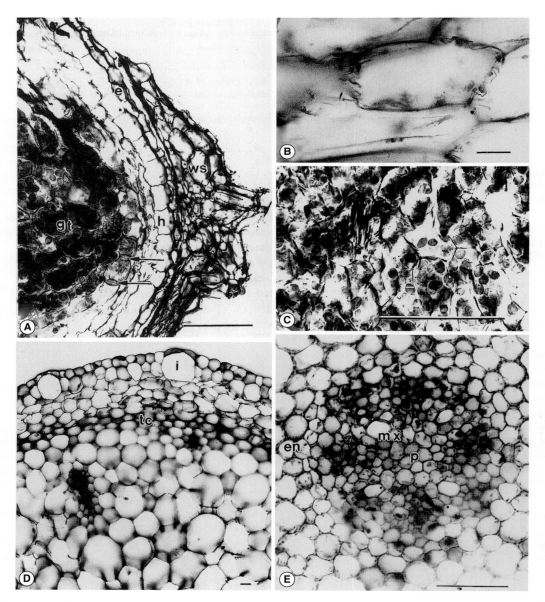

Fig. 75. Gastrodieae. *Uleiorchis ulei*. (A–C) Rhizome. (A) TS with warty scale (ws) external to epidermis (e), hypodermis (h) subtending epidermis, endodermis with thickened casparian strip (arrows) bordering ground tissue (gt). (B) Pleated cortical water-storage cells. (C) Spiranthosomes in ground-tissue cells. (D) TS stem with circular idioblasts (i) subtending epidermis and band of thick-walled cells (tc) at cortex/ground tissue boundary. (E) TS filiform root showing tetrarch vascular cylinder (mx = metaxylem), thick-walled endodermal cells (en), and thin-walled pith cells (p). Scale bars: A,C–E = 100 μm, B = 10 μm. Reprinted from Stern (1999).

Subfamily Epidendroideae 139

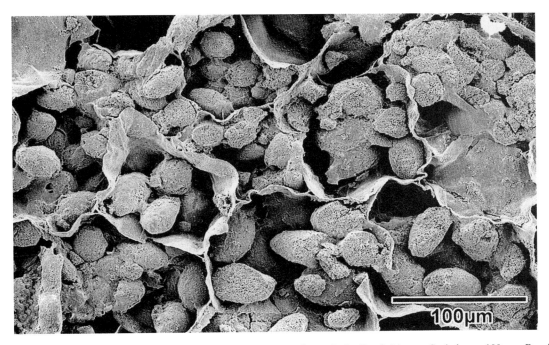

Fig. 76. Gastrodieae. *Uleiorchis ulei*. SEM with ovoid spiranthosomes in cortical cells of rhizome. Scale bar = 100 μm. Reprinted from Stern (1999).

(1999), the limited structural information available has come from research on mycorrhizal relationships between Gastrodieae and the roots of host plants.

Tribe Malaxideae Lindl.

Malaxideae are small-flowered plants, terrestrial or epiphytic, pseudobulbous or cormous, autotrophic, often with creeping rhizomes. Roots are velamentous and usually mycorrhizal. Leaves are thin to fleshy, plicate or conduplicate, dorsiventral or iridiform, alternate or distichous, articulate or not. Malaxideae are widely distributed in the Old and New World tropics extending northwards into temperate regions. They are absent from New Zealand and occur only in northern tropical Australia. There are 13 genera, two of which, *Hammarbya* and *Risleya*, are monospecific; *Malaxis* has 300 species and *Liparis* 320. A few species of small-flowered terrestrials of *Liparis* and *Malaxis* are sometimes grown in hobby collections. *Malaxis* species have been used medicinally in the Himalayas.

The information on this tribe has been compiled entirely from the literature but the names now accepted have been used where possible. Where no data are given for any organ or tissue, it is because none was available.

Leaf surface

Hairs present in *Crepidium* species (as *Microstylis*) and *Liparis duthiei*, sunken in *Crepidium dentatum* (as *Microstylis philippinensis*) and *Oberonia wightiana*; leaves glabrous in some *Oberonia* and *Stichorkis* species. Glands sunken in epidermis in *Crepidium rheedii* (as *Microstylis rheedii*) and *Liparis walkeriae*. **Epidermis**: adaxial cells often polygonal (Figs 77A, 78A). **Stomata** abaxial (Figs 77B, 78B), except on both surfaces in terete or equitant leaves, e.g. *Oberonia* species; tetracytic in most taxa, tetracytic and anomocytic in *Crepidium acuminatum* (as *Microstylis wallichii*), *Dienia cylindrostachya* (as *Microstylis cylindrostachya*), *Liparis resupinata*, and *Oberonia ensifolia*, with up to seven subsidiary cells in *Crepidium dentatum*; two subsidiary cells in *Liparis nervosa* (as *L. elata*) and anomocytic in some *Liparis* species.

Leaf TS

Cuticle thin in *Crepidium purpureum* (as *Microstylis purpurascens*), *Dienia ophrydis* (as *Microstylis congesta*), and *Liparis liliifolia*, moderately thick to thick in other taxa, adaxial 14–21 μm thick and abaxial 10–14 μm thick and tuberculate in *Stichorkis viridiflora* (as *Liparis longipes*).

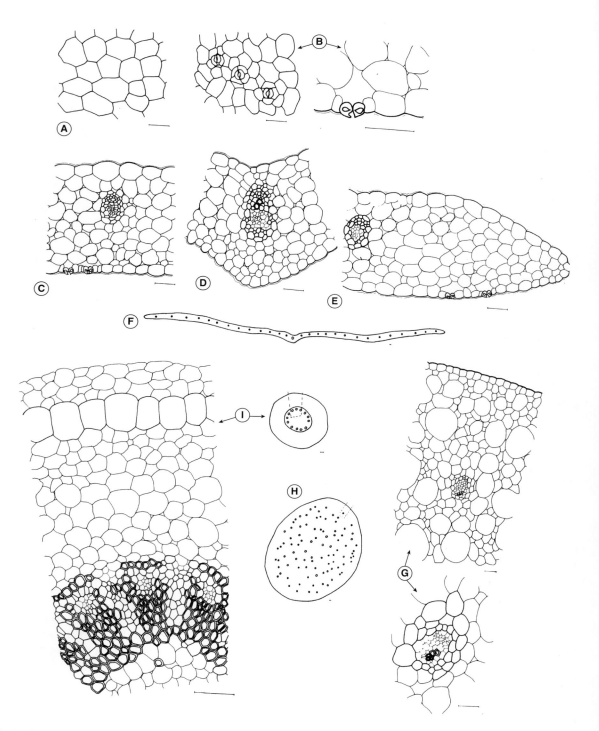

Fig. 77. Malaxideae. *Liparis platyglossa*. (A–F) Leaf. (A) Adaxial surface with polygonal cells. (B) Abaxial surface showing tetracytic and anomocytic stomata, and cells with curvilinear walls and irregular outlines. (C–E) TS with superficial stomata, homogeneous mesophyll, and collateral vascular bundles. (F) TS with one row of vascular bundles. (G,H) TS stem lacking sharply defined cortex, with pronounced water-storage cells, and scattered collateral vascular bundles. (I) TS root with 1–3-layered velamen, thin-walled anticlinal exodermal cells, broad cortex, endodermal cells with O-thickened walls opposite phloem and thin walls opposite xylem, vascular strands surrounded by thin-walled sclerenchyma cells, and thin-walled pith cells.

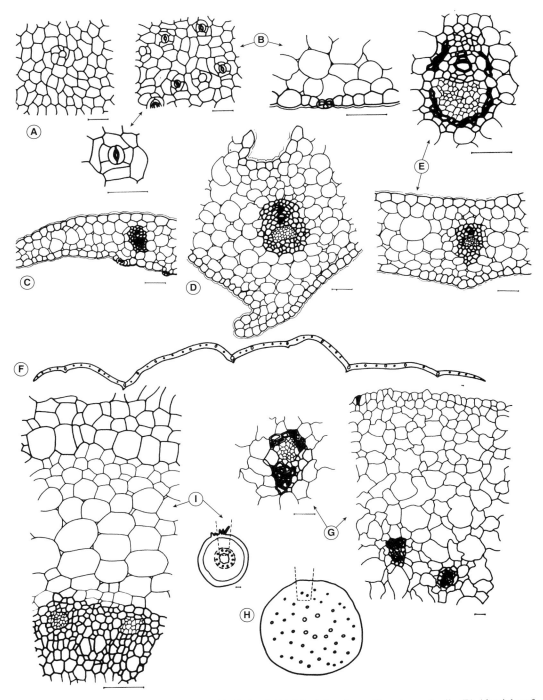

Fig. 78. Malaxideae. *Crepidium maximowiczianum*. (A–F) Leaf. (A) Adaxial surface with polygonal cells. (B) Abaxial surface cells with curvilinear walls and anomocytic/tetracytic stomata, also TS stoma. (C–E) TS with moderately thick cuticle, homogeneous mesophyll, 1-layered adaxial hypodermis, and superficial stomata. (F) TS with one row of vascular bundles. (G,H) TS stem with scattered vascular bundles having marked ad- and abaxial sclerenchyma caps, lack of vascular bundle sheath, and ground-tissue cells with variable outlines. (I) TS root with 3- or 4-layered velamen, thin-walled anticlinal exodermal cells, thin-walled endodermal cells only slightly thickened over phloem, large oval to polygonal, thin-walled cortical cells, and vascular tissue embedded in thin-walled sclerenchyma cells.

Epidermis: cells usually thin-walled; cells small, convexly domed in *Oberonia* species, large in *Crepidium purpureum*; irregularly triradiate idioblasts present in *Oberonia pachyrachis*. **Stomata** superficial. **Hypodermis**: absent in most species; adaxial, cells palisade-like in *Stichorkis viridiflora* (as *Cestichis pendula*); cells with helical thickenings. **Mesophyll** homogeneous (Figs 77C–E, 78C–E), except heterogeneous with tall, narrow palisade cells in *Crepidium metallicum* (as *Microstylis metallica*); axially or tangentially elongated water-storage cells with fine to broad spiral or reticulate thickenings scattered in mesophyll, water-storage cells spirally thickened in species of *Liparis*, *Oberonia*, and *Stichorkis*, reticulately thickened in some species of *Oberonia* and *Stichorkis*. Raphides in mesophyll cells of *Hammarbya*, *Liparis liliifolia*, *Oberonia* species, and *Stichorkis viridiflora*. **Vascular bundles** collateral (Figs 77C–E, 78C–E), in a single series (Figs 77F, 78F), except in two to three series in species with equitant leaves, e.g. *Oberonia*; weakly developed; sclerenchyma consisting of large, wide-lumened cells strongly developed over phloem, weakly developed over xylem. Bundle sheath cells parenchymatous, containing chloroplasts, in *Crepidium rheedii* and *Liparis walkeriae*. **Stegmata** absent.

Stem/pseudobulb TS

Epidermis: cells thin-walled (Figs 77G, 78G), covered by thin cuticle. **Cortex**: outer layers often of thin-walled chlorophyllous cells (Figs 77G, 78G) followed by 1-layered sclerenchymatous ring; inner layers including large cells, reticulately thickened in *Oberonia wightiana* and spiral or reticulate water-storage cells in *Liparis* and *Stichorkis* species, elongated in TS or LS; such cells lacking in *Crepidium calophyllum* (as *Microstylis scottii*) and *C. chlorophrys* (as *Microstylis chlorophrys*). **Endodermis**: 1-layered, of thin-walled cells with casparian strips, noted in *Crepidium rheedii* and *Liparis walkeriae*. **Ground tissue** of small, starch-containing cells and larger cells often containing starch or mucilage. Sclerenchyma ring lacking in pseudobulbs of *Hammarbya* (as *Malaxis paludosa*) and *Stichorkis* species, but present in inflorescence axis of *Hammarbya*. **Vascular bundles** collateral, xylem and phloem sheaths of wide-lumened cells present in *Stichorkis*, absent in *Malaxis* species; vascular bundles closely packed in stem centre of *Stichorkis luteola* (as *Liparis luteola*), scattered in *Crepidium maximowiczianum* (Fig. 78H), *Hammarbya*, *Liparis platyglossa* (Fig. 77H), *S. parviflora* (as *L. parviflora*), and *S. viridiflora* (as *L. pendula*). Raphides noted in *Crepidium calophyllum*, *C. chlorophrys*, *Stichorkis parviflora*, and *S. viridiflora*. **Stegmata** absent.

Root TS

Velamen 1-layered in *Crepidium acuminatum* (as *Microstylis wallichii*), *C. rheedii* (as *Microstylis rheedii*), *Hammarbya*, *Hippeophyllum*, and *Orestias* species, and in *Oberonia imbricata*, *O. pumilio* (as *O. treubii*), *O. wightiana*, and *Stichorkis latifolia* (as *Liparis latifolia*); 1-, 2-, or 3-layered in *Crepidium metallicum* (as *Microstylis metallica*) and most *Liparis* (Fig. 77I), and *Stichorkis* (as *Liparis*) species, 2-layered in *Dienia ophrydis* (as *Malaxis longifolia*) and *Crepidium taurinum* (as *Malaxis taurina*), 2- or 3-layered in *Oberonia brunoniana*, *O. pachyrachis*, and some *Liparis* and *Malaxis* species, 5-layered in *Oberonia acaulis* (as *O. myriantha*); cells lacking spiral thickenings in most genera (many with *Malaxis* type of velamen of Porembski and Barthlott 1988); walls with pits of different sizes in *Dienia*, many *Liparis*, *Malaxis*, and *Stichorkis* species; walls with small pits in *Crepidium taurinum*, *Hippeophyllum scortechinii*, *Liparis formosana*, *Orestias*, and *Stichorkis parviflora* (*Calanthe* type of Porembski and Bartlott). Root hairs often present. **Tilosomes** absent. **Exodermis**: cells frequently large, only slightly thickened; Ω-type thickenings noted in *Liparis duthiei*, *Oberonia pumilio*, and *O. wightiana*. **Cortex** with spirally thickened cells in *Liparis liliifolia*, *Oberonia acaulis*, *Stichorkis latifolia*, and *S. luteola* (as *Liparis luteola*); mucilage cells present in cortex of *O. platyglossa*; fungal hyphae present in some species. **Endodermis** 1-layered (Figs 77I, 78I), cells O-thickened in *Liparis platyglossa*, *Oberonia wightiana*, *Stichorkis luteola*, U-thickened in *Crepidium acuminatum*, *Oberonia brunoniana*, *O. pachyrachis*, thin-walled with casparian strips in *Liparis liliifolia* and *L. loeselii*. **Pericycle** 1-layered, cells thin-walled. **Vascular cylinder** 8-arch in *Liparis bowkeri*, 12-arch in *L. liliifolia*; 12–16-arch in *Crepidium acuminatum*, 14-arch in *Liparis nervosa*, vascular tissue in these species surrounding pith of small, thin-walled cells; vascular cylinder 5-arch in *Oberonia pachyrachis*, 7-arch in *Stichorkis luteola*, 9-arch in *Crepidium calophyllum* (as *Malaxis scottii*), 7-arch in *Oberonia wightiana*, and 9-arch in *Stichorkis longipes*; vascular tissue surrounding lignified pith in these species.

Material reported

Crepidium (5 leaf and root, as *Malaxis* or *Microstylis*), *Dienia* (1 leaf as *Microstylis congesta*, 1 root as *Malaxis longifolia*), *Hammarbya* (1, as *Malaxis paludosa*), *Hippeophyllum* (2 root), *Liparis* (3 leaf, 12 root), *Malaxis* (4 root), *Oberonia* (6 leaf, 8 root), *Stichorkis* (3 leaf, 7 root, as *Liparis*).

Reports from the literature

Burr and Barthlott (1991): root cortex of *Liparis* and *Microstylis*. Camus (1908, 1929): general anatomy of *Liparis* and *Malaxis*. Chatin (1856, 1857): vegetative anatomy of *Liparis*. Czapek (1909): root, stem, leaf anatomy of *Liparis* and *Oberonia*. Dietz (1930): root and tuber anatomy of *Liparis* and *Malaxis*. Dressler (1993): survey of systematics of orchid family. Duruz (1960): stomata of *Liparis*. Fan et al. (2000): mycorrhiza of *Liparis*. Fuchs and Ziegenspeck (1925, 1927a): stem, root, and mycorrhiza of *Liparis* and *Malaxis* (including *Hammarbya*). Guttenberg (1968): root anatomy of *Oberonia*. Hadley and Williamson (1972): mycorrhiza of *Liparis*. Holm (1904, 1927): root and leaf anatomy of *Liparis*. Huber (1921): mycorrhiza of *Liparis*. Isaiah et al. (1991): leaf, pseudobulb, root anatomy of *Liparis duthiei*. Katiyar et al. (1986): mycorrhiza of *Liparis* and *Oberonia*. Kaushik (1983): vegetative anatomy of *Liparis* and *Oberonia*. Khasim and Mohana Rao (1986): leaf and root anatomy of *Oberonia brunoniana*. Kim and Kim (1986): anatomy of five species of *Liparis*. Kozhevnikova and Vinogradova (1999): pseudobulb anatomy of *Malaxis*. Krüger (1883): aerial organs of *Liparis*. Leitgeb (1864b): aerial root anatomy of *Liparis* and *Oberonia*. Meinecke (1894): aerial root anatomy of *Liparis*, *Microstylis*, and *Stichorkis*. Möbius (1887): leaf anatomy of *Cestichis* (= *Liparis*), *Microstylis* and *Oberonia*. Mohana Rao et al. (1989): stem and root anatomy of *Liparis*. Møller and Rasmussen (1984): stegmata absent in *Liparis*, *Malaxis*, and *Oberonia*. Mulay and Panikkar (1956): velamen in *Liparis* and *Microstylis rheedii* (= *Crepidium*). Mulay et al. (1954, 1956): root anatomy of *Liparis* and *Microstylis rheedii* (= *Crepidium*). Olatunji and Nengim (1980): tracheoids of *Liparis*. Pirwitz (1931): 'storage tracheids' in *Liparis* and *Malaxis*. Porembski and Barthlott (1988): velamen in *Hippeophyllum*, *Liparis*, *Malaxis*, *Oberonia*, and *Orestias*. Pridgeon (1987): velamen and exodermis in *Hammarbya*, *Liparis*, *Malaxis*, and *Oberonia*. Pridgeon et al. (1983): tilosomes in root of *Liparis* and *Malaxis*. Pridgeon et al. (2005): comprehensive review of the systematics of Malaxideae. Rao (1998): leaf anatomy of *Liparis*, *Malaxis*, and *Oberonia*. Rasmussen and Whigham (2002): root and mycorrhiza of *Liparis*. Raunkiaer (1895–9): root and stem anatomy of *Hammarbya* and *Sturmia* (= *Liparis*). Roth and Merida de Bifano (1979): leaf anatomy of *Liparis*. Sanford and Adanlawo (1973): velamen and exodermis of *Liparis*, *Malaxis*, *Microstylis*, *Oberonia*, and *Stichorkis*. Sasikumar (1973a,b): root anatomy of *Liparis walkeriae* and *Microstylis rheedii* (= *Crepidium*). Sasikumar and Navalkar (1973 [1974]): anatomy of *Liparis walkeriae* and *Microstylis rheedii*. Singh (1981): stomata of *Liparis*, *Malaxis*, and *Oberonia*. Singh (1986): root anatomy of *Malaxis*. Solereder and Meyer (1930): review of orchid anatomy and original observations on *Malaxis* (including *Hammarbya*), *Microstylis*, *Oberonia*, *Stichorkis*, and *Sturmia* (= *Liparis*). Tatarenko (1995): mycorrhiza of *Liparis*. Tominski (1905): leaf anatomy of *Liparis*, *Microstylis*, and *Oberonia*. Vij et al. (1991): leaf epidermis of *Liparis*. Weltz (1897): stem anatomy of *Liparis* and *Microstylis* (= *Crepidium*). Williams (1979): stomata of *Liparis*.

Taxonomic notes

According to Porembski and Barthlott (1988) the *Malaxis* type of velamen predominates in this tribe, that is, a velamen of one or two layers with cells lacking helical thickenings. Many *Liparis* and *Malaxis* species cited in this description fit the definition of the *Malaxis* type. However, some species of *Liparis* have a multilayered velamen and *Oberonia acaulis* a 5-layered velamen. Porembski and Barthlott concede that some species of *Oberonia* have a peculiar multilayered velamen that does not fit into any of their types of velamina. That may indicate 'a possibly polyphyletic nature of Malaxideae' (Porembski and Barthlott 1988).

Tribe Neottieae Lindl.

Neottieae are terrestrial, sometimes mycotrophic herbs with clustered, coralloid, fleshy roots lacking a velamen. Plants may be leafless, as in *Aphyllorchis* and *Limodorum*, or have spiral, convolute, or plicate, non-articulate leaves scattered along the stem. Species are widespread, occurring in the Old World across Eurasia from Great Britain to the East Indian islands and through much of northern, central, and western Africa, and in the New World from Alaska through north-eastern North America into northern Mexico and northern South America. Neottieae are at home in arctic regions, temperate zones, and subtropical and tropical areas of the world. There are seven genera and about 140 species. Several species of *Cephalanthera*, *Epipactis*, and *Neottia* are in cultivation.

Epipactis has been used medicinally in North America and Europe and *Listera* (now in *Neottia*) in Spain.

The information on this tribe has been compiled entirely from the literature and the line drawings, but the names now accepted have been used where possible. *Listera* is included in *Neottia* in Pridgeon et al. (2005). Where no data are given for any organ or tissue, it is because none was available.

Leaf surface

Hairs present on both surfaces in *Cephalanthera rubra*, secretory, with rounded or elongated terminal cell; hairs 1- or 2-celled in *C. longifolia* (as *C. ensifolia*), 2–4-celled in *Epipactis* and *Neottia* species. **Epidermis**: anticlinal cell walls of *Neottia* species undulating, adaxial curved and abaxial undulating in *N. ovata* (as *Listera ovata*), wavy to straight in *Cephalanthera* (Fig. 79A,B) and *Epipactis* (Fig. 80A,B) species. **Stomata** abaxial in most taxa (Figs 79B, 80B); numerous abaxially and few adaxially in *C. rubra* and *Neottia ovata*; on both surfaces in *E. atrorubens* (as *E. rubiginosa*); stomata anomocytic (Figs 79B, 80B) in all except *Palmorchis* where two subsidiary cells present. Stomata absent from both surfaces of sheathing leaves in *Neottia nidus-avis*.

Leaf TS

Cuticle ridged in *Cephalanthera rubra* and adaxially in *Epipactis helleborine* (as *E. latifolia*) and *E. palustris*. **Epidermis**: cells elongated, papillate in *Cephalanthera* and adaxially in *Epipactis*. **Stomata** superficial in *Cephalanthera*, *Epipactis* (Figs 79C, 80E), and *Neottia ovata*. **Mesophyll** homogeneous in *Cephalanthera* and *Epipactis* (Figs 79D, 80C–E), cells elongated; 5–7 layers of cells, slightly transversely elongated, containing chlorophyll and some cells with raphides in *Cephalanthera rubra*; 5–8 layers, adaxial cells compact, abaxial layers of large, irregularly shaped cells, raphides rare in *Neottia ovata*; eight or nine layers of thin-walled cells with small intercellular spaces, raphides rare or absent in *N. nidus-avis*; water-storage cells noted in *Epipactis* species and *N. ovata*. **Vascular bundles** in one row (Figs 79G, 80F), associated with surrounding sclerenchyma in *Cephalanthera* and *Epipactis* (Figs 79E,F, 80C,D); bundles lacking sclerenchyma or collenchyma, surrounded by parenchyma in *N. nidus-avis* and *N. ovata*. **Stegmata** with conical silica bodies in *Cephalanthera longifolia*, absent in *C. damasonium*, *Epipactis*, and *Neottia*.

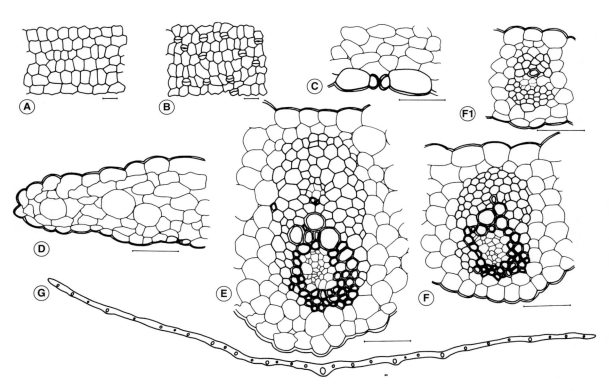

Fig. 79. Neottieae. ***Cephalanthera damasonium* leaf**. (A) Adaxial surface with thin-walled polygonal cells. (B) Abaxial surface with thin-walled polygonal cells and anomocytic stomata. (C) TS to show superficial stoma. (D–F1) TS with 1-layered epidermis, homogeneous mesophyll, collateral vascular bundles with phloem sclerenchyma, and circular raphide idioblasts. (G) TS with one row of vascular bundles.

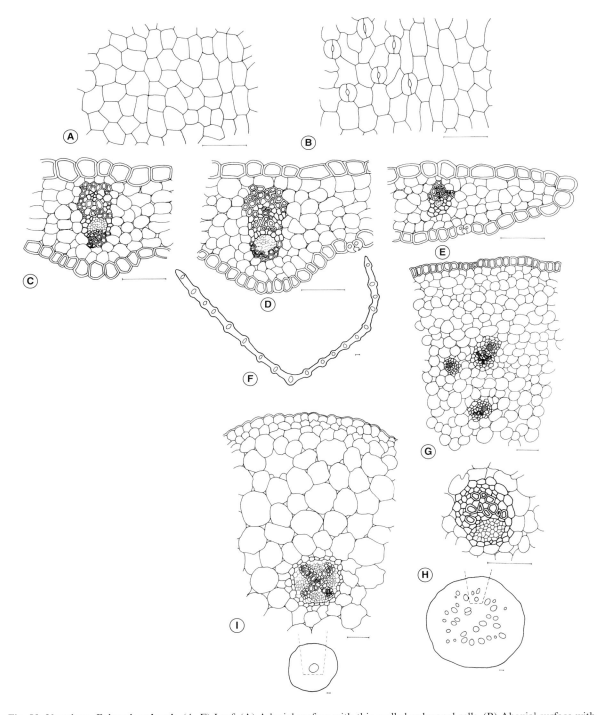

Fig. 80. Neottieae. *Epipactis palustris*. (A–F) Leaf. (A) Adaxial surface with thin-walled polygonal cells. (B) Abaxial surface with thin-walled polygonal, elongated cells, and anomocytic stomata. (C–E) TS with thick-walled ad- and abaxial epidermal cells, collateral vascular bundles with phloem and xylem sclerenchyma. (F) TS with one row of vascular bundles. (G,H) TS stem with scattered collateral vascular bundles, groundmass tissue of rounded cells, lack of distinct cortex. (I) TS root with 1-layered epidermis, 1-layered exodermis with small thin-walled cells, wide cortex with thin-walled rounded cells and triangular intercellular spaces, thin-walled endodermal cells, and tetrarch vascular system.

Stem TS

Outline ridged or undulating in *Cephalanthera* species. **Hairs** on young stem in *Aphyllorchis pallida*; secretory hairs with rounded or slightly elongated terminal cell on upper part of stem in *Cephalanthera rubra*; hairs 250–400 μm long in *Neottia nidus-avis* and 3–6-celled in *N. ovata*, terminal cell enlarged with granular yellow contents. **Cuticle** thick in *Cephalanthera* and *Epipactis*. **Epidermis**: cells in *Epipactis* isodiametric, all walls thin, walls vaulted externally and internally; epidermal cells in *Cephalanthera* thick-walled externally, thin-walled internally. **Stomata** numerous in *C. rubra*, present in *Aphyllorchis* and *Limodorum abortivum*. **Cortex**: cells in *Aphyllorchis*, *Cephalanthera*, and *Epipactis* thin-walled with prominent intercellular spaces. Chlorophyll present in outer stem layers of *Cephalanthera* and *Epipactis*. Cortex 3–5-layered with raphides and a little chlorophyll in *Limodorum*, 4–6-layered in *Neottia nidus-avis*, 3- or 4-layered with intercellular spaces and air canals in *N. ovata*. Sclerenchyma ring between cortex and ground tissue 3- or 4-layered of narrow-lumened fibres in *Aphyllorchis* and most *Cephalanthera* species, up to 9-layered in *C. rubra*, 5- or 6-layered of thick-walled fibres in *Epipactis*, 7–9-layered in *Limodorum*, 3–8-layered in *N. ovata*, and 4–6-layered in *N. nidus-avis* (Fig. 81A,B). Inner ground-tissue cells in *Cephalanthera* and *Epipactis* rounded (Fig. 80G), thin-walled, with intercellular spaces. **Vascular bundles**: larger bundles scattered in ground tissue in *Aphyllorchis*, *Cephalanthera*, and *Epipactis* (Fig. 80H); smaller bundles attached to one another, resulting in broadened sclerenchyma ring at these places. Bundles in circle in sclerenchyma ring in *Neottia nidus-avis* (Fig. 81B), bundles touching ring or separated from it by a few layers of cells in *N. ovata*. Ground tissue parenchymatous (Figs 80G, 81A,B), inner ground tissue free of vascular bundles. Sclerenchymatous cap at phloem poles or forming sheath around bundles; in *Aphyllorchis* bundles surrounded by incomplete sheath of fibres. Sclerenchyma bridge between xylem and phloem absent (Fig. 80G). *Limodorum* differing in not having vascular bundles in regular ring and outer bundles larger than inner. Central parenchyma with rare raphide bundles in *C. rubra*; centre sometimes slightly lacunose in *Neottia nidus-avis* and sometimes hollow in *C. rubra*. **Stegmata** absent.

Root TS

Root hairs not seen in *Neottia nidus-avis*; numerous in *N. ovata*. Piliferous layer cells with suberized walls in *Cephalanthera rubra* and *Limodorum*. **Velamen** absent. **Epidermis**: cells thin-walled (Fig. 80I), except outer walls thickened in *N. nidus-avis* (Fig. 81C). Hyphae in *Aphyllorchis* can be traced from root hairs into deeper tissues. **Exodermis** with small passage cells in *Aphyllorchis*. **Cortex**: cells thin-walled; cells often harbouring fungi except in *Epipactis purpurata* (as *E. violacea*); outer cells 2- or 3-layered with raphide/mucilage cells (absent from other parts of root) in *Aphyllorchis*. Cortex with raphides and starch granules in *C. rubra* and *Neottia ovata* and in outer cortex of *Limodorum*; raphides rare in *Neottia nidus-avis*; mucilage cells in middle cortex of *Limodorum* and some in *N. ovata*; starch in *Epipactis gigantea*, *N. bifolia* (as *Listera australis*), and *N. cordata* (as *L. cordata*). **Endodermis**: cells with casparian strips in *Aphyllorchis*, *Cephalanthera*, *Epipactis*, *Limodorum*, and *Neottia nidus-avis*. Endodermal cells O-thickened in *Cephalanthera damasonium* (as *C. alba*), *C. longifolia*, *C. rubra*, *Epipactis atrorubens* (over phloem), *E. gigantea*, *E. helleborine*, *E. purpurata* (as *E. violacea*), *Listera* species, and *Limodorum*; thin-walled in *E. palustris* (Fig. 80I) and *E. atrorubens* (as *E. rubiginosa*). **Pericycle** cells thin-walled with thick-walled cells opposite phloem clusters except in *C. rubra* where most cells thin-walled. **Vascular cylinder**: xylem stellate, 5-arch in *Aphyllorchis*, 4–11-arch in *Cephalanthera*, 4–8-arch in *Epipactis*, 9–25-arch in *Limodorum*, 3–6-arch in *Neottia* (Fig. 81E); xylem cells extending into centre of root. Xylem elements scattered in *Limodorum* and *N. nidus-avis* (Fig. 81D); phloem poorly developed in *C. rubra* and *N. ovata*. **Pith** cells thick-walled in *Aphyllorchis*.

Material reported

Aphyllorchis (1), *Cephalanthera* (6), *Epipactis* (6), *Limodorum* (1), *Neottia* (4, including 3 *Listera* spp.).

Reports from the literature

Abadie et al. (2006): mycorrhiza of *Cephalanthera*. Aybeke (2012): vegetative anatomy of three *Cephalanthera* species, *Epipactis helleborine*, *Limodorum abortivum*, and *Neottia nidus-avis*. Bernard (1902): developmental anatomy of *Neottia*. Bernardos et al. (2004a,b): leaf and stem anatomy of *Epipactis*. Borsos (1977, 1980, 1982a,b, 1983): leaf and stem anatomy of *Cephalanthera damasonium*, *C. longifolia*, and *C. rubra*. Burgeff (1909, 1959): mycorrhiza of *Limodorum* and *Neottia*. Camus (1908, 1929): general anatomy of *Cephalanthera* species, *Limodorum abortivum*, *Neottia nidus-avis*, and *N. ovata* (as *Listera ovata*). Champagnat and Loiseau (1975): root development in *Neottia*. Chatin (1857): anatomy of *Epipactis*. Chodat and Lendner (1896): mycorrhiza of *Neottia cordata* (as *Listera cordata*). Cyge (1930): leaf anatomy of *Cephalanthera rubra*, *Epipactis helleborine*

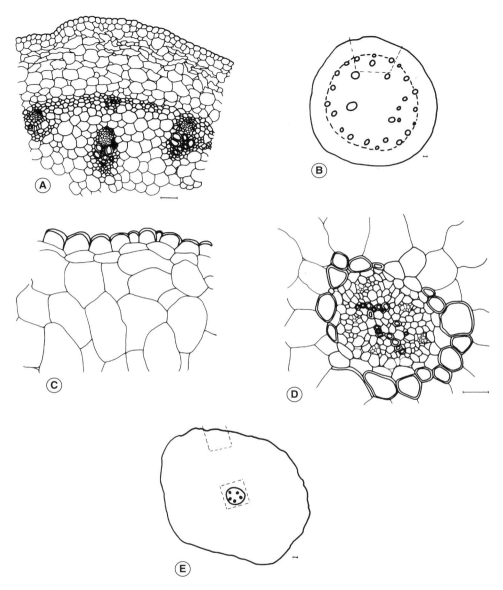

Fig. 81. Neottieae. *Neottia nidus-avis*. (A,B) TS stem with thick-walled epidermal cells, several-layered stratified cortex, and collateral vascular bundles in a ring. (C–E) TS root with epidermal cells having thickened outer walls, periclinal exodermal cells, irregularly shaped thin-walled parenchymatous cortical cells, pentarch vascular cylinder surrounded by endodermal cells with O-thickened walls, and vascular elements embedded in a parenchymatous matrix.

(as *E. latifolia*), and *Neottia ovata* (as *L. ovata*). Danilova and Barmicheva (1990): root phloem of *Neottia*. Dörr and Kollmann (1969): mycorrhiza of *Neottia*. Dressler (1993): taxonomic review of Neottieae including a phylogenetic diagram of the tribe. Drude (1873): vegetative anatomy of *Neottia nidus-avis*. Falkenberg (1876): stem anatomy of *Cephalanthera longifolia* (as *C. pallens*) and *Epipactis palustris*. Filipello Marchisio et al. (1985):

mycorrhiza of *Cephalanthera*. Fricke (1926): leaf and bract anatomy of *Epipactis*. Fuchs and Ziegenspeck (1925, 1926): mycorrhizal relationships and root and stem anatomy of *Cephalanthera*, *Epipactis* (including *Helleborine*), *Limodorum*, *Listera*, and *Neottia*. Groom (1895–7): stem and root anatomy of *Aphyllorchis pallida*. Guillaud (1878): stem anatomy of *Epipactis palustris*. Guttenberg (1968): root and mycorrhiza of *Epipactis atrorubens*. Holm (1904): root anatomy of *Cephalanthera austiniae* (as *Chloraea austiniae*), *Epipactis gigantea*, *Neottia bifolia* (as *Listera australis*), and *N. cordata* (as *L. cordata*). Holm (1927): leaf anatomy of *Cephalanthera austiniae* (as *C. oregana*). Jakubska-Busse and Gola (2010): leaf anatomy of *Epipactis*. Johow (1889): anatomy of *Neottia*. Jurcak (1999): anatomy of *Epipactis*. Kasapligil (1961): leaf anatomy of *Limodorum*. Latr et al. (2008): mycorrhiza of *Cephalanthera longifolia*. Lee and Kim (1986): leaf epidermis of *Cephalanthera erecta*, *C. falcata*, and *C. longibracteata*. MacDougal (1899a): anatomy of *Cephalanthera*. Magnus (1900): mycorrhiza of *Neottia*. Meinecke (1894): aerial root anatomy of *Epipactis*. Möbius (1886, 1887): leaf and stem anatomy of *Cephalanthera rubra*, *C. longifolia* (as *C. grandiflora*), *Epipactis atrorubens*, *E. palustris*, *Limodorum*, and *Neottia ovata* (as *Listera ovata*). Molisch (1920): stegmata in *Cephalanthera*. Mollberg (1884): mycorrhiza of *Cephalanthera*, *Epipactis*, and *Neottia*. Møller and Rasmussen (1984): stegmata present in *Cephalanthera longifolia* leaf, absent in *C. damasonium*. Nieuwdorp (1972): mycorrhiza of *Epipactis*, *Limodorum*, *Listera*, and *Neottia*. Porembski and Barthlott (1988): simple rhizodermis in *Cephalanthera*, *Epipactis*, and *Neottia* (including *Listera*). Pridgeon et al. (2005): comprehensive systematic review of tribe Neottieae. Prillieux (1856): anatomy of *Neottia nidus-avis*. Raunkiaer (1895–9): anatomy of *Epipactis palustris* and *N. cordata* (as *Listera*). Salmia (1989a,b): mycorrhiza and general anatomy of *Epipactis helleborine*. Scrugli et al. (1991): mycorrhiza of *Limodorum trabutianum*. Singh (1981): stomata in *Epipactis helleborine* (as *E. latifolia*). Solereder and Meyer (1930): survey of orchid anatomy, including original observations on *Epipactis palustris* and *N. ovata* (as *Listera*). Staudermann (1924): hairs in *Cephalanthera*, *Epipactis*, and *Listera*. Tatarenko (1995): mycorrhiza of *Cephalanthera* and *Neottia*. Vij et al. (1991): leaf epidermis of *Cephalanthera longifolia* (as *C. ensifolia*), *Epipactis gigantea*, *E. helleborine*, and *E. veratrifolia*. Voigt (1991): mycorrhiza of *Epipactis*. Wahrlich (1886): mycorrhiza of *Epipactis*. Weiss (1880): root and rhizome anatomy of *Epipactis* and *Listera*. Weltz (1897): stem anatomy of *Cephalanthera longifolia* (as *C. pallens*) and *Epipactis palustris*. Williams (1979): stomata in *Epipactis* and *Palmorchis*.

Taxonomic notes

Anatomical literature on Neottieae is sparse and poorly integrated. Entire cohorts of features are neglected or almost so, e.g. detailed studies of leaf anatomy and root structure. Relatively few neottioids have been examined thoroughly in contrast to reports and interpretations of floral morphology, molecular studies, cytogenetics, palynology, phytochemistry, ecology, etc. Until there are more complete anatomical analyses, it is doubtful any meaningful taxonomic conclusions can be drawn.

Tribe Nervilieae Dressler

Plants of Nervilieae are photosynthetic or mycotrophic terrestrial herbs produced from tubers, rhizomes, or coralloid underground stems. Tubers are spherical or ellipsoid. There is a solitary plicate, non-articulate hairy or glabrous leaf in *Nervilia* borne erect or parallel to the substrate, or leaves are reduced, sheath-like, and achlorophyllous. The stem is erect, lacking chlorophyll in *Epipogium*, *Silvorchis*, and *Stereosandra*; leaves are reduced to buff or whitish scales. Nervilieae are confined to the Old World from western Europe to the tropics and subtropics of Africa, Madagascar, mainland Asia, the Malay Archipelago, the Philippines, Australia, New Guinea, and south-west Pacific ocean islands. There are only four genera: *Nervilia* with about 60 species, *Epipogium* of three species; *Silvorchis* and *Stereosandra* are monospecific. *Nervilia* is used medicinally in Taiwan.

Information on this tribe has been compiled entirely from the literature but the names now accepted have been used where possible. Where data are lacking for any given tissue or organ, it is because none was available. Observations are based on *Epipogium* and *Nervilia*, including species wrongly assigned by Moreau to *Pogonia*.

Leaf surface (Nervilia only)

Lateral veins protruding on both surfaces or only abaxially. **Hairs** absent or sticky, unicellular, on adaxial or both surfaces; numerous, long. multicellular in *N. plicata* (as *N. purpurea*). **Cuticle** wavy, dentate, or smooth. **Epidermis**: cells polygonal on both surfaces. **Stomata** abaxial, tetracytic, or anomocytic with 3–5 subsidiary cells.

Leaf TS (Nervilia only)

Epidermis 1-layered, adaxial cells polygonal, oblong, rectangular or square; papillose in *N. aragoana*. **Mesophyll** homogeneous, cells spherical and polygonal, or slightly heterogeneous, with one adaxial layer of oblong palisade cells and circular, thin-walled spongy cells, some containing mucilage, raphide bundles, or solitary crystals. **Vascular bundles** in 1–3 rows, each bundle surrounded by an arc or ring of collenchyma; collenchyma also present as girders in midrib and protruding lateral veins. **Stegmata**: none.

Petiole TS (Nervilia)

Outline arcuate to elliptical. **Epidermis**: one layer of elliptical to circular cells. **Mesophyll**: thin-walled cells, elliptical and circular, some containing mucilage and raphides. **Vascular bundles** from six to 23 in one series or peripheral series plus scattered inner bundles, with collenchyma sheaths. Bilaterally symmetrical in *N. leguminosarum* with slightly flattened adaxial surface, bearing tufted hairs; vascular bundles in closed arc near centre of petiole with two or more bundles in centre.

Inflorescence axis TS (Epipogium)

Hairs and **stomata** absent. **Cuticle** thin. **Epidermis**: cells narrow, elongated, thin-walled. **Cortex**: *c*.25 layers of very thin-walled parenchyma cells with well-developed intercellular spaces (Fig. 82); some cells containing mucilage and raphides. Sclerenchyma sheath absent. **Vascular bundles** in several circles (Fig. 82), bundles lacking sclerenchyma; starch in cells surrounding vascular bundles. Central cavity. **Stegmata** absent.

Rhizome TS (Epipogium, Nervilia)

Scale leaves adpressed to rhizome in *Epipogium roseum* (as *E. nutans*), bearing elongated hairs with thin, non-cuticularized walls and often containing fungal hyphae. **Epidermis**: cells tangentially elongated; bearing unicellular hairs in *Nervilia crociformis* (as *N. monantha*) and papillae bearing groups of hairs in *N. aragoana*. **Cortex**: cells elliptical and parenchymatous. Fungal hyphae in epidermis and cortex of *N. crociformis* and *N. aragoana*, not in *Epipogium*. **Endodermis**: one layer of oblong to slightly polygonal cells; cells with prominent casparian strips in *N. crociformis*. Cortex and ground-tissue cells containing starch granules, some cells with mucilage, raphide bundles, or solitary crystals. **Vascular cylinder** lacking sclerenchyma ring; vascular bundles collateral, from six to over 30, in peripheral ring and scattered in centre. **Stegmata** absent.

Tuber TS (Nervilia)

Weakly cutinized tubers are storage organs formed at ends of rhizome or stolon. Surface irregular, warty, bearing tiny roots. Non-functional **stomata** with two subsidiary cells occurring next to hairs on scale leaves. Structure similar to rhizome except endodermis not easily distinguishable. **Central cylinder** surrounded by an 11- or 12-layered cortex with wide-lumened cells bearing starch granules; large water-storage cells in *N. aragoana*. **Vascular bundles** collateral, numerous, distributed in ground tissue of central cylinder; outer bundles arranged in a circle at fixed distance from epidermis; inner bundles scattered.

Root TS (Nervilia)

Velamen absent. **Epidermis**: rectangular to irregularly polygonal cells bearing root hairs. **Cortex**: cells very thin-walled, parenchymatous, containing starch granules, some also with mucilage, raphide bundles, or solitary crystals. **Endodermis**: one layer of rectangular cells with casparian strips. **Vascular cylinder** 4–8(–15)-arch. **Pith** central. Roots lacking in some species, e.g. *Epipogium roseum* and *Nervilia crociformis*.

Material reported

The treatment of Nervilieae was prepared from only a few published sources varying in numbers of taxa yielding the report above. Most significant of these sources were those of Moreau (1912), Namba, Lin et al. (1981a,b), and Burgeff (1932). Moreau published the anatomy of *Nervilia* under the mistaken identity of *Pogonia*.

Reports from the literature

Burgeff (1932): vegetative anatomy and mycorrhiza of *Epipogium* and *Nervilia aragoana*. Dietz (1930): tuber anatomy of *Epipogium roseum* (as *E. nutans*), *Nervilia aragoana*, and *N. crociformis* (as *Pogonia crispata*). Fuchs and Ziegenspeck (1927): stem anatomy of *Epipogium*. Groom (1895–7): vegetative anatomy of *E. roseum* (as *E. nutans*). Hermans et al. (2007): checklist of orchids of Madagascar including *Nervilia*. Johow (1889): anatomy of *Epipogium*. Møller and Rasmussen (1984): stegmata absent from *E. aphyllum* and *N. plicata*. Moreau (1912): anatomical and developmental anatomy of *Nervilia* misconstrued as species of *Pogonia* from Madagascar. Namba, Lin et al. (1981a,b): anatomy of *Nervilia* species, especially *N. aragoana* and *N. plicata* (as *N. purpurea*), prepared for drug use; organization of text is confusing. Pirwitz (1931): 'storage tracheids' in *Epipogium*. Porembski and Barthlott (1988): simple rhizodermis in *N. bicarinata* and *N. infundibulifolia*. Pridgeon et al. (2005): comprehensive review of tribe Nervilieae. Rao (1998): leaf anatomy of *N. aragoana* and *N. crociformis*. Raunkiaer (1895–9): anatomy of

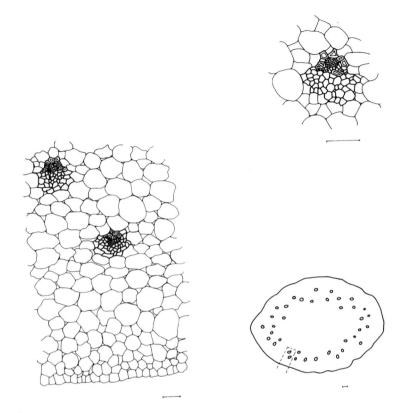

Fig. 82. Nervilieae. *Epipogium roseum*. TS stem with scattered collateral vascular bundles.

Epipogium. Reinke (1873): rhizome anatomy of *Epipogium*. Roy, Yagame et al. (2009): mycorrhiza of *Epipogium*. Sasikumar (1973a): stem anatomy of *N. crociformis* (as *N. monantha*). Solereder and Meyer (1930): survey of orchid anatomy. Wahrlich (1886): mycorrhiza of *Epipogium*.

Taxonomic notes

Anatomical data are too sparse to elaborate any taxonomy.

Tribe Podochileae Pfitzer

Podochileae are usually epiphytic and lithophytic herbs with velamentous, wiry, hairy roots. They are rhizomatous, pseudobulbous or slender-stemmed. Leaves are conduplicate, leathery, distichous, sometimes laterally flattened and fleshy, articulated or non-articulated at the leaf base. The tribe is divided into two subtribes, Eriinae and Thelasinae, with 29 genera widely dispersed from India and Sri Lanka in the west to China, the Philippines, South East Asia, the Malay Archipelago, south-western Pacific islands, Australia, and New Zealand. *Stolzia* occurs on tropical Africa. *Phreatia* is the largest genus with 190 species; *Notheria* and *Ridleyella* are monospecific. Species of *Ceratostylis*, *Cryptochilus*, and *Eria* appear in some hobby collections.

Information on this tribe has been compiled mainly from the literature and the line drawings, but the names now accepted have been used where possible. Where data are missing for any tissue or organ it is because none was available.

Leaf surface

Hairs ad- and abaxial, with one or two sunken basal cells (Figs 84B,C, 87C); 3–5 cells arising from base in *Eria*,

Subfamily Epidendroideae 151

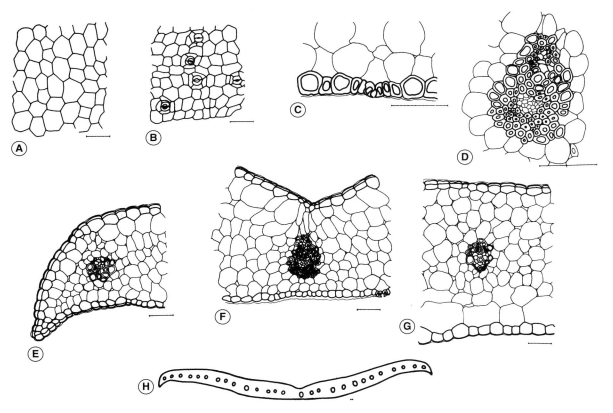

Fig. 83. Podochileae. *Cryptochilus luteus* leaf. (A) Adaxial epidermis with polygonal, curvilinear/straight-sided cell walls. (B) Abaxial epidermis with polygonal, curvilinear/straight-sided cell walls and tetracytic stomata. (C) TS with superficial stoma and thick-walled epidermal cells. (D) TS collateral vascular bundle with thick-walled bundle sheath cells, phloem and xylem sclerenchyma groups. (E–G) TS with one layer of thin-walled ad- and abaxial hypodermal cells, and homogeneous mesophyll. (H) TS with one row of vascular bundles.

branched or tufted according to taxon; two stiff hair cells branching from base in *Trichotosia*, one 2–6 times length of other (further details in Rosinski 1992); hairs adaxial in *Conchidium muscicola* (as *Eria*), abaxial in *E. javanica* (as *E. stellata*). **Epidermis**: cells polygonal (Figs 83A,B–87A,B), thin-walled in *Appendicula angustifolia* (as *A. monoceras*), *Cryptochilus luteus* (Fig. 83A,B), *Eria ornata*, and *E. rosea*. **Stomata** abaxial; whole surface abaxial in terete leaves of *Campanulorchis pellipes* (as *Eria*), *E. pannea*, and *Octarrhena*; tetracytic in all taxa (Figs 83B–86B), occasional stomata with five or six subsidiary cells (Fig. 87B), frequently cyclocytic in *Pseuderia* (Fig. 123D, p. 124).

Leaf TS
Cuticle: adaxial 1–5(–10) µm thick, abaxial 3–5 µm thick in *Eria*, adaxial 3–6 µm thick, abaxial 3–5 µm thick in

Trichotosia; cuticle covering epidermal anticlinal walls adjacent to hair bases. **Epidermis**: adaxial cells in *Conchidium muscicola* occupying one-third of leaf thickness; abaxial cells thick-walled in many species (Figs 83C–87C). **Stomata** superficial (Figs 83C–87C). **Hypodermis**: large water-storage cells in one layer in *Octarrhena*, in one or several ad- and abaxial layers in *Eria* species (Fig. 84E–G), except *E. nutans* and *E. paniculata*, in one or two layers in *Appendicula angustifolia* (as *A. monoceras*), in several adaxial and one abaxial layer(s) in *Thelasis*, in several adaxial layers in *Trichotosia*; absent in *Pseuderia*. **Fibre bundles** in one row in mesophyll in *Appendicula*; absent in *Pseuderia*. **Mesophyll** heterogeneous or homogeneous in *Eria* (Fig. 84E–G); heterogeneous in *Octarrhena*, *Phreatia* (Fig. 85E–G), and *Thelasis*; homogeneous, composed of parenchyma with large intercellular spaces in *Conchidium muscicola*,

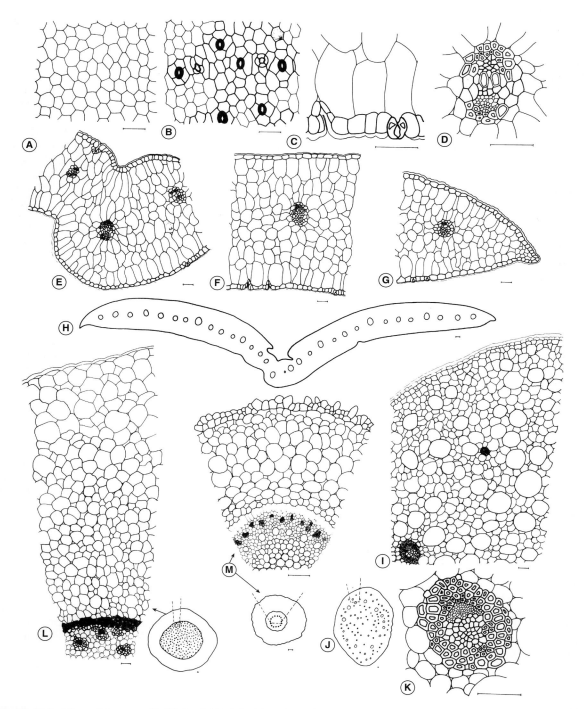

Fig. 84. Podochileae. *Eria rosea*. (A–H) Leaf. (A) Adaxial surface with polygonal cells and curvilinear/straight-sided cell walls. (B) Abaxial surface with polygonal cells, mainly straight-sided cell walls, and tetracytic stomata. (C) TS with superficial stoma and thick cuticle. (D) TS collateral vascular bundle with thin-walled bundle sheath cells, xylem and phloem sclerenchyma caps. (E–G) TS with superficial stomata, ad- and abaxial thin-walled hypodermal cells, and homogeneous mesophyll. (H) TS lamina with one row of vascular bundles. (I–K) TS pseudobulb with 1-layered hypodermis, scattered vascular bundles, mainly circular water-storage cells. (L) TS rhizome, with broad cortex and scattered vascular bundles. (M) TS root with 1-layered velamen, thin-walled exo- and endodermal cells, thin-walled cortical and pith cells.

Cryptochilus (Fig. 83E–G), *Ridleyella* (Fig. 86D–F), and *Stolzia* (Fig. 87E–G), and of elongated oval cells and large water-storage cells in *Trichotosia*; subepidermal cells polygonal, remainder ±oval in *Pseuderia*, lacking water-storage cells. Idioblasts in hypodermis and mesophyll in most taxa, including mucilage cells and raphide cells. **Vascular bundles** in one row or sometimes two rows (Figs 83D,H–85D,H, 86G, 87H). **Stegmata** with spherical silica bodies occurring in species of *Appendicula*, *Ceratostylis*, *Eria*, and *Podochilus*; conical in *E. javanica*; absent in *Conchidium*, *Phreatia*, *Porpax*, *Pseuderia*, *Stolzia*, *Thelasis*, and *Eria muscicola* (now in *Conchidium*).

Stem TS

Hypodermis: 1-layered in pseudobulb of *Eria rosea* (Fig. 84I). **Ground tissue** thin-walled in *Appendicula angustifolia*, *Eria bicristata*, and *E. ovata* (both now in *Pinalia*), with intercellular spaces; water-storage cells in pseudobulb of *E. rosea* (Fig. 84I); outer cortical cells thick- or very thick-walled in *Pseuderia*, decreasing in thickness towards central ground tissue (Fig. 88A), scattered gelatinous fibres present. Sclerenchyma ring between cortex and ground tissue in rhizome of *E. rosea* (Fig. 84L). **Vascular bundles** in a circle and scattered (Figs 84J,L, 86H, 87I); larger bundles encircled by sclerenchyma sheath (Fig. 84K), smaller bundles with sclerenchyma only around phloem. Thin-walled bundle sheath cells in rhizome of *E. rosea* (Fig. 84L). **Stegmata**: spherical silica bodies aggregated around phloem in *A. angustifolia* and adjacent to xylem sclerenchyma in *E. bicristata*; silica bodies in both genera absent from bridge sclerenchyma. Spherical silica bodies also seen in two

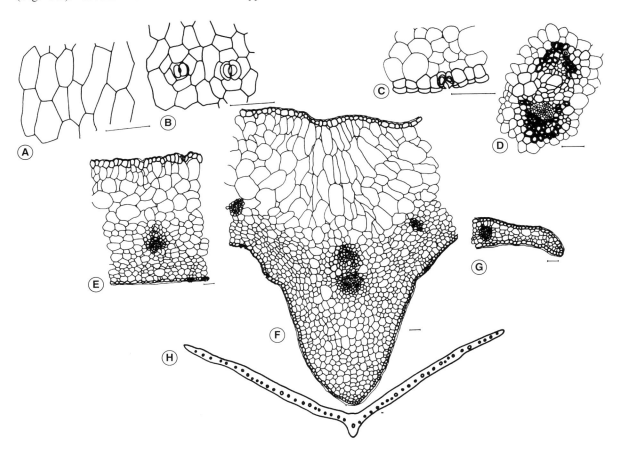

Fig. 85. Podochileae. *Phreatia micrantha* leaf. (A) Adaxial epidermis with elongated polygonal cells and curvilinear cell walls. (B) Abaxial surface with polygonal cells, curvilinear cell walls and tetracytic stomata. (C) TS showing superficial stoma and thick cuticle. (D) TS collateral vascular bundle with phloem and xylem sclerenchyma. (E–G) TS with adaxial, 1-layered hypodermis, extensive adaxial mesophyll in part with upright cells, remainder homogeneous, and one row of vascular bundles. (H) TS with one row of vascular bundles.

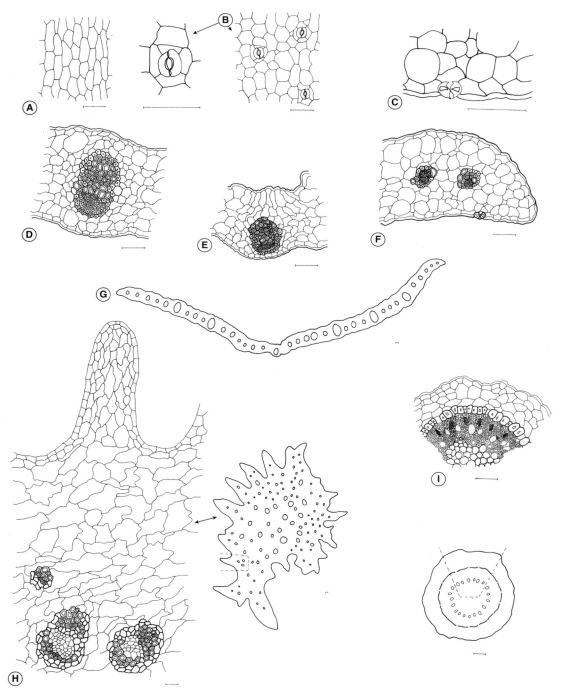

Fig. 86. Podochileae. *Ridleyella paniculata*. (A–G) Leaf. (A) Adaxial surface with elongated polygonal cells and curvilinear cell walls. (B) Abaxial surface with more or less isodiametric polygonal cells with curvilinear cell walls and tetracytic stomata. (C) TS with superficial stoma and thick cuticle. (D–F) TS with homogeneous mesophyll, ad- and abaxial hypodermises, collateral vascular bundles with thick-walled bundle sheath cells, xylem and phloem sclerenchyma. (G) TS with one row of wider and narrower vascular bundles. (H) TS stem with irregularly shaped ground-tissue cells, scattered collateral vascular bundles with thin-walled bundle sheath cells, and xylem and phloem sclerenchyma caps. (I) TS root with thin-walled cortical cells, O-thickened endodermal cells, vascular elements embedded in thick-walled sclerenchyma cells, and thin-walled pith cells.

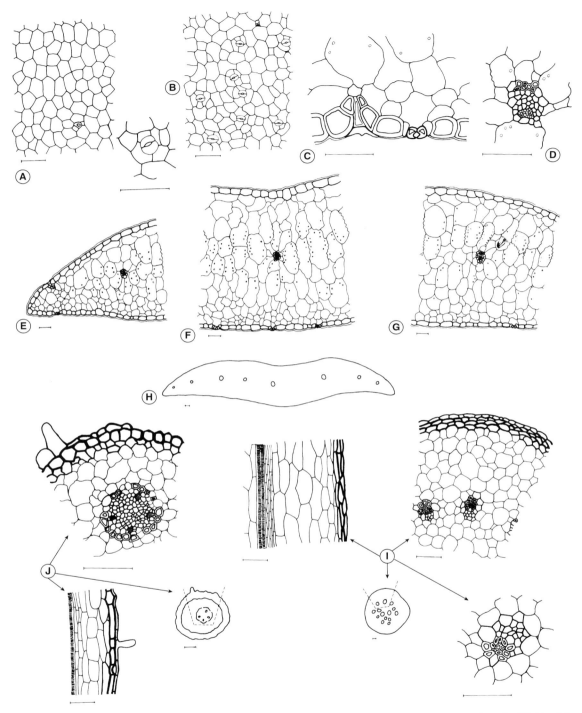

Fig. 87. Podochileae. *Stolzia repens*. (A–H) Leaf. (A) Adaxial surface with polygonal cells and curvilinear, thin walls. (B) Abaxial surface with polygonal cells and curvilinear walls, mainly tetracytic stomata. (C) TS with superficial stoma, thick-walled epidermal cells and hair base. (D) TS vascular bundle. (E–G) TS with heterogeneous/homogeneous mesophyll and superficial stomata. (H) TS with one row of vascular bundles. (I) TS and LS stem with scattered vascular bundles, thick-walled outer cortical cells, and thin-walled ground-tissue cells. (J) TS and LS root with one layer of thick-walled velamen cells, exodermis with moderately thick-walled cells, strongly thickened endodermal cell walls, and thin-walled cortical and pith cells.

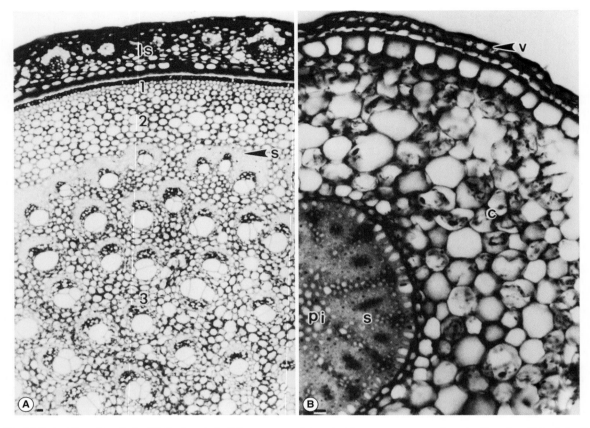

Fig. 88. Podochileae. *Pseuderia*. (A) *P. platyphylla*, TS stem, ground tissue lacking water-storage cells, outer tissue band is leaf sheath (ls), zone 1 cells subtending epidermis, zone 2 cells subtending zone 1 cells external to sclerenchyma (s) ring of vascular bundles, and zone 3 cells constituting central cauline core with vascular bundles. (B) *P. smithiana*, TS root with 2-layered velamen (v) subtended by thin-walled almost isodiametric hypodermal cells; cortex consisting of thin-walled parenchyma cells; vascular tissues embedded in sclerenchyma (s) surrounding thick-walled parenchymatous pith cells (pi). Scale bars = 10 μm. Reprinted from Morris et al. (1996).

other *Appendicula* species, *E. myristiciformis*, and rhizome of *Ceratostylis*; none seen in *Conchidium muscicola*, *Phreatia*, *Podochilus*, *Porpax*, *Pseuderia*, and *Stolzia*.

Root TS

Velamen in both subtribes of *Calanthe*-type (Figs 84M, 86I, 87J); 1-layered in *Bryobium eriaeoides* (as *Eria*), *Ceratostylis*, *Epiblastus*, *Eria japonica* (as *E. arisanensis*), *Eria lasiopetala* (as *E. albidotomentosa*), *E. laniceps*, and *E. rosea* (Fig. 84M), *Mediocalcar*, *Stolzia* (Fig. 87J), and *Thelasis*, 1- or 2-layered in *Appendicula* and *Phreatia*, 2-layered in *Eria ornata*, 2- or 3-layered in *Pseuderia* (Fig. 88B), 2- or 4-layered in *Trichotosia*, 3- or 4-layered in *E. javanica*, and 3- or more-layered in *Conchidium nanum* (as *Eria*). **Tilosomes** absent. **Exodermis**: cells with ∩-thickenings in *Appendicula*, *Ceratostylis*, *Conchidium nanum*, *Phreatia*, and *Trichotosia velutina*; slightly U-shaped thickenings in *Pseuderia*. **Cortex** parenchymatous (Fig. 84M), with water-storage cells in middle region in *Eria coronaria* and *E. crassicaulis* and with scattered large raphide-containing cells in *Pseuderia*. **Endodermis** with 1–3 thick-walled cells alternating with 3–5 passage cells in *Eria ornata* (now in *Callostylis*), *Ridleyella paniculata* (Fig. 86I), and *Stolzia repens* (Fig. 87J); cells thick-walled in *E. japonica* (as *E. arisanensis*). **Vascular cylinder** in *Eria* species and *R. paniculata* (Fig. 86I) with 6–22 xylem rays, with 11–18 xylem rays in *Pseuderia*. **Pith**: central parenchyma thin-walled (Figs 84M, 86I, 87J); except thick-walled in *Pseuderia* (Fig. 88B); pith cells in *E. javanica* rich in starch; pith in *E. crassicaulis* and *E. paniculata* (now in *Mycaranthes*) composed of lignified fibres.

Material reported

Eriinae: *Appendicula* (1), *Ceratostylis* (1), *Epiblastus* (1), *Eria* (27 for leaf, including species now transferred to *Bryobium*, *Campanulorchis*, *Conchidium*, and *Mycaranthes*), *Mediocalcar* (1), *Podochilus* (1), *Porpax* (2), *Pseuderia* (6), *Stolzia* (2), *Trichotosia* (8 for leaf).

Thelasinae: *Octarrhena* (1), *Phreatia* (2), *Ridleyella* (1), *Thelasis* (1).

Leaf description based mainly on Rosinski (1992); full details for *Pseuderia* in Morris et al. (1996).

Reports from the literature

Andersen et al. (1988): developmental anatomy of *Eria*. Chiang and Chou (1971): root anatomy of *Eria*. Dressler (1993): taxonomic review of the tribe segregating *Ridleyella* as basis of a separate subtribe, Thelasinae. Dressler and Cook (1988): conical silica bodies in *Eria javanica*. Guttenberg (1968): root anatomy of *Ceratostylis*. Hadley and Williamson (1972): mycorrhiza of *Appendicula*. Isaiah et al. (1990): vegetative anatomy of *Eria*. Janczewski (1885): root anatomy of *Eria*. Katiyar et al. (1986): mycorrhiza of *Eria* and *Phreatia*. Kaushik (1983): vegetative anatomy of *Eria* (including *Mycaranthes*). Kuttelwascher (1964, 1965 [1966]): aerial root anatomy of *Eria*. Leitgeb (1864b): aerial root anatomy of *Eria* (including *Pinalia*) and *Trichotosia*. Löv (1926): leaf anatomy of *Eria*. Meinecke (1894): aerial root anatomy of *Eria*. Möbius (1887): leaf anatomy of *Eria* (including species now in *Pinalia*, also *Tainia stellata*), and *Thelasis*. Møller and Rasmussen (1984): stegmata and silica bodies in *Appendicula*, *Ceratostylis*, *Eria* (also *Conchidium*, *Pinalia*), *Phreatia*, *Podochilus*, *Porpax*, *Stolzia*, and *Thelasis*. Morris et al. (1996): anatomy of subtribe Dendrobiinae, including *Pseuderia*. Mulay et al. (1958): velamen in *Eria*. Neubauer (1978): inflorescence axis idioblasts in *Eria*. Oudemans (1861): aerial root anatomy of *Eria*. Porembski and Barthlott (1988): velamen in many genera of Podochileae. Prasad (1960): velamen in *Eria*. Pridgeon (1987): velamen in several genera of Podochileae. Pridgeon et al. (1983): tilosomes absent in *Appendicula*, *Ceratostylis*, and *Eria*. Pridgeon et al. (2005): comprehensive review of systematics of Podochileae. Raciborski (1898): root anatomy of *Eria*. Rosinski (1992): leaf anatomy of *Eria* (including segregated genera) and *Trichotosia*. Singh (1981, 1986): stomata and root anatomy of *Eria*. Solereder and Meyer (1930): review of work on orchid anatomy with original observations on *Appendicula* and *Eria* (now *Conchidium*). Staudermann (1924): hairs of *Eria*. Subrahmanyam (1974): velamen in pseudobulb of *Eria*. Tominski (1905): leaf anatomy of *Eria* (now *Conchidium*) and *Octarrhena*. Vij et al. (1991): leaf epidermis of *Eria*. Weltz (1897): stem anatomy of *Appendicula* and *Eria*. Williams (1979): subsidiary cells in several taxa of Podochileae.

Taxonomic notes

Spherical silica bodies characterize Podochileae, except for *Eria javanica* where conical silica bodies predominate and the absence of stegmata in *Conchidium muscicola*, *Phreatia*, *Porpax*, *Pseuderia*, *Stolzia*, and *Thelasis*. That silica body shape and presence of stegmata are genetically determined traits has been reported by Møller and Rasmussen (1984) based upon consistent environmental factors in the several groups. That in *Eria* both conical and spherical silica bodies occur implies the genus may be polyphyletic (Dressler and Cook 1988). The occurrence in the tribe as a whole of both spherical and conical silica bodies and the lack of silica bodies in several genera may indicate a polyphyletic condition in this taxon.

Tribe Sobralieae Pfitzer

Plants of tribe Sobralieae are wide-ranging in tropical America, distributed from Mexico and the Caribbean islands throughout South America south to southern Brazil. Sobralieae are especially diverse in the Andes. There are only three genera, the largest of which are *Elleanthus* and *Sobralia* each with about 100 species. *Sertifera* has only six or seven species. Plants are epiphytic, lithophytic, or terrestrial herbs without pseudobulbs but with reed-like stems, sometimes up to 10 m tall in *Sobralia*. Leaves are distichous, plicate, and articulate. *Sobralia* is frequently grown by hobbyists for its large, colourful, but short-lived *Cattleya*-like flowers. This report is based partly on the PhD dissertation of Kurt M. Neubig at the University of Florida and Eastern Illinois University.

Leaf surface

Hairs three or four cells long, sunken, on both leaf surfaces in *Elleanthus*, only on lower surface in *Sobralia* (Fig. 89A); occasionally two or three hairs arising simultaneously from same point. **Epidermis**: cells rectangular. **Stomata** abaxial, tetracytic or with two subsidiary cells.

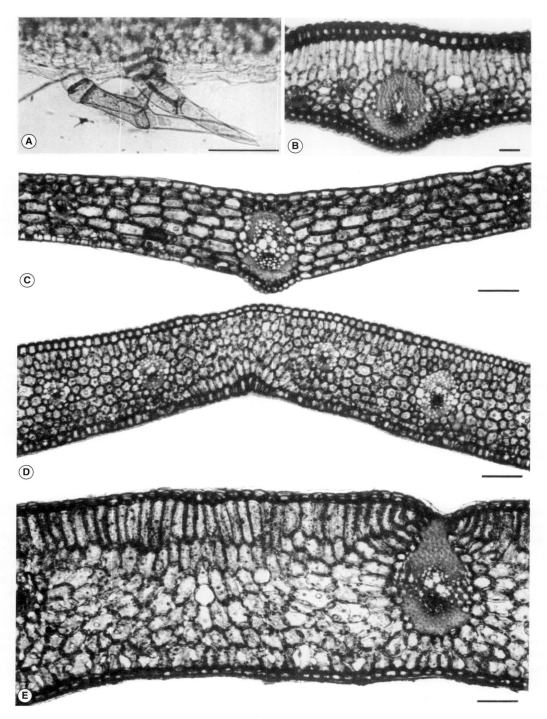

Fig. 89. Sobralieae. *Sobralia* and *Sertifera* leaf. (A) *Sobralia warszewiczii*, TS abaxial, showing several-celled, pointed epidermal hairs. (B) *S. callosa*, TS with heterogeneous mesophyll and vascular tissue surrounded by a multilayered sclerenchyma band. (C) *S. recta*, TS vascular bundle with ad- and abaxial sclerenchyma bands and homogeneous mesophyll. (D) *S. portillae*, TS with heterogeneous mesophyll and one row of vascular bundles. (E) *Sertifera colombiana*, TS with heterogeneous mesophyll, vascular bundles with multilayered sclerenchyma sheath showing adaxial extension reaching adaxial surface. All scale bars = 100 μm. Photographs by K.M. Neubig.

Leaf TS

Epidermis: cells rectangular (Fig. 89C), periclinal (Fig. 89B), or isodiametric (Fig. 89D). **Stomata**: outer ledges thin, inner ledges thick. Substomatal chambers small or large, often irregularly shaped. **Fibre bundles** present in *Elleanthus*, *Sertifera*, *Sobralia ciliata*, *S. crispissima*, *S. dichotoma*, *S. rosea*, *S. theobromina*, and *S. warszewiczii*. **Mesophyll** heterogeneous in *E. ensatus*, *E. stolonifer*, and *Sertifera* (Fig. 89E); homo- or heterogeneous in *Sobralia* (Fig. 89A–D). **Vascular bundles** collateral in one row (Fig. 89D); sclerenchyma crescents at both xylem and phloem poles; embedded in thick-walled parenchyma enclosed by sickle-shaped fibres. **Stegmata** absent.

Stem TS

Cuticle vaulted, very thin. **Epidermis**: cells isodiametric (Fig. 90C,D), outer walls vaulted, thickened. Sclerenchyma ring of three or four layers of polygonal fibres between outer and inner ground tissue. **Ground tissue**: outer ring with rounded cells and large intercellular spaces, chloroplasts present in cells of outermost layers; inner ring with thin-walled cells containing cruciate starch granules and large intercellular spaces (Fig. 90C). **Vascular bundles** inside sclerenchyma ring, scattered in ground tissue and fused with ring (Fig. 90A,B). Bundles with xylem and phloem caps of narrow- to wide-lumened, thick-walled fibres (Fig. 90E). **Stegmata** absent.

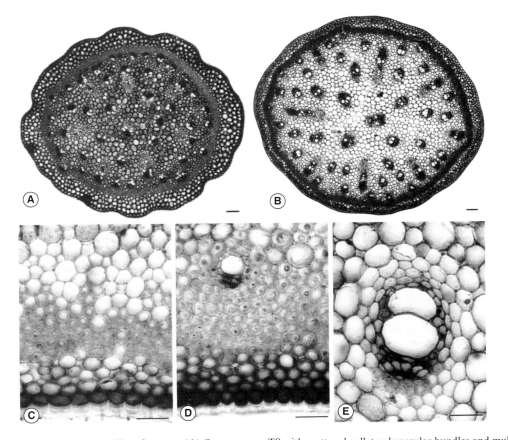

Fig. 90. Sobralieae. *Sobralia* and *Elleanthus* **stem.** (A) *S. mucronata*, TS with scattered collateral vascular bundles and multilayered sclerenchyma band surrounding ground tissue comprising thick-walled parenchyma cells. (B) *Elleanthus longibracteatus*, TS showing scattered collateral vascular bundles, multilayered sclerenchyma band surrounding parenchymatous ground tissue with thin-walled cells. (C) *S. ciliata*, TS having thick-walled more or less isodiametric epidermal cells covered by thick cuticle, portion of sclerenchyma band, cortical tissue with rounded cells and thin-walled, almost circular ground-tissue cells. (D) *S. mandonii*, TS with isodiametric epidermal cells, thick-walled cortical cells, large metaxylem cell of vascular bundle, and circular ground-tissue cells. (E) *S. ciliata*. TS showing details of collateral vascular bundle having two large metaxylem cells, sieve cells, thin-walled bundle sheath cells, angular intercellular spaces among rounded ground-tissue parenchyma cells. All scale bars = 100 μm. Photographs by K.M. Neubig.

Root TS

Velamen two, three, four, or six cell layers wide in *Sobralia*; cells with thick parallel bars; cell walls frequently perforated; two, three, or five layers in *Elleanthus*. **Tilosomes** spongy. **Exodermis** 1-layered with thin-walled passage cells alternating with ∩- and/or O-thickened cell walls. **Cortex**: parenchyma relatively wide in *S. liliastrum* and *S. macrantha*; cell walls spirally thickened in *Elleanthus*. **Endodermis**: 1-layered, cells with O-thickened walls (Fig. 91C,D). **Vascular cylinder** 6–40-arch; *S. liliastrum* with 15 xylem rays alternating with large phloem strands and centre of root woody; *S. macrantha* with 14 xylem rays alternating with small phloem strands and central parenchyma thin-walled. **Pith** heterogeneous (Fig. 91B) or homogeneous (Fig. 91A), heterogeneous sclerenchymatous, homogeneous parenchymatous towards centre.

Material reported

Elleanthus (11), *Sertifera* (1), *Sobralia* (17, including *Epilyna*). Details in Neubig (2012).

Reports from the literature (all for *Sobralia*, except where stated)

Baker (1972): leaf anatomy of *Elleanthus linifolius*. Benzing et al. (1982): root structure of *S. macrantha*. Dietz (1930): underground organs. Dressler (1993): treated tribe Sobralieae as subtribe Sobraliinae in tribe Epidendreae adding genus *Epilyna*, a synonym of *Elleanthus*. Engard (1944): velamen and exodermis. Fellerer (1892): aerial roots. Leitgeb (1864b): aerial roots. Meinecke (1894): aerial roots. Möbius (1887): leaf anatomy of *Elleanthus kermesinus* and *Sobralia macrantha*. Møller and

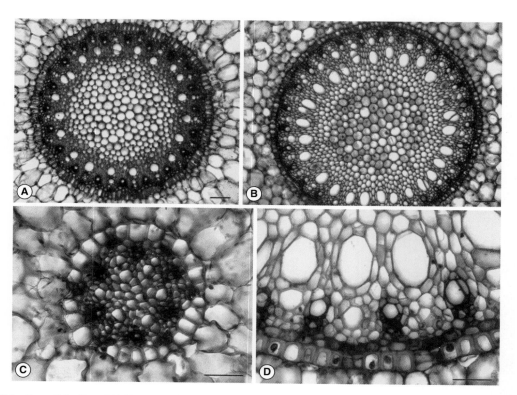

Fig. 91. Sobralieae. *Sobralia* and *Epilyna* root. (A) *S. wilsoniana*, TS vascular cylinder with alternating xylem and phloem strands embedded in a multilayered sclerenchyma band, surrounding circular-celled, homogeneous parenchymatous pith; endodermal cell walls are O-thickened. (B) *S. dichotoma*, TS vascular cylinder showing alternating xylem and phloem strands embedded in thick-walled parenchyma cells surrounding heterogeneous pith having outer ring of small cells and central mass of large circular cells. (C) *Epilyna hirtzii*, TS vascular cylinder with 6-arch xylem rays in the angles of which are phloem strands surrounded by endodermis of O-thickened somewhat anticlinal cells and no pith. (D) *S. dichotoma*, TS arc of vascular tissue with alternating xylem and phloem strands embedded in thick-walled parenchyma cells and surrounded by an endodermis with O-thickened cell walls and thin-walled passage cells opposite xylem strands. All scale bars = 100 μm. Photographs by K.M. Neubig.

Rasmussen (1984): stegmata absent in *Elleanthus longibracteatus* and *Sobralia macrantha*. Neubig (2012): anatomy and systematics of Sobralieae. Oudemans (1861): aerial roots. Porembski and Barthlott (1988): velamen in *Elleanthus* and *Sobralia*. Pridgeon (1987): velamen and exodermis in *Elleanthus* and *Sobralia*. Pridgeon et al. (1993): tilosomes in *Elleanthus* and *Sobralia*. Pridgeon et al. (2005): comprehensive systematic treatment of tribe Sobralieae. Richter (1901): aerial roots. Solereder and Meyer (1930): review of orchid anatomy and original observations on leaf anatomy of *Sobralia macrantha*. Weltz (1897): stem anatomy. Williams (1979): stomata in *Elleanthus* and *Sobralia*.

Taxonomic notes

Sobralia and *Elleanthus*, each with about 100 species, are the largest genera of Sobralieae. *Sobralia* flowers are ephemeral, difficult to preserve as pressed specimens, hence the classification of the genus is still unsatisfactory.

Epilyna has been submerged in *Elleanthus* (Pridgeon et al. 2005) because the floral structure of the former contains few features to differentiate it from the latter. According to B.S. Carlsward (personal communication), however, there appear to be features of vegetative anatomy that tend to separate *Epilyna* from *Elleanthus*: a biseriate velamen, a flattened stem, absence of lacunae between vascular bundles in the leaf sheath, conduplicate vernation, and an abruptly exserted foliar midrib.

Tribe Triphoreae Dressler

Triphoreae consist of two subtribes: Diceratostelinae Summerh. and Triphorineae (Dressler) Szlach. *Diceratostele* of tropical West Africa is a monospecific genus comprising subtribe Diceratostelinae. Triphorinae consists of three American genera: *Monophyllorchis* with two species, *Psilochilus* with nine species, and *Triphora* with about 19 species. *Monophyllorchis* is found from Nicaragua to Colombia, Venezuela, and Ecuador; *Psilochilus* grows on islands of the Caribbean, in Central America,

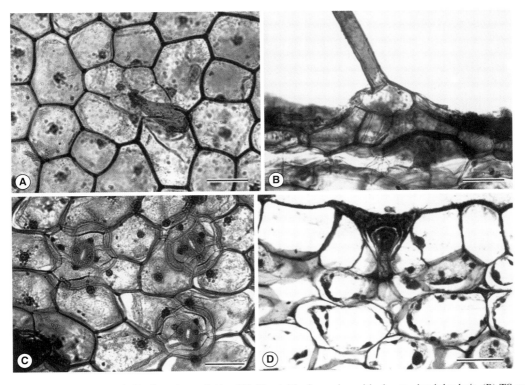

Fig. 92. Triphoreae. (A,B) *Monophyllorchis microstyloides*. (A) Abaxial leaf scraping with clavate glandular hair. (B) TS root with hair borne on velaminal buttress. (C,D) *Psilochilus* cf. *macrophyllus*. (C) Abaxial surface leaf scraping exhibiting both anisocytic and tetracytic stomata. (D) TS leaf showing sunken glandular 3-celled hair. Scale bars: A,C,D = 50 μm, B = 100 μm. Reprinted from Carlsward and Stern (2009).

and tropical South America, and *Triphora* ranges from the eastern USA and southern Ontario, Canada, to the Caribbean islands, through Central America into central South America. These are reed-like, terrestrial, mycorrhizal herbs, with or without roots, sometimes with fleshy or knobby tuberoids, with convolute or plicate, non-articulate leaves that are sometimes bract-like.

Leaf surface

Hairs on both surfaces, glandular; unicellular, sunken in *Diceratostele*, 2-celled in *Monophyllorchis* (Fig. 92A), 3-celled, sunken in *Psilochilus* (Fig. 92D), 3-celled in *Triphora* (Fig. 93A). **Cuticle** smooth. **Epidermis**: adaxial and abaxial cells polygonal with straight-sided and curvilinear walls (Fig. 92C). **Stomata** abaxial, mostly tetracytic in *Diceratostele*, *Monophyllorchis* (Fig. 92C), and *Psilochilus*; largely anomocytic in *Triphora*.

Leaf TS

Cuticle <2.5 μm thick. **Epidermis**: cells variable, squarish, oval, or periclinal (Fig. 93B). **Stomata** superficial,

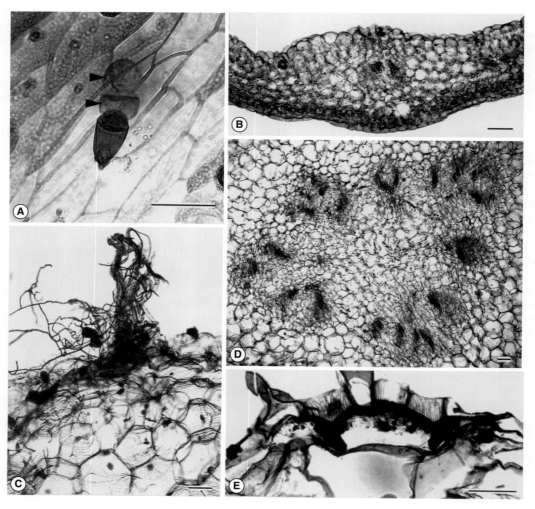

Fig. 93. Triphoreae. *Triphora gentianoides*. (A) Leaf scraping with 3-celled hair (arrowheads); basal cells thin-walled and clear, apical cell thicker-walled and secretory. (B) TS leaf showing two tracheary strands in midvein. (C) TS tuberculate subterranean stem having tufts of unicellular hairs and tangled hyphae. (D) TS vascular cylinder in subterranean stem with U-shaped and paired vascular bundles. (E) TS root with pectinated velaminal cell walls. Scale bars: A,C–E = 50 μm, B = 100 μm. Reprinted from Carlsward and Stern (2009).

substomatal chambers small to moderate; outer ledges small, inner ledges minute or absent. **Hypodermis**: absent, except 1-layered in *Triphora*. **Fibre bundles** absent. **Mesophyll** homogeneous (Fig. 93B), raphide idioblasts circular, intercellular spaces triangular. **Vascular bundles** collateral, in one row; sclerenchyma and **stegmata** absent, except with conical silica bodies adjacent to vascular sclerenchyma in *Diceratostele*. Bundle sheath cells thin-walled with chloroplasts.

Stem TS (Triphora described separately)

Hairs bicellular in *Diceratostele*, absent in *Monophyllorchis*, sunken in *Psilochilus*. **Cuticle** <2.5 µm thick, except moderately thick in *Diceratostele*; smooth. **Hypodermis**: absent. **Cortex**: cells thin-walled, rounded, with chloroplasts; intercellular spaces triangular. Sclerenchyma band (Fig. 94A) present between cortex and ground tissue in *Diceratostele* and *Psilochilus*; absent in *Monophyllorchis*. **Endodermis** 1-layered in *Diceratostele*; absent in other genera. **Vascular bundles** in a ring in *Monophyllorchis* and *Psilochilus*; smaller bundles in a ring in *Diceratostele*, larger bundles scattered in ground tissue. **Stegmata** with conical rough-surfaced silica bodies present in *Diceratostele*; absent in other genera.

Triphora

Stems aerial and subterranean. Aerial and subterranean stem surfaces tuberculate, subterranean stems with tufts of unicellular **hairs** borne on tubercles (Fig. 93C). **Cuticle** in aerial stems <2.5 µm thick, absent in subterranean stems. **Epidermis**: cells thin-walled, rounded in aerial stems, square in subterranean stems. **Hypodermis**: 1-layered in aerial stems, absent in underground stems. **Cortex**: cells thin-walled, rounded to angular, five or six cells wide in aerial stems, 15 cells wide in terrestrial stems. Cells with dead fungal masses and pelotons in subterranean stems; intercellular spaces triangular. **Endodermis** and **pericycle** absent. **Vascular bundles** in a ring, bundles U-shaped, lobed and paired in underground stems (Fig. 93D). Sclerenchyma and stegmata absent.

Root TS

Velamen: two or three cells wide in *Monophyllorchis*, 1-layered in *Psilochilus* (Fig. 94C), two cells wide in *Triphora* (Fig. 93E). Root hairs in velamenal buttresses in *Monophyllorchis* (Fig. 92B), arising internally and adjacent to exodermal passage cells in *Psilochilus* (Fig. 94C). **Tilosomes** absent. **Exodermal** cells thin-walled (Fig. 94C), except thickened in *Diceratostele*. **Cortex**: cells thin-walled, 20 or more cells wide in *Monophyllorchis* and *Triphora* with hyphae and dead fungal masses; 3-layered in *Psilochilus* (Fig. 94C), outer layer 1–3 cells wide, cells fungus-free (Fig. 94C), middle layer two or four cells wide, fungus-filled; inner layer five or six cells wide, fungus-free; parenchymatous in *Diceratostele*, cells containing cruciate starch granules. **Endodermis** 1-layered, cells thin-walled, except thickened on radial and inner surfaces in *Diceratostele*. **Pericycle** 1-layered, cells thick-walled opposite phloem groups, thin-walled opposite xylem strands in *Monophyllorchis*; cells O-thickened in *Diceratostele* and *Psilochilus*; all thin-walled in *Triphora*. **Vascular cylinder**: xylem and phloem groups embedded in parenchyma and alternating in *Monophyllorchis* and *Triphora*; metaxylem organized as a solid scalloped ring in *Psilochilus* with phloem present in the embayments (Fig. 94B); embedded in sclerenchyma in *Diceratostele*; 20–22-arch in *Diceratostele*, 8-arch in *Monophyllorchis*, 33-arch in *Triphora*. **Pith** parenchymatous (Fig. 94B).

Material examined

Diceratostele (1), *Monophyllorchis* (1), *Psilochilus* (1), *Triphora* (2). Full details appear in Stern, Morris et al. (1993c) and Carlsward and Stern (2009).

Reports from the literature

Carlsward and Stern (2009): anatomical review of Triphorinae sans *Diceratostele*. Dressler (1993): comprehensive treatise on phylogeny and classification of Orchidaceae. Holm (1927): leaf anatomy of *Triphora trianthophora* (as *T. pendula*). Pridgeon et al. (2005): detailed taxonomic discussion of Triphoreae including Triphorinae and Diceratostelinae. Stern, Morris et al. (1993c): anatomical review of Cranichideae (as Spiranthoideae) including *Diceratostele*. Williams (1979): stomata of *Monophyllorchis microstyloides*.

Taxonomic notes

Dressler (1993) treated *Diceratostele* in tribe Diceratosteleae. Pridgeon et al. (2005) used data from molecular sequences to indicate a sister-group relationship between *Diceratostele* and the three genera in Triphorinae. They joined *Diceratostele* in subtribal rank with these genera in tribe Triphoreae. Anatomical evidence is somewhat equivocal about this union. *Monophyllorchis*, *Psilochilus*, and *Triphora* are devoid of both sclerenchyma and stegmata that are present in leaves and stems of *Diceratostele*. Moreover, an endodermis is present in stems of *Diceratostele* but not in stems of the other three genera. The vascular cylinder is embedded in parenchyma in roots of *Monophyllorchis* and *Triphora* but in sclerenchyma in

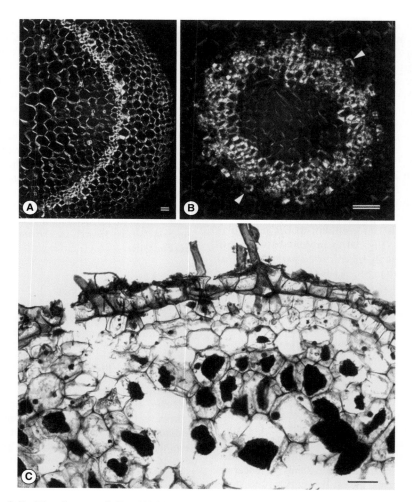

Fig. 94. Triphoreae. *Psilochilus* cf. *macrophyllus*. (A) TS stem with sclerenchyma band between cortex and ground tissue under polarized light. (B) TS root with encircling scalloped metaxylem vascular cylinder and endodermal cells with O-thickened walls (arrowheads) under polarized light. (C) Hairs arising endogenously below velamen, usually opposite a passage cell of exodermis (note the 3-layered cortex and digested hyphae of the mycorrhiza). Scale bars = 100 μm. Reprinted from Carlsward and Stern (2009).

Diceratostele. The relationships postulated by Pridgeon et al. bear further examination in view of these anatomical variations.

Tribe Tropidieae (Pfitzer) Dressler

Tropidieae are terrestrial, upright, somewhat woody, rhizomatous herbs. Leaves are non-articulate, plicate, ovate, elliptical to lanceolate with sheathing bases. The only two genera, *Tropidia* and *Corymborkis*, are widely distributed in Old and New Worlds. *Tropidia* does not occur in Africa, but is found on some Pacific islands; a single species ranges from Florida and the Caribbean islands into South and Central America and the Galapagos Islands. *Corymborkis* grows in central and southern Africa, Indian Ocean and Caribbean islands, Central and South America to Argentina (Pridgeon et al. 2005).

Leaf surface

Hairs on both surfaces unicellular, glandular; bases sunken, tips bulbous. **Epidermis**: anticlinal cell walls deeply sinuate in *Tropidia* on both surfaces. **Stomata** abaxial, tetracytic (Fig. 95), few anomocytic in *Tropidia*.

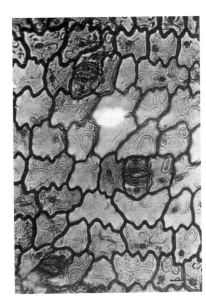

Fig. 95. Tropidieae. *Tropidia curculigoides*. Abaxial leaf epidermis with tetracytic stomata and sinuous anticlinal cell walls. Scale bar = c.10 μm. Reprinted from Stern et al. (1993c).

Leaf TS

Cuticle smooth, thin, on both surfaces. **Epidermis**: cells rectangular, elliptical, oval, thin-walled. **Stomata** superficial, substomatal chambers small, outer ledges small, inner ledges minute. **Hypodermis** and **fibre bundles** absent. **Mesophyll** homogeneous, three or four cells wide, wider near major veins. **Vascular bundles** collateral (Fig. 96A), in one row, surrounded by sclerenchyma. Bundle sheath cells thin-walled, except thick-walled in *C. veratrifolia*. **Stegmata** with rough-surfaced conical silica bodies (Fig. 96B,C) surrounding vascular bundle sclerenchyma. Raphide bundles in elongated, thin-walled idioblasts.

Stem TS

Hairs and **Stomata** absent. **Epidermis**: cells elliptical to rectangular, thick-walled in *Tropidia*, thin-walled in *C. veratrifolia*, outer walls appreciably thickened in *C. forcipigera*. **Ground tissue**: subepidermal parenchyma band two cells wide in *T. curculigoides* (as *T. graminea*), cells thin-walled; 6–8 to many cells wide in *T. polystachya* (Fig. 96D), subtending sclerenchyma many cells wide, cell walls thick to very thick, stratified; 10–13 cells wide in *C. forcipigera*, cells thin-walled (Fig. 96E); 2–4 cells wide in *C. veratrifolia*, thick-walled; subtending sclerenchyma 6–8 cells wide, cell walls thick, stratified in *C. forcipigera*; one or two cells wide in *C. veratrifolia*, cell walls moderately thick,

stratified. **Endodermoid layer** identified only in *C. veratrifolia*. Central ground-tissue cells in *Tropidia* somewhat thick-walled, parenchymatous, cells polygonal to rounded; cruciate starch granules present; intercellular spaces triangular; vascular bundles absent. Central ground-tissue cells in *Corymborkis* thick-walled, rounded; cruciate starch granules present; intercellular spaces triangular; vascular bundles present, collateral, in two or three irregularly disposed series. **Vascular bundles**: in *Tropidia* and *C. forcipigera* outer bundles enclosed by sclerenchyma band (Fig. 96E); in *C. veratrifolia* phloem bundle cap continuous with sclerenchyma band, xylem free. Vascular bundle sclerenchyma only as caps at phloem poles in *Tropidia*, at both poles in *Corymborkis*. **Stegmata** with rough-surfaced, conical silica bodies along outer margin of sclerenchyma band in *Tropidia* and *C. veratrifolia*; absent from vascular bundle sclerenchyma in both genera. Raphide idioblasts in outer ground-tissue parenchyma in *Tropidia*.

Root TS

Velamen 4- to 5-layered in *C. veratrifolia*, cells without spiral thickenings; velamen present in *T. curculigoides* (as *T. graminea*). **Tilosomes** absent. **Exodermis**: cell walls O-thickened in *T. curculigoides* (Fig. 97A), passage cells intermittent, thin-walled; exodermis absent in *Corymborkis* (Fig. 97B). **Cortex**: two regions in *Tropidia* (Fig. 97D), outer region 4–7 cells wide, cell walls very thick, stratified, cells tangentially flattened, inner region three or four cells wide, cells thick-walled, unstratified, cruciate starch granules present. Cortex in *Corymborkis* homogeneous (Fig. 97C), eight or 10 or more cells wide, cells thin-walled, cruciate starch granules and fungal pelotons present. **Endodermis** 1-layered, cell walls O-thickened opposite phloem, walls stratified in *Tropidia*. **Pericycle** 1- or 2-layered, O-thickened opposite phloem, walls stratified in *Tropidia*. **Vascular tissue** embedded in sclerenchyma (Fig. 97C). Vascular cylinder with alternating xylem and phloem elements, 23-arch in *Tropidia*, 6–11-arch in *Corymborkis* (Fig. 97C). Protoxylem radially disposed, metaxylem tangentially oriented in *Tropidia*. Xylem arms extending well into pith in *Tropidia* and *C. forcipigera* (Fig. 97C), confined to perimeter of vascular cylinder in *C. veratrifolia*. **Pith** in *Tropidia* parenchymatous, cells polygonal, fairly thick-walled; cruciate starch granules and small intercellular spaces present. Pith in *C. forcipigera* narrow, cells few, angular, parenchymatous, thick-walled; cruciate starch granules present, intercellular spaces absent. Pith in *C. veratrifolia* broad; cells polygonal, parenchymatous, somewhat thick-walled to sclerotic in different specimens; cruciate starch granules and intercellular spaces present. Raphide bundles in

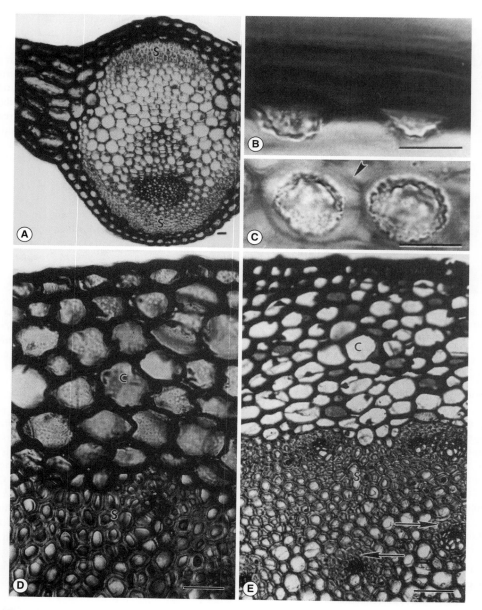

Fig. 96. Tropidieae. *Tropidia* **and** *Corymborkis*. (A) *T. curculigoides*, TS foliar midrib with ad- and abaxial sclerenchyma (S) and homogeneous chlorenchyma. (B,C) *Corymborkis forcipigera*. (B) Conical silica bodies in stegmata in side view associated with vascular bundle sclerenchyma of leaf. (C) Conical silica bodies in stegmata (arrowhead) of leaf in surface view. (D) *T. polystachya*, TS stem showing parenchymatous (c) and subtending sclerenchymatous (s) ground tissue. (E) *C. forcipigera*, TS stem with wide parenchyma band (c) subtending epidermis; outermost vascular bundles embedded in sclerenchyma, and sclerenchyma associated with vascular bundles (arrows). Scale bars = *c*.10 μm. Reprinted from Stern et al. (1993c).

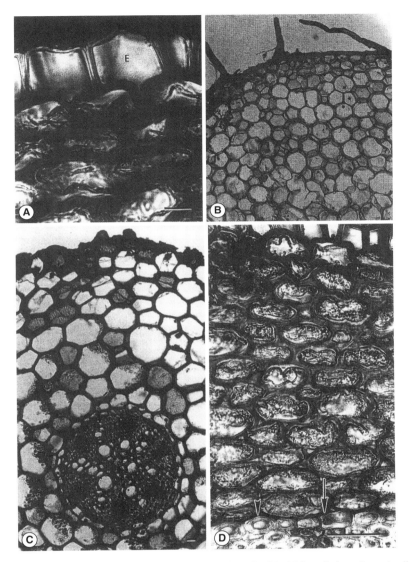

Fig. 97. Tropidieae. *Tropidia* and *Corymborkis* root. (A) *T. curculigoides*, TS with thick-walled exodermal cells. (B) *C. veratrifolia*, TS, velamen and exodermis absent, ×95. (C) *C. forcipigera*, TS, velamen and exodermis absent, endodermis with O-thickened cell walls. (D) *T. curculigoides*, TS showing 2-layered cortex without intercellular spaces in outer layer, thick-walled endodermal cells (arrowhead) and thin-walled passage cells (arrow). Scale bars = $c.10$ μm. Reprinted from Stern et al. (1993c).

saccate, thin-walled idioblasts among thick-walled cortical cells in *Tropidia*; in cortical idioblasts in *C. veratrifolia*, not seen in *C. forcipigera*.

Material examined

Corymborkis (3), *Tropidia* (2). Full details appear in Stern, Morris et al. (1993c).

Reports from the literature

Dressler (1993): discussions of relationships of tribe Tropideae. Porembski and Barthlott (1988): velamen in *Corymborkis veratrifolia*. Pridgeon et al. (2005): comprehensive analysis of tribe Tropidieae. Rasmussen (1977): taxonomic revision of *Corymborkis*. Stern, Morris et al. (1993c): vegetative anatomy of Spiranthoideae.

Taxonomic notes

Tropideae have been treated as a tribe by Dressler (1993), parallel with tribe Diceratosteleae, in his subfamily Spiranthoideae. On the other hand, Pridgeon et al. (2005) have maintained Tropidieae, but as a member of subfamily Epidendroideae, removed from its former placement in subfamily Spiranthoideae. Diceratosteleae have been reduced by them to subtribe Diceratostelinae in tribe Triphoreae, subfamily Epidendroideae, and Dressler's tribe Cranichideae has been moved to subfamily Orchidoideae owing to recent molecular and embryological studies indicating a position as sister to orchidoid tribe Diurideae.

Tribe Xerorchideae P.J.Cribb

The sole genus, *Xerorchis*, consists of two species occurring in Brazil, the Guyanas, Venezuela, and Peru; there is no information on the anatomy (Pridgeon et al. 2005).

Tribe Cymbidieae Pfitzer

Cymbidieae comprise predominantly sympodial epiphytes, terrestrials, lithophytes, and rarely leafless mycotrophs with a cormous or pseudobulbous structure. Leaves are various, one to several, plicate, conduplicate, convolute, soft or fleshy to hard coriaceous, and articulate or not. Cymbidieae are pantropical: Catasetinae and Coeliopsidinae are tropical American, Eulophinae and Cymbidiinae occur in the Old World, Cyrtopodiinae and Eriopsidinae each with a single genus are present in the New World tropics, Maxillariinae are restricted to the Neotropics as are Oncidiinae and Stanhopeinae. Vargasiellinae has two species, one each in Venezuela and Peru, and Zygopetalinae occur throughout the American tropics. Many genera of Cymbidieae comprise highly desired ornamental species, namely *Cymbidium*, *Oncidium*, *Stanhopea*, *Catasetum*, *Maxillaria*, and *Lycaste*.

Subtribe Catasetinae Schltr.

Plants in subtribe Catasetinae comprise a distinctive tropical American group distributed from Florida and the Antilles to Mexico, northern Argentina and Bolivia. *Catasetum* and *Cycnoches* are almost unique among orchids because of the unisexual flowers in most species. In these genera the pollinaria are ejected forcefully when the proper floral trigger is moved by the pollinator so that they strike it and adhere to its body. There are seven genera. Catasetums and galeandras are grown by hobbyists. *Catasetum* has been used medicinally in Guyana, Paraguay, and Japan.

Leaf surface

Hairs on both surfaces, 3-celled, sunken, except 5-celled in *Galeandra batemanii*; absent in *Grobya*. **Stomata** abaxial; anomocytic and tetracytic in *Catasetum* (Fig. 98A, Pl. 2C), *Clowesia*, *Dressleria*, and *Mormodes*; tetracytic in *Grobya*; tetracytic and actinocytic in *Galeandra*; actinocytic in *Cycnoches* (Fig. 98B). **Epidermis**: cells polygonal (Pl. 2B).

Leaf TS

Cuticle smooth, 2.5–50.0 μm thick. **Epidermis**: cells isodiametric and periclinal (Fig. 98C). **Stomata** superficial, substomatal chambers small; large in *Clowesia glaucoglossa* and *Dressleria*, outer ledges small in *Galeandra*, small to moderate in *Clowesia russelliana*, *Cycnoches*, *Dressleria*, *Grobya*, and some *Mormodes* species; large in *Clowesia thylaciochila*, *C. glaucoglossa*, and *Mormodes sinuata*; inner ledges minute to absent, except moderate in *Galeandra*, prominent in *Clowesia glaucoglossa*. **Hypodermis**: present ad- and abaxially in *Galeandra* and *Grobya*. **Fibre bundles** in ad- and abaxial rows with a few scattered mid-mesophyll in *Catsetum* (Fig. 98C), *Clowesia russelliana*, *C. thylaciochila*, *Cycnoches lehmannii*, *Dressleria*, *Galeandra*, *Grobya*, and *Mormodes*; randomly distributed in *Cycnoches chlorochilon* and *C. warszewiczii*. **Mesophyll** homogeneous (Fig. 98C), cells thin-walled, variously shaped; crystalliferous idioblasts thin-walled, circular; mucilaginous idioblasts with a peripheral coil of fine cellulosic filaments in *Cycnoches* (Fig. 98D) randomly distributed in the mesophyll. **Vascular bundles** collateral, in one row, sclerenchyma caps at both xylem and phloem poles, except surrounding conductive tissue in *Galeandra*. **Stegmata** bearing conical, rough-surfaced silica bodies associated with vascular bundle sclerenchyma and all fibre bundles (Pl. 2B), except those of *Grobya* and *Mormodes*. Chlorophyllous bundle sheath cells thin-walled, mostly not well organized, discontinuous, or absent, except in *Catasetum fimbriatum*, *C. punctatum*, and *Clowesia glaucoglossa* where forming a continuous ring around vascular bundle.

Pseudobulb TS

Hairs and **stomata** absent. **Cuticle** smooth, 2.5–22.5 μm thick (Fig. 99A). **Epidermis**: cells with strongly thickened external walls tapering internally towards thin inner wall

Subfamily Epidendroideae 169

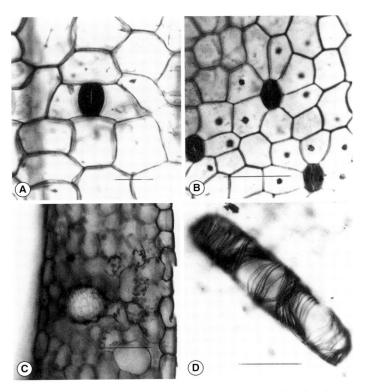

Fig. 98. Catasetinae. *Catasetum* **and** *Cycnoches* **leaf**. (A) *Catasetum planiceps*, abaxial surface showing tetracytic stoma and polygonal epidermal cell walls. (B) *Cycnoches chlorochilon*, abaxial surface with characteristic actinocytic stomata. (C) *Catasetum punctatum*, TS showing fibre bundle (left), raphide idioblast (right), and homogeneous mesophyll. (D) *Cycnoches lehmannii*, mucilaginous idioblast with spiral cell wall banding. Scale bars: A,D = 25 μm, B,C = 50 μm. Reprinted from Stern and Judd (2001).

(Fig. 99A). Ground tissue with numerous, thin-walled, variously shaped polygonal, sometimes chlorophyllous, often starch-containing (Pl. 1C,D) assimilatory cells surrounding fewer, larger, thin-walled, enucleate, globose, water-storage cells (Pl. 2A). Water-storage cells with finely branched helical thickenings in *Grobya* (Fig. 99C). Idioblasts with spiral cell wall thickenings and mucilaginous cells with peripheral coils of thread-like cellulosic material present in *Cycnoches* and with banded walls in *Mormodes*. **Vascular bundles** collateral, numerous, scattered in *Catasetum*, *Cycnoches*, and other genera; phloem associated with thick-walled sclerenchyma (Fig. 99D); xylem subtended by thick-walled parenchyma (Fig. 99D) in these genera and others; sclerenchyma only at phloem poles in *Galeandra* and *Grobya*. **Stegmata** next to phloem sclerenchyma with rough-surfaced conical silica bodies in *Catasetum*, *Clowesia*, *Dressleria*, and smooth-appearing silica bodies in *Cycnoches*; stegmata absent in *Mormodes*.

Root TS

Velamen 6–12 cells wide in *Catasetum* and *Clowesia*, 5–10 cells wide in *Cycnoches*, 9–12 in *Dressleria* (Fig. 100E), 6–8 in *Galeandra*, 6–10 in *Grobya*, and 5–9 cells wide in *Mormodes*. Cells thin-walled with fine, coalescing spiral thickenings (Fig. 100A); epivelamen cells isodiametric and periclinal, endovelamen cells anticlinal (Fig. 100F). **Hairs** single-celled. **Exodermis**: cell walls thin (Fig. 100B), except outer and radial walls thickened in *Grobya*; cells polygonal, isodiametric, sometimes anticlinal. **Cortex**: cells thin-walled, 8–15 cells wide with thin coalescing bands of cell wall material in *Catasetum atratum* (Fig. 100C) and *C. callosum*; water-storage cells present in *C. expansum*; masses of dead fungi and pelotons present in some specimens; 6–9 cells wide with branched bars in *Clowesia russelliana*; 7–13 cells wide in *Cycnoches* with mucilaginous spirally banded cells (Fig. 99B); 18–20 cells wide with broad-banded, branched, and coalescent thickenings in *Dressleria*; 7–11 cells wide with branched bars in

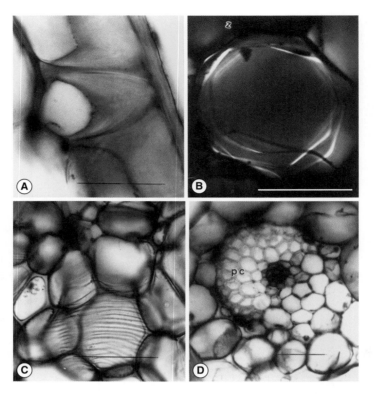

Fig. 99. Catasetinae. (A) *Catasetum pileatum*, TS pseudobulb with thickened cuticle, epidermis, and thick-walled epidermal cells. (B) *Cycnoches chlorochilon*, TS root loose strips of coils of cell wall material in mucilaginous idioblast, polarized. (C) *Grobya galeata*, TS pseudobulb, cells marked with strips of cell wall material. (D) *Clowesia glaucoglossa*, TS pseudobulb vascular bundle with xylem lacking sclerenchyma cap (pc). Scale bars: A,C = 100 μm, B,D = 25 μm. Reprinted from Stern and Judd (2001, 2002).

Galeandra baueri but not *G. batemanii*, 11–15 cells wide with broad branched bars in *Grobya*, 10–13 cells wide in *Mormodes*, cells with broad branched thickenings in *M. aromatica* and *M. sinuata*. **Endodermis** 1-layered, cells periclinal and isodiametric, cell walls O-thickened opposite phloem, thin-walled opposite xylem in most taxa; thin-walled throughout in *Catasetum tabulare*, *Clowesia glaucoglossa* (Fig. 100D), *Galeandra*, and *Grobya*. **Pericycle** 1-layered, cells periclinal and isodiametric, O-thickened opposite phloem, thin-walled opposite xylem; thin-walled throughout in *Catasetum tabulare*. **Vascular cylinder**: xylem and phloem alternating; 11–17-arch in *Catasetum*, 8–13-arch in *Clowesia*, 12–17-arch in *Cycnoches*, 24-arch in *Dressleria*; 11- or 12-arch in *Galeandra*, 12-arch in *Grobya*, 12- or 13-arch in *Mormodes*. Vascular tissue embedded in thick-walled parenchyma in some *Catasetum* species, *Clowesia*, *Dressleria*, *Mormodes aromatica*. and *M. paraensis*, but in sclerenchyma in *Catasetum expansum* and *Mormodes sinuata*.

Pith parenchymatous, cells thin-walled almost circular in *Catasetum*, *Cycnoches*, *Dressleria*, and *Mormodes*; thick-walled parenchyma in *Clowesia*. Intercellular spaces triangular.

Material examined

Catasetum (9), *Clowesia* (3), *Cycnoches* (4), *Dressleria* (1), *Galeandra* (2), *Grobya* (1), *Mormodes* (3). Full details appear in Stern and Judd (2001) and, for *Grobya*, Stern and Judd (2002).

Reports from the literature

Barthlott (1976): velamen of *Catasetum*. Barthlott and Capesius (1975): velamen of *Catasetum*. Burr and Barthlott (1991): root anatomy of Catasetinae. Chatin (1856, 1857): vegetative anatomy of *Catasetum*. Dressler (1993): phylogeny and classification of Orchidaceae. Holst

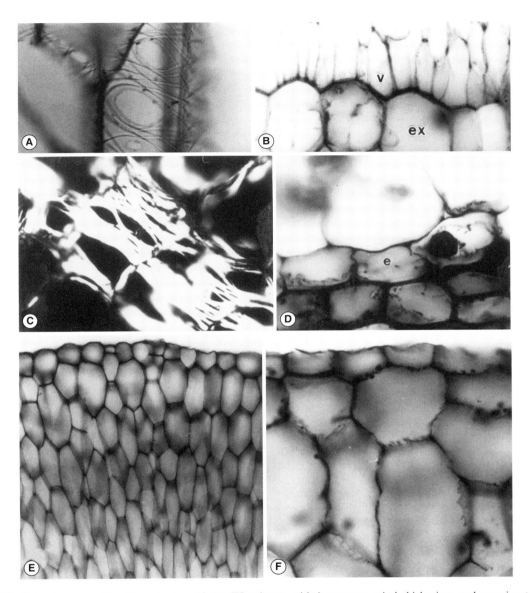

Fig. 100. **Catasetinae root.** (A) *Clowesia glaucoglossa*, TS velamen with interwoven spiral thickenings and prominent loops. (B) *Dressleria dilecta*, exodermal cells (ex) significantly wider than adjacent cells of velamen (v). (C) *Catasetum atratum*, TS cortical cells showing branched bars of cell wall thickenings, polarized. (D) *Clowesia glaucoglossa*, TS with thin-walled endodermal cells (e). (E) *Dressleria dilecta*, TS velamen showing isodiametric epivelamen cells and anticlinal endovelamen cells. (F) *Clowesia russelliana*, TS velamen with periclinal epivelamen cells and clearly delineated endovelamen cells. Scale bars: A = 12 μm, B,C,F = 50 μm, D,E = 25 μm. Reprinted from Stern and Judd (2001).

(1999): monograph of *Catasetum*. Khayota (1995): leaf and root anatomy of *Grobya*. Krüger (1883): anatomy of aerial organs of *Mormodes*. Leitgeb (1864b): aerial root anatomy of *Catasetum*. Möbius (1887): leaf anatomy of *Galeandra* and *Mormodes*. Neubauer (1978): inflorescence idioblasts of *Catasetum*, *Cycnoches*, and *Mormodes*. Oliveira and Sajo (1999, 2001): leaf and stem anatomy of *Catasetum*. Oudemans (1861): aerial root anatomy of *Mormodes*. Pedroso-de-Moraes et al. (2012): root anatomy of 12 species of Catasetinae. Porembski and Barthlott (1988): velamen of Catasetinae. Pridgeon (1987): velamen and exodermis of Catasetinae. Pridgeon and Chase (1998): phylogenetics of Catasetinae. Romero (1990): phylogenetic relationships of Catasetinae. Stern and Judd (2001): comparative anatomy and systematics of Catasetinae. Stern and Judd (2002): comparative anatomy of Cymbidieae, including *Galeandra* and *Grobya*. Weltz (1897): axis anatomy of *Catasetum*. Williams (1976, 1979): stomata in *Catasetum*, *Cycnoches*, and *Grobya*.

Taxonomic notes

Dressler's (1993) Catasetinae included five genera: *Catasetum*, *Clowesia*, *Cycnoches*, *Dressleria*, and *Mormodes*. Pridgeon et al. (2009) included these same genera in Catasetinae plus *Grobya* and *Galeandra*. They recognized that adding the last two genera to Catasetinae 'may seem questionable' except for the homoblastic pseudobulbs, i.e. pseudobulbs with several internodes, that are shared by these taxa. Anatomically, these taxa are much alike except that hairs in leaves of the first five genera of Catasetinae are 3-celled, 5-celled in *Galeandra*, and absent in *Grobya*. Sclerenchyma occurs at both xylem and phloem poles in *Catasetum* and *Grobya*, but it surrounds vascular bundles in *Galeandra*. Stegmata with conical silica bodies associated with fibre bundles and vascular bundle sclerenchyma occur in *Catasetum*, *Clowesia*, *Cycnoches*, *Dressleria*, and *Galeandra* but are absent along fibre bundles in *Grobya*. In these features we have a rather mixed bag. However, molecular evidence demonstrates that *Galeandra* and *Grobya* are successively sister to remaining Catasetinae and form a clade with 100% jackknife support (Freudenstein et al. 2004; Pridgeon et al. 2009).

Subtribe Coeliopsidinae Szlach.

Coeliopsidinae comprise three tropical American genera, *Coeliopsis*, *Lycomormium*, and *Peristeria*, that range from Costa Rica south to Bolivia and Brazil. *Coeliopsis* has only a single species, *Lycomormium* has five, and *Peristeria* has 13. These are mostly epiphytic herbs with pseudobulbs that may bear up to four convolute, articulate, petiolate leaves. *Peristeria* is occasionally grown by hobbyists.

Leaf surface
Hairs 3-celled, glandular, sunken, on both leaf surfaces.
Stomata abaxial, tetracytic.

Leaf TS
Cuticle smooth. **Stomata** superficial. **Hypodermis**: absent. **Fibre bundles** in a single row with the vascular bundles. **Mesophyll** homogeneous. **Vascular bundles** in one row, collateral. **Stegmata** with conical silica bodies lining fibre bundle and vascular bundle sclerenchyma.

Pseudobulb TS
Trichomes and **stomata** absent. **Epidermis**: cells isodiametric, thick-walled, turbinate, bases bulbous. **Endodermis** and **pericycle** absent. Ground tissue of larger, dead water-storage cells surrounded by smaller, living, thin-walled assimilatory cells with cruciate starch granules. **Fibre bundles** present in *Coeliopsis*. **Vascular bundles** collateral, scattered. **Stegmata** with conical silica bodies associated with fibre bundles in *Coeliopsis* and with phloic sclerenchyma in vascular bundles of *Coeliopsis*, *Lycomormium*, and *Peristeria*.

Root TS
Velamen four or five cells wide in *Coeliopsis*, three or four cells wide in *Lycomormium*, and 5–11 cells wide in *Peristeria*; cells with fine, spiral, anticlinal wall thickenings. **Tilosomes** absent. **Exodermis**: cells square with uniformly thickened walls in *Coeliopsis*, thin-walled in *Lycomormium* and *Peristeria*; passage cells intermittent. **Cortex**: pelotons, hyphae, and dead cell masses present in cells in middle layers of cortex and raphides in unmodified cells. **Endodermis** 1-layered, cells U-thickened. **Vascular cylinder** 7-arch in *Coeliopsis*, 15-arch in *Lycomormium*, and up to 15-arch in *Peristeria*. **Pith** cells sclerotic in *Coeliopsis*, parenchymatous in *Lycomormium* and *Peristeria*.

Material examined

Coeliopsis (1), *Lycomormium* (1), *Peristeria* (2). Full details appear in Stern and Whitten (1999).

Reports from the literature

Dressler (1993): taxonomy of subtribe Stanhopeinae including genera of Coeliopsidinae. Mulay and Panikkar (1956): velamen of *Peristeria*. Porembski and Barthlott (1988): velamen of *Peristeria*. Pridgeon et al. (2009): systematics of subtribes Ceoliopsidinae and Stanhopeinae including discussion of establishment of Coeliopsidinae and its removal from Stanhopeinae *sensu lato*. Stern and Whitten (1999): anatomy of plants in subtribe Stanhopeinae including genera of subtribe Coeliopsidinae.

Szlachetko (1995): establishment of subtribe Coeliopsidinae with five genera but his circumscription included features not present in all five genera. Whitten et al. (2000): relationships of Maxillarieae and Stanhopeinae, molecular evidence. Williams (1976, 1979): stomata of *Coeliopsis* and *Lycomormium*.

Taxonomic notes

See discussion under Stanhopeinae.

Subtribe Cymbidiinae Benth.

Cymbidiinae are epiphytic, lithophytic, or terrestrial, autotrophic or rarely mycoheterotrophic herbs with leaves borne on ovoid to fusiform pseudobulbs. Leaves are coriaceous, plicate or conduplicate, articulate with sheathing bases. Plants occur exclusively in the Old World from the Himalaya to Japan and south to Indochina, the Philippines, New Guinea, and Australia. There are 11 genera. *Ansellia* has a single species indigenous to sub-Saharan Africa and monospecific *Imerinaea* is endemic to Madagascar. Species and hybrids of *Cymbidium* are grown commercially and by orchid fanciers. *Grammatophyllum* is sometimes seen in hobby collections and may be grown outside as a terrestrial under tropical conditions.

Ansellia is used medicinally in parts of Africa; *Cymbidium* species are widely used medicinally in India, eastern Asia, and Australia; *Grammatophyllum* is noted as medicinal in Indonesia.

Leaf surface

Hairs present adaxially in *Acriopsis* and *Thecostele*; scarce adaxially, absent abaxially in *Ansellia*, absent in *Cymbidium*, *Dipodium*, *Grammatophyllum*, and *Porphyroglottis*. **Stomata** predominantly tetracytic; abaxial in *Ansellia* (Fig. 101A), *Cymbidium* (Fig. 101B), *Dipodium paludosum*, *Grammatophyllum* (occasionally adaxial), *Graphorkis*, *Porphyroglottis*, and *Thecostele*; amphistomatal in *Cymbidium canaliculatum* and *Dipodium pictum*.

Leaf TS

Cuticle smooth, 2.5 μm thick overall; 10 μm in *Cymbidium finlaysonianum*, 2.0–5.0 μm in other *Cymbidium* species. **Epidermis**: cells mostly periclinal (Figs 101C, 102B), all walls thickened in *Ansellia*, *Graphorkis*, and *Porphyroglottis*. **Stomata** superficial, somewhat depressed in *Dipodium* and several *Cymbidium* species. Substomatal chambers small in most *Cymbidium* species, *Grammatophyllum speciosum*, *Graphorkis*, *Porphyroglottis*, and *Thecostele*; small to absent in *Acriopsis*; moderate to large in *Cymbidium canaliculatum* and *C. aliciae*; moderate in *Ansellia*; large in *Dipodium* and *Grammatophyllum scriptum*. Outer ledges moderate, inner ledges apiculate in *Ansellia* and *Porphyroglottis*; outer ledges large in several species of *Cymbidium*; inner ledges small in most *Cymbidium* species; outer ledges large, inner ledges apiculate in *Grammatophyllum speciosum*, both apiculate in *G. scriptum*; outer ledges moderate to large, inner ledges small to minute in *Dipodium*; outer and inner ledges moderate in *Graphorkis*; outer ledges small, inner ledges minute in *Thecostele*. **Hypodermis**: ad- and abaxial in *Ansellia*, *Dipodium*, *Grammatophyllum*, *Graphorkis*, and *Porphyroglottis*. **Fibre bundles** absent in *Ansellia*; in four different positions in *Cymbidium* and absent in one species of *Cymbidium* (*C. goeringii*); in single ad- and abaxial series in *Acriopsis*, *Dipodium*, *Grammatophyllum* (Fig. 101C), *Graphorkis*, *Porphyroglottis*, and *Thecostele*; also mid-mesophyll in *Dipodium*, *Porphyroglottis*, and many throughout mesophyll in *Grammatophyllum*. **Mesophyll** homogeneous (Figs 101C, 102A,B), but heterogeneous in some species of *Cymbidium*, cells thin-walled with triangular intercellular spaces; layer of water-storage cells between ad- and abaxial layers of smaller cells in *Acriopsis*. **Vascular bundles** collateral in a single row (Figs 101C, 102B), sclerenchyma at xylem and phloem poles, except surrounding the circular phloem pole in *Cymbidium* (Fig. 102D). **Stegmata** with conical silica bodies in *Acriopsis* and *Thecostele*; associated with all fibre bundles (Fig. 101B) except in *Cymbidium aliciae*. Bundle sheath cells thin-walled (Fig. 102D), variously thin- and thick-walled in *Cymbidium* and *Grammatophyllum speciosum*, sometimes with chloroplasts in *Ansellia*, *Cymbidium madidum*, and *Grammatophyllum*; stegmata replacing bundle sheath cells in *Cymbidium madidum*.

Pseudobulb TS (Acriopsis, Ansellia, Cymbidium, Dipodium, Grammatophyllum, Graphorkis, Thecostele)

Hairs and **stomata** absent. **Cuticle** smooth, from less than 2.0 μm to 42.0 μm thick. **Epidermis**: cell walls evenly thickened in *Ansellia* and *Grammatophyllum scriptum*; in *Acriopsis*, some species of *Cymbidium*, *Graphorkis*, and *Thecostele* outer and lateral walls thickest, and strongly ∩-thickened in most *Cymbidium* species (Fig. 103A) and *Grammatophyllum speciosum*. Solitary rhombic crystals in epidermal cells of *Grammatophyllum speciosum* (Fig. 103B). **Cortex**: small fibre bundles with stegmata in cortex of *Dipodium pictum* (but not *D. paludosum*). Ground tissue with larger, circular and oval, thin-walled, empty water-storage cells embedded among smaller, circular and oval assimilatory cells, some containing cruciate starch

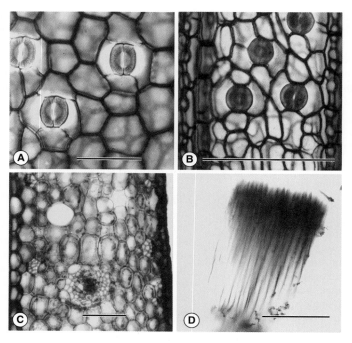

Fig. 101. Cymbidiinae. (A) *Ansellia africana*, abaxial leaf surface with tetracytic stomata. (B) *Cymbidium canaliculatum*, abaxial leaf surface leaf with tetracytic stomata and files of stegmata associated with fibres. (C) *Grammatophyllum scriptum*, TS leaf with homogeneous mesophyll, ad- and abaxial rows of fibre bundles, and raphide idioblast. (D) *Cymbidium ensifolium*, TS ground tissue of pseudobulb to show raphide crystals. Scale bars: A–C = 50 µm, D = 12 µm. (A) Reprinted from Stern and Judd (2002); (B–D) reprinted from Yukawa and Stern (2002).

granules or raphides (Fig. 101D). **Vascular bundles** collateral, scattered, numerous, with sclerenchyma at xylem and phloem poles; sclerenchyma only at phloem poles in *Acriopsis*, *Cymbidium*, *Dipodium*, *Grammatophyllum*, and *Thecostele*. Xylem of *Cymbidium* subtended or surrounded by thick-walled parenchyma. Vascular bundles of *Graphorkis* collateral, but consisting of a circular mass of starch-containing parenchyma cells enclosing an arcuate several-cell-wide band of sclerenchyma cells extending around phloem segment of vascular tissue but separated from it by parenchyma cells surrounding the vascular tissue (Fig. 103C). **Stegmata** with conical, rough-surfaced silica bodies lying next to phloem sclerenchyma only, except on all sides of sclerenchyma arc in *Graphorkis*, and along fibre bundles of *Cymbidium bicolor*; silica bodies in *Dipodium* small, not well-formed.

Root TS

Velamen cells thin-walled with fine spiral thickenings (Fig. 103D); eight or nine cells wide in *Acriopsis*, 6–8 cells wide in *Ansellia*, 3–16 cells wide in *Cymbidium*, 3–15 cells wide in *Dipodium*, 2–6 cells wide in *Grammatophyllum*, six or seven cells wide in *Graphorkis*, cells thick-walled, five or six cells wide in *Porphyroglottis*, and 6–10 cells wide in *Thecostele*; root hairs unicellular throughout. **Tilosomes** baculate in *Ansellia*, *Cymbidium* (Fig. 104A), *Dipodium*, and *Grammatophyllum*; webbed in *Porphyroglottis*, absent in *Acriopsis*, *Graphorkis*, and *Thecostele*. **Exodermis**: cells with thickened outer and lateral walls in *Ansellia* and *Dipodium paludosus*, all walls thickened in *Dipodium pictum*; walls thin in *Acriopsis*, *Cymbidium* (Fig. 104B), *Graphorkis*, *Porphyroglottis*, and *Thecostele* (Fig. 103D). **Cortex**: cells mostly circular, thin-walled, 9–11 cells wide in *Acriopsis*, 10 cells wide in *Ansellia*, seven or eight cells wide in *Cymbidium*, 8–13 cells wide in *Dipodium*, 8–18 cells wide with branched, banded wall thickenings in *Grammatophyllum*, seven or eight cells wide with branched bars in *Graphorkis*, 9–11 cells wide in *Porphyroglottis*, 10–12 cells wide in *Thecostele* (Fig. 103D). Scattered, empty, thick-walled angular cells occurring in *Ansellia* and *Porphyroglottis*. **Endodermis** 1-layered

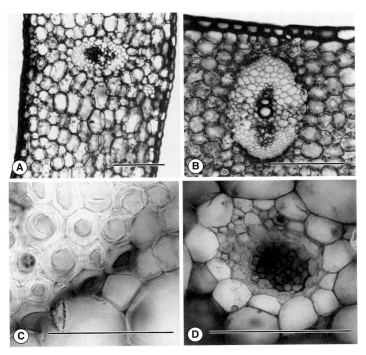

Fig. 102. **Cymbidiinae.** (A) *Thecostele alata*, TS leaf with homogeneous mesophyll, ad- and abaxial rows of fibre bundles. (B) *Ansellia africana*, TS foliar vascular bundle with sclerenchyma caps at xylem and phloem poles. (C) *Cymbidium atropurpureum*, TS root with stegmata among pericycle cells and dislodged silica body. (D) *C. hookerianum*, TS leaf with vascular bundle showing circular pole of phloem surrounded by sclerenchyma. Scale bars: A = 100 μm, B,D = 50 μm, C = 25 μm. Reprinted from Yukawa and Stern (2002).

(Fig. 104C), except 2-layered in part in *Grammatophyllum speciosum*; cells isodiametric, thin-walled throughout in *Ansellia*, *Graphorkis*, and *Thecostele*. O-thickened adjacent to phloem in *Acriopsis*, *Cymbidium*, *Dipodium*, and *Grammatophyllum speciosum*, O-thickened throughout in *Grammatophyllum scriptum*. **Pericycle** similar to endodermis; cells thin-walled opposite xylem; some cells converted to stegmata in several *Cymbidium* species (Fig. 102C). **Vascular cylinder** with radial rows of xylem alternating with strands of phloem cells; 16-arch in *Acriopsis*, 15-arch in *Ansellia*, variously polyarch in *Cymbidium* (Fig. 104C), 16–18-arch in *Dipodium*, 10–24-arch in *Grammatophyllum*, 6-arch in *Graphorkis*, 12-arch in *Porphyroglottis*, 17-arch in *Thecostele*. Vascular tissue embedded in parenchyma, except in thick-walled sclerenchyma in *Acriopsis* and most *Cymbidium* species. Medullary bundles present in *Dipodium*. **Pith** parenchymatous with triangular intercellular spaces; cells polygonal in *Graphorkis* and *Porphyroglottis*, lacking intercellular spaces.

Material examined

Acriopsis (1), *Ansellia* (1), *Cymbidium* (21), *Dipodium* (2), *Grammatophyllum* (2), *Graphorkis* (1), *Porphyroglottis* (1), *Thecostele* (1). Full details appear in Stern and Judd (2001, 2002), Yukawa and Stern (2002).

Reports from the literature

Barthlott and Capesius (1975) [1976]: velamen of *Ansellia*, *Cymbidium*, and *Graphorkis*. Burr and Barthlott (1991): velamen of *Acriopsis* and *Graphorkis*. Champagnat et al. (1966): tuber of *Cymbidium*. Chatin (1856, 1857): vegetative anatomy of *Cymbidium*. Czapek (1909): stem and root anatomy of *Acriopsis* and *Grammatophyllum*. Das and Paria (1992): stomata of *Cymbidium*. Du Puy (1986): leaf anatomy of *Cymbidium*. Du Puy and Cribb (1988): leaf anatomy of *Cymbidium*. Fan et al. (1999, 2000): mycorrhiza of *Cymbidium*. Goh et al. (1992): mycorrhiza of *Grammatophyllum*. Groom (1893):

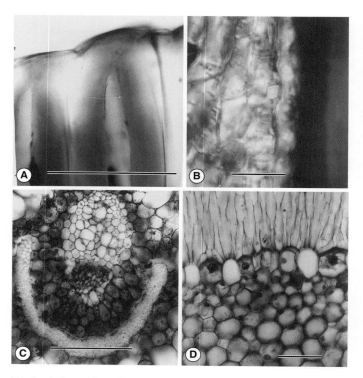

Fig. 103. **Cymbidiinae.** (A) *Cymbidium chloranthum*, TS pseudobulb showing characteristic thick-walled epidermal cells. (B) *Grammatophyllum speciosum*, TS pseudobulb with rhomboid crystals in epidermal cells. (C) *Graphorkis lurida*, TS pseudobulb vascular bundle with hippocrepiform arc of sclerenchyma enclosing parenchyma cells facing adjacent phloem. (D) *Thecostele alata*, TS root showing anticlinally oriented cells of inner velamen, exodermal cells, and outer cortical cells; long cells of exodermis with larger nuclei than those in passage cells. Scale bars: A,B = 25 μm, C = 100 μm, D = 50 μm. (A,B,D) Reprinted from Yukawa and Stern (2002); (C) reprinted from Stern and Judd (2002).

root anatomy of *Grammatophyllum speciosum*. Guttenberg (1968): root anatomy of *Cymbidium* and *Grammatophyllum*. Hadley and Williamson (1972): mycorrhiza of *Cymbidium*. Holtermann (1902): leaf anatomy of *Cymbidium*. Kaushik (1983): vegetative anatomy of *Cymbidium*. Khayota (1995): leaf and root anatomy of many specimens of *Ansellia africana* and *A. gigantea* (now in *A. africana*), also other genera of Cymbidiinae. Krüger (1883): anatomy of aerial organs of *Cymbidium*. Leitgeb (1864b): aerial root anatomy of *Ansellia* and *Cymbidium*. Li et al. (2002): leaf anatomy of *Cymbidium sinense*. Lin and Namba (1982): vegetative anatomy of *Cymbidium*. Löv (1926): leaf anatomy of *Cymbidium*. Mayer et al. (2008): leaf and root anatomy of *Cymbidium*. McLuckie (1922): mycorrhiza of *Dipodium*. Metzler (1924): leaf anatomy of *Cymbidium*. Möbius (1887): leaf anatomy of *Cymbidium* and *Eulophiopsis* (now *Graphorkis*). Moreau (1913): pseudobulb anatomy of *Cymbidium*. Nieuwdorp (1972): mycorrhiza of *Cymbidium*. Noel (1974): velamen of *Ansellia*. Olatunji and Nengim (1980): tracheoidal idioblasts of *Ansellia* and *Graphorkis*. Porembski and Barthlott (1988): velamen of Cymbidiinae. Raciborski (1898): root anatomy of *Acriopsis*. Sanford and Adanlawo (1973): root anatomy of *Ansellia* and *Graphorkis*. Singh (1981): stomata of *Cymbidium*. Singh (1986): root anatomy of *Cymbidium*. Solereder and Meyer (1930): review of orchid anatomy with original observations on leaf of *Cymbidium*. Stern and Judd (2001): anatomy and systematics of Catasetinae including outgroups *Acriopsis* and *Thecostele*. Stern and Judd (2002): comprehensive account of vegetative anatomy and systematics of Cymbidieae. Sun (1995): SEM of leaves of *Cymbidium*. Tominski (1905): leaf anatomy of *Cymbidium*. Vij et al. (1991): leaf epidermis of *Cymbidium*. Warcup (1981): mycorrhiza of *Cymbidium* and *Dipodium*. Weltz (1897): axis anatomy of *Ansellia* and *Cymbidium*. Williams

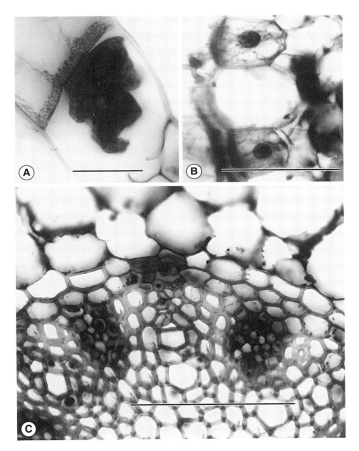

Fig. 104. Cymbidiinae. *Cymbidium* root. (A) *C. ensifolium*, TS with baculate tilosomes in velamen cells opposite passage cell of exodermis. (B) *C. chloranthum*, TS to show nucleated exodermal passage cells. (C) *C. kanran*, TS showing thickened endodermal cell walls opposite phloem poles and thin-walled endodermal cells opposite xylem poles. Scale bars: A = 25 μm, B,C = 50 μm. Reprinted from Yukawa and Stern (2002).

(1976, 1979): stomata of *Ansellia*, *Cymbidium*, and *Grammatophyllum*. Wu et al. (2005): mycorrhiza of *Cymbidium goeringii*. Xu et al. (2009): mycorrhiza of *Cymbidium goeringii*. Ye et al. (1992): leaf anatomy of *Cymbidium sinense*. Yukawa and Stern (2002): vegetative anatomy and systematics of *Cymbidium*.

Taxonomic notes

The unusual occurrence of stegmata in the pericycle of roots in several species of *Cymbidium* (Yukawa and Stern 2002) and in several genera of Maxillarieae (Holtzmeier et al. 2002) plus the presence of baculate tilosomes in several genera of Cymbidiinae are evidence of a possibly close relationship between members of subtribe Cymbidiinae and tribe Maxillarieae. Molecular evidence (Cameron et al. 1999) also suggested a close relationship between members of these two taxa.

Subtribe Cyrtopodiinae Benth.

Members of subtribe Cyrtopodiinae are usually pseudobulbous terrestrial or lithophytic herbs. The several leaves are distichous, coriaceous, plicate, and eventually deciduous. There is only a single genus, *Cyrtopodium*, with 47 species spread in North America from southern Florida, Caribbean islands, and Mexico through Central America and southwards in South America to southern Brazil, Bolivia, and Argentina. Several showy species appear in cultivation, namely *C. andersonii* and *C. punctatum*.

Leaf surface

Hairs: sunken 3-celled hairs on both surfaces of leaves in *C. andersonii* but absent in *C. punctatum*. **Stomata** tetracytic, basically abaxial, sometimes on both surfaces in *C. andersonii*.

Leaf TS

Cuticle smooth, less than 2.5 μm thick. **Epidermis**: cells more or less periclinal on both surfaces. Cell walls generally thin. **Stomata** superficial. **Hypodermis**: absent. **Fibre bundles** in one row abaxially, alternating with vascular bundles. **Mesophyll** cells thin-walled, crowded; intercellular spaces few, tiny. Crystalliferous idioblasts circular in TS, short, saccate, blunt-ended in LS. **Vascular bundles** collateral, in one series, joined laterally by thick-walled parenchyma cells. Sclerenchyma at both xylem and phloem poles. **Stegmata** with conical, rough-surfaced silica bodies along both xylem and phloem margins and around fibre bundles. Bundle sheath cells thin-walled, chlorophyllous.

Pseudobulb TS

Hairs and **stomata** absent. **Cuticle** smooth, 7.5–10.0 μm thick. **Epidermis**: cell walls thickest externally, tapering to thin inner walls. **Ground tissue** cells thin-walled; smaller circular and oval assimilatory cells surrounding fewer, much larger, empty water-storage cells. Raphide-containing cells unspecialized. **Vascular bundles** collateral, numerous, small, scattered. Sclerenchyma only at phloem pole; xylem subtended by thick-walled parenchyma cells. **Stegmata** with conical, rough-surfaced silica bodies adjacent to phloem sclerenchyma.

Root TS

Velamen 10–24 cells wide. Epivelamen cells isodiametric and periclinal. Endovelamen cells peripherally isodiametric grading to anticlinally oriented cells internally. Cells thin-walled with fine reticulate thickenings. **Tilosomes** lamellate. **Exodermis**: cells polygonal; outer walls slightly thickened, tapering inwardly to thin inner walls. Passage cells intermittent; ladder-like bars present on lateral walls of long cells. **Cortex** 7–11 cells wide; cells thin-walled, circular and oval. Scattered, empty, angular cells with thickened birefringent cell walls, possibly specialized water-storage cells. Banded cell wall thickenings absent. Masses of dead hyphae in *C. punctatum*. **Endodermis** 1-layered, cells isodiametric and periclinal, O-thickened opposite phloem, thin-walled opposite xylem. **Pericycle** cells as endodermis. **Vascular cylinder** 15- and 17-arch. Xylem in radial rows alternating with elliptical strands of phloem cells. Vascular tissue surrounded by thin-walled parenchyma cells. **Pith** parenchymatous, cells thin-walled, circular and oval. Pith and cortical cells in *C. andersonii* appearing chlorophyllous.

Material examined

Cyrtopodium (2). Full details in Stern and Judd (2001).

Reports from the literature

Benzing and Pridgeon (1983): leaf trichomes. Burr and Barthlott (1991): velamen-like tissue in root cortex. Dressler (1993): taxonomic descriptions of Cyrtopodiinae including 12 genera. Freudenstein and Rasmussen (1999): morphology and orchid relationships. Kutschera et al. (1997): root anatomy. Leitgeb (1864b): anatomy of aerial roots. Möbius (1887): leaf anatomy. Oudemans (1861): anatomy of aerial roots. Porembski and Barthlott (1988): velamen. Pridgeon (1987): velamen and exodermis. Pridgeon et al. (2009): comprehensive taxonomic description of Cyrtopodiinae. Solereder and Meyer (1930): review of orchid anatomy. Stern and Judd (2001, 2002): comparative vegetative anatomy and systematics of Catasetinae and Cymbidieae. Weltz (1897): axis anatomy. Williams (1976, 1979): stomata.

Taxonomic notes

Pridgeon et al. (2009), based on DNA phylogenetic studies, reduced Dressler's 1993 Cyrtopodiinae of 12 genera to a single genus, *Cyrtopodium*, that has no close relatives at the generic level. Indeed, *Cyrtopodium* stands away from 11 of these 12 genera in its possession of lamellate tilosomes (Stern and Judd 2002). There are no publications on the infrageneric phylogenetic relationships of *Cyrtopodium*–Cyrtopodiinae. Anatomy has nothing to offer to explain these.

Subtribe Eriopsidinae Szlach.

Eriopsidinae are epiphytic, lithophytic, or terrestrial, sympodial herbs with 2–4 coriaceous, conduplicate leaves produced from a pseudobulb. There are only five species widely distributed from Honduras, Guatemala, and Belize in Central America south to the western Andes in Peru and in the Amazonian lowlands of much of northern and central South America. *Eriopsis biloba* is found in some hobby collections.

Little is known of the vegetative anatomy except that the stomata are confined to the abaxial surface of the leaf in *E. sceptrum* (as *E. helenae*), with two subsidiary cells (Williams 1976, 1979); the root in *Eriopsis* species has a cymbidioid velamen of five or seven layers with long, slender wall thickenings and wide pores in non-thickened parts and 'tracheoidal' idioblasts occur in the cortex (Porembski and Barthlott 1988). Pridgeon et al. (2009) present a review of the subtribe.

Subtribe Eulophiinae Lindl.

Eulophiinae are pantropical, but largely Old World terrestrials, lithophytes, and epiphytes, cormous or pseudobulbous plants with spiral or solitary articulate leaves. There are nine genera, four of which, *Cymbidiella*, *Eulophiella*, *Grammangis*, and *Paralophia*, are endemic to Madagascar. *Pteroglossaspis* is included in *Eulophia* by Pridgeon et al. (2009), but is described separately here.

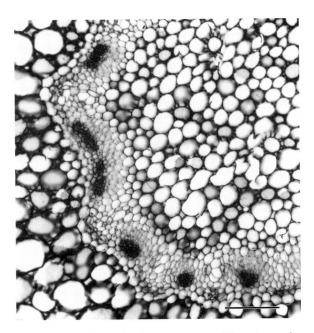

Fig. 105. Eulophiinae. *Geodorum purpureum*. TS root, vascular cylinder with undulating margin. Scale bar = 100 μm. Reprinted from Stern and Judd (2002).

Oeceoclades maculata, *Cymbidiella pardalina*, *Grammangis ellisii*, and *Eulophia guineensis* are grown in specialized hobby collections.

Eulophia has been noted as medicinal in Africa, India, and Indonesia.

Leaf surface

Hairs ad- and abaxial, 3-celled, sunken in *Eulophia*, glandular, sunken in *Oeceoclades*, present in *Eulophiella* and *Pteroglossaspis ecristata* (now *Eulophia ecristata*), rare in *Cymbidiella*, none in *Geodorum* and *Grammangis*. **Stomata** tetracytic (Figs 106B–108B) to anomocytic, abaxial in most *Eulophia* species, *Geodorum*, *Grammangis*, and *Oeceoclades saundersiana*, amphistomatal in *Cymbidiella*, *Eulophiella*, *O. maculata* (Fig. 108A,B), and *Pteroglossaspis*. **Epidermis**: cells polygonal (Figs 106A,B–108A,B).

Leaf TS

Cuticle smooth, <2.5 μm thick in *Cymbidiella* and *Grammangis*, 2.5 μm thick in *Eulophiella*, <2.5 μm thick to thicker in *Geodorum*, *Oeceoclades* (Fig. 108D–F), and *Pteroglossaspis*, <2.5 μm to 10 μm thick in *Eulophia*. **Epidermis**: cells isodiametric in *Eulophia* (Fig. 106D–F), *Eulophiella*, and *Grammangis*, periclinal in *Geodorum* and *Oeceoclades*, and isodiametric and periclinal in *Cymbidiella* and *Pteroglossaspis*. **Stomata** somewhat sunken in *Eulophia petersii*, but superficial in other species of *Eulophia* (Figs 106C, 107C), superficial to sunken in *Grammangis*, superficial in other genera (Fig. 108C). Outer ledges moderate in *Cymbidiella*, *Eulophiella*, and *Oeceoclades*, moderate to large with prominent cuticular horns in *Grammangis*, small in *Geodorum* and *Pteroglossaspis*; inner ledges absent, small or minute in most *Eulophia* species, *Geodorum*, *Oeceoclades*, and *Pteroglossaspis*, apiculate in *Cymbidiella* and *Eulophiella*, moderate to large in *Grammangis*, large in *Eulophia macrobulbon*. **Hypodermis**: absent except 1-layered, ad- and abaxial, in *Eulophia petersii*. **Fibre bundles** in one abaxial row in *Eulophia callichroma*, *E. gracilis*, *E. macra* (Fig. 107F), *E. petersii*, and *Geodorum*, in ad- and abaxial rows in *Cymbidiella*, *Eulophiella*, *Eulophia andamanensis* (as *E. keithii*), *Grammangis*, and *Oeceoclades*, absent in *Eulophia guineensis*, *E. macrobulbon*, and *Pteroglossaspis*. **Mesophyll** homogeneous (Figs 106D–F–108D–F); cells thin-walled, crowded, oval and almost circular; intercellular spaces few, tiny, triangular; raphide idioblasts present in *Cymbidiella*, *Eulophiella*, *Geodorum*, *Grammangis*, and crystalliferous idioblasts in *Pteroglossaspis*. **Vascular bundles** collateral (Figs 106F, 107F), in one row (Figs 106G–108G); sclerenchyma associated with xylem and phloem in larger

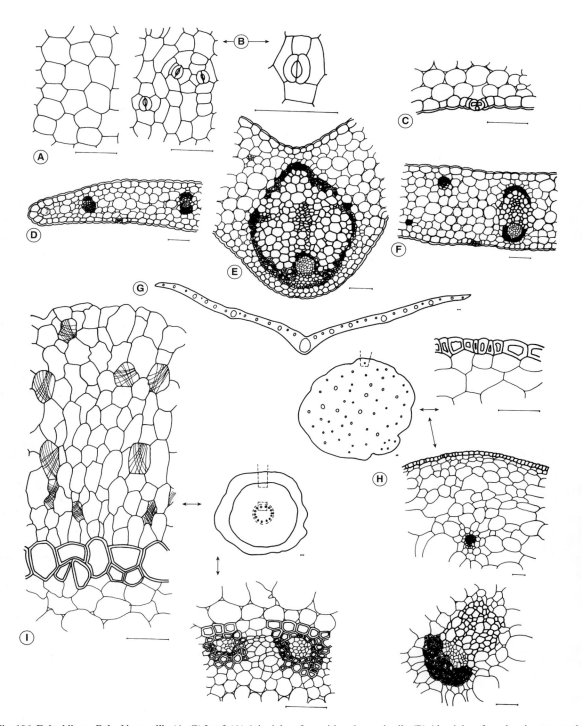

Fig. 106. Eulophiinae. *Eulophia gracilis*. (A–G) Leaf. (A) Adaxial surface with polygonal cells. (B) Abaxial surface showing tetracytic stomata and polygonal cells with curvilinear walls. (C) TS showing superficial stoma. (D–F) TS with homogeneous mesophyll, collateral vascular bundles in a single row each with xylem and phloem sclerenchyma caps and midrib with vascular tissue surrounded by thick-walled bundle sheath cells. (G) TS showing one row of vascular bundles. (H) TS stem with scattered collateral vascular bundles and thick-walled epidermal cells. (I) TS root with multilayered velamen cells with thin cell wall bands, thick-walled exodermal cells, thin-walled cortical cells, O-thickened more or less isodiametric endodermal cells, and vascular tissue surrounded by thick-walled cells.

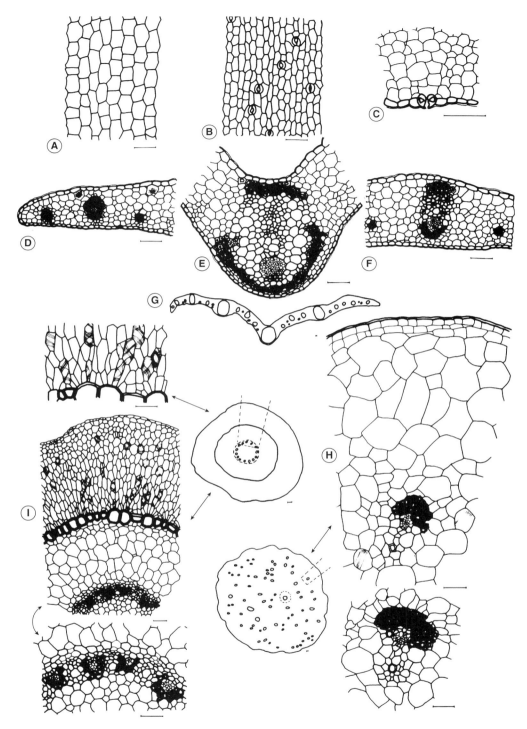

Fig. 107. Eulophiinae. *Eulophia macra*. (A–G) Leaf. (A) Adaxial surface with polygonal cells. (B) Abaxial surface with more or less rectangular cells and tetracytic stomata. (C) TS showing superficial stoma. (D–F) TS with homogeneous mesophyll, cells containing what appear to be crystal sand, collateral vascular bundles with strong xylem and phloem sclerenchyma caps in a single row. (G) TS with one row of vascular bundles. (H) TS stem with scattered collateral vascular bundles and dense phloem sclerenchyma. (I) TS root with multilayered velamen containing thick-walled barred idioblasts, velamen cell walls with thin bands, O-thickened isodiametric exodermal cells, thin-walled cortical cells, thin-walled endodermal cells, and vascular tissue surrounded by sclerenchyma.

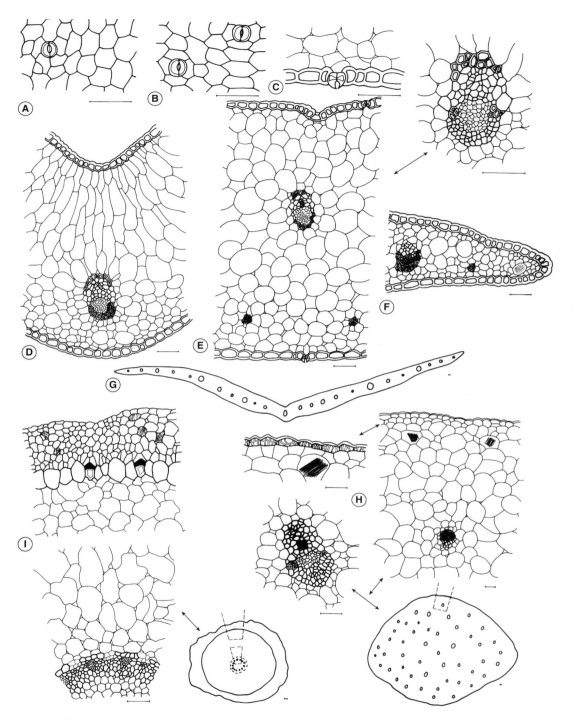

Fig. 108. Eulophiinae. *Oeceoclades maculata*. (A–G) Leaf. (A) Adaxial surface showing polygonal cells with curvilinear cell walls and tetracytic stoma. (B) Abaxial surface with polygonal cells and tetracytic stomata. (C) TS with superficial stoma and thick-walled epidermal cells. (D–F) TS with thick-walled epidermal cells, basically homogeneous mesophyll but upright cells adaxial to midrib, collateral vascular bundles in a single row, and tiny abaxial fibre bundles. (G) TS showing one row of vascular bundles. (H) TS stem showing epidermis with barred cell walls, scattered collateral vascular bundles, and raphide idioblasts. (I) TS root with several-layered velamen having cells with barred walls, thin-walled exodermal cells, tilosomes associated with exodermal passage cells, thin-walled cortical cells, thin-walled endodermal cells, and vascular tissue surrounded by parenchyma cells.

bundles, with only phloem in smaller bundles; midvein in *Eulophia*, *Geodorum*, and *Oeceoclades* flanked laterally by a pair of smaller bundles on either side, by one bundle in *Pteroglossaspis*; midvein consisting of one bundle in *Cymbidiella*, *Eulophiella*, and *Grammangis*. **Stegmata** with conical rough-surfaced silica bodies along fibre bundles, and xylem and phloem sclerenchyma in vascular bundles of *Cymbidiella*, *Eulophia*, *Geodorum*, *Grammangis*, *Oeceoclades*, and *Pteroglossaspis*, only along phloem sclerenchyma in *Eulophiella*. Bundle sheath cells thin-walled.

Pseudobulb/stem/corm TS

Hairs and **stomata** none. **Cuticle** smooth, 5.0–20.0 μm thick in *Eulophia* (Fig. 107H), 12.5 μm thick in *Geodorum*, 17.5–25.0 μm thick in *Grammangis*, 5.0–7.5 μm thick in *Oeceoclades*, c.2.5 μm thick in *Pteroglossaspis*. **Epidermis**: cells conical in *Grammangis*, isodiametric in *Oeceoclades*, mostly square to rectangular in *Pteroglossaspis*; walls variably thickened in *Eulophia*, all walls thickened in *Geodorum* and *Grammangis*. **Cortex** absent. Groundtissue cells thin-walled, circular; larger water-storage cells surrounded by smaller assimilatory cells sometimes with cruciate starch granules in *Eulophia* and *Grammangis*; **spiranthosomes** in assimilatory cells of *E. gracilis* and *E. petersii*; large water-storage cells surrounded by smaller assimilatory cells, some of them containing spiranthosomes, in *Geodorum*, *Oeceoclades*, and *Pteroglossaspis*. **Vascular bundles** collateral, numerous, scattered (Figs 106H–108H), sclerenchyma only at phloem poles in *Eulophia* (Figs 106H, 107H) and *Geodorum*; lacking in *Pteroglossaspis*. **Stegmata** with conical rough-surfaced silica bodies occurring next to phloem sclerenchyma in *Eulophia*, *Geodorum*, *Grammangis*, and *Oeceoclades*; absent in *Pteroglossaspis*.

Root TS

Velamen six cells wide in *Cymbidiella*, 7–14 cells wide in *Eulophia* (Figs 106I, 107I), two or three cells wide in *Eulophiella*, five or six cells wide in *Geodorum*, 7–13 cells wide in *Grammangis*, 8–13 cells wide in *Oeceoclades* (Fig. 108I), two or four cells wide in *Pteroglossaspis*; cell walls with intermeshing network of fine spiral thickenings (Fig. 106I); broad in *Eulophiella*; unicellular hairs present throughout. **Tilosomes** baculate in *Cymbidiella* and *Grammangis*, webbed in *Eulophia* and *Oeceoclades* (Fig. 108I), absent in *Eulophiella*, *Geodorum*, and *Pteroglossaspis*. **Exodermis**: outer cell walls thickened in *Eulophia callichroma*, *E. gracilis*, *E. macrobulbon*, and *E. petersii*; all walls thin in *Cymbidiella*, *Eulophia guineensis*, *E. andamanensis* (as *E. keithii*), *Grammangis ellisii*, *Oeceoclades saundersiana*, and *Pteroglossaspis*; outer and lateral walls thickened in *Eulophiella*, *Geodorum*, *Grammangis spectabilis*, and *O. maculata*. **Cortex**: cells thin-walled in most taxa, 10–12 cells wide in *Cymbidiella*, 7–14 cells wide in *Eulophia* (Fig. 107I), 11–13 cells wide in *Eulophiella*, 10–16 cells wide in *Grammangis*, 12 or 13 cells wide in *Geodorum* and *Pteroglossaspis*, 8–12 cells wide in *Oeceoclades* (Fig. 108I). A meshwork of broad-banded spiral thickenings occurring in cortical cells of *Eulophia callichroma* and broad, branched thickenings in *Grammangis spectabilis*. Dead hyphal masses and pelotons in almost all cortical cells. Raphide bundles seen in *Eulophia callichroma* (Pl. 3B,C). **Endodermis** 1-layered, cells isodiametric, walls O-thickened opposite phloem (Figs 106I, 107I, Pl. 3A), thin-walled opposite xylem; all walls thin in *Cymbidiella* and *Grammangis ellisii*. **Pericycle** 1-layered, like endodermis. **Vascular cylinder** 10-arch in *Cymbidiella*, 12–17-arch in *Eulophia* (Pl. 2D), 18-arch in *Eulophiella*, 16-arch with undulating margin in *Geodorum* (Fig. 105), 14–19-arch in *Grammangis*, 8–12-arch in *Oeceoclades maculata* (Fig. 108I), 14-arch in *O. saundersiana*, 8-arch in *Pteroglossaspis*; xylem in radial rows alternating with clusters of phloem cells. **Pith** cells thin-walled (Figs 106I–108I), but with moderately thick walls in *Geodorum* and thin and thick walls in *Eulophiella*, cells mostly circular, intercellular spaces triangular.

Material examined

Cymbidiella (1), *Eulophia* (6), *Eulophiella* (1), *Geodorum* (1), *Grammangis* (2), *Oeceoclades* (2), *Pteroglossaspis* (1, now included in *Eulophia*). Full details appear in Stern and Judd (2001, 2002).

Reports from the literature

Burr and Barthlott (1991): root anatomy of *Grammangis*. Das and Paria (1992): stomata of *Eulophia*. Dietz (1930): underground organs of *Eulophia*. Dressler (1993): comprehensive review of orchid systematics. Freudenstein et al. (2000): broad study of mitochondrial DNA and relationships within Orchidaceae. Grace (1999): velamen in 40 *Eulophia* species and two *Oeceoclades* species. Guttenberg (1968): root anatomy of *Eulophia*. Kaushik (1983): anatomy of *Eulophia*. Khayota (1995): thesis on *Ansellia*, also including leaf and root anatomy of *Acrolophia*, *Cymbidiella*, *Eulophiella*, and *Grammangis*. Lakshminarayana and Venkateswarlu (1950): velamen in *Eulophia graminea*. Möbius (1887): leaf anatomy of *Eulophia*. Møller and Rasmussen (1984): stegmata in leaf and stem of *Eulophia andamanensis*. Moreau (1913): pseudobulb anatomy of *Grammangis*. Olatunji and

Nengim (1980): tracheoids in *Eulophia* and *Oeceoclades*. Porembski and Barthlott (1988): velamen in most genera of Eulophiinae. Pridgeon (1987): velamen and exodermis in *Eulophia*, *Grammangis*, and *Oeceoclades*. Pridgeon et al. (1983): tilosomes present in *Eulophia keithii* (now *E. andamanensis*) and *Oeceoclades maculata*, absent in *Grammangis ellisii*. Pridgeon et al. (2009): comprehensive review of subtribe Eulophiinae. Puri (1971): tuber anatomy of *Eulophia hormusjii* (now *E. dabia*). Rao (1988): leaf anatomy of *Eulophia* and *Geodorum*. Sarin (1960): root anatomy of *Eulophia herbacea*. Sasikumar (1973a, 1974): root anatomy of *Eulophia*. Solereder and Meyer (1930): review of orchid anatomy. Stern and Judd (2001): anatomy and systematics of Catasetinae including *Eulophia* and *Pteroglossaspis*. Stern and Judd (2002): anatomy and systematics of Cymbidieae including *Cymbidiella*, *Eulophiella*, *Geodorum*, *Grammangis*, and *Oeceoclades*. Tominski (1905): leaf anatomy of *Eulophia*. Vij et al. (1991): leaf epidermis of *Eulophia*. Weltz (1897): axis anatomy of *Eulophia*. Williams (1979): stomata in *Cymbidiella*, *Eulophia*, and *Eulophidium* (now in *Oeceoclades*).

Taxonomic notes

Cladograms developed by Stern and Judd (2000) do not support the monophyly of Eulophiinae. The placement of *Pteroglossaspis* (now = *Eulophia*) in Eulophiinae is especially divergent. Freudenstein et al. (2000) present results supporting a close relationship between *Eulophia* and *Oeceoclades*. Anatomical data, however, do not indicate this relationship.

Subtribe Maxillariinae Benth.

Maxillariinae are epiphytic and lithophytic, usually sympodial plants with distichous, variously shaped,

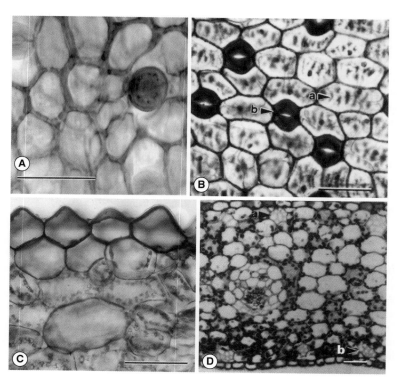

Fig. 109. Maxillariinae and Zygopetalinae leaf. (A) *Xylobium colleyi*, epidermis to show glandular hair in face view. (B) *Mormolyca rufescens*, abaxial epidermis showing granular ridges perpendicular to long axis of cell (a), anomocytic stomata, and ±circular guard cell pairs (b). (C) *Benzingia caudata*, TS papillate epidermis. (D) *Brasiliorchis picta*, TS with adaxial (a) and abaxial (b) hypodermal fibre bundles without water-storage cells. Scale bars: A = 50 μm, B–D = 100 μm. (A,C) Reprinted from Stern et al. (2004); (B,D) reprinted from Holtzmeier et al. (1998).

thin to succulent articulate leaves. Pseudobulbs are heteroblastic, one- to several-leaved, often subtended by foliaceous sheathing bracts. There are 29 genera and more than 720 species restricted to the Neotropics with centres of diversity in Andean South America and the Brazilian Atlantic rain forest. Plants of interest to hobbyists include species of *Anguloa*, *Bifrenaria*, *Lycaste*, *Maxillaria*, *Neomoorea*, *Scuticaria*, *Xylobium*, and the strangely flowered *Mormolyca* and *Trigonidium*. *Maxillaria tenuifolia* flowers are distinguished by their coconut-like fragrance. *Maxillaria* species have been used medicinally in Mexico.

Leaf surface

Hairs glandular (Fig. 109A), on both surfaces; multicellular in *Brasiliorchis*, *Camaridium*, *Maxillaria*, *Mormolyca* (Fig. 111C), *Scuticaria*, and *Trigonidium*; single-celled in *Anguloa*, *Bifrenaria*, *Cryptocentrum*, *Cyrtidiorchis*, *Lycaste*, *Neomoorea*, *Rudolfiella*, *Sudamerlycaste*, *Teuscheria*, and *Xylobium* (Fig. 109A), with bulbous tips, ad- and abaxial, sunken, one cell arising from a basal cell, not exserted; anastomosing cell wall thickening bands in surrounding crypt cells of *Maxillaria endresii* and *M. ringens*. **Epidermis**: cells rectangular and periclinal, elongated parallel with veins in *Cryptocentrum roseans* (as *Anthosiphon*) and *Cryptocentrum gracillimum*; polygonal, isodiametric in *Cyrtidiorchis*; walls straight-sided and curvilinear (Fig. 109A,B). **Stomata** abaxial (Figs 109B, 110A), except on both surfaces in *Cryptocentrum gracillimum*; mainly tetracytic, but both tetracytic and anomocytic in *Brasiliorchis porphyrostele* (as *Maxillaria*) and *Camaridium cucullatum* (as *Maxillaria*) and some *Maxillaria* and *Mormolyca* species.

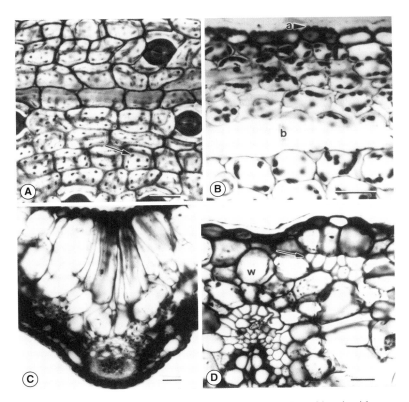

Fig. 110. Maxillariinae. *Maxillaria* **and** *Mormolyca* **leaf**. (A) *Maxillaria nasuta*, abaxial epidermis with tetracytic stomata and small scattered papillae (arrow). (B) *Mormolyca rufescens*, LS showing rugulose projections of adaxial epidermal cell walls (a) and elongate mesophyll idioblast (b). (C) *Maxillaria nasuta*, TS showing adaxial anticlinal cells of midvein. (D) *Mormolyca ringens*, TS illustrating adaxial fibre bundles (arrow) scattered among water-storage cells (w). Scale bars = 100 μm. Reprinted from Holtzmeier et al. (1998).

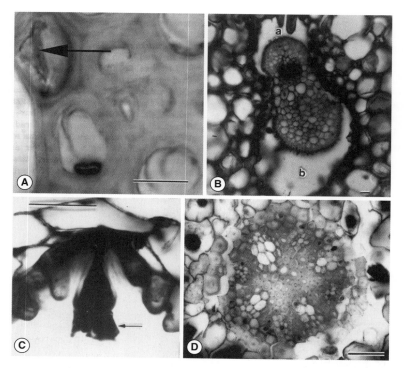

Fig. 111. Maxillariinae and Oncidiinae. (A) *Anguloa x ruckeri*, TS root showing stegma with a conical silica body (arrow). (B) *Maxillaria ringens*. TS pseudobulb showing vascular bundle with phloem sclerenchyma and lacuna (a) and xylem sclerenchyma and lacuna (b). (C) *Mormolyca rufescens*, TS lamina showing sunken multicellular hair with apical cell remains (arrow). (D) *Notylia ramonensis*, TS root with tetrarch vascular system, xylem and phloem surrounded by sclerenchyma, and thick-walled pith cells. Scale bars: A–C = 100 μm, D = 50 μm. (A) Reprinted from Stern et al. (2004); (B,C) reprinted from Holtzmeier et al. (1998); (D) reprinted from Stern and Carlsward (2006).

Leaf TS

Cuticle 2.5–19.4 μm thick in most taxa, up to 25.0 μm thick in *Scuticaria*; smooth, except granular in *Maxillaria* and *Mormolyca* species (Figs 109B, 110B) and striated in *Maxillaria nasuta*. **Epidermis**: cells isodiametric and periclinal; papillate in a few *Maxillaria* and *Mormolyca* species. **Stomata** superficial, except sunken in two *Bifrenaria* species, *Maxillaria ringens*, and *Rudolfiella*; substomatal chambers small; outer ledges small to moderate to very large in some *Maxillaria* species and *Mormolyca ringens*, ruminate in some taxa; inner ledges obscure, apiculate, but large in *Scuticaria*. **Hypodermis**: absent in *Anguloa*, *Bifrenaria*, *Cryptocentrum roseans*, *Cyrtidiorchis*, *Lycaste*, *Neomoorea*, *Rudolfiella*, *Sudamerlycaste*, *Teuscheria*, and *Xylobium*; present in *Cryptocentrum gracillimum* as an adaxial row of few-celled fibre bundles interrupted by chlorenchyma cells; as an adaxial row of water-storage cells in *Maxillaria nasuta* and *Mormolyca ringens* (Fig. 110D); as fibre bundles scattered amidst water-storage cells in *Camaridium* and several *Maxillaria* species; and as subepidermal bundles of fibres among chlorenchyma cells in *Brasiliorchis picta* (Fig. 109D), *B. porphyrostele*, and *Trigonidium*. **Fibre bundles** in one row in mesophyll in *Anguloa virginalis*, *Bifrenaria harrisoniae*, *Lycaste deppei*, and *Rudolfiella*, in a ring in *Scuticaria*, in ad- and abaxial rows in *Cryptocentrum roseans*, *Teuscheria*, and *Xylobium*, absent in *Sudamerlycaste*. **Mesophyll** homogeneous (Fig. 109D), cells anticlinally elongated adaxial to midrib in *Maxillaria nasuta* (Fig. 110C), narrow in *Cryptocentrum roseans*, broad in *Cyrtidiorchis*; adaxial two-thirds with wide, circular thin-walled water-storage cells in most species of *Maxillaria* and in *Mormolyca* (Fig. 110D), abaxial one-third of very small circular and oval, closely packed, thin-walled chloroplast-containing cells; leaf isobilateral in *Cryptocentrum gracillimum* with palisade layer

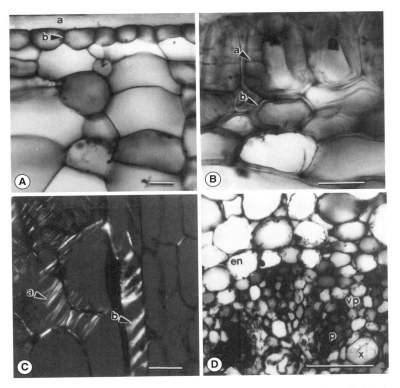

Fig. 112. Maxillariinae. (A) *Mormolyca rufescens*, TS pseudobulb with thick cuticle (a) and thin-walled epidermal cells (b). (B) *Maxillaria endresii*, TS pseudobulb showing extremely thick-walled epidermal cells (a) and continuous layer of dead cells with thickened walls (b) just below epidermis. (C) *Mormolyca rufescens*, LS root polarized to show tenuous wall thickenings of velamen cells (a), banded wall thickenings of dead exodermal cells (b). (D) *Brasiliorchis porphyrostele*, TS root with thin-walled endodermal cells (en), vascular parenchyma (vp), xylem (x), and phloem (p). Scale bars = 100 μm. Reprinted from Holtzmeier et al. (1998).

developed towards both surfaces, separated by a row of vascular bundles; intercellular spaces triangular; heterogeneous in *Bifrenaria harrisoniae*, *Hylaeorchis minuta* (as *B. minuta*), *Rudolfiella*, and *Scuticaria*. Crystalliferous idioblasts circular, saccate in LS. **Vascular bundles** collateral (Fig. 109D), in one row in most taxa; in one or two rows in *Maxillaria*, 1–3 rows in *Mormolyca* and two or three rows in *Neomoorea*; sclerenchyma at both poles, except in *Cryptocentrum gracillimum* where present at phloem pole, absent at xylem pole in smaller bundles; phloem sclerenchyma more strongly developed than xylem sclerenchyma. **Stegmata** with conical, rough-surfaced silica bodies adjacent to vascular bundle sclerenchyma and fibre bundles, except absent in *Cyrtidiorchis*. Bundle sheath cells thin-walled (Fig. 109D), sometimes absent, obscure, or incomplete; prominent in *Mormolyca ringens* and *Scuticaria*.

Pseudobulb TS

Hairs occasional, multicellular, sunken. **Stomata** absent. **Cuticle** thick, very thick in *Mormolyca rufescens* (Fig. 112A), smooth. **Epidermis**: cells various, polygonal and periclinal to isodiametric (Fig. 112A); walls very thick in *Maxillaria endresii* (Fig. 112B). **Ground tissue** with many smaller, thin-walled, circular to oval assimilatory cells surrounding fewer, larger, circular to oval water-storage cells with birefringent walls. Lacunae in ground tissue occurring next to phloem sclerenchyma of vascular bundles in most species of *Maxillaria* (Fig. 111B), *Mormolyca*, and *Trigonidium*. **Vascular bundles** collateral (Fig. 111B), scattered; phloem sclerenchyma prominent, xylem sclerenchyma absent in *Bifrenaria harrisoniae* and *Mormolyca rufescens*, usually associated only with largest bundles. **Stegmata** with rough-surfaced conical silica bodies occurring adjacent to vascular

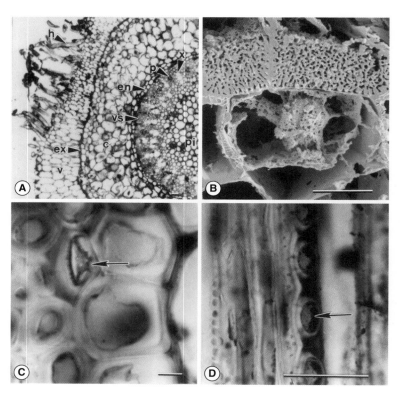

Fig. 113. Maxillariinae root. (A) *Mormolyca hedwigiae*, TS showing root hairs (h), multilayered velamen (v), exodermis (ex), vascular sclerenchyma (vs), cortex (c), xylem (x), phloem (p), and pith (pi). (B) *Mormolyca ringens*, TS (SEM) with spongy tilosomes along exodermal cells. (C,D) *Brasiliorchis picta*. (C) TS showing conical silica body in stegma (arrow). (D) LS showing row of stegmata containing conical silica bodies (arrow) associated with vascular sclerenchyma. Scale bars: A,D = 100 μm, B,C = 10 μm. Reprinted from Holtzmeier et al. (1998).

bundle sclerenchyma. Chloroplasts and cruciate starch granules occurring in assimilatory cells.

Root TS

Velamen 2–9(–11) cells wide (Fig. 113A); cell walls somewhat thickened in *Cryptocentrum gracillimum*, *Cyrtidiorchis*, and *Teuscheria*; spiral thickenings coarse in *Cyrtidiorchis* and *Scuticaria*, fine in most genera (Fig. 112C). **Tilosomes** lamellate in *Anguloa*, *Cyrtidiorchis*, and *Lycaste*, baculate in *Brasiliorchis*, *Camaridium*, and some *Maxillaria* species, spongy in *Mormolyca* species (Fig. 113B), plaited in *Trigonidium*, plaited or baculate in *Cryptocentrum roseans*, absent in *Bifrenaria*, *Rudolfiella*, *Scuticaria*, and *Teuscheria*. **Exodermis**: cells isodiametric in *Cryptocentrum gracilipes*, squarish in *Bifrenaria* and *Cyrtidiorchis*, polygonal in *Maxillaria*, *Mormolyca*, and *Trigonidium*; cell walls U-thickened in *Cryptocentrum gracilipes*, heavily O-thickened in *Cyrtidiorchis*; O-thickened in *Anguloa ruckeri*, *Lycaste deppei*, and *Xylobium leontoglossum*; ∩-thickened in *Brasiliorchis*, *Maxillaria*, *Mormolyca*, and *Trigonidium*, thin in other taxa (Fig. 113A). Birefringent bands in cells of some *Maxillaria* species, *Mormolyca ringens*, and *Trigonidium*. **Cortex**: cells circular to oval (Fig. 113A), walls thickened in *Cryptocentrum gracilipes* and *Cyrtidiorchis*; dead polygonal cells with somewhat thickened walls scattered in most taxa. **Endodermis** 1-layered (Fig. 112D), cells isodiametric, polygonal, or periclinal; cell walls O-thickened opposite phloem in *Anguloa*, *Cryptocentrum gracilipes*, *Lycaste*, *Neomoorea*, *Trigonidium*, and *Xylobium*; cell walls U-thickened in *Brasiliorchis*, *Cyrtidiorchis*, *Maxillaria* and *Mormolyca* species, *Scuticaria*, and *Teuscheria*; thin-walled opposite xylem. **Pericycle** cells polygonal in *Brasiliorchis*, *Camaridium*, *Maxillaria*, *Mormolyca*, periclinal in *Cryptocentrum gracilipes*, isodiametric in

Cyrtidiorchs; cell walls O-thickened opposite phloem. **Vascular cylinder** 17–26-arch in *Brasiliorchis* (Fig. 112D), 16-arch in *Camaridium cucullatum*, 6-arch in *Cryptocentrum gracilipes*, 20-arch in *Cyrtidiorchis*, 9–26-arch in *Maxillaria*, 13–24-arch in *Mormolyca*, and 12- or 13-arch in *Trigonidium*; xylem and phloem alternating around circumference of vascular cylinder (Fig. 112D); vascular tissue embedded in sclerenchyma in most *Maxillaria* species and in *Trigonidium*. **Stegmata** with conical silica bodies opposite phloem in *Anguloa* (Fig. 111A), *Bifrenaria*, *Brasiliorchis picta* (Fig. 113C,D) (as *Maxillaria*), *Cyrtidiorchis*, *Lycaste*, *Neomoorea*, and *Xylobium*; absent in *Rudolfiella*, *Scuticaria*, and *Teuscheria*; apparently absent in *Maxillaria* species, *Mormolyca*, and *Trigonidium*. Stegmata replace pericycle cells in roots of some orchid species. **Pith** cells sclerified in *Cryptocentrum gracilipes*; sclerenchymatous or parenchymatous in *Camaridium*, *Maxillaria*, and *Mormolyca* (Fig. 113A), parenchymatous in *Trigonidium*.

Material examined

Anguloa (4), *Bifrenaria* (5 + 1 as *Stenocoryne*), *Brasiliorchis* (2, as *Maxillaria*), *Camaridium* (1, as *Maxillaria*), *Cryptocentrum* (2 + 1 as *Anthosiphon*), *Cyrtidiorchis* (1), *Hylaeorchis* (1, as *Bifrenaria minuta*), *Lycaste* (11, + 2 transferred to *Sudamerlycaste*), *Maxillaria* (6, + 2 transferred to *Brasiliorchis*, 1 to *Camaridium*, 4 to *Mormolyca*), *Mormolyca* (1, + 4 as *Maxillaria*), *Neomoorea* (1), *Rudolfiella* (1), *Scuticaria* (1), *Sudamerlycaste* (2, as *Lycaste*, formerly *Ida*), *Teuscheria* (1), *Trigonidium* (1), *Xylobium* (4, + 1 transferred to *Maxillaria*). Full details appear in Holtzmeier et al. (1998), Stern, Judd et al. (2004). Species names have been updated in accord with those in the World Checklist of Selected Plant Families (2012).

Reports from the literature

Barthlott and Capesius (1975) [1976]: velamen of *Bifrenaria* and *Mormolyca*. Carnevali (1996): systematics and anatomy of *Cryptocentrum*. Chatin (1856, 1857): vegetative anatomy of *Maxillaria*. Davies (1999): leaf anatomy of *Heterotaxis* and *Maxillaria* species. Holtzmeier et al. (1998): vegetative anatomy of *Maxillaria*, *Mormolyca*, and *Trigonidium*. Krüger (1883): aerial organs of *Maxillaria*. Leitgeb (1864b): aerial root anatomy of *Bifrenaria*, *Camaridium*, *Lycaste*, *Maxillaria*, *Trigonidium*, and *Xylobium*. Löv (1926): leaf anatomy of *Lycaste* and *Maxillaria*. Metzler (1924): leaf anatomy of *Ornithidium*. Möbius (1887): leaf anatomy of *Lycaste*, *Maxillaria*, *Ornithidium*, *Scuticaria*, and *Xylobium*. Moreau (1913): pseudobulb of *Maxillaria*. Neubauer (1978): inflorescence idioblasts of *Lycaste*. Ojeda (2003): anatomy and phylogeny of *Heterotaxis* and *Ornithidium*. Oudemans (1861): aerial root anatomy of *Anguloa* and *Lycaste*. Parrilla Diaz and Ackerman (1990): root anatomy of *Maxillaria*. Porembski and Barthlott (1988): velamen of *Anguloa*, *Bifrenaria*, *Lycaste*, *Maxillaria*, *Mormolyca*, *Rudolfiella*, and *Xylobium*. Pridgeon et al. (2009): taxonomic treatment of Maxillariinae. Rǔgina et al. (1987): leaf and root anatomy of *Maxillaria* and *Ornithidium*. Solereder and Meyer (1930): general survey and original observations on leaf of *Anguloa*, *Bifrenaria*, *Lycaste*, *Maxillaria*, and *Ornithidium*. Stern and Judd (2002): systematics and anatomy of Cymbidieae and *Maxillaria*, *Mormolyca*, and *Trigonidium*. Stern, Judd et al. (2004): systematics and anatomy of Maxillarieae. Toscano de Brito (1994): leaf anatomy of *Maxillaria*. Weltz (1897): axis anatomy of *Bifrenaria*, *Lycaste*, *Maxillaria*, *Ornithidium*, *Trigonidium*, and *Xylobium*. Went (1895): root anatomy of *Ornithidium*. Williams (1976, 1979): stomata of *Anguloa*, *Cryptocentrum*, *Lycaste*, *Maxillaria*, *Mormolyca*, *Neomoorea*, *Teuscheria*, *Trigonidium*, and *Xylobium*.

Taxonomic notes

Maxillariinae are a morphologically diverse group of plants. A cladistic analysis (Whitten et al. 2000) has demonstrated that the monophyly of the group is only weakly supported. A later study (Whitten et al. 2007) demonstrated that *Maxillaria* is polyphyletic, and that all minor genera of core Maxillariinae are embedded in *Maxillaria* s.l. As currently understood, *Maxillaria* is not defined by a uniform set of character states and is a 'rag bag' taxon. Anatomically, however, there is evidence to show the monophyly of the group in the appearance of several almost consistent features: (1) superficial stomata with a couple of exceptions, (2) foliar fibre bundles, (3) homogeneous mesophyll with a few exceptions, (4) collateral vascular bundles in leaves mostly in one row, (5) stegmata with conical rough-surfaced silica bodies in leaves and pseudobulbs, (6) stomata absent in pseudobulbs, (7) tilosomes in roots of most genera, and (8) the unusual occurrence of stegmata in roots of several genera. Interestingly, *Anguloa*, *Bifrenaria*, *Lycaste*, *Maxillaria* (at least one species), *Neomoorea*, and *Xylobium*, in which these stegmata occur, form the backbone of a part of Dressler's 1993 Lycastinae.

Subtribe Oncidiinae Benth.

Oncidiinae constitute a vegetatively diverse group of tropical American plants. They may be epiphytes, lithophytes, or terrestrials, pseudobulbous or pseudobulbless

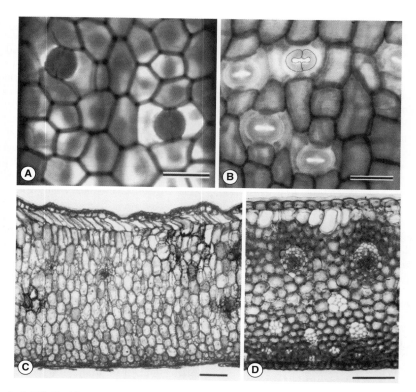

Fig. 114. Oncidiinae leaf. (A) *Oncidium venustum*, epidermis with tetracytic stomata and polygonal straight-sided cell walls. (B) *Trichopilia marginata*, epidermis with C-shaped guard cells, outlined in pair at upper middle. (C) *Rodriguezia lanceolata*, TS with homogeneous mesophyll and 1-layered adaxial hypodermis. (D) *Cochlioda noezliana*, TS showing 3-layered homogeneous mesophyll, adaxial hypodermis, two abaxial fibre bundle rows, collateral vascular bundles, and ±isodiametric adaxial epidermal cells. Scale bars: A,B = 50 μm, C = 200 μm, D = 100 μm. Reprinted from Stern and Carlsward (2006).

with short to long slender stems. Leaves are equitant, distichous, conduplicate, articulate or not, cylindric, isobilateral, or flattened laterally. Habit varies from short-lived delicate twig epiphytes to robust long-lived plants. Oncidiinae include about 70 genera and about 1600 species distributed in the New World tropics from Florida and northern Mexico, the Caribbean islands through Central America into Peru, Bolivia, and northern Argentina. Many species of Oncidiinae are favourites of orchid hobbyists especially in the genera *Brassia*, *Miltonia*, *Miltoniopsis*, *Oncidium*, *Psychopsis*, *Rossioglossum*, and *Tolumnia*. Some of these form the basis of numerous hybrids beloved of orchid enthusiasts. *Oncidium* has been noted as medicinal in Venezuela.

Leaf surface

Hairs absent throughout, except sunken hairs on both surfaces of *Telipogon* sp. (as *Stellilabium*). **Epidermis**: cells polygonal on both leaf surfaces (Fig. 114A,B); cells rectangularly elongate in *Hintonella* and *Ornithocephalus ecuadorensis* (as *Sphyrastylis*); cell walls straight-sided or curvilinear (Fig. 114A,B). **Stomata** abaxial, except amphistomatal in *Oncidium venustum*, *Trichocentrum albococcineum*, *T. microchilum* (as *Oncidium*), *T. stramineum* (as *Oncidium*), and possibly other species. Stomata present on all surfaces of isobilateral leaves, e.g. *T. lacerum* (as *Oncidium stipitatum*), where exposed surfaces are abaxial. Stomatal apparatuses basically tetracytic; guard cells typically reniform, but C-shaped in *Trichopilia marginata* (Fig. 114B).

Leaf TS

Cuticle smooth, grooved in *Psychopsis papilio*, papillate in *Trichocentrum albococcineum* and *Zygostates apiculata* (as *Dipteranthus planifolius*); generally 2.5–5.0 μm

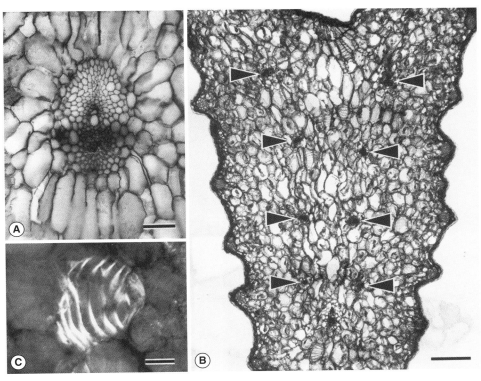

Fig. 115. Oncidiinae leaf. (A) *Ionopsis utricularioides*, TS with collateral vascular bundle and radiating thin-walled elongated bundle sheath cells. (B) *Tolumnia velutina*, isobilateral lamina with vascular bundles (arrowheads) occurring in opposite pairs. (C) *Trichocentrum microchilum*, TS water-storage cell with polarized banded wall thickenings. Scale bars: A = 100 μm, B = 200 μm, C = 50 μm. Reprinted from Stern and Carlsward (2006).

thick, up to 25 μm thick in *Trichocentrum microchilum*; <2.5 μm thick in *Telipogon* (including *Stellilabium*). **Epidermis**: cells mostly periclinal and isodiametric (Fig. 114D). **Stomata** usually superficial, but sunken to slightly sunken in a number of species and slightly raised in other species. Substomatal chambers usually very small to moderate, small to large in *Oncidium venustum*, moderately large in *Miltoniopsis roezlii* (as *M. santanaei*), large in *Psychopsis papilio*, *Trichocentrum lacerum*, and *T. microchilum*. Stomatal ledges often poorly developed, outer ledges small to large, inner ledges vestigial to apiculate. **Hypodermis**: 1-layered adaxially in most species (Fig. 114C,D), except *Pachyphyllum* sp. and *Plectrophora* sp., 1-layered abaxially in some species, absent in *Cuitlauzina candida* (as *Palumbina candida*), *Hintonella*, *Ornithocephalus* (including *Sphyrastylis*), *Phymatidium*, *Psychopsis papilio*, *Telipogon* (including *Stellilabium*), *Trichocentrum albococcineum*, *T. lacerum*, *T. microchilum*, *T. stramineum*, *Trichoceros*, and *Zygostates* (including *Dipteranthus*). Hypodermis having only parenchyma cells (Fig. 114C,D), some of them assimilatory, or alternating series of parenchyma cells and thick-walled sclerenchyma cells. **Fibre bundles** generally present abaxially (Figs 114D, 116A); absent in *Brachtia andina*, *Cuitlauzina pulchella* (as *Osmoglossum pulchellum*), *Hintonella*, *Ionopsis utricularioides*, *Mesospinidium* sp., *Ornithocephalus* (including *Sphyrastylis*), *Pachyphyllum* sp., *Phymatidium*, *Telipogon* (including *Stellilabium*), *Trichocentrum albococcineum*, *Trichoceros*, and *Zygostates* (including *Dipteranthus*); rare adaxially, present in *Chelyorchis ampliata* (as *Oncidium*), *Cuitlauzina candida*, *C. pulchella*, *Oncidium boothianum*, *O. sphacelatum*, and *Trichocentrum stramineum*. **Mesophyll** predominantly homogeneous (Fig. 114C,D), except heterogeneous in *Chelyorchis ampliata*, *Cuitlauzina pendula* (Fig. 116A), *Notylia barkeri*, *Psychopsis papilio*, *Trichocentrum microchilum*, and *Trichoceros tupaipi*. Angular water-storage cells occurring in mesophyll of

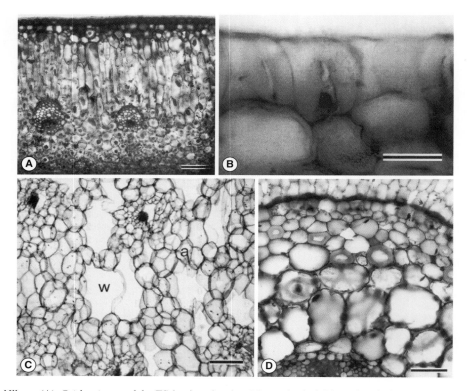

Fig. 116. Oncidiinae. (A) *Cuitlauzina pendula*, TS lamina showing 1-layered adaxial hypodermis, heterogeneous mesophyll, and collateral vascular bundle. (B) *C. pendula*, TS pseudobulb epidermal cells with highly thickened walls. (C) *Cochlioda noezliana*, TS pseudobulb showing ground tissue with water-storage cells (w), assimilatory cells (a), and scattered vascular bundles. (D) *Cuitlauzina pulchella*, TS root showing 3-layered cortex with idioblastic thick-walled cells surrounded by thin-walled parenchyma cells, triangular intercellular spaces, velamen cells, ∩-thickened exodermal cell walls, and O-thickened endodermal cell walls. Scale bars: A,D = 100 μm, B = 30 μm, C = 200 μm. Reprinted from Stern and Carlsward (2006).

some species, sometimes with banded or spirally-thickened walls (Fig. 115C). Mesophyll 3-layered in *Cochlioda noezliana* (Fig. 114D). Thin-walled crystalliferous idioblasts up to 182 μm in diameter occupying large part of leaf in *Telipogon* sp. **Vascular bundles** collateral (Figs 114D, 115A, 116A), in one row, except in two or three rows in *Ionopsis utricularioides* and *Trichocentrum microchilum*; linear series of bundles and paired bundles with phloem facing abaxial epidermis in *Ornithocephalus ecuadorensis* (as *Sphyrastylis*) and *Tolumnia velutina* (Fig. 115B). Sclerenchyma usually adjacent to xylem and phloem in larger bundles; in different species sclerenchyma sometimes predominating at xylem pole, sometimes at phloem pole; smallest bundles often lacking sclerenchyma completely. **Stegmata** with conical rough-surfaced silica bodies occurring along vascular bundle and fibre bundle sclerenchyma, except none along fibre

bundles in *Miltoniopsis roezlii*; absent in *Hintonella*, *Ornithocephalus ecuadorensis*, *O. inflexus*, *Pachyphyllum* sp., *Telipogon* sp., *Trichocentrum albococcineum*, and *Zygostates* (including *Dipteranthus*). Bundle sheaths variable, discontinuous, ill-defined, or absent, cells thin-walled and chlorenchymatous. In *Cochlioda noezliana* thick- and thin-walled cells occurring in same sheath, thin-walled radiating in *Ionopsis utricularioides*.

Stem TS

Hairs and **stomata** absent. **Cuticle** smooth, 2.5 μm thick, up to 37 μm thick in *Cochlioda noezliana* and *Oncidium boothianum*; <2.5 μm thick in *Phymatidium*. **Epidermis**: cells mostly periclinal, tiny, bead-like in *Pachyphyllum* sp. Epidermal cell walls usually thin, but thickened overall in *Cuitlauzina pendula* (Fig. 116B) almost to exclusion of cell lumen, outer walls thick in *Zygostates apiculata* (as *Dipteranthus*). **Cortex** present in pseudobulbs of *Aspasia* and *Brassia bidens*, in rhizomes of *Aspasia lunata*, *Gomesa crispa*, *Oncidium baueri*, and *Warmingia* sp., and in stems of *Lockhartia lunifera* and *Telipogon pulcher*. **Endodermis** and **pericycle** absent throughout, except for 1–3-layered endodermis of O-thickened, oval cells surrounding ground tissue of rhizome in *Gomesa crispa*. **Ground tissue** in pseudobulbs similar throughout, consisting of many variably shaped, smaller assimilatory cells surrounding fewer, large water-storage cells (Fig. 116C), the cell walls of which may be banded or spirally thickened in some species. **Vascular bundles** in pseudobulbs collateral, scattered, numerous; in a ring of 11 bundles in *Zygostates apiculata* and a ring of peripheral bundles with others scattered in ground tissue in *Phymatidium* and *Telipogon pulcher*; sclerenchyma surrounding entire bundle in *Oncidium baueri*, occurring only at phloem pole in several species and at both poles of larger bundles in other species. **Stegmata** with conical, rough-surfaced silica bodies associated with phloem sclerenchyma in almost all species, with phloem and xylem sclerenchyma in several species; absent in *Oncidium baueri*, *Pachyphyllum* sp., *Phymatidium*, *Telipogon pulcher*, and *Zygostates apiculata*.

Root TS

Velamen mostly 4–7 cells wide ranging from one or two cells wide in *Telipogon* sp. to two cells in *Macradenia* and to 16 cells in *Mesoglossum*. Epi- and endovelamen cell orientation variable. Cell wall thickenings mostly tenuous; wall thickenings spiral and looped or coarse. **Tilosomes** mostly absent; lamellate in *Cuitlauzina candida*, *Oncidium lineoligerum* (as *O. stenotis*), and *Trichopilia turialbae*. **Exodermis**: cell wall thickenings variable, usually thin-walled. Passage cells intermittent. **Cortex** 3–5 cells wide in *Comparettia speciosa* and *Ornithophora radicans* (as *Sigmatostalix*) to more than 20 cells wide in *Macradenia* sp.; cells thin-walled. Middle cortical region of almost all species consisting of typical thin-walled, more or less circular parenchyma cells; angular cells with greatly thickened birefringent walls occurring singly or clustered in small groups in outer layers of this region (Fig. 116D) and infested with fungal pelotons and dead hyphal masses. Angular water-storage cells with broad-branched cell wall bands occurring among parenchyma cells of middle region of cortex in some species. **Endodermis** and **pericycle**: cells variously oriented; most isodiametric with O-thickened walls. **Vascular cylinder** 4-arch in *Notylia ramonensis* (Fig. 111D), 5-arch in *Leochilus johnstonii* and *Telipogon* sp. to 27-arch in *Brachtia andina*. Vascular tissue embedded in thick- or thin-walled sclerenchyma (Fig. 111D); in parenchyma in *Ornithophora radicans*. **Pith** cells mostly thin-walled, parenchymatous, circular to oval; some species having thick to somewhat thickened cell walls and angular cells (Fig. 111D).

Material examined

Ada (1), *Aspasia* (1), *Brachtia* (1), *Brassia* (1), *Caucaea* (1, as *Oncidium*), *Chelyorchis* (1, as *Oncidium*), *Cischweinfia* (1), *Cochlioda* (1), *Comparettia* (1), *Cuitlauzina* (3, including *Osmoglossum* 1, *Palumbina* 1), *Erycina* (1), *Gomesa* (1), *Helcia* (1), *Hintonella* (1), *Ionopsis* (1), *Leochilus* (1), *Lockhartia* (2), *Macradenia* (1), *Macroclinium* (1, as *Notylia*), *Mesoglossum* (1), *Mesospinidium* (1), *Mexicoa* (1), *Miltonia* (1), *Miltoniopsis* (1), *Notylia* (2), *Odontoglossum* (1, as *Symphyglossum*), *Oncidium* (5), *Ornithocephalus* (10, including *Sphyrastylis* 4), *Ornithophora* (1, as *Sigmatostalix*), *Otoglossum* (1), *Pachyphyllum* (1), *Phymatidium* (1), *Plectrophora* (1), *Psychopsis* (1), *Rhynchostele* (2, including *Lemboglossum* 1), *Rodriguezia* (1), *Rossioglossum* (1), *Solenidium* (1), *Telipogon* (2, including *Stellilabium* 1), *Tolumnia* (3), *Trichocentrum* (5, including *Oncidium* 4), *Trichoceros* (2), *Trichopilia* (1), *Trizeuxis* (1), *Warmingia* (1), *Zygostates* (2, including *Dipteranthus* 1). For full details see Stern and Carlsward (2006), Stern, Judd et al. (2004).

Reports from the literature

Bentham (1881): definition of subtribe Oncidiinae including several genera now recognized in the subtribe, but with the exclusion of *Cryptarrhena*. Chase (1987): pollinarium morphology of *Oncidium*, *Odontoglossum*, and allied genera relative to their systematics. Chase and Palmer (1989): discussion of monophylesis of *Trichocentrum* based on molecular studies. Möbius (1887): leaf anatomy of many genera of Oncidiinae. Neubig et al. (2012): generic recircumscriptions in Oncidiinae. Pridgeon et al. (1983): description of tilosomes, their distribution and classification. Pridgeon et al. (2009): comprehensive treatment of taxonomy and phylogeny of Oncidiinae. Sandoval-Zapotitla and Terrazas (2001): comparative leaf anatomy of *Trichocentrum* and related taxa. Sandoval-Zapotitla et al. (2010a): phylogeny of Oncidiinae, partly based on leaf anatomy. Sandoval-Zapotitla et al. (2010b): mineral inclusions in Oncidiinae. Stern and Carlsward (2006): vegetative anatomical study of Oncidiinae. Stern, Judd et al. (2004): vegetative anatomy of Maxillarieae, including several genera now recognized in Oncidiinae. Toscano de Brito (1998):

comprehensive review of leaf anatomy in subtribe Ornithocephalinae including genera now recognized in Oncidiinae. Toscano de Brito (2001): systematic review of the *Ornithocephalus* group in Oncidiinae. Weltz (1897): stem anatomy of *Ada, Aspasia, Brassia, Lockhartia, Miltonia, Odontoglossum, Oncidium, Rodriguezia, Sigmatostalix* (now in *Ornithophora*), and *Trichopilia*. Williams et al. (2001): molecular systematics of Oncidiinae. Williams et al. (2005): molecular systematics of *Telipogon* and its allies.

Additional literature

Ayensu and Williams (1972): leaf anatomy and relationships of *Palumbina* (now in *Cuitlauzina*) and *Odontoglossum*. Barthlott and Capesius (1975): velamen of *Oncidium*. Capesius and Barthlott (1975): velamen of *Oncidium* and *Rodriguezia*. Chase (1986): monograph of *Leochilus*, with brief leaf and root anatomy. Chatin (1856, 1857): leaf, stem, root anatomy of *Oncidium*. Chiron and Guirad (2008): leaf anatomy of *Baptistonia* (now in *Oncidium*). Colleta and Silva (2007): leaf anatomy of *Psygmorchis* (now in *Erycina*) and *Ornithocephalus*. Duruz (1960): stomata of *Odontoglossum* and *Oncidium*. Dycus and Knudson (1957): velamen of *Oncidium*. Engard (1944): root anatomy of *Brassia* and *Oncidium*. Goh (1975): leaf anatomy of *Oncidium*. Hofmann (1930): leaf anatomy of *Oncidium ascendens* (now in *Trichocentrum*). Izaguirre de Artucio (1972): leaf anatomy of *Capanemia*. Janczewski (1885): root anatomy of *Oncidium*. Kraft (1949): velamen of *Oncidium*. Krüger (1883): aerial organs of *Brassia, Miltonia*, and *Oncidium*. Kuttelwascher (1964, 1965) [1966]: aerial root anatomy and development of *Brassia* and *Oncidium*. Leitgeb (1864b): aerial root anatomy of *Brassia, Cyrtochilum, Notylia, Oncidium*, and *Rodriguezia* (including *Burlingtonia*). Löv (1926): leaf anatomy of *Miltonia* and *Oncidium*. Meinecke (1894): aerial root anatomy of *Rodriguezia* and *Trichopilia*. Metzler (1924): leaf anatomy of *Oncidium*. Moreau (1913): pseudobulb anatomy of *Odontoglossum* and *Oncidium*. Neubauer (1978): idioblasts in axis of *Miltonia, Odontoglossum, Oncidium*, and *Trichopilia*. Oliveira and Sajo (1999, 2001): leaf and stem anatomy of *Miltonia*. Oudemans (1861): aerial root anatomy of *Brassia, Oncidium*, and *Rodriguezia* (including *Burlingtonia*). Parrilla Diaz and Ackerman (1990): root anatomy of *Comparettia, Ionopsis, Leochilus*, and *Oncidium*. Pfitzer (1877): silica bodies in *Oncidium*. Pirwitz (1931): 'storage tracheids' in *Oncidium*. Porembski and Barthlott (1988): velamen in many genera of Oncidiinae. Raciborski (1898): root anatomy of *Rhynchostele*. Richter (1901): aerial root anatomy of *Oncidium*. Rojas Leal (1993): leaf anatomy of *Mesoglossum, Odontoglossum, Oncidium*, and *Rhynchostele*. Rŭgina et al. (1987): leaf and root anatomy of *Oncidium*. Rüter and Stern (1994): root anatomy of *Oncidium*. Sandoval-Zapotitla (1993): leaf anatomy of *Cuitlauzina pendula*. Schimper (1884): root anatomy of *Oncidium*. Silva et al. (2006): leaf anatomy of *Oncidium*. Solereder and Meyer (1930): original observations on *Brassia, Oncidium*, and *Rodriguezia*. Tischler (1910): root anatomy of *Oncidium*. Wejksnora (1971): leaf and stem anatomy of *Oncidium*. Williams (1974, 1979): leaf anatomy and stomata, including many genera of Oncidiinae.

Taxonomic notes

There is considerable anatomical homogeneity among Oncidiinae despite the great diversity of vegetative features. Bentham (1881) included *Zygostates, Ornithocephalus, Phymatidium*, and *Chytroglossa* in Oncidiinae (as Oncidieae). Dressler (1993) and Toscano de Brito (1998), however, removed them into subtribe Ornithocephalinae, but Williams et al. (2001), based on molecular studies, supported the inclusion of these genera in Oncidiinae.

Since Bentham circumscribed Oncidiinae there have been many treatments in which the subtribe has been both broadened (Dressler 1993) and narrowed (Szlachetko 1995), showing great diversity in chromosome number and vegetative features. Recent phylogenetic studies based on DNA data (Williams et al. 2001, 2005) have demonstrated that two other former subtribes, namely Ornithocephalinae and Telipogoninae, are nested within Oncidiinae. There are no accounts in the literature on the anatomy of members of Telipogoninae, but Toscano de Brito (1998, 2001) has reported on leaf anatomy of Ornithocephalinae and 'related subtribes': cuticle thin to thick, hairs absent, epidermal cells isodiametric or elongated, anticlinal cell walls straight, stomata superficial, tetracytic; leaves hypostomatal, except amphistomatal in *Hintonella, Hofmeisterella, Rauhiella*, and *Thysanoglossa*; hypodermis absent; mesophyll homogeneous; fibre bundles absent; stegmata absent, but present in *Chytroglossa, Phymatidium falcifolium, Platyrhiza*, and *Rauhiella*; vascular bundles collateral with xylem and phloem sclerenchyma caps. Leaf anatomy of Ornithocephalinae reveals that fibre bundles and hypodermises are absent in these as are stegmata, except in *Chytroglossa, Phymatidium falcifolium, Platyrhiza*, and *Rauhiella*. Thus, these features support Toscano de Brito's Ornithocephalinae that are in distinct contrast to leaf anatomy of other Oncidiinae where fibre bundles, hypodermis, and stegmata are almost constant features.

In their study of leaf anatomy of 33 genera of Oncidiinae, Sandoval-Zapotitla et al. (2010a,b) included one species each of seven genera not described above. Additional data for these genera are as follows: Hairs present adaxially in *Capanemia*, *Cyrtochiloides*, *Cyrtochilum*, *Systeloglossum*, and *Zelenkoa*, absent in *Fernandezia* and *Macroclinium*; cuticle with adaxial and abaxial papillae in *Capanemia* only; stomata abaxial; hypodermis present adaxially; fibre bundles present abaxially in all genera except *Fernandezia*; mesophyll homogeneous; vascular bundles in one row; stegmata with conical silica bodies associated with abaxial fibre bundles in *Cyrtochiloides* and *Cyrtochilum* and with vascular bundle sclerenchyma in all except *Fernandezia*.

Sandoval-Zapotitla et al. (2010b) showed the value of mineral inclusions in leaves in characterizing clades in the subtribe. Their clade D (*Fernandezia*, *Hintonella*, *Ornithocephalus*, *Pachyphyllum*, *Telipogon*) lacked crystals apart from raphides, and completely lacked fibre bundles and stegmata. Data from the present study support these findings, except that stegmata occur in vascular sclerenchyma in *Telipogon pulcher*, a species not studied by Sandoval-Zapotitla et al.

Subtribe Stanhopeinae Benth.

Stanhopeinae are tropical American epiphytes with fragrant flowers pollinated exclusively by euglossine bees. Plants are pseudobulbous with 1–4 plicate, articulate leaves. There are 20 genera including seven that are monospecific. *Stanhopea* is the largest genus with about 55 species. Flowers of some Stanhopeinae display the most startlingly bizarre pollination systems. Owing to the beauty, waxy texture, pervasive aromas, and intricate floral structure, species of several genera, namely *Stanhopea*, *Gongora*, and *Coryanthes*, are widely grown by orchid fanciers, but the short-lived flowers of *Stanhopea* persuade some growers to avoid cultivating these.

Leaf surface

Hairs uniseriate, 3-celled, on both surfaces, proximal two cells submerged in a crypt. **Epidermis**: cells polygonal, anticlinal walls straight-sided or curvilinear (Fig. 117A,B). **Stomata** abaxial, basically tetracytic, sometimes anomocytic, cyclocytic, hexacytic, and anisocytic (Fig. 117A,B).

Leaf TS

Cuticle *c.*2.5 μm thick, often smooth. **Epidermis**: cells thin-walled (Fig. 117C), surfaces rounded to somewhat flattened, mostly isodiametric. **Stomata** superficial, to raised in *Paphinia neudeckeri*; substomatal chambers small to moderate and large in some species; outer ledges small to moderate, large in *Paphinia neudeckeri* and *Sievekingia reichenbachiana*; inner ledges minute or lacking. **Hypodermis**: absent. **Fibre bundles** surrounded by stegmata (Fig. 117C) with conical silica bodies present in all genera except *Acineta*. **Mesophyll** homogeneous, tending towards heterogeneity in some species of *Stanhopea*; cells thin-walled, rounded to oval to nearly circular. Crystalliferous idioblasts with raphides. Thin-walled water-storage cells occurring in *Gongora*, thick-walled in *Lueddemannia pescatorei*. **Vascular bundles** collateral (Fig. 117C), in a single series, each surrounded by thin-walled, chloroplast-containing bundle sheath cells. Sclerenchymatous bundle caps adjacent to xylem and phloem poles. **Stegmata** with conical rough-surfaced silica bodies associated with vascular sclerenchyma.

Pseudobulb TS

Hairs absent, except sunken in *Acineta densa*. **Cuticle** smooth (Fig. 118E), 5.0–17 μm thick in different genera. **Epidermis**: cells very thick-walled, usually turbinate (Fig. 118A), isodiametric, elliptical, or periclinal in different genera. **Stomata** absent. **Endodermis** and **pericycle** absent. **Ground tissue**: cells of outer 1–4 layers small, polygonal, thin-walled, without intercellular spaces, except thick-walled in *Polycycnis gratiosa*. Bulk of ground tissue composed of large, thin-walled, dead, rounded to angular water-storage cells surrounded by a matrix of small, thin-walled, living, rounded to polygonal assimilatory cells with cruciate starch granules; intercellular spaces triangular. Ground tissue of *Paphinia neudeckeri* with two components: (1) peripheral layer of loosely organized smaller assimilatory cells and large water-storage cells and (2) central mass of closely packed small assimilatory cells among which are larger water-storage cells. **Fibre bundles** usually absent but present in *Houlletia* species, *Paphinia neudeckeri*, *Schlimmia* species, and *Sievekingia reichenbachiana*. **Vascular bundles** collateral (Fig. 118C), scattered, and in *Gongora* greatly elongated, tapering, sock-like away from phloem (Fig. 118B); sclerenchyma phloic; xylem sclerenchyma present only in larger bundles of some species, absent in others. **Stegmata** with conical, rough-surfaced silica bodies (Fig. 118D) always accompanying phloic and xylic sclerenchyma, except for *Sievekingia* species.

Root TS

Velamen width ranging from four or five cells in *Kegeliella atropilosa* up to 11–13 cells in *Lueddemannia pescatorei*

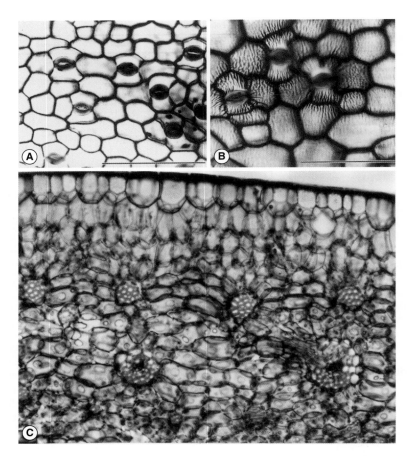

Fig. 117. Stanhopeinae leaf. (A) *Acineta densa*, abaxial surface with tetracytic stomata. Scale bar = 100 μm. (B) *Embreea rodigasiana*, abaxial surface with tetracytic stomata and cuticular corrugations. Scale bar = 100 μm. (C) *Stanhopea ruckeri*, TS showing heterogeneous mesophyll and fibre bundles; note primary pit fields in chlorenchyma cells; ×320. (A,B) Reprinted from Stern and Whitten (1999); (C) photograph by W.L. Stern.

and 12–15 cells in *Acineta densa*; cells thin-walled with fine anticlinal spirals. Hairs unicellular, bulbous-based. **Tilosomes** absent. **Exodermis**: cells polygonal (Fig. 119A), thin-walled; dead cells having tenuous scalariform thickenings along axial walls; passage cells intermittent. **Cortex** five cells wide in *Soterosanthus shepheardii* to ±15 cells in *Acineta densa*. Scattered thick-walled, enucleate, angular cells occurring in middle portion of cortex; cells with pelotons and dead cell masses also in this region. **Endodermis** 1-layered (Fig. 119B,C), cell walls U- or O-thickened opposite phloem, thin-walled opposite xylem. **Pericycle** 1-layered, 2- or 3-layered in *Stanhopea pulla*, cell walls O-thickened opposite phloem, thin-walled opposite xylem. **Vascular tissue** varying from 5-arch in *Soterosanthus shepheardii* to 20-arch in *Acineta densa*. Xylem and phloem components surrounded by sclerenchyma (Fig. 119C) and alternating around periphery of vascular cylinder. **Pith** cells thin-walled, polygonal to circular, parenchymatous, with intercellular spaces, or thick-walled and angular without intercellular spaces (Fig. 119C).

Material examined

Acineta (1), *Braemia* (1), *Cirrhaea* (1), *Coryanthes* (1), *Embreea* (1), *Gongora* (4), *Horichia* (1), *Houlletia* (2), *Kegeliella* (1), *Lueddemannia* (1), *Paphinia* (1), *Polycycnis* (1), *Schlimmia* (2), *Sievekingia* (2), *Soterosanthus* (1),

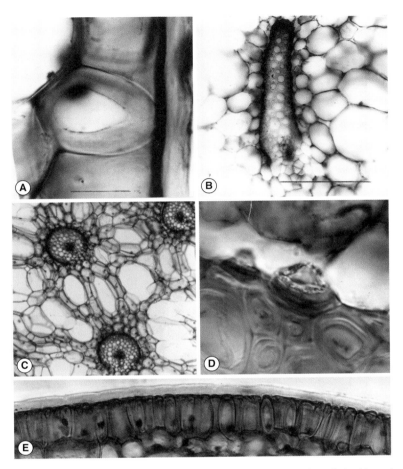

Fig. 118. **Stanhopeinae pseudobulb.** (A) *Sievekingia reichenbachiana*, TS with turbinate, thick-walled epidermal cell. Scale bar = 10 μm. (B) *Gongora truncata*, TS vascular bundle with digitiform sclerenchyma extension leading from phloem at lower right margin of bundle. Scale bar = 100 μm. (C) *Stanhopea anfracta*, TS showing scattered collateral vascular bundles in ground mass of assimilatory and water-storage cells, ×80. (D) *S. tricornis*, TS with rough-surfaced conical silica body in stegma associated with fibres of vascular bundle cap, ×2500. (E) *S. ruckeri*, TS showing thick cuticle, thick-walled epidermal cells, and edge of parenchymatous ground tissue, ×330. (A,B) Reprinted from Stern and Whitten (1999); (C,D) photographs by W.L. Stern.

Stanhopea (17). Full details appear in Stern and Whitten (1999) and Stern and Morris (1992).

Reports from the literature

Burgeff (1909): mycorrhiza of *Coryanthes*. Burr and Barthlott (1991): root anatomy of *Gongora*. Curry, McDowell, Judd, and Stern (1991): osmophores and systematics of *Stanhopea*. Curry, Stern, and McDowell (1989): osmophores in *Stanhopea anfracta* and *S. pulla*. Foroughbakhch et al. (2008): leaf epidermis and taxonomy of *Stanhopea* species. Krüger (1883): anatomy of aerial organs of *Stanhopea*. Leitgeb (1864b): aerial root anatomy of *Cirrhaea*, *Gongora* (including *Acropera*), *Houlletia*, and *Stanhopea*. Link (1849–51): aerial root anatomy of *Stanhopea*. Möbius (1887): leaf anatomy of *Acineta*, *Gongora*, and *Stanhopea*. Moreau (1913): pseudobulb anatomy of *Acineta*, *Gongora*, and *Stanhopea*. Neubauer (1978): inflorescence axis idioblasts of *Gongora* and *Polycycnis*. Oliveira and Sajo (1999, 2001): leaf and stem anatomy of *Stanhopea*. Pfitzer (1877): silica bodies in *Stanhopea*. Porembski and Barthlott (1988): velamen of *Acineta*, *Coryanthes*, *Gongora*, *Sievekingia*, and *Stanhopea*. Rŭgina et al. (1987): leaf and root anatomy

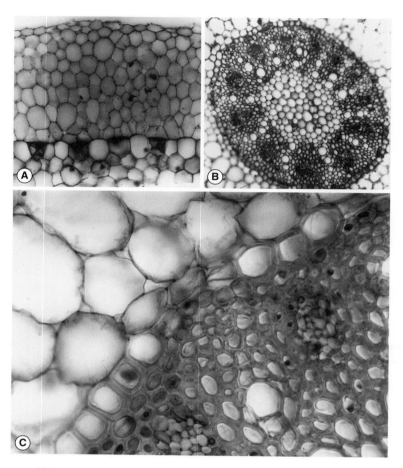

Fig. 119. **Stanhopeinae root.** (A) *Sievekingia reichenbachiana*, TS velamen to illustrate periclinal epivelamen cells, isodiametric to anticlinal endovelamen cells, and angular cortical cells. Scale bar = 100 μm. (B) *Polycycnis gratiosa*, TS vascular cylinder showing spoke-like organization of alternating xylem and phloem strands surrounding parenchymatous pith and thin-walled endodermal cells. Scale bar = 100 μm. (C) *Stanhopea pulla*, TS edge of vascular cylinder showing O-thickened walls of endodermal cells, two- or three-cell-wide pericycle, thin-walled cortical cells with triangular intercellular spaces, xylem and phloem strands, and thick-walled pith cells, ×960. (A,B) Reprinted from Stern and Whitten (1999); (C) photograph by W.L. Stern.

of *Gongora* and *Stanhopea*. Solereder and Meyer (1930): survey of orchid anatomy and original observations of leaf anatomy of *Gongora* and *Stanhopea*. Stern, Curry, and Pridgeon (1987): osmophores of *Stanhopea*. Stern and Morris (1992): vegetative anatomy of *Stanhopea*. Stern and Whitten (1999): vegetative anatomy of subtribe Stanhopeinae. Tischler (1910): root anatomy of *Stanhopea*. Weltz (1897): axis anatomy of *Gongora* and *Stanhopea*. Whitten, Williams, and Chase (2000): relationships of Maxillarieae and Stanhopeinae, molecular evidence. Williams (1976, 1979): stomata of *Acineta, Gongora, Kegeliella, Lueddemannia,* and *Stanhopea*.

Taxonomic notes

Despite the wonderful variation in floral details of Stanhopeinae and the amazing array of features associated with fragrance production and pollination strategies, especially in *Stanhopea*, the vegetative anatomy of these plants is unremarkable taxon to taxon and of little taxonomic utility (Stern and Morris 1992; Stern and Whitten 1999). Molecular research, on the contrary, has supported generic limits in Stanhopeinae. Differences in floral morphology among Stanhopeinae have prompted workers to split this taxon into two subtribes, Coeliopsidinae and Stanhopeinae *sensu stricto*, that molecular studies have shown are sister groups (Szlachetko 1995; Whitten et al. 2000).

Subtribe Vargasiellinae C. Schweinf. ex G.A. Romero & Carnevali

Plants of Vargasiellinae are terrestrial, sympodial herbs with elongated, decumbent, slender, leaf sheath-covered stems. Leaves are membranaceous to subcoriaceous, distichous, plicate, articulate. Vargasiellinae are represented by two South American species endemic to the tepuis of Venezuela and Pasco in eastern Peru. There are no reports on the vegetative anatomy. Vargasiellas are not in cultivation.

Subtribe Zygopetalinae Schltr.

Plants in this subtribe are either epiphytic or terrestrial herbs with or without pseudobulbs. Leaves are distichous, plicate or conduplicate, dorsiventrally flattened, articulated or not, often distinctly petiolate, membranous to coriaceous or fleshy. There are 35 genera, which occur strictly in the American tropics from southern Mexico and Caribbean islands to Brazil, Paraguay, Bolivia, and northern Argentina. The greatest diversity in numbers of genera and species occurs along the Andean chain of mountains of north-western South America. Many species are desirable ornamentals grown by orchid fanciers, among which are *Pescatoria cerina*, *Warczewiczella amazonica* (formerly *Cochleanthes amazonica*), *Chaubardia heteroclita*, *Huntleya burtii*, *Zygopetalum crinitum*, and *Z. mackaii*.

Leaf surface

Hairs with uniseriate cells, basal cells sunken. **Epidermis**: cells polygonal (Figs 120A, 121B) to rectangular (Figs 120B, 121A) and isodiametric, anticlinal walls straight-sided or curvilinear. **Stomata** abaxial, tetracytic (Figs 120B, 121B).

Leaf TS

Cuticle smooth, *c*.2.5 μm thick, ranging to 7.5 μm thick in species of *Chaubardia*, *Dichaea*, *Huntleya*, and *Otostylis*. **Epidermis**: cells papillate in *Benzingia* (as *Ackermania*) (Fig. 109C), isodiametric and periclinal in most taxa (Figs 120E–G, 121E–G). **Stomata** superficial (Figs 120C, 121C), substomatal chambers small; outer ledges small to moderate, inner ledges obscure to apiculate. **Hypodermis**: abaxial in *Koellensteinia altissima* only. **Fibre bundles** usually absent, present ad- and abaxially in *Koellensteinia altissima*, *Otostylis lepida*, *Zygopetalum crinitum*, and *Z. intermedium*; absent in *Z. mackaii*. **Mesophyll** homogeneous (Figs 120E–G, 121E–G), cells thin-walled, crowded, mostly oval; cruciate starch granules present in some cells. Crystalliferous idioblasts with raphide bundles present. Water-storage cells with secondary cell wall thickenings scattered in mesophyll. **Vascular bundles** collateral (Figs 120D–G, 121D–G), in one series (Figs 120H, 121H). Sclerenchyma caps at both xylem and phloem poles of each bundle but absent from smallest bundles. All sclerenchyma associated with **stegmata** bearing conical, rough-surfaced silica bodies. Bundle sheaths poorly defined or absent, cells thin-walled, usually chlorophyllous, sometimes with cruciate starch granules.

Stem TS

Hairs and **stomata** absent. **Cuticle** smooth, ranging from 2.5 μm to 37 μm thick. **Epidermis**: cell walls heavily C- to U-thickened facing cuticle (Fig. 121J), thin-walled internally; entirely thin-walled in *Dichaea* and *Warrea*. **Cortex** in *Dichaea* sharply defined, chlorenchymatous, not in other genera. Ground-tissue cells thin-walled, variably shaped; smaller chloroplast-containing assimilatory cells, some with cruciate starch granules, surrounding fewer, larger circular to oval water-storage cells. Endodermis and pericycle absent. **Vascular bundles** collateral (Fig. 121I,J), scattered; sclerenchyma associated with xylem and phloem poles, only with phloem poles in smaller bundles. **Fibre bundles** with stegmata and conical silica bodies scattered in ground tissue of *Otostylis lepida*. In *Dichaea camaridoides* and *D. chasei* outer vascular bundles embedded in a layer of thick-walled sclerenchyma cells. **Stegmata** with conical, rough-surfaced silica bodies associated with sclerenchyma of vascular bundles, but not in *Promenaea xanthina*. Raphide idioblasts present in unmodified cells.

Root TS

Velamen three or four cells wide in *Cryptarrhena*, four or five cells wide in most genera, up to 15 cells in *Warrea warreana*; cells thin-walled with tenuous spiral thickenings, except somewhat thickened in *Benzingia caudata* (as *Ackermania*) and *Galeottia negrensis*. **Tilosomes** absent, except lamellate in *Zygopetalum intermedium* var. *peruvianum*. **Exodermis**: cells polygonal (Fig. 120I), thin-walled; passage cells intermittent, lateral walls of long cells with ladder-like thickenings. **Cortex** 5–19 cells wide; cells thin-walled (Fig. 120I), mostly circular; angular idioblasts with thick, birefringent walls present; cell

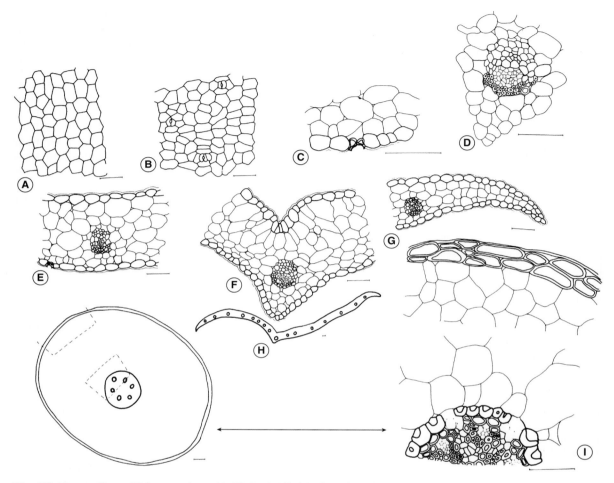

Fig. 120. Zygopetalinae. *Dichaea muricata*. (A–H) Leaf. (A) Adaxial epidermis with straight-sided polygonal cells. (B) Abaxial leaf surface with squarish cells and tetracytic stomata. (C,D) TS showing collateral vascular bundles and superficial stoma. (E–G) TS with collateral vascular bundles and homogeneous mesophyll. (H) TS with one row of vascular bundles. (I) TS root showing 2-layered exodermis, thin-walled cortical cells, heavily U-thickened endodermal cell walls, and vascular cylinder with xylem and phloem strands alternating within a sclerenchymatous matrix.

walls with branched, banded thickenings in *Chaubardia heteroclita*, *Koellensteinia graminea*, and *Neogardneria murryana*. Chloroplasts and cruciate starch granules present in some species; cells often infested with hyphae, pelotons, and dead fungal masses. **Endodermis** 1-layered (Figs 120I, 121K), cells isodiametric, periclinal in some species; U-thickened opposite phloem, very heavily so in *Benzingia* (as *Ackermania*), *Chondroscaphe escobariana*, and *Dichaea muricata* (Fig. 120I), O-thickened in *Pescatoria violacea*, *Promenaea stapelioides* (Fig. 121K), *Koellensteinia altissima*, and *Warrea warreana*, thin-walled in *Cryptarrhena lunata* and *Paradisanthus micranthus*. **Vascular cylinder** 6-arch in *Dichaea muricata* (Fig. 120I), *Paradisanthus micranthus*, and *Zygosepalum lindenii*, 8-arch in *Promenaea stapelioides* (Fig. 121K), up to 18-arch in *Otostylis lepida*; xylem and phloem alternating around circumference; vascular tissue embedded in thick-walled sclerenchyma cells, except thin-walled in *Paradisanthus micranthus* and *Pescatoria cerina*. **Pith** cells circular, thin-walled, except thick-walled in *Benzingia caudata* (as *Ackermania*), *Batemannia colleyi*, *Cryptarrhena*, *Dichaea muricata* (Fig. 120I), *Promenaea lentiginosa*, and *P. stapelioides* (Fig. 121K). Intercellular spaces triangular.

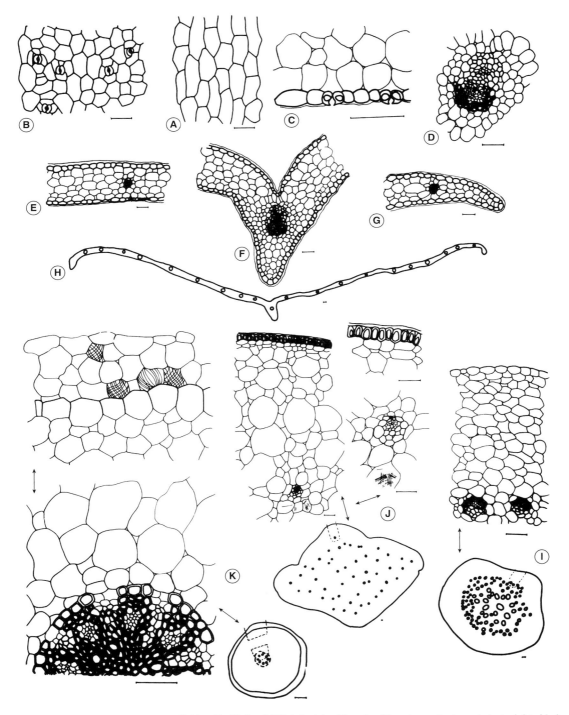

Fig. 121. Zygopetalinae. *Promenaea stapelioides*. (A–H) Leaf. (A) Adaxial epidermis with polygonal, elongated, straight-sided and curvilinear cell walls. (B) Abaxial epidermal surface with ±straight-sided cell walls and tetracytic stomata. (C,D) TS showing superficial stomata and collateral vascular bundle. (E–G) TS with homogeneous mesophyll and collateral vascular bundles. (H) TS with one row of vascular bundles. (I) TS rhizome with O-thickened epidermal cell walls, thin-walled cortical cells, and scattered collateral vascular bundles. (J) TS pseudobulb with thick-walled epidermal cells and vascular bundles scattered throughout most of ground tissue. (K) TS root showing vascular cylinder with O-thickened endodermal cell walls, alternating xylem and phloem strands, and thin-walled cortical cells with secondary cell wall strips.

Material examined

Aganisia (2), *Batemannia* (3), *Benzingia* (2, as *Ackermania*), *Chaubardia* (2), *Chaubardiella* (2), *Chondrorhyncha* (8, 1 sp. now *Stenotyla*), *Chondroscaphe* (1), *Cochleanthes* (1, 2 spp. now in *Warczewiczella*), *Cryptarrhena* (2), *Dichaea* (7), *Galeottia* (3), *Huntleya* (2), *Kefersteinia* (6), *Koellensteinia* (3), *Neogardneria* (1), *Otostylis* (1), *Pabstia* (1), *Paradisanthus* (1), *Pescatoria* (4, including *Bollea*), *Promenaea* (2), *Stenia* (2), *Stenotyla* (1, as *Chondrorhyncha lendyana*), *Warczewiczella* (2, as *Cochleanthes* spp.), *Warrea* (1), *Zygopetalum* (3), *Zygosepalum* (2). Full details appear in Stern et al. (2004).

Reports from the literature

Dressler (1993): systematics of Orchidaceae. Duruz (1960): stomata in *Zygopetalum crinitum*. Hering (1900): stem anatomy of *Dichaea*. Krüger (1883): aerial organs of *Huntleya* and *Zygopetalum*. Leitgeb (1864b): aerial root structure in *Zygopetalum* species. Meinecke (1894): aerial root anatomy of *Warczewiczella*, *Zygopetalum crinitum*, and *Z. mackayi*. Möbius (1887): leaf anatomy of *Chondrorhyncha*, *Colax* (now = *Pabstia*), *Dichaea*, *Warrea*, and *Zygopetalum*. Moreau (1913): pseudobulb structure in *Zygopetalum gauthieri*. Moreira et al. (2009): leaf and root anatomy of *Dichaea cogniauxiana*. Nelson et al. (2005): leaf anatomy of *Zygopetalum*. Neubauer (1978): inflorescence axis of *Chondrorhyncha*. Oliveira and Sajo (1999, 2001): leaf and stem anatomy of *Dichaea*. Parrilla Diaz and Ackerman (1990): root anatomy of *Cochleanthes* and *Dichaea*. Porembski and Barthlott (1988): velamen in several genera of Zygopetalinae. Pridgeon (1987): velamen in several genera of Zygopetalinae. Pridgeon et al. (1983): tilosomes absent in Zygopetalinae except present in *Zygopetalum intermedium* var. *peruvianum*. Silva et al. (2006): leaf anatomy of *Zygopetalum*. Solereder and Meyer (1930): review of orchid anatomy including original observations on *Zygopetalum* leaf. Stern, Judd et al. (2004): systematic and comparative anatomy of Maxillarieae including 26 genera of Zygopetalinae. Weltz (1897): axis anatomy of *Promenaea* and *Zygopetalum mackayi*. Williams (1979): stomata in several genera of Zygopetalinae.

Taxonomic notes

Dressler (1993) included *Cryptarrhena* in the monogeneric subtribe Cryptarrheninae in tribe Maxillarieae, but Pridgeon et al. (2009), based on DNA sequence data developed by Whitten et al. (2000, 2005), concluded *Cryptarrhena* is a monophyletic genus probably best included in Zygopetalinae. *Cryptorrhena* is supported anatomically as a member of Zygopetalinae. Genera *Benzingia*, *Chaubardia*, *Chaubardiella*, *Chondrorhyncha*, *Cochleanthes*, *Dichaea*, *Huntleya*, *Pescatoria*, and *Stenia* may be related because they are not distinguishable anatomically and in the DNA-based phylogenetic analysis of Whitten et al. (2005) these genera are members of a derived clade within Zygopetalinae; this specialized clade is unified by the loss of pseudobulbs.

Tribe Dendrobieae Endl.

Dendrobieae are mainly epiphytes with robust upright stems of one to many internodes; others have slender wiry stems, and still others have foreshortened stems hidden by small imbricated leaves. They vary widely in vegetative morphology. Leaves may be laterally flattened and conduplicate, some are grass-like and thin, still others are coriaceous and fleshy, terete, lunate, or falcate, persistent or deciduous. Of the several segregate genera, *Dendrobium* and *Bulbophyllum* are the only ones now recognized.

Dendrobium species range widely in the Old World tropics and subtropics, but are absent from Africa and Madagascar. They are prominent in India, the Philippines, southern China, New Guinea, northern Australia, Pacific islands, and Indonesia. They also occur in Korea and Japan. Dendrobiums are widely grown as ornamentals, chiefly as hybrids of *Dendrobium bigibbum* (= *D. phalaenopsis*). *Dendrobium* species have been used medicinally throughout East and South East Asia and Australia.

Bulbophyllum is the largest orchid genus with more than 1900 species. Most orchid genera are unknown anatomically. *Bulbophyllum*, on the contrary, has been studied by several researchers and most of the available information results from these efforts. Many bulbophyllums are bizarre plants with flowers sometimes borne on fleshy, flattened rachises. Plants vary in size from *B. pygmaeum* and *B. minutissimum*, which are among the smallest species in the orchid family, to other species in which the leaves may be more than 1 m long. Bulbophyllums are epiphytes, most species of which inhabit the rain or cloud forests of South East Asia although the genus is pantropical. There are one or two often fleshy articulate leaves from a pseudobulb of a single node. Pseudobulbs are often widely separated on an elongated rhizome. All bulbophyllums are worth considering for cultivation, many because of their curious structure. Because many species have flowers with obnoxious or offensive odours, growers should be cautious about selecting species to introduce. *Bulbophyllum* species have been used medicinally in Asian countries from India to China and Japan.

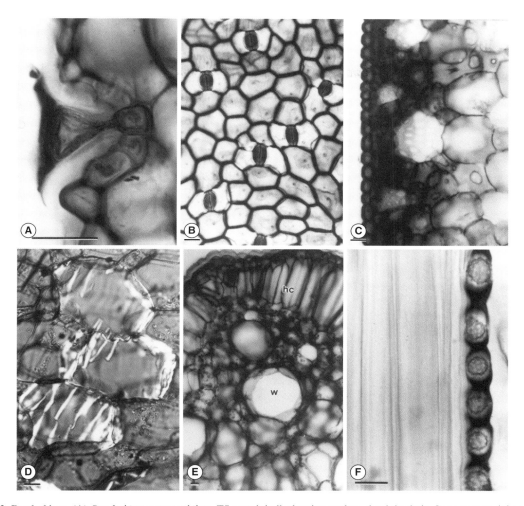

Fig. 122. Dendrobieae. (A) *Dendrobium potamophilum*, TS pseudobulb showing sunken glandular hair. Opaque material spreading from apex is apparently the exudate. (B–F) Leaf. (B) *D. albosanguineum*, abaxial epidermis with paracytic stomata. (C) *D. acinaciforme*, TS showing alternating larger and smaller abaxial fibre bundles. (D) *D. potamophilum*, TS with polarized helical thickening bands of water-storage cells. (E) *D. rigidum*, TS with elongated hypodermal cells (hc) and unpleated water-storage cells (w). (F) *D. albosanguineum*, LS with spherical silica bodies in stegmata adjacent to vascular bundle fibres. Scale bars = 10 μm. Reprinted from Morris et al. (1996).

Dendrobium Sw.

Leaf
See below for sections *Rhizobium* and *Aporum*.

Leaf surface
Hairs ambifacial, uniseriate, uni- to bicellular in most species, paired in *D. crumenatum* (Fig. 123A), tricellular in *D. angustifolium* and *D. pallens* and mostly so in *D. senile*; glandular, base sunken in epidermal crypt, walls of crypt cells birefringent with anastomosing thickening bands in *D. speciosum*. **Cuticle** smooth to irregular. **Epidermis**: cells polygonal, anticlinal walls straight-sided and curvilinear, sinuous in a few species. **Stomata** usually abaxial, but ab- and adaxial in *D. canaliculatum*; mostly tetracytic (Fig. 123A), but paracytic in some species, including section *Cadetia* (Fig. 122B).

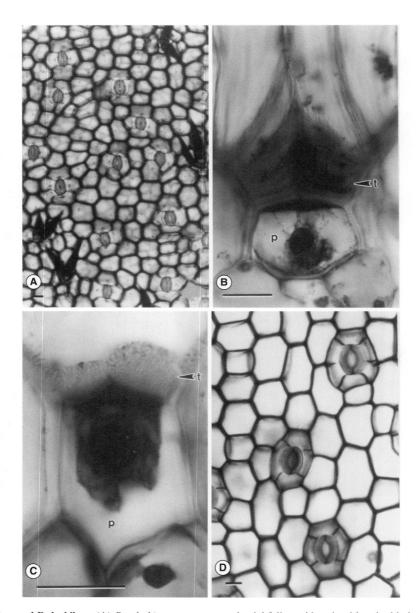

Fig. 123. Dendrobieae and Podochileae. (A) *Dendrobium crumenatum*, abaxial foliar epidermis with paired hairs (lower left). (B) *D. malbrownii*, TS root exodermis with passage cell (p) subjacent to tilosome (t) in adjoining velamen cell. (C) *D. salaccense*, TS root exodermis with tilosome (t) adjacent to exodermal passage cell (p). (D) *Pseuderia platyphylla*, abaxial foliar epidermis showing tetracytic and cyclocytic stomata. Scale bars = 10 μm. Reprinted from Morris et al. (1996).

Leaf TS

Cuticle thin to moderately thick. **Epidermis**: cells elliptical, square, rectangular (Fig. 122C), and hexagonal. **Stomata** superficial, substomatal chambers usually small, large in some species; outer ledges large, inner ledges minute. **Hypodermis**: present (Fig. 122E) or absent; ad- or abaxial or both, of one or more layers; cells with helical wall thickenings in species of section *Cadetia*;

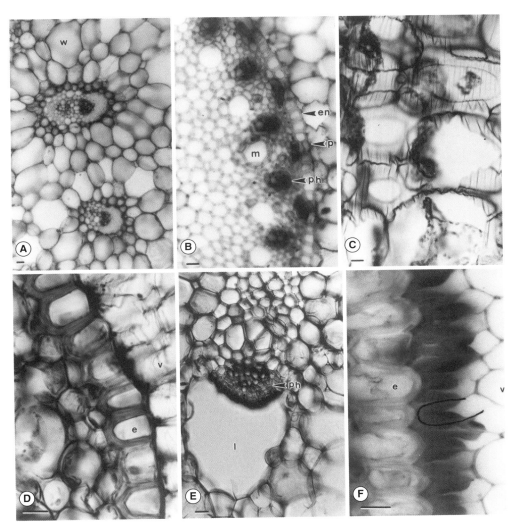

Fig. 124. Dendrobieae. (A) *Dendrobium jonesii*, TS pseudobulb ground tissue with collateral vascular bundles and smooth water-storage cell walls (w). (B) *D. salaccense*, TS root with O-thickened endodermal cell walls (en), endodermal passage cells (p) opposite xylem strands, alternating clusters of xylem and phloem (ph), and wide metaxylem cells (m). (C) *D. pachyphyllum*, TS lamina showing pleated water-storage cell walls. (D) *D. pachyphyllum*, TS root showing helical thickening bands of velamen cell walls (v), and U-thickened exodermal cell walls (e). (E) *D. macraei*, TS pseudobulb with phloem group (ph) of vascular bundle adjacent to lacuna (l). (F) *D. schoeninum*, TS root with V- and U-thickened walls of innermost velamen cells (v) adjacent to exodermis (e). The inked line outlines a velamen cell within which the thickening forms. Scale bars = 10 μm. Reprinted from Morris et al. (1996).

parenchyma cells sometimes alternating with fibre bundles. **Fibre bundles**: present, ad- or abaxial or both, in a few species (Fig. 122C), but absent in most species (including section *Cadetia*); fibre bundles with adjacent stegmata bearing spherical, rough-surfaced silica bodies in species of section *Desmotrichum* (except *D. comatum*), and *D. arachnoideum*, *D. crumenatum*, *D. cymbidioides*, *D. guttulatum*, and *D. malbrownii*. **Mesophyll**: homogeneous in most species, heterogeneous in section *Cadetia* and several other taxa; water-storage cells with helical thickenings present in *D. munificum*, *D. pachyphyllum* (Fig. 124C), and section *Cadetia* (Fig. 122D). Raphide

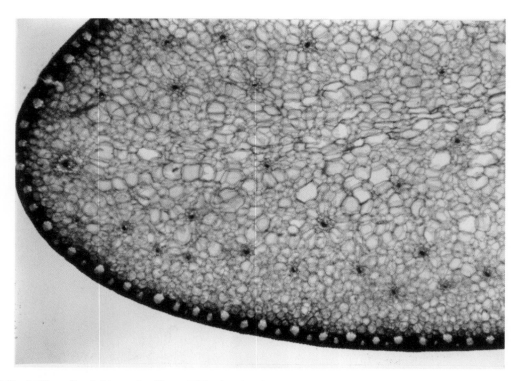

Fig. 125. Dendrobieae. *Dendrobium acinaciforme*. TS lamina showing major vascular bundle at leaf tip, rows of peripheral vascular bundles on either side of incipient lacuna, and peripheral alternating larger and smaller abaxial fibre bundles; mesophyll homogeneous including irregularly shaped water-storage cells, ×192. Reprinted from Carlsward et al. (1997).

idioblasts present in section *Cadetia* and some other taxa; other crystal types occasionally present. **Vascular bundles** collateral, in one row, phloic and xylic sclerenchyma present. **Stegmata** with spherical, rough-surfaced silica bodies present opposite vascular sclerenchyma (Fig. 122F), but absent in *D. senile*. Bundle sheath cells thin-walled in most species to somewhat thick-walled in a few species.

Sections *Rhizobium* and *Aporum*

Leaves in species of sections *Rhizobium* and *Aporum* are unique in tribe Dendrobieae because exposed surfaces are for the most part abaxial. Species in section *Rhizobium* stand out because of their fleshy, usually terete or subterete leaves and lack of fleshy stems. Plants in section *Aporum* have coriaceous, equitant, and laterally flattened leaves borne in fascicles on upright or pendent stems arising from a short rhizome.

Hairs uniseriate, bicellular, and glandular on all surfaces in both sections. **Epidermis** mainly abaxial on exposed surfaces. Adaxial epidermis exposed in groove of leaves in *D. rigidum* and *D. toressae*, but in most species of section *Rhizobium* adaxial epidermis internal and completely surrounded by mesophyll cells, forming an epithelium enclosing a canal represented as a lacuna in TS, the cells of which are covered by a cuticle and subtended by a hypodermis (Fig. 126). **Stomata** abaxial only; paracytic in species of section *Aporum* and tetracytic in section *Rhizobium*. **Hypodermis**: 1–3-layered adaxially, 1- or 2-layered abaxially in section *Rhizobium*, 1–3-layered abaxially in section *Aporum*; cell walls thin, some with banded thickenings in *D. toressae* only. **Fibre bundles** absent in section *Rhizobium* but occurring abaxially and with stegmata containing rough-surfaced, spherical silica bodies in section *Aporum*. **Mesophyll**: tendency to zonation in most species; central region comprising a reticulum of assimilatory cells and water-storage cells in section *Rhizobium*; water-storage cells thin-walled in most species, but thick-walled in *D. rigidum* and *D. teretifolium*, thin-walled with banded thickenings in *D. toressae*. **Vascular bundles** collateral, occurring in two or

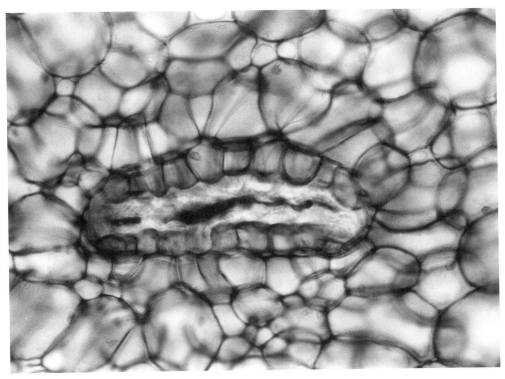

Fig. 126. Dendrobieae. *Dendrobium linguiforme*. TS isobilateral lamina showing internal adaxial epidermis surrounded by cells of hypodermis; these forming an epithelium enclosing an elliptical canal represented as a lacuna in TS; epidermal cells are covered by cuticle, ×1100. Reprinted from Stern, Morris, and Judd (1994).

more rings or facing rows (Fig. 125), between which is a suture or incipient suture in species of section *Aporum* and a lacuna or incipient lacuna in species of section *Rhizobium*; a major vascular bundle present towards margin of each leaf tip. **Stegmata** containing rough-surfaced spherical silica bodies associated with phloem sclerenchyma in all species of section *Rhizobium* except *D. rigidum* where stegmata lacking; also absent in *D. acinaciforme* of section *Aporum*.

Stem TS

Hairs absent, except present in a few species of section *Cadetia* (Fig. 122A) and *D. senile*. **Cuticle** moderately thick, smooth. **Epidermis**: cells square, rectangular, hexagonal, or rounded; thick-walled. **Stomata** absent. **Ground tissue**: parenchymatous, bi- or trizonate, or non-zonate; outer one to several layers of very thick-walled, polygonal, parenchymatous cells; inner region composed of smaller, thin-walled, parenchymatous cells interspersed with larger, dead, thin-walled water-storage cells, often with 'pleated' outlines; water-storage cells with helical wall thickenings present in inner region of zonate ground tissue in some species of section *Cadetia*, *D. amplum*, *D. cymbidioides*, *D. treacherianum* (as *D. lyonii*), and *D. munificum*; water-storage cells absent in *D. cunninghamii*, *D. malbrownii*, and *D. salaccense*. Raphide bundles sometimes present in thin-walled ground-tissue cells. **Vascular bundles** collateral (Fig. 124A), concentrated in inner regions of stems with zonate parenchyma and sometimes forming a ring between outer zones; randomly dispersed in non-zonate ground tissue. Phloic and xylic sclerenchyma present (Fig. 124A). Phloem subtended by a large lacuna in *D. macraei* (Fig. 124E). **Stegmata** bearing spherical rough-surfaced silica bodies present in association with phloic sclerenchyma, but absent in *D. cunninghamii*, *D. malbrownii*, *D. senile*, and all species of section *Rhizobium* (except *D. lichenastrum*).

Root TS

Velamen: 2–17-layered with variably shaped cells and fine or relatively wide, helical, anastomosing thickening bands; in *D. schoeninum* innermost walls of velamen cells next to exodermis strongly thickened. **Tilosomes** usually absent, but present in *D. amplum*, *D. arachnoideum*, *D. cymbidioides*, *D. fariniferum*, *D. malbrownii* (Fig. 123B), *D. salaccense* (Fig. 123C), and *D. treacherianum* (as *D. lyonii*). **Exodermis**: dead cells O-, U- (Fig. 124D,F), or ∩-thickened; walls laminated or unlaminated; anticlinal walls with oval bordered pits in many species; anticlinal walls with tenuous scalariform bars present in sections *Cadetia*, *Desmotrichum*, and *Diplocaulobium*, and in *D. pachyphyllum*, absent in other species. **Cortex**: cells of outermost one or two layers polygonal, other cells rounded or oval, thin-walled, parenchymatous. Raphide bundles in some cortical cells. **Endodermis** 1-layered, cells isodiametric or anticlinal, cell walls O-thickened opposite phloem sectors (Fig. 124B), thin-walled opposite xylem sectors. **Pericycle** configured like endodermis. **Vascular cylinder** 6–8- or 10-arch in species of section *Cadetia* and 5–27-arch in other species; conductive cells embedded in sclerenchyma; xylem and phloem elements alternating around periphery of vascular cylinder (Fig. 124B). **Pith** sclerenchymatous in section *Cadetia* and sclerenchymatous or parenchymatous in other species (Fig. 124B). **Stegmata** absent.

Bulbophyllum Thouars.

Leaf surface

Hairs with sunken bases present on both surfaces, more numerous abaxially than adaxially. Hairs glandular in *B. dasypetalum*, *B. gracile* (as *B. thouarsii*), and *B. makoyanum*. **Epidermis**: cells polygonal (Figs 127A,B–129A,B). **Stomata** tetracytic or anomocytic (Figs 127B, 128B), abaxial only; 'floating' in *B. viridiflorum* owing to dissolution of anticlinal walls of subsidiary cells.

Leaf TS

Cuticle thick in some species (Fig. 129C), thin in others (Figs 127C, 128C); considerably thickened in *B. molossus* (as *B. sessiliflorum*), papillate on abaxial surface in *B. tuberculatum*. **Epidermis** 1-layered; cells small, flat-surfaced (Figs 127D–F, 128D–F); radial and inner tangential walls thickened, oil droplets in cells of both epidermises. In *B. hirtum* adaxial epidermal cells large, reaching one-sixth of leaf thickness and serving for water storage. **Hypodermis**: cells generally polyhedral (Fig. 129D–F); spirally thickened in *B. gracile*; mostly 1-layered ab- and adaxially with thin walls; adaxially only in *B. griffithii*, *B. penicillium*, and *B. reptans*, abaxially only in *B. khasyanum*. Subepidermal fibres present ad- and abaxially in *B. pumilum* (Fig. 128C–F). **Fibre bundles** absent. **Mesophyll** heterogeneous in *B. clandestinum*, *B. makoyanum*, *B. pumilum* (Fig. 128D–F), *B. retiusculum*, *B. saltatorium* var. *calamarium* (as *B. calamarium*; Fig. 127D–F), *B. vaginatum* (as *B. whiteanum*), and other species (Fig. 129D–F); homogeneous in *B. leopardinum* and other species. Water-storage cells with spiral thickenings occurring among mesophyll cells in *B. gracile*, *B. reptans*, *B. retiusculum*, and additional species; these cells elongated, sac-like, and parallel to midvein; in some other species these cells isodiametric; some larger hypodermal cells also storing water as in *B. khasyanum*. Mucilage cells, raphides, and fibres occurring in hypodermis of *B. gibbosum*, *B. multiflorum*, and *B. virescens* (as *B. ericssonii*); small druses in mesophyll cells of *B. acutiflorum*; oil droplets in mesophyll cells of *B. saltatorium* var. *calimarium*. **Vascular bundles** usually in one row (Figs 127G–129G); bundles lying at juncture of upright and spongy cells in species with heterogeneous mesophyll. Bundles surrounded or capped by 2- or 3-layered fibrous sheath, containing cells in which lumen almost absent in some species (Figs 127C–129C); fibres more numerous at phloem pole than at xylem pole. Sclerenchyma bridge of thick-walled cells present in *B. ambrosia* (as *B. watsonianum*), *B. virescens*, and other species. **Stegmata** absent.

Pseudobulb TS

Cuticle thin in *B. leopardinum* and other species, but thick to very thick in some species (Fig. 129H). **Epidermis**: cells periclinal (Fig. 127I) or squarish (Fig. 128I), with weakly vaulted outer walls in *B. ambrosia*. Cells beneath epidermis 3- or 4-layered, colourless, parenchymatous, sometimes collenchymatously thickened and mucilaginous. When spiral cells present in leaf, also present in pseudobulb parenchyma. **Cortex** and ground-tissue cells parenchymatous, polygonal; in *B. penicillium*, *B. reptans*, and *B. triste* some larger cells and water-storage cells of cortex (Figs 127I, 128I, 129H) and ground tissue with spirals; some cortical cells mucilaginous. **Vascular bundles** typically scattered in cortex and ground tissue (Figs 127I, 128I, 129H), but some species having a central ring of small bundles. In *B. gibbosum* parenchymatous cortex of polyhedral cells with spirals outside sclerenchymatous ring and large parenchyma cells with spirals inside ring. Larger vascular bundles may be surrounded by complete sclerenchymatous sheath; smaller bundles having few sclerenchyma cells at xylem pole and more pronounced sclerenchyma at phloem pole.

Rhizome TS

Cuticle generally thick, very thick in *B. clavatum* and *B. occulatum*; thick in *B. acutiflorum* and *B. lobbii*.

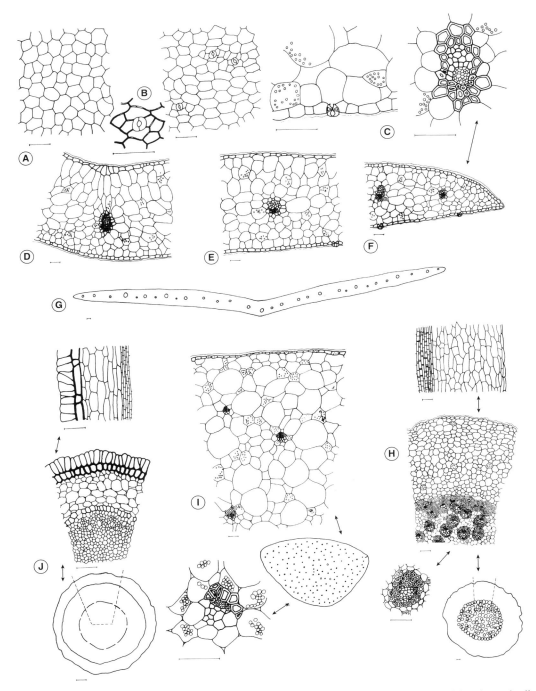

Fig. 127. Dendrobieae. *Bulbophyllum saltatorium* var. *calamarium*. (A–G) Leaf. (A) Adaxial epidermis with polygonal cells mostly with curvilinear walls. (B) Abaxial epidermis with polygonal cells and mostly curvilinear walls, stomata tetracytic and anomocytic. (C) TS with superficial stoma, oil droplets in mesophyll cells, and collateral vascular bundle. (D–F) TS with heterogeneous/homogeneous mesophyll, oil droplets in cells, and collateral vascular bundles with thin-walled bundle sheath cells. (G) TS with one row of vascular bundles. (H) TS and LS rhizome with parenchymatous oil-bearing cells, collateral scattered vascular bundles, sclerenchyma of outer bundles fused to form a band or ring. (I) TS pseudobulb showing water-storage cells of cortex, cells with oil droplets, and scattered vascular bundles. (J) TS and LS root with 1-layered velamen of thin-walled, narrow anticlinal cells, exodermal cells with O-thickened walls, parenchymatous cortex, endodermis with thin-walled cells, vascular elements embedded in sclerenchyma, and parenchymatous, thin-walled pith cells.

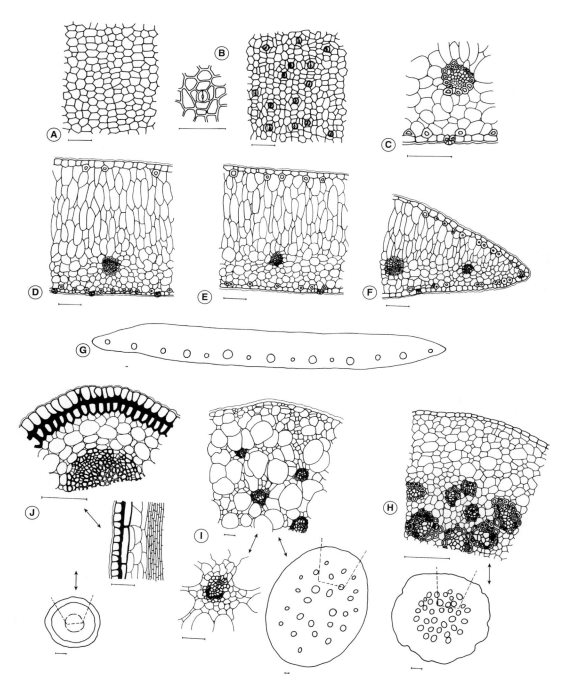

Fig. 128. Dendrobieae. *Bulbophyllum pumilum*. (A–G) Leaf. (A) Adaxial epidermis with polygonal cells, mostly with curvilinear walls. (B) Abaxial epidermis with polygonal cells and curvilinear and straight walls and tetracytic stomata. (C) TS with superficial stoma and subepidermal fibres. (D–F) TS showing heterogeneous mesophyll, collateral vascular bundles situated at juncture of upright cells and underlying spongy cells, and ad- and abaxial row of single thick-walled fibres directly under epidermis. (G) TS with one row of vascular bundles. (H) TS rhizome with scattered vascular bundles in central cylinder. (I) TS pseudobulb having scattered collateral vascular bundles, oval to irregularly formed water-storage cells, ground tissue of thin-walled parenchyma cells. (J) TS and LS root with 1-layered velamen of thin-walled anticlinal cells with thickened internal walls, O-thickened exodermal cells, thin-walled parenchymatous cortical cells, O-thickened endodermal cells alternating with thin-walled cells opposite xylem, and thick-walled pith cells.

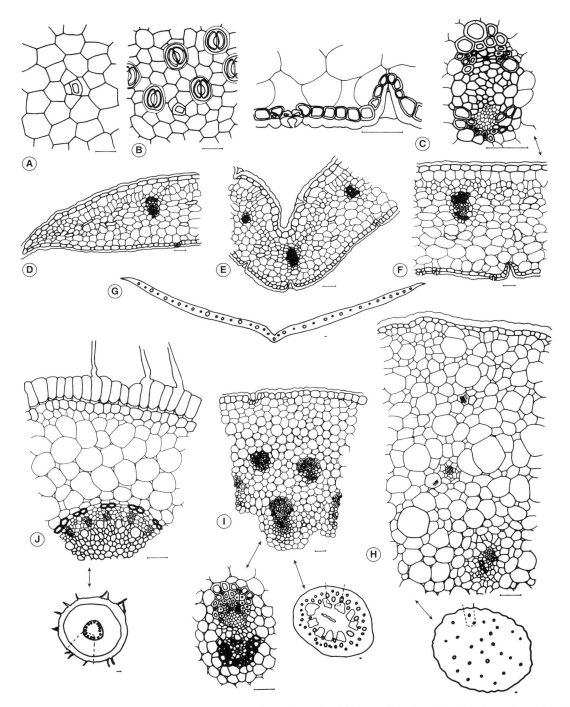

Fig. 129. Dendrobieae. *Bulbophyllum* sp. (as *Monomeria barbata*). (A–G) Leaf. (A) Adaxial epidermis with polygonal straight-sided thin-walled cells. (B) Abaxial epidermal cells polygonal with straight-sided walls, and anomocytic stomata. (C) TS with superficial stoma, thick cuticle, sunken hair base, epidermal cells with O-thickened walls, and collateral vascular bundle. (D–F) TS with heterogeneous mesophyll, 1-layered hypodermis, and collateral vascular bundles. (G) TS showing one row of vascular bundles. (H) TS pseudobulb with scattered collateral bundles, raphide bundles in idioblasts, thin-walled ground-tissue, and water-storage cells. (I) TS rhizome with scattered vascular bundles. (J) TS root with 1-layered velamen, thin-walled, anticlinal U-thickened cells, thick-walled endodermal cells alternating with thin-walled cells opposite xylem, wide metaxylem cells, and thick-walled parenchymatous pith cells.

Epidermis: cells elongated in axial direction; radial and tangential walls evenly thickened, inner and outer walls weakly vaulted; radial and tangential walls thick in *B. careyanum*. **Cortex** often consisting of few cell layers; cells thick-walled and narrow-lumened. Spiral cells occurring in ground tissue of *B. gibbosum*, *B. lobbii*, *B. penicillium*, *B. reptans*, and *B. triste*. Most *Bulbophyllum* species examined having a 1-layered sclerenchyma cylinder at border of cortex and ground tissue, e.g. *B. clandestinum*, *B. gibbosum*, and *B. makoyanum*; two layers of sclerenchyma present in *B. gamblei*, four in *B. acutiflorum*, and six in *B. virescens*. **Vascular bundles** scattered in central parenchyma cylinder (Figs 128H, 129I), the peripheral ones adjacent to inner border of sclerenchyma ring (Fig. 127H). All bundles with a 1-layered xylem sheath and most with two or three layers of sclerenchyma forming a complete phloem sheath; a sclerenchyma bridge present in vascular bundles of *B. acutiflorum* and *B. clavatum*.

Root TS

Velamen 1-layered (Figs 127J–129J), cells isodiametric, lacking ornamentation. Root hairs unicellular (Fig. 129J). **Tilosomes** broadly lamellate in most bulbophyllums; spongy in *B. micholtzianum* and *B. ornatissimum*. **Exodermis**: cell walls O-thickened in most *Bulbophyllum* species (Figs 127J, 128J), Ո-thickened in *B. medusae*, Ո- and O-thickened in *B. imbricatum* (as *B. kamerunense*), unthickened in *Bulbophyllum* sp. (as *Monomeria barbata*; Fig. 129J). **Cortex** 4- or 5-layered with spirally thickened columnar or sac-like water-storage cells in *B. penicillium*, *B. reptans*, and *B. triste*. **Endodermis**: cells barrel-shaped (Figs 127J, 129J); walls O-thickened. **Pericycle** cells sclerenchymatous with thin-walled passage cells opposite xylem sectors. **Vascular cylinder** polyarch; xylem and phloem components alternating (Figs 127J–129J); *B. griffithii* 10-arch, *B. membranifolium* (as *B. maculosum*) 7–13-arch, *B. odoratissimum* 20–66-arch, *B. reptans* 9–10-arch, *B. triste* 6-arch, *B. viridiflorum* 6–12-arch. Xylem and phloem elements embedded in thick-walled cells in some species. **Pith** sclerotic in some taxa.

Contrary to other studies, Kaushik (1982) [1983] describes the epidermis of the root as multiseriate, 1–3-layered in *B. reptans*, *B. triste*, *B. griffithii*, and *B. khasyanum* and 4- or 5-layered in *B. penicillium*. Kaushik records these as velamen layers and states the epivelamen is generally ruptured in mature roots. The anatomical account for roots of *Cirrhopetalum caudatum* and *C. caespitosum* (now = *B. scabratum*) describes a velamen of one or two layers. Other researchers note the velamen of *Bulbophyllum* to be 1-layered!

Material examined for *Dendrobium*

Dendrobium (94, + 5 *Cadetia* spp.). Full details appear in Carlsward et al. (1997), Morris et al. (1996), Stern et al. (1994).

Material reported for *Bulbophyllum*

All available anatomical data refer to species of *Bulbophyllum* (including *Cirrhopetalum* and *Megaclinium*). For information on numbers of species please refer to references in Reports from the literature. There is no anatomical information in the literature on *Chaseela*, *Codonosiphon*, *Monosepalum*, *Saccoglossum*, *Tapeinoglossum*, and other genera now included in *Bulbophyllum*.

Reports from the literature

Adams (2011): systematics of Dendrobiinae through molecular and phylogenetic analyses, especially of Australian taxa. Bechtel et al. (1992): manual of cultivated orchids. Carlsward et al. (1997): comparative anatomical study of thick-leaved species in two sections of *Dendrobium*. Dressler (1993): phylogeny and classification of orchids. Kraenzlin (1910): classification of Dendrobiinae. Morris et al. (1996): vegetative anatomical and systematic analysis of plants in subtribe Dendrobiinae. Pridgeon et al. (2014): comprehensive survey of Dendrobieae. Schlechter (1918): description and classification of taxa in Dendrobiinae largely from New Guinea. Solereder and Meyer (1930): survey of orchid anatomy with original data on leaf of *B. dasypetalum*, stem of *Cadetia thouarsii*, and leaf, stem, and root of *Dendrobium*. Stern et al. (1994a): anatomical analysis of thick-leaved species in section *Rhizobium* of *Dendrobium*. Yukawa and Uehara (1996): anatomy and phylogeny of Dendrobiinae, chloroplast DNA phylogeny and anatomical features in this subtribe; an attempt to clarify the evolutionary processes causing the diversity in this subtribe and the anatomical characters that contributed to the plasticity of vegetative organs. Yukawa et al. (1990): leaf surface structure in some dendrobiums. Yukawa et al. (1992): stomata in *Dendrobium* and their systematic importance.

Additional literature (*Dendrobium* including *Cadetia*, *Ephemerantha*, *Epigeneium*, and *Flickingeria*)

Ando (1987): variegated leaf anatomy of cultivars. Barthlott (1976): velamen. Barthlott and Capesius (1975): velamen. Chatin (1856, 1857): vegetative anatomy. Chiang (1970): root anatomy. Chiang and Chou (1971): root. Conran et al. (2009): fossil leaf of *Dendrobium*.

Curtis (1917): vegetative anatomy. Dycus and Knudson (1957): velamen. Engard (1944): root. Fan et al. (2000): mycorrhiza. Goh (1975): leaf. Goh et al. (1992): mycorrhiza. Gupta et al. (1970) [1972]: vegetative anatomy of drug plant, as *Desmotrichum*. Hadley and Williamson (1972): mycorrhiza. Isaiah and Mohana Rao (1992): vegetative anatomy of *D. jenkinsii*. Janse (1897): mycorrhiza. Katiyar et al. (1986): mycorrhiza. Kaushik (1982) [1983]: vegetative anatomy, including *Ephemerantha*. Khasim and Mohana Rao (1989): general anatomy of four species. Kraft (1949): velamen. Krüger (1883): aerial organs. Kuttelwascher (1964), (1965) [1966]: root. Lancaster (1910) [1911]: mycorrhiza. Leitgeb (1864b): aerial root, also as *Sarcopodium*. Li et al. (1989): leaf epidermis of drug plants. Meinecke (1894): aerial root. Mejstrik (1970): root of *D. cunninghamii*. Möbius (1887): leaf. Mohana Rao et al. (1989): vegetative anatomy. Mollberg (1884): mycorrhiza. Moraes et al. (2007): root. Moreau (1913): pseudobulb. Morisset (1964): aerial root. Namba and Lin (1981a,b): drug plant anatomy, also *Ephemerantha* and *Epigeneium*. Oudemans (1861): aerial root. Porembski and Barthlott (1988): velamen, also *Flickingeria*. Richter (1901): aerial root. Schelpe and Stewart (1990): species of *Dendrobium* in cultivation. Seidenfaden (1980): *Flickingeria* and *Epigeneium* in Thailand. Seidenfaden (1985): *Dendrobium* species in Thailand. Seidenfaden (1992a): Dendrobiinae in Indochina. Seidenfaden (1992b): Dendrobiinae in Malaysia and Singapore. Singh (1981): stomata. Singh (1986): root. Takahashi et al. (1965): anatomy of 'Chukanso' and three *Dendrobium* species. Tominski (1905): leaf. Weltz (1897): axis. Went (1895): root. Williams (1979): stomata, also *Cadetia*. Wu et al. (2009): leaf and stem anatomy and molecular studies of drug plants.

Additional literature (*Bulbophyllum* including *Cirrhopetalum* and *Megaclinium*)

Chatin (1856, 1857): vegetative anatomy of *B. careyanum*. Chiang and Chou (1971): root anatomy. Curtis (1917): anatomy of *B. pygmaeum* and *B. tuberculatum*. Duruz (1960): stomata of *B. lobbii*. Goh (1975): leaf anatomy of *B. vaginatum*. Guttenberg (1968): root anatomy of *B. penduliscapum* and *B. lobbii*. Kaushik (1982) [1983]: vegetative anatomy of seven species. Leitgeb (1864b): velamen in *Cirrhopetalum wallichii*. Metzler (1924): leaf anatomy of *B. coriaceum* and *B. lopezianum*. Möbius (1887): leaf anatomy of several species of *Bulbophyllum* and *Cirrhopetalum*. Mohana Rao and Khasim (1987a): vegetative anatomy of *B. dyerianum* (now = *B. rolfei*), *B. andersonii*, and *B. leopardinum*. Mohana Rao, Khasim, and Isaiah (1991): anatomy of leaf, stem, and root of *B. fischeri*. Møller and Rasmussen (1984): lack of stegmata in *Bulbophyllum* species. Moreau (1913): pseudobulb of *B. occultum* (also *Genyorchis*). Neubauer (1978): inflorescence axis idioblasts in *Megaclinium*. Olatunji and Nemgin (1980): tracheoidal idioblasts in 13 species of *Bulbophyllum* (also *Genyorchis*). Oliver (1930): leaf anatomy of *B. tuberculatum*. Oudemans (1861): aerial root anatomy of *Bulbophyllum* species. Porembski and Barthlott (1988): velamen and tilosomes of five *Bulbophyllum* species, *Monomeria barbata*, and *Pedilochilus* species. Pridgeon, Stern, and Benzing (1983): tilosomes in *Bulbophyllum* and *Cirrhopetalum*. Rao (1998): leaf anatomy of five *Bulbophyllum* species. Sanford and Adanlawo (1973): velamen and exodermis. Singh (1981): stomata in *C. viridiflorum* (now = *B. viridiflorum*), *C. maculosum* (now = *B. umbellatum*), and *B.* sp. Singh (1986): root anatomy of *B. odoratissimum*, *C. maculosum*, and *C. viridiflorum*. Singh and Singh (1974): stomata of *B. odoratissimum*. Sprenger (1904): comprehensive study of anatomy of numerous species of *Bulbophyllum* (including *Cirrhopetalum*, *Bulbophyllaria*, and *Megaclinium*). Tominski (1905): leaf anatomy of *C. thwaitesii* (now = *B. thwaitesii*). Weltz (1897): stem anatomy of eight species. Went (1895): root of *Cirrhopetalum*. Williams (1979): stomata in Bulbophyllinae.

Taxonomic notes

Dendrobium and *Bulbophyllum* represent two large and complex groups of orchids, each of which has been associated with, presumably, related genera (Dressler 1993): *Dendrobium* with *Cadetia*, *Diplocaulobium*, *Epigeneium*, *Flickingeria*, and *Pseuderia* (Eriinae!) and *Bulbophyllum* with *Cirrhopetalum*, *Genyorchis*, *Drymoda*, *Hapalochilus*, *Monomeria*, and *Sunipia*. Molecular and cladistic studies have persuaded researchers that the minor genera, listed above, are embedded in *Dendrobium* and *Bulbophyllum*, respectively, and are so treated in Pridgeon et al. (2014).

There remain, however, outstanding anatomical and other differences between *Dendrobium* and *Bulbophyllum*, among which are: spherical, rough-surfaced silica bodies in stegmata of *Dendrobium* whereas *Bulbophyllum* lacks stegmata entirely, stems of *Dendrobium* feature one to several internodes while those of *Bulbophyllum* have only a single internode, velamina in *Dendrobium* have two to several layers but *Bulbophyllum* velamina are 1-layered. Geographically *Dendrobium* is confined to the Old World, whereas *Bulbophyllum* is pantropical.

The clear anatomical differences between *Dendrobium* and *Bulbophyllum* indicate that further analysis needs to be done with these two taxa in order to resolve the ambivalence bound up in the Pridgeon et al. (2014) hypothesis.

Tribe Vandeae Lindl.

Vandeae consist of monopodial, mostly epiphytic plants. They exhibit a wide variety of habits from plants with elongated stems and well-developed photosynthetic leaves to those with abbreviated stems and caducous leaves, to those with abbreviated stems and small non-photosynthetic scale leaves.

Subtribe Adrorhizinae Schltr.

This subtribe comprises three genera: *Adrorhizon*, *Bromheadia*, and *Sirhookera*. It is distributed from southern India and Sri Lanka through Myanmar (Burma), Thailand, Indochina, Malaysia, and Indonesia north to the Philippines, south to New Guinea and Queensland in northern Australia. There are 26 species divided into two sections in *Bromheadia*. There is only one species in *Adrorhizon* and there are two species in *Sirhookera*. Herbs are epiphytic or terrestrial, occasionally lithophytic, with creeping rhizomes, conduplicate flattened or terete leaves that are distichous, articulate, stiffly herbaceous, and coriaceous. *Bromheadia finlaysoniana* appears to be the only species in cultivation.

Leaf surface

Hairs absent in *Bromheadia aporoides*, the only species studied personally; not recorded for other taxa. **Epidermis**: cells polygonal in *B. aporoides*. **Stomata** tetracytic, covering all surfaces in *B. aporoides*, subsidiary cells often divided perpendicularly to length of guard cells; stomata confined to abaxial surface in *Adrorhizon purpurascens*, with large cuticular ledges.

Leaf TS

Blade terete, elliptical in TS in *B. aporoides*, dorsiventral in *B. finlaysoniana*, *Adrorhizon*, and *Sirhookera lanceolata*. **Cuticle** smooth, 15 μm thick, intruding between epidermal cells in *B. aporoides*; thick adaxially and moderately thick abaxially in *Adrorhizon* and *B. finlaysoniana*. **Epidermis**: cells conical, outer and lateral walls thickened, inner walls thin in *Bromheadia*; cells small, thick-walled in *A. purpurascens*; adaxial cells large in *S. lanceolata*. **Stomata** sunken, guard cells lying at base of cuticle-lined pit between subsidiary cells in *Bromheadia*. **Hypodermis**: very thick-walled fibres, elongated in LS, present ad- and abaxially in *Adrorhizon*. **Fibre bundles** alternating regularly with vascular bundles around periphery of leaf in *B. aporoides*. **Mesophyll**: heterogeneous in *Adrorhizon* and *B. aporoides*; band of 5–8 layers of subepidermal palisade tissue in *B. aporoides*, and central mass of tissue comprising empty, irregularly shaped water-storage cells; adaxial palisade and abaxial spongy tissue with intercellular spaces in *Adrorhizon*; slightly heterogeneous in *B. finlaysoniana*, adaxial cells rounded and abaxial cells oblong to rectangular; homogeneous in *S. lanceolata*, with scattered smaller hyaline cells. **Vascular bundles** collateral, xylem facing inwards, phloem outwards in *B. aporoides*; phloem associated with massive sclerenchyma contacting upper epidermis; xylem subtended by modest layer of sclerenchyma. Vascular bundles small, in one row in *Adrorhizon*, *B. finlaysoniana*, and *S. lanceolata*; 3- to 4-layered sclerenchyma cap at phloem pole, 1- to 2-layered cap at xylem pole, and sheath of thick-walled parenchyma in *Adrorhizon*. **Stegmata** with conical, rough-surfaced silica bodies associated with phloem sclerenchyma and fibre bundle sclerenchyma in *B. aporoides*. Raphide idioblasts absent.

Root TS

Velamen 2-layered in *Bromheadia alticola*, *B. finlaysoniana*, and *B. palustris* (now *B. finlaysoniana*); inner layer of cells thin-walled with delicate reticulate thickenings. Velamen 3-layered in *Adrorhizon*, 5-layered in *Sirhookera lanceolata*, cell walls without helical thickenings. Root hairs in *B. finlaysoniana* may be reduced to papillae. **Exodermis** with elongated cells and small passage cells. **Tilosomes** absent. **Cortex** mycorrhizal with localized zones of intense infection in *B. finlaysoniana*; large intercellular spaces in cortex of *B. finlaysoniana* and small clusters of lignified cells with reticulate thickenings present.

Material examined

Bromheadia aporoides (leaf); full details in Stern and Judd (2002). Information on other taxa from literature.

Reports from the literature

Goh (1975): leaf anatomy of *B. finlaysoniana*. Goh et al. (1992): mycorrhiza of *Bromheadia*. Groom (1893): root anatomy of *B. alticola* and *B. finlaysoniana* (as *B. palustris*). Hadley and Williamson (1972): mycorrhiza of *B. finlaysoniana*. Porembski and Barthlott (1988): velamen structure in *Adrorhizon purpurascens*, *B. finlaysoniana*, and *Sirhookera lanceolata*. Rao (1998): leaf anatomy of *S. lanceolata*. Stern and Judd (2002): leaf anatomy of *B. aporoides*. Tominski (1905): leaf anatomy of *A. purpurascens*.

Taxonomic notes

There is not sufficient anatomical information to draw any taxonomic inferences.

Subtribes Angraecinae Summerh. and Aeridinae Pfitzer

These subtribes are composed of predominantly Old World plants inhabiting Africa, Madagascar, and tropical Asia, with two genera in tropical America. There are almost 2000 species and approximately 158 genera, 50 of which are in Africa and Madagascar. Leaves are distichous, conduplicate, terete or semi-terete, reduced to small, non-photosynthetic scales, or absent. The species of Africa and Madagascar are generally white-flowered, fragrant at night, and are pollinated by moths. Because Vandeae are monopodial, studies of the stem would have necessitated sacrificing the entire plant. Hence, anatomical studies have been restricted mostly to leaf and root preparations. These orchids are widely grown as ornamentals by hobbyists and many hybrids have been produced. Among the favourite species in cultivation are *Aerangis fastuosa*, *A. luteoalba* var. *rhodosticta*, *Aeranthes grandiflora*, *Aerides multiflora*, *Angraecum infundibulare*, *A. leonis*, *A. magdalenae*, *A. scottianum*, *Chiloschista lunifera*, *Papilionanthe teres*, *P. vandarum*, *Phalaenopsis sanderiana*, *P. schilleriana*, *P. stuartiana*, *P. violacea*, *Vanda coerulea*, *V. tessellata*, and *V. tricolor*.

Angraecum has been used as a remedy in equatorial Africa; *Acampe* has been recorded as medicinal in eastern Asia, including Hong Kong; *Aerides* species have been used medicinally in India, Himalayas, and Burma; *Phalaenopsis* species from the Philippines and Japan have been used as medicines; *Vanda* species have been noted as medicinal in India, Sri Lanka, and Burma.

Leaves

Well-formed photosynthetic leaves occur in most angraecoids. However, leaves are completely lacking in the mature state of *Chiloschista*, *Dendrophylax*, *Microcoelia*, and *Taeniophyllum*; some species are leafy and others leafless in *Campylocentrum* and *Phalaenopsis*. Planar leaves occur in most Angraecinae; terete or semi-terete leaves occur among species of *Angraecum*, *Bolusiella*, *Podangis*, and *Ypsilopus*.

Leaf surface

Hairs multicellular, glandular, basal cell sunken in an epidermal crypt. Hairs absent in *Bolusiella*, *Calyptrochilum emarginatum*, *Campylocentrum*, *Chamaeangis*, *Neobathiea*, *Ossiculum*, *Phalaenopsis*, *Summerhayesia*, and *Vanda*. **Stomata** paracytic, mostly abaxial; ad- and abaxial in some species of *Aerangis*, *Aeranthes*, *Amesiella*, *Ancistrorhynchus*, *Angraecopsis*, *Bolusiella*, *Chamaeangis*, *Diaphananthe*, *Mystacidium*, *Rangaeris*, *Rhipidoglossum*, *Sphyrarhynchus*, *Trichoglottis*, and *Tridactyle*.

Leaf TS

Cuticle smooth to ridged (Fig. 132B,C), 1.25–23.7 μm thick. **Epidermis**: cells periclinal to isodiametric (Figs 130B, 132A–C). **Stomata** superficial, outer ledges thin, inner ledges thick, cuticular horns prominent. **Hypodermis**: adaxial in *Aeranthes*, *Bonniera* (now *Angraecum*), *Cribbia*, and *Podangis*; ad- and abaxial in *Ancistrorhynchus*, *Angraecum* (Figs 130B, 132A), *Campylocentrum micranthum*, *Eggelingia*, *Jumellea*, *Listrostachys*, *Neofinetia*, *Plectrelminthus*, *Solenangis*, *Summerhayesia*, *Tridactyle*, *Vanda*, *Ypsilopus* (Fig. 130C), also ad- and abaxial in some species of *Bolusiella* and *Rangaeris*; sometimes absent. **Fibre bundles** usually absent; present in mesophyll of some species of *Aeranthes*, *Angraecum*, and *Ossiculum*. **Mesophyll** homogeneous (Fig. 130A,C), but heterogeneous (Figs 130B, 132A,C) in some species of *Aeranthes*, *Ancistrorhynchus*, *Angraecum* (Fig. 130B), *Calyptrochilum*, *Campylocentrum micranthum*, *Jumellea*, *Listrostachys* (Fig. 132C), *Mystacidium*, *Rangaeris*, *Rhipidoglossum*, *Tridactyle*, and *Ypsilopus*; heterogeneous throughout in *Bonniera* (now *Angraecum*), *Eggelingia*,

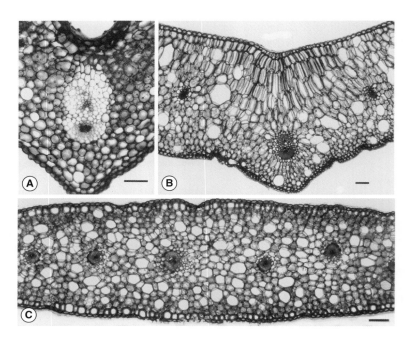

Fig. 130. **Angraecinae leaf.** (A) *Cribbia brachyceras*, TS mid-leaf with homogeneous mesophyll and collateral vascular bundle. (B) *Angraecum dives*, TS with heterogeneous mesophyll, collateral midrib vascular bundle, and water-storage cells. (C) *Ypsilopus viridiflorus*, TS with homogeneous mesophyll, water-storage cells, one row of collateral vascular bundles, and ad- and abaxial hypodermises. Scale bars: A,B = 100 μm, C = 200 μm. Reprinted from Carlsward et al. (2006a).

Lemurella, *Lemurorchis*, *Plectrelminthus*, *Podangis*, *Sobennikoffia*, and *Summerhayesia*. Water-storage cells with smooth or pitted walls, some with banded (Pl. 3D, 4C) or spiral thickenings (Fig. 131A). **Vascular bundles** collateral (Fig. 130A–C), in one row, except two rows in *Angraecum teres*. Sclerenchyma associated with xylem and phloem. Bundle sheath cells distinct, chlorophyllous. **Stegmata** containing rough-surfaced, spherical silica bodies associated with xylem and phloem sclerenchyma; only with phloem in some species of *Angraecum*, *Neobathiea*, *Ossiculum*, *Tridactyle*, and *Ypsilopus*; stegmata absent in *Amesiella*.

Root TS

Velamen 1–7 cells wide, mostly two or three cells (Pl. 4D), sometimes absent in *Chiloschista*, *Lemurorchis*, and *Taeniophyllum* (including *Microtatorchis*). Epivelamen cells radially elongate and isodiametric with fine anastomosing radial wall thickenings; endovelamen cells angular and isodiametric to radially elongate; wall thickenings often absent. Root hairs thin-walled, single-celled. **Exodermis**: cells radially elongate (Figs 131E, 133A,D) to isodiametric. Walls of long cells O-thickened to U-thickened (Fig. 131E) to slightly ∩-thickened (Pl. 4A). Exodermal proliferations occurring where velamen lost; these superficially similar cellularly to velamen cells, consisting of large, empty, lignified (Figs 131E, 133A) or suberized cell walls. Aeration units (analogous to stomata in leaves by permitting gaseous exchange with the outside atmosphere) composed of one or two exodermal cells with thin inner tangential walls and usually with two differentially thickened cortical cells below (Fig. 133C,D). **Tilosomes** absent. **Cortex**: cells thin-walled, chlorenchymatous, isodiametric (Fig. 133B) to radially elongate. Water-storage cells with smooth to pitted walls (Fig. 131C); birefringent banded cell wall thickenings occurring in species of many genera (Fig. 131B) and uniformly thickened walls in *Dendrophylax porrectus* (Fig. 131D). **Endodermis** 1-layered; cells isodiametric to radially elongate, with O-thickened walls opposite phloem strands and thin-walled cells opposite xylem rays. **Pericycle** 1-layered; cells thin-walled opposite xylem, thick-walled opposite phloem. **Vascular cylinder** 6–28-arch (Fig. 133B, Pl. 4B), of alternating strands of

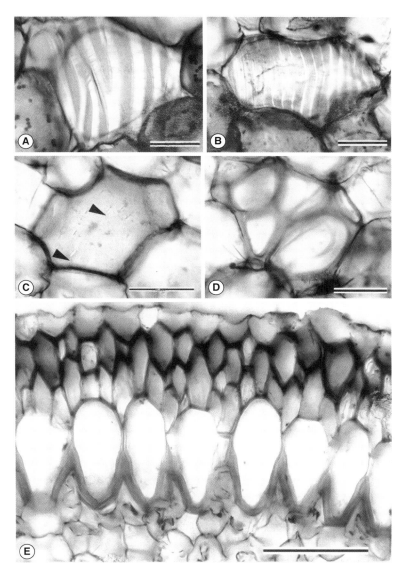

Fig. 131. Angraecinae. (A) *Cryptopus paniculatus*, TS leaf with spirally thickened water-storage cell wall. (B) *Calyptrochilum christyanum*, TS root with evenly banded water-storage cell wall. (C) *Jumellea walleri*, TS root with water-storage cell and pitted wall (arrowheads). (D) *Dendrophylax porrectus*, TS root showing water-storage cells with uniformly thickened walls. (E) *Beclardia macrostachya*, TS root with U-thickened exodermal cells. Scale bars: A–D = 50 μm, E = 100 μm. Reprinted from Carlsward et al. (2006a).

xylem and phloem. Vascular elements embedded in sclerenchyma or parenchyma. **Pith** usually sclerenchymatous.

Material examined

Subtribe Angraecinae: *Aerangis* (11), *Aeranthes* (5), *Ancistrorhynchus* (3), *Angraecopsis* (3), *Angraecum* (19), *Beclardia* (1), *Bolusiella* (3), *Bonniera* (1, now *Angraecum*), *Calyptrochilum* (2), *Campylocentrum* (5), *Chamaeangis* (4), *Cribbia* (2), *Cryptopus* (2), *Cyrtorchis* (5), *Dendrophylax* (7), *Diaphananthe* (4), *Eggelingia* (1), *Eurychone* (1), *Jumellea* (7), *Lemurella* (1), *Lemurorchis* (1), *Listrostachys* (1), *Microcoelia* (13), *Microterangis* (1), *Mystacidium* (3), *Neobathiea* (1), *Oeonia* (1),

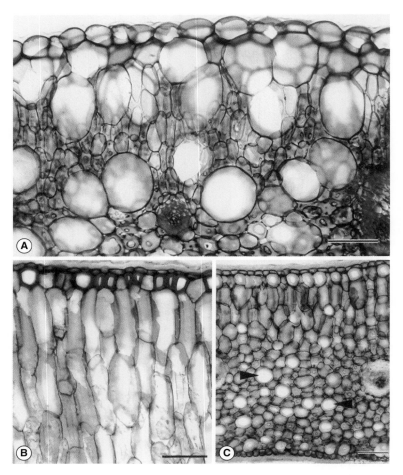

Fig. 132. **Angraecinae leaf.** (A) *Angraecum conchiferum*, TS showing adaxial hypodermis composed of water-storage cells and water-storage cells scattered in mesophyll. (B) *Jumellea walleri*, TS with upright mesophyll cells and adaxial hypodermis of thick-walled fibres. (C) *Listrostachys pertusa*, TS with heterogeneous mesophyll containing scattered fibrous idioblasts (arrowheads) and thick-walled hypodermal cells. Scale bars = 100 μm. Reprinted from Carlsward et al. (2006a).

Oeoniella (1), *Ossiculum* (1), *Plectrelminthus* (1), *Podangis* (1), *Rangaeris* (4), *Rhipidoglossum* (8), *Sobennikoffia* (2), *Solenangis* (3), *Sphyrarhynchus* (1), *Summerhayesia* (1), *Tridactyle* (8), *Ypsilopus* (2).

Subtribe Aeridinae: *Acampe* (1), *Amesiella* (1), *Chiloschista* (4), *Neofinetia* (1), *Phalaenopsis* (2), *Taeniophyllum* (4, + *Microtatorchis* 1), *Trichoglottis* (1), *Vanda* (1). Full details appear in Carlsward, Stern, and Bytebier (2006a).

Reports from the literature

Benzing et al. (1983): study of shootlessness in Orchidaceae and its relationship to orchids as epiphytes including *Campylocentrum* (and as *Harrisella*), *Polyradicion*, and *Rangaeris*. Cameron et al. (1999): phylogenetic analysis of Orchidaceae based on molecular research. Carlsward (2004): comparison of anatomy and molecular systematic studies of Vandeae. Carlsward, Stern, and Bytebier (2006a): comparative anatomy and systematics of angraecoids. Carlsward, Whitten, Williams, and Bytebier (2006b): molecular phylogenetics of Vandeae and evolution of leaflessness. Chase et al. (2003): molecular data applied to orchid systematics and phylogeny. Dressler (1993): phylogeny and classification of Orchidaceae. Jonsson (1981): monograph of leafless *Microcoelia*, including root anatomy of 26 species. Møller and Rasmussen (1984): orchid stegmata. Porembski and

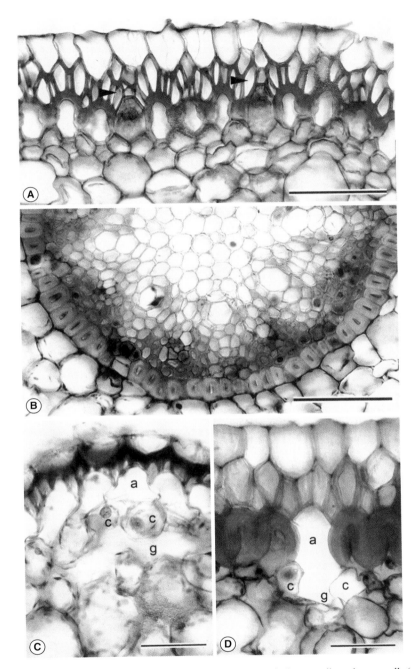

Fig. 133. Angraecinae root. (A) *Solenangis clavata*, TS with radially elongate epivelamen cells and cover cells (arrowheads) arranged above passage cells of ∩-thickened exodermal cells. (B) *Jumellea sagittata*, TS with thick-walled endodermal cells. (C) *Campylocentrum fasciola*, TS showing aeration units: (a) aeration unit; (c) modified cortical cells; (g) cortical region of gas exchange. (D) *Chamaeangis sarcophylla*; caption as in (C). Scale bars: A,B = 100 μm, C,D = 50 μm. (A,C,D) Reprinted from Carlsward et al. (2006a); (B) photograph by B.S. Carlsward.

Barthlott (1988): velamen in orchids, including several genera of angraecoids. Pridgeon (1987): velamen and exodermis of orchid roots including *Aerangis, Aeranthes, Angraecum, Cyrtorchis, Eurychone, Jumellea, Plectrelminthus,* and *Tridactyle.* Pridgeon et al. (1983): survey of tilosomes in Orchidaceae. Schlechter (1918): a new classification of African angraecoid orchids. Solereder and Meyer (1930): systematic anatomy of orchids and original observations on *Angraecum* (and as *Macroplectrum*) and *Listrostachys.*

Additional literature

Arends and Stewart (1989): leaf of *Aerangis gracillima.* Barthlott and Capesius (1975) [1976]: velamen of *Acampe, Microcoelia, Rhynchostylis,* and *Vanda.* Baruah (2001): vegetative anatomy of *Vanda teres.* Benzing (1982), Benzing and Friedman (1981), and Benzing et al. (1983): root anatomy and mycorrhizae of *Campylocentrum, Dendrophylax* (and as *Harrisella* and *Polyradicion*), and *Phalaenopsis* (as *Kingidium*). Bloch (1935): root anatomy of *Aerides* and *Renanthera.* Burgeff (1959): mycorrhizae of *Phalaenopsis* and *Vanda.* Capesius and Barthlott (1975): velamen in *Angraecum.* Chatin (1856, 1857): leaf and stem anatomy of *Aerides* and *Saccolabium.* Chiang and Chow (1971): root anatomy of *Haraella, Phalaenopsis, Saccolabium, Sarcanthus,* and *Thrixspermum.* Clements (1988): mycorrhizae of *Phalaenopsis.* Costantin (1885): root anatomy of *Vanda.* Curtis (1917): vegetative anatomy of *Sarcochilus.* Czapek (1909): root anatomy of *Luisia.* Dixon (1894): vegetative anatomy of *Vanda teres.* Duruz (1960): stomata of *Vanda.* Dycus and Knudson (1957): velamen of *Phalaenopsis* and *Vanda.* Engard (1944): root anatomy of *Aerides, Phalaenopsis,* and *Vanda.* Fan et al. (2000): mycorrhizae of *Vanda.* Gasson and Cribb (1986): leaf anatomy of *Ossiculum.* Goebel (1922): aerial root anatomy of *Phalaenopsis* and *Taeniophyllum.* Goh (1975): leaf anatomy of *Phalaenopsis, Thrixspermum,* and *Vanda.* Goh et al. (1992): mycorrhiza of *Chiloschista, Doritis,* and *Vanda.* Guttenberg (1968): root anatomy of *Arachnis* and *Trichoglottis.* Hadley and Williamson (1972): mycorrhiza of *Arachnis, Thrixspermum,* and *Vanda.* Heim (1945): aerial roots of *Phalaenopsis schilleriana.* Hering (1900): stem anatomy of many genera of angraecoids. Janczewski (1885): root anatomy of *Aeranthes, Phalaenopsis,* and *Sarcanthus.* Jayeola and Thorpe (2000): SEM leaf of *Calyptrochilum.* Jeyamurthy et al. (1990): leaf and root anatomy of *Chiloschista lunifera.* Jonsson (1979): root anatomy of *Taeniophyllum.* Katiyar et al. (1986): mycorrhiza of *Sarcanthus.* Kaushik (1982 [1983], 1983): vegetative anatomy of *Aerides, Camarotis,*

Cleisostoma, Gastrochilus, Luisia, Phalaenopsis (as *Kingidium*), *Rhynchostylis, Vanda,* and *Vandopsis.* Khasim and Mohana Rao (1986): leaf and root anatomy of *Ascocentrum.* Kraft (1949): velamen of *Phalaenopsis* and *Vanda.* Krüger (1883): aerial organs of *Renanthera, Saccolabium, Sarcanthus,* and *Vanda.* Leitgeb (1864b): aerial root anatomy of *Aerides, Angraecum, Arachnis, Cottonia, Phalaenopsis, Renanthera, Saccolabium, Sarcanthus, Thrixspermum* (as *Dendrocolla*), and *Vanda.* Mahyar (1988): leaf epidermis of *Renanthera coccinea.* Meinecke (1894): aerial root anatomy of *Aerides* and *Angraecum.* Metzler (1924): leaf anatomy of *Aerides, Mystacidium, Renanthera,* and *Vanda.* Möbius (1887): leaf anatomy of *Acampe, Aeranthes, Aerides, Angraecum, Listrostachys, Rhynchostylis, Saccolabium, Sarcanthus,* and *Vanda.* Mohana Rao et al. (1989): vegetative anatomy of *Luisia.* Müller (1900): root anatomy of *Taeniophyllum.* Musampa Nseya and Arends (1995): vegetative anatomy of *Cyrtorchis.* Napp-Zinn (1953): aerial root anatomy of *Phalaenopsis* and *Vanda.* Olatunji and Nengim (1980): tracheoid idioblasts in several genera of angraecoids. Oudemans (1861): aerial root anatomy of *Aerides, Arachnis, Angraecum. Renanthera, Rhynchostylis, Saccolabium, Sarcanthus,* and *Vanda.* Palla (1889): aerial root anatomy of *Angraecum* (possibly *Campylocentrum*) and *Dendrophylax* (as *Polyrrhiza*). Parrilla Diaz and Ackerman (1990): root anatomy of *Campylocentrum* and *Dendrophylax* (as *Harrisella*). Pirwitz (1931): 'storage tracheids' in *Aerides* and *Vanda.* Prillieux (1879): aerial root anatomy of *Aerides, Renanthera,* and *Vanda.* Raciborski (1898): root anatomy of *Trichoglottis.* Ramanujam and Rao (1998): leaf idioblasts of *Papilionanthe* and *Vanda.* Rao (1998): leaf anatomy of several genera of angraecoids. Richter (1901): aerial root anatomy of *Aerides.* Sanford and Adanlawo (1973): root anatomy of several genera of angraecoids. Schimper (1884, 1888): root anatomy of *Aeranthes.* Senthilkumar et al. (1998): mycorrhiza of *Papilionanthe subulata.* Singh (1981): stomata in *Aerides, Luisia, Renanthera, Rhynchostylis,* and *Vanda.* Singh (1986): root anatomy of *Aerides, Rhynchostylis,* and *Vanda.* Sohma (1954): mycorrhiza of *Sarcochilus japonicus.* Sulistiarini (1986): leaf anatomy of *Luisia latipetala.* Tischler (1910): root anatomy of *Taeniophyllum.* Tominski (1905): leaf anatomy of *Aerides, Cottonia, Luisia, Saccolabium, Sarcochilus,* and *Vanda.* Wahrlich (1886): mycorrhiza of *Vanda.* Wallach (1938–9): aerial root of *Phalaenopsis.* Watthana (2007): leaf anatomy of *Pomatocalpa.* Went (1895): root anatomy of *Phalaenopsis* (and as *Polychilos*), *Taeniophyllum,* and *Vanda.* Williams (1979): stomata of *Aerangis, Aeranthes, Angraecopsis, Angraecum, Chamaeangis, Diaphananthe, Jumellea, Oeonia, Podangis,* and *Rangaeris.*

Taxonomic notes

Traditionally Vandeae have been divided into three subtribes (Dressler 1993): Aeridinae, Angraecinae, and Aerangidinae. As more recently constituted (Pridgeon et al., 2014), however, Vandeae are considered to comprise five subtribes: Adrorhizinae, Aeridinae, Agrostophyllinae, Angraecinae, and Polystachyinae. Adrorhizinae contain three genera, *Bromheadia*, *Adrorhizon*, and *Sirhookera*; Aeridinae have about 103 genera (Dressler 1993); Agrostophyllinae contain two genera, *Agrostophyllum* and *Earina*; Angraecinae comprise 19 genera (Dressler 1993); and Polystachyinae have two genera, *Hederorkis* and *Polystachya* (including *Neobenthamia*). The genera in Adrorhizinae and Polystachyinae were contained in Dressler's tribe 'Epidendreae II' and *Agrostophyllum* and *Earina* were listed in subtribe Glomerinae in tribe 'Epidendreae II'. *Bromheadia*, basis of the monogeneric subtribe Bromheadiinae, was placed in tribe Cymbidieae. Recent molecular studies (Carlsward 2004; Carlsward et al. 2006) have shown Angraecinae to include members of Aerangidinae. Thus, Vandeae, *sensu stricto*, have only two subtribes, Aeridinae and Angraecinae. These constitute the angraecoid orchids with about 50 genera in Africa and Madagascar and two genera from tropical America (Carlsward 2004; Carlsward et al. 2006).

Looking more closely at the anatomy of the newly established Vandeae, we find structure in the two subtribes Angraecinae and Aeridinae to agree anatomically in that fibre bundles are usually absent in the mesophyll, that stegmata usually contain rough-surfaced spherical silica bodies associated with xylem and phloem sclerenchyma, that tilosomes are absent in roots, and that the pith of roots is generally sclerenchymatous. Thus we may conclude the anatomy in these two subtribes is substantially similar.

Considering the anatomy of genera in subtribes Adrorhizinae, Agrostophyllinae, and Polystachyinae we note: a homogeneous mesophyll in *Sirhookera*, *Agrostophyllum*, and *Polystachya* but a heterogeneous mesophyll in *Adrorhizon* and *Earina*; fibre bundles in leaves of *Bromheadia*, *Agrostophyllum*, and *Earina* but none in *Polystachya*; stegmata with conical silica bodies in *Bromheadia*, *Agrostophyllum*, and *Earina*, but none in *Polystachya*, and large, webbed tilosomes in *Polystachya*, but none in *Adrorhizon*, and unknown in *Agrostophyllum* and *Earina*. There are some sharp anatomical differences among the genera in these three subtribes and between these three subtribes and subtribes Angraecinae and Aeridinae. Thus, relationships of subtribes in the newly established Vandeae may need to be re-evaluated based upon the anatomical differences among them.

Subtribe Agrostophyllinae Szlach.

Subtribe Agrostophyllinae consists of two genera: *Agrostophyllum* and *Earina*. These are epiphytic, terrestrial, occasionally lithophytic, rhizomatous herbs. Leaves are conduplicate, articulate, distichous, narrow in *Agrostophyllum*, narrow and broad in *Earina*, distributed evenly along the stem in *Agrostophyllum*, clustered in overlapping basal sheaths in *Earina*. *Agrostophyllum* consists of 50 or 60 species distributed in the Old World tropics from the Seychelles in the Indian Ocean and tropical Asia to Samoa in the Pacific Ocean with the centre of distribution in New Guinea. They are absent from Australia. *Earina* has six or seven species on the south Pacific Ocean islands of New Zealand, New Caledonia, Solomon Islands, Vanuatu, eastward to Fiji and Samoa. There are no records of these plants in cultivation.

Agrostophyllum has been used medicinally in eastern Asia.

Leaf surface

Epidermis: cells rectangular and polygonal in *Agrostophyllum*. **Stomata** in both genera abaxial, tetracytic in *Agrostophyllum*, tetracytic to cyclocytic in *Earina*.

Leaf TS

Cuticle in *Earina* conspicuously thick. **Fibre bundles** occurring in mesophyll of both genera. **Mesophyll** in *Agrostophyllum* homogeneous with water-storage cells; heterogeneous in *Earina*. **Vascular bundles** in *Agrostophyllum* with fibrous cap at phloem pole; in *Earina* surrounded by sclerenchyma sheath. **Stegmata** with conical silica bodies present in leaves of both genera.

Stem TS (Earina)

Cuticle thicker than height of subtending epidermal cells. **Cortex** surrounded internally by complete or incomplete sclerenchyma sheath. **Vascular bundles** included in sclerenchyma ring and also scattered among starch-containing, oval, ground-tissue cells. Vascular bundle phloem associated with a fibre layer that may be absent from xylem. **Fibre bundles** present in ground tissue.

Stegmata with conical silica bodies occurring in rhizomes and stems of *Agrostophyllum* but not in *Earina*.

Root TS

Velamen in *Agrostophyllum* 2–4 cell layers wide and 3–5 in *Earina*; cells in *Earina* isodiametric and containing hyphae. **Exodermis** in *Agrostophyllum* consisting of large, thick-walled cells and narrow, thin-walled passage cells. In *Earina* cells with radial and external walls thickened alternating with smaller passage cells. **Cortex**: some cells reticulately thickened in *Agrostophyllum*; in *Earina* large inner cortical cells reticulately thickened while central cell walls spirally thickened. Hyphae and raphides occurring in cortical cells of *Earina*. **Endodermis**: cells of *Agrostophyllum* with thickened radial and inner walls alternating with thin-walled cells; in *Earina* cells with thick, striated, cutinized walls alternating with smaller passage cells. **Vascular cylinder** with 11–13 strands of alternating xylem and phloem in *Earina*.

Material reported

Anatomical observations have been recorded from the literature: *Agrostophyllum* from Khasim and Mohana Rao (1986), *Earina* from Conran et al. (2009), Curtis (1917), and Oliver (1930), stegmata of both genera from Møller and Rasmussen (1984).

Reports from the literature

Conran et al. (2009): leaf anatomy of fossil *Earina fouldenensis* and extant *E. autumnalis* and *E. mucronata*. Curtis (1917): anatomy of *E. autumnalis* (as *E. suaveolens*) and *E. mucronata*. Katiyar et al. (1986): mycorrhiza of *Agrostophyllum*. Khasim and Mohana Rao (1986): leaf and root anatomy of *Agrostophyllum callosum*. Lancaster (1910): mycorrhiza of *Earina*. Møller and Rasmussen (1984): stegmata and silica bodies in *Agrostophyllum* and *Earina*. Oliver (1930): leaf and root anatomy of *E. autumnalis*. Porembski and Barthlott (1988): velamen in *Agrostophyllum* and *Earina*. Pridgeon et al. (2014): comprehensive overview of *Agrostophyllum* and *Earina*. St George (1994): Pacific island distribution of *Earina*. Schuiteman (1997): monograph on *Agrostophyllum* section *Appendiculopsis*.

Taxonomic notes

Earnia fouldenensiwith with abaxial tetra- to cyclocytic stomata has been described from Early Miocene strata from New Zealand together with *Dendrobium winikaphyllum* (Dendrobieae). These represent the first credible vegetative orchid fossils.

There is only one major difference in anatomy between *Agrostophyllum* and *Earina*: mesophyll in the former is homogeneous, heterogeneous in the latter.

Schuiteman (1997) confuses 'stegmata' with silica bodies. Silica bodies are deposits contained within stegmata or silica cells. Stegmata are not spherical or conical, the silica bodies are.

He states all *Agrostophyllum* species have umbonate (read conical) stegmata, and he notes, Podochilinae should all have spherical stegmata (read stegmata have spherical silica bodies). Yet, we find some species in Podochilinae to have conical silica bodies (*Eria javanica*) and others spherical silica bodies (*Appendicula* and *Podochilus*) (Møller and Rasmussen 1984). Schuiteman favours the position of *Agrostophyllum* in Podochilinae despite the heterogeneous state of silica bodies in that subtribe. Other features will have to be examined to establish the true place of *Agrostophyllum* and *Earina*.

Earina, the second genus in Agrostophyllinae, has conical silica bodies in the leaf, as does *Agrostophyllum*. Schuiteman favours placement of *Earina* in Podochilinae and indicates the 'strange paniculate inflorescences' in that genus resemble those in *Agrostophyllum* section *Doichodesme*. The question remains in the placement of these two genera: how much weight to place on the shape of silica bodies in establishing the relationships of these two genera.

Subtribe Polystachyinae Pfitzer

Polystachyinae are pantropical. *Polystachya*, the largest genus with about 180 species, ranges in the east from the Philippines to India, Sri Lanka, Madagascar, and sub-Saharan Africa and in the west from Mexico, southern Florida, and Caribbean islands to Peru, Bolivia, and Brazil. The lone species of *Neobenthamia* (now *Polystachya benthamia*) grows in Tanzania, and *Hederorkis*, with two species, is found on some Indian Ocean islands. Plants are primarily epiphytic, but may grow on rock and soil substrates. Stems are slender, as in *P. benthamia*, or form pseudobulbs. Some African species of *Polystachya* section *Dendrobianthe* have an affinity for the fibrous root mantle surrounding stems of *Vellozia* (Velloziaceae) and *P. microbambusa* grows on the columnar root masses of the bog sedge, *Afrotrilepis pilosa* (Cyperaceae).

Specimens of *Hederorkis* were unavailable for study. The anatomical outline of the subtribe rests on *Polystachya* of which good material was available. The anatomical observations below are based primarily on the MSc thesis (2006) research of James M. Heaney, University of Florida, Gainesville.

Leaf surface

Leaves planar, or isobilateral in *Polystachya* section *Aporoides*. **Hairs**: bicellular, clavate hairs present on both surfaces of *Polystachya*; unicellular, elongate, falcate hairs present abaxially on *P. zambesiaca*. **Cuticle** granular, ridged. **Epidermis**: cells polygonal to tabular. Crystals present in abaxial cells of *P. affinis*, *P. concreta* (as *P. tessellata*), *P. fallax*, *P. masayensis*, *P. supfiana*, *P. valentina*, and *P. zambesiaca*, in adaxial cells of *P. concreta*, *P. foliosa*, *P. fulvilabia*, *P. modesta*, and *P odorata*; and in ad- and abaxial cells of *P. benthamia*, *P. goetzeana*, *P. minima*, and *P. ottoniana*. **Stomata** abaxial, except ad- and abaxial in *P. foliosa*; mainly tetracytic, but tetracytic and anomocytic in *P. affinis*, *P. bella*, *P. canaliculata*, *P. leonensis*, *P. longiscapa*, *P. maculata*, and other species.

Leaf TS

Cuticle thin to 2 µm thick; up to 2.5 µm in *P. caespitifica*, *P. canaliculata*, *P. concreta*, *P. odorata*, *P. shega*, and *P. vulcanica*; up to 9 µm in *P. elegans*. **Epidermis**: cells thin-walled, isodiametric to rectangular and periclinal; conical to papillate in *P. calluniflora* and *P. hastata*, often anticlinal at adaxial groove. Adaxial cells with banded walls in *P. masayensis* and *P. ottoniana*. **Hypodermis**: adaxial cell layers mostly one or two (Fig. 134A); three in *P. fulvilabia*, *P. galeata*, *P. supfiana*, and *P. vulcanica*; three or four in *P. adansoniae*; four or five in *P. coriscensis*, five or six in *P. elegans*; abaxial cell layers none or one in *P. adansoniae*, *P. coriscensis*, *P. elegans*, *P. foliosa*, *P. longiscapa* (Fig. 134B), *P. mystacioides*, and *P. polychaete*. Cells isodiametric to anticlinal, thin-walled, mucilage-containing and water-storage cells, frequently with banded cell walls. **Fibre bundles** absent. **Mesophyll** homogeneous (Fig. 134A–D); cells thin-walled, isodiametric; mucilage idioblasts thin-walled, generally scattered (Fig. 134A–D), or concentrated in *P. mystacioides*. Banded water-storage cells occurring in *P. adansoniae*, *P. caespitifera*, *P. campyloglossa*, *P. canaliculata*, *P. cultriformis*, *P. fusiformis*, *P. galeata*, *P. longiscapa*, *P. maculata*, *P. minima*, *P. shega*, *P. spatella*, and *P. vulcanica*; absent from other species. **Vascular bundles** in one row (Fig. 134A–C), except in two rows in *P. mystacioides*. Sclerenchyma present at xylem and phloem poles; phloem sclerenchyma predominantly thick-walled, xylem sclerenchyma thin-walled. Bundle sheath cells undifferentiated, merging with mesophyll cells, chlorenchymatous; smaller than surrounding mesophyll cells in *P. minima*. **Stegmata** absent.

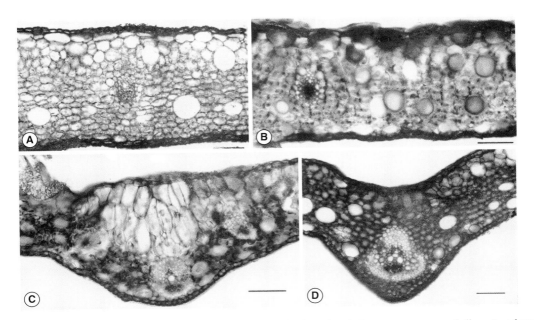

Fig. 134. Polystachyinae leaf. (A) *Polystachya concreta*, TS with adaxial hypodermis, homogeneous mesophyll, scattered mucilage-containing idioblasts, collateral vascular bundle, and thin-walled bundle sheath cells. (B) *P. longiscapa*, TS showing adaxial hypodermis with mucilage-containing water-storage idioblasts, homogeneous mesophyll, collateral vascular bundle, and thin-walled, chlorenchymatous bundle sheath cells. (C) *P. benthamia*, TS, mesophyll homogeneous, except with upright cells adaxial to midrib, collateral vascular bundle, mucilage cells scattered throughout mesophyll. (D) *P. cultriformis*, TS with homogeneous mesophyll. Scale bars = 100 µm. Photographs by B.S. Carlsward.

Stem TS

Stems mainly pseudobulbous, but reed-like in *P. benthamia*. **Hairs** mostly absent; bicellular in *P. elegans*. **Cuticle** 4–29 μm thick; smooth, but mostly granular in *P. benthamia*, *P. goetzeana*, *P. minima*, *P. shega*, *P. spatella*, and species in *Polystachya* section *Polychaete*. **Epidermis**: cells ∩-thickened with scalariformly thickened walls in *P. affinis*; banded in species of *Polystachya* section *Calluniflora*, *P. caespitifera*, *P. concreta*, *P. coriscensis*, and *P. galeata*. **Hypodermis**: 1-layered, but one or two cells wide in species of *Polystachya* section *Calluniflora*, *P. bella*, *P. eurychila*, *P. goetzeana*, *P. heckmanniana*, *P. minima*, *P. ottoniana*, and *P. shega*, one to three cells wide in *P. adansoniae*, *P. modesta*, *P. odorata*, and *P. spatella*; absent in *P. benthamia*, *P. fallax*, *P. lawrenceana*, and *P. seticaulis*. **Ground tissue**: mucilage cells thin-walled, scattered; banded water-storage cells present in *P. calluniflora*, *P. hastata*, *P. ottoniana*, *P. piersii*, species of *Polystachya* sections *Cultriformes* and *Dendrobianthe*, absent in *P. adansoniae*, *P. goetzeana*, *P. heckmanniana*, *P. minima*, *P. polychaete*, *P. seticaulis*, and *P. shega*. **Vascular bundles**: sclerenchyma associated with xylem and phloem, phloem sclerenchyma mostly thicker-walled than xylem sclerenchyma. Banded water-storage cells often surrounding vascular bundle sclerenchyma.

Root TS

Velamen width varying widely (Fig. 135A–C) from two or three cells in *P. johnstonii* and *P. valentina* to six or seven cells in *P. pubescens*. Epivelamen cells thin-walled (Fig. 135A,B), periclinal; epi- and endovelamen cells with fine anastomosing wall thickenings. **Tilosomes** large, webbed (Fig. 135D). **Exodermis**: cells thin-walled, ∩-thickened (Fig. 135B,C), very thick-walled in *P. fallax*, *P. golungensis*, *P. maculata*, *P. masayensis*, *P. modesta*, *P. piersii*, and some species in *Polystachya* section *Polychaete*. **Cortex**: parenchyma cells isodiametric (Fig. 135A,B); mucilage cells thin-walled, isodiametric, scattered; water-storage cells isodiametric, scattered, walls banded. **Endodermis**: cells O-thickened opposite phloem; all thin-walled in *P. concreta*, *P. foliosa*, and *P. modesta*. **Pericycle** cells thick-walled opposite phloem, but thin-walled opposite phloem in *P. concreta* and *P. odorata*. **Vascular cylinder**:

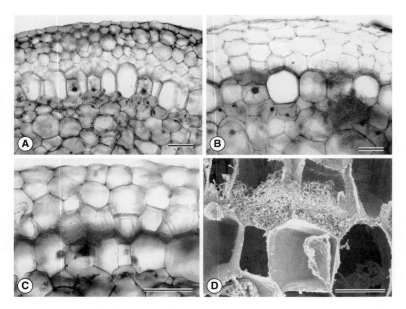

Fig. 135. Polystachyinae root. (A) *Polystachya benthamia*, TS with 4- or 5-layered velamen, epivelamen cells periclinal, endovelamen cells polygonal to rounded, cells thin-walled, evenly thickened, exodermal cells radially elongate. (B) *P. concreta*, TS with 4-layered velamen, epivelamen cells periclinal, exodermal cells almost isodiametric, ∩-thickened, cortical cells isodiametric, rounded, thin-walled. (C) *P. cultriformis*, TS with angular thin-walled endovelamen cells (epivelamen cells appear to be absent), exodermal cells ∩-thickened, isodiametric. (D) *P. longiscapa*, SEM of velamen/exodermal region showing webbed tilosomes. Scale bars: A–C = 50 μm, D = 23 μm. (A–C) Photographs by B.S. Carlsward; (D) reprinted from Carlsward et al. (2006a).

11–14-arch, conductive elements embedded in parenchyma or sclerenchyma. **Pith** comprising isodiametric, thin-walled parenchyma cells with scalariform thickenings in *P. piersii* and *P. pubescens*; banded in *P. adansoniae*, *P. benthamia*, *P. concreta*, *P. coriscensis*, *P. dendrobiiflora*, and *P. modesta*; cells thick-walled in *P. johnstonii* and *P. valentina*. Cruciate starch granules present or absent.

Material examined

Polystachya (59, including *Neobenthamia*, 1). Full details appear in Carlsward et al. (2006a), Heaney (2006).

Reports from the literature (all *Polystachya*)

Carlsward, Stern, and Bytebier (2006a): leaf and root anatomy of many species, including *P. benthamia* (as *Neobenthamia*). Cribb (1978): *Polystachya* in Africa. Dressler (1993): phylogeny and classification of Orchidaceae. Heaney (2006): comparative anatomy and systematics of Polystachyinae. Kraenzlin (1926): monograph of *Polystachya*. La Croix and la Croix (1997): account of all species of *Polystachya* in cultivation with sectional comparisons. Meinecke (1894): aerial root anatomy. Möbius (1887): leaf anatomy. Møller and Rasmussen (1984): stegmata absent from all parts of *P. concreta*. Moreau (1913): pseudobulb anatomy. Olatunji and Nengim (1980): tracheoids. Parrilla Diaz and Ackerman (1990): root anatomy. Podzorski and Cribb (1979): revision of *Polystachya* section *Cultriformes*. Porembski and Barthlott (1988): velamen in three species. Rolfe (1897): *Polystachya* in tropical Africa. Sanford and Adanlawo (1973): root anatomy. Senghas (1996): study of subtribe Polystachyinae. Tominski (1905): leaf anatomy. Weltz (1897): axis anatomy. Williams (1979): stomata of *Polystachya* (including *Neobenthamia*).

Taxonomic notes

Based on anatomical features, Polystachyinae are monophyletic: fibre bundles are lacking and stegmata are not associated with vascular bundle sclerenchyma. Thin-walled mucilage cells are consistently present in mesophyll, foliar hypodermis, and ground tissue of stems and roots. An abaxial hypodermis is lacking in most species of *Polystachya* (including *Neobenthamia*). Banded water-storage cells are present in ground tissue of some pseudobulbs, in cortex of roots, and in mesophyll.

TABLES OF DIAGNOSTIC CHARACTERS

These lists refer only to taxa studied; in some tribes very few taxa have been studied. Values in brackets are ones less common.

Table 1. Leaf. Cuticle and hairs

Taxon	Cuticle	Hairs				Papillate
	Patterned in some taxa	Ab- or adaxial in some taxa	Both or all surfaces in some taxa	Uni- or multicellular or present (+)	Glandular or non-glandular	In some taxa
Apostasioideae						
Cypripedioideae		+		+	g, ng	
Diseae	patt			(uni, multi)	(g, ng)	
Acianthinae	patt					
Caladeniinae	patt			multi	g, ng	+
Cryptostylidinae	patt					+
Diuridinae	patt					(1 sp.)
Drakaeinae	patt					+
Megastylidinae	patt					(1 genus)
Prasophyllinae	patt	ad		uni, multi	g, ng	
Thelymitrinae	patt					+ 1 per cell
Orchideae	patt		+	(uni)	(ng)	
Chloraeeae						
Cranichideae	patt			(uni 1 sp.)		(1 sp.)
Pogonieae						(1 sp.)
Vanilleae						
Arethusinae				(+)		
Coelogyninae			+	multi		
Calypsoeae			+	2 cells	g, ng?	
Collabieae	patt			+, (uni)		
Bletiinae	patt			multi		
Chysinae		ad ± ab	+	(multi)		
Coeliinae			+	uni		
Laeliinae	patt			(+1 genus)	(g)	
Pleurothallidinae			+	multi 1–4 cells	g	
Ponerinae	(patt)	(ad 1 sp.)		(+ 1 sp.)		
Malaxideae				+		
Neottieae	patt		+	uni, multi	g, ng	
Nervilieae	patt	ad	+	uni, multi		
Podochileae			+	multi		
Sobralieae		ab	+	multi 3 or 4		
Triphoreae			+	uni, multi 2 or 3	g	
Tropidieae			+	uni	g	
Catasetinae			+	multi 3,(5)	ng?	
Coeliopsidinae			+	multi 3	g	
Cymbidiinae		ad		+		
Cyrtopodiinae			+	multi 3		
Eulophiinae			+	multi	g, ng?	
Maxillariinae	(patt 2 gen)		+	uni, multi	g, ng?	
Oncidiinae			(1 sp.)	(+ in 1 sp.)		
Stanhopeinae			+	multi 3	g?	
Zygopetalinae				multi		

Table 1. continued

Taxon	Cuticle	Hairs				Papillate
	Patterned in some taxa	Ab- or adaxial in some taxa	Both or all surfaces in some taxa	Uni- or multicellular or present (+)	Glandular or non-glandular	In some taxa
Dendrobieae			+	1–2, 2, 3	g, (ng)	
Adrorhizinae						
Aeridinae & Angraecinae	patt			multi	g	
Agrostophyllinae						
Polystachyinae	patt	ab	+	uni, (multi, 2 cells)	(g)	

Table 2. Leaf. Stomata and hypodermis
Stomata abaxial, superficial are main character states in all tribes; other values apply to some taxa.

Taxon	Stomata					Hypodermis
	Ab- and adaxial	Sunken or raised	Mainly anomocytic	Mainly tetracytic	Paracytic or 2 subs	Ad- or abaxial or both; no. layers
Apostasioideae	+			+ (or anom)		
Cypripedioideae	+			+ (or anom)		
Diseae	+	(sl. rais)	+			
Acianthinae			+			
Caladeniinae	+		+			(2 or 3 ad in 1 gen)
Cryptostylidinae			+			
Diuridinae	+		+			
Drakaeinae	+		+			
Megastylidinae	+		+			
Prasophyllinae			+			
Thelymitrinae	+		+			
Orchideae	(2 gen)		+			
Chloraeeae			+			
Cranichideae	(2 gen)		various	various	various	
Pogonieae	+					
Vanilleae	(few spp.)	(sunk, rais)	(2 gen)	+ (or & anom)		1 ad & ab or ab
Arethusinae	(2 spp.)		+			
Coelogyninae		sunk		+		1 (2) ad & ab or ad
Calypsoeae	(1 gen)		(1 gen)	+ (or & anom)		(1 or 2 ab)
Collabieae	(3 spp.)			+		
Bletiinae	(+)				(+)	3 or 4 ad or 1 ab
Chysinae				(+)	+ (or tet)	
Coeliinae				+	+	1 ad or ad & ab
Laeliinae	(+)	sunk	(+)	+ (or & anom)		(1 or 2 ad ± ab)
Pleurothallidinae		(rais)		+ (or other)		1–2 (–11) ad & 1 ab
Ponerinae		sunk		+		1 (2) ad
Malaxideae	+			+ (or & anom)		(ad in 1 sp.)
Neottieae	(+)		+		(2 subs in 1gen)	
Nervilieae			+ (or tet)	+ (or anom)		
Podochileae	+			+		1 to several ad ± ab
Sobralieae				+ (or 2 subs)	(2 subs)	
Triphoreae				+ (or anom)		(1 in 1 gen)

Table 2. continued

Taxon	Stomata					Hypodermis
	Ab- and adaxial	Sunken or raised	Mainly anomocytic	Mainly tetracytic	Paracytic or 2 subs	Ad- or abaxial or both; no. layers
Tropidieae				+		
Catasetinae			+ (& tet)	+ (& others)		(ad & ab)
Coeliopsidinae			+			
Cymbidiinae	(+)	(sl. sunk)		+		ad & ab
Cyrtopodiinae	+			+		
Eulophiinae	+	(sl. sunk)		+		(1 ad & ab in 1 sp.)
Maxillariinae		(sunk)		+ (or & anom)		(ad)
Oncidiinae	+	sunk, (rais)		+		1 ad (or 1 ab)
Stanhopeinae		(rais 1 sp.)		+ (& others)		
Zygopetalinae				+		(ab in 1 gen)
Dendrobieae	(1 sp.)			+ (or para)	(+)	1–3 ad ± ab
Adrorhizinae	+	sunk		+		(ad & ab in 1 gen)
Aeridinae & Angraecinae	(+)				+	1 or 2 or more ad & ab (or ad)
Agrostophyllinae				+ (or ± cyclo)		
Polystachyinae				+ (or & anom)		1–6 ad (± 1 ab)

Table 3. Leaf. Mesophyll, vascular bundles, inclusions
Vascular bundles in one row in most taxa.

Taxon	Mesophyll			Vascular	Silica	Crystals
	Homogeneous, heterogeneous	Palisade layers	Fibre bundles	Bundles (if not 1 row)	Conical, spherical	Druses, crys, raphides, rods
Apostasioideae	ho				c	ra
Cypripedioideae	ho, het	(2 ad)			(c)	
Diseae	ho, (het)					ra
Acianthinae	ho, (het 1 gen)	(2 ad in 1 gen)				d, ra
Caladeniinae	ho, het					ra
Cryptostylidinae	ho, het	(1–3)				ra
Diuridinae	ho, het	1–2				ra, ro
Drakaeinae	(ho 2 spp.), het					ra
Megastylidinae	ho, het					ra
Prasophyllinae	ho, het	(1–2 ab)		1 row/ring		ra, ro
Thelymitrinae	ho, het	1–5 ad ± ab, (ab)		1–2 rows/rings		ra, ro
Orchideae	ho					crys
Chloraeeae	ho					crys
Cranichideae	ho, (het 1 gen)					ra, crys
Pogonieae	ho					ra
Vanilleae	ho					ra, crys
Arethusinae	ho		(1 sp.)			(d, ra)
Coelogyninae	ho, (het)	1 or 2	(+)	(2 in 2 taxa)	(c)	d, ra, crys
Calypsoeae	ho		(1 sp.)		(c)	ra
Collabieae	ho				(c)	ra
Bletiinae	ho		+		(c)	ra, crys
Chysinae	ho				c	

Table 3. continued

Taxon	Mesophyll			Vascular	Silica	Crystals
	Homogeneous, heterogeneous	Palisade layers	Fibre bundles	Bundles (if not 1 row)	Conical, spherical	Druses, crys, raphides, rods
Coeliinae	–, het	2 or 3			pres	ra in epid
Laeliinae	ho, (het)		+	(3 rows in 1 gen)	c	crys
Pleurothallidinae	ho, het		(1 sp.)		(c 1 gen)	ra
Ponerinae	ho, het	ad			c	d, ra
Malaxideae	ho, (het 1 sp.)			(2 or 3 rows)		ra
Neottieae	ho, (het)				(c 1 sp.)	ra
Nervilieae	ho, (het)	(1 ad)		1, 2, 3 rows		ra, crys
Podochileae	ho, het		(1 sp.)	(2 rows)	(s),(c 1 sp.)	ra
Sobralieae	ho, het		+			
Triphoreae	ho				(c 1 gen)	ra
Tropidieae	ho				c	ra
Catasetinae	ho		ad & ab		c	crys
Coeliopsidinae	ho		1 row		c	
Cymbidiinae	ho, (het)		ad & ab		c	crys
Cyrtopodiinae	ho		1 ab row		c	crys
Eulophiinae	ho		ad ± ab		c	ra, (crys)
Maxillariinae	ho, (het)		ad & ab	(1–3 rows)	c	crys
Oncidiinae	ho, (het)		(ab)	(2 or 3 rows)	(c)	crys
Stanhopeinae	ho, (het 1 gen)		+		c	ra
Zygopetalinae	ho		(+)		c	ra
Dendrobieae	ho, (het 1 gen)		(ad ± ab)	(2+ rows/rings)	s	(d), ra, crys
Adrorhizinae	(ho 1 gen), het	(5–8 in 1 sp.)	(1 sp.)	(2 rows in 1 sp.)	(c 1 sp.)	
Aeridinae & Angraecinae	ho, (het)		(+)	(2 rows in 1 sp.)	s	ra
Agrostophyllinae	ho, het		+		c	
Polystachyinae	ho			(2 rows in 1 sp.)		crys in epid

Table 4. Root. Velamen

Hyphens indicate range of values, not necessarily a continuous series.

Taxon	Velamen					
	1–5 layers	Up to 10 layers	Up to 15 layers	Up to 20–25 layers	Absent in some taxa	Entirely absent
Apostasioideae	(in 2 spp.)					
Cypripedioideae		1, 6–10			several abs	
Diseae	+					
Acianthinae						abs
Caladeniinae	+				(abs 1 gen)	
Cryptostylidinae	+					
Diuridinae	+					
Drakaeinae	+					
Megastylidinae	+					
Prasophyllinae	+				some abs	
Thelymitrinae	+					
Orchideae	+				many abs	

Table 4. continued

Taxon	Velamen					
	1–5 layers	Up to 10 layers	Up to 15 layers	Up to 20–25 layers	Absent in some taxa	Entirely absent
Chloraeeae	+					
Cranichideae	(+)				many abs	
Pogonieae	+					
Vanilleae	+				(abs 2 gen)	
Arethusinae	+					
Coelogyninae		1–10				
Calypsoeae	+					
Collabieae			1, 2–13			
Bletiinae	+					
Chysinae	+					
Coeliinae	+					
Laeliinae		3–10				
Pleurothallidinae	+					
Ponerinae		3–8				
Gastrodieae						abs
Malaxideae	+					
Neottieae						abs
Nervilieae						abs
Podochileae	+					
Sobralieae		2–6				
Triphoreae	+					
Tropidieae	+					
Catasetinae			5–12			
Coeliopsidinae			3–11			
Cymbidiinae				2–16		
Cyrtopodiinae				10–24		
Eulophiinae			2–14			
Maxillariinae			2–9(–11)			
Oncidiinae				(1–2–)4–7(–16)		
Stanhopeinae			4–15			
Zygopetalinae			3–15			
Dendrobieae				1–17		
Adrorhizinae	+					
Aeridinae & Angraecinae		1–3(–7)				
Agrostophyllinae	+					
Polystachyinae		2–7				

Table 5. Root. Tilosomes
Tilosomes absent except where listed.

Taxon	Spongy	Lamellate	Webbed	Bacculate	Discoid	Meshed	Unspecified
Apostasioideae							
Cypripedioideae		(+)					
Diseae							
Acianthinae							

Table 5. continued

Taxon	Spongy	Lamellate	Webbed	Bacculate	Discoid	Meshed	Unspecified
Caladeniinae							
Cryptostylidinae							(in 1 sp.)
Diuridinae							
Drakaeinae							
Megastylidinae							
Prasophyllinae							
Thelymitrinae							
Orchideae							
Chloraeeae							
Cranichideae		+		+			
Pogonieae							
Vanilleae							
Arethusinae							
Coelogyninae	+						
Calypsoeae							
Collabieae							
Bletiinae							
Chysinae		+					
Coeliinae		+					
Laeliinae	(+)	+					
Pleurothallidinae	+	+			(+)	(1 sp.)	
Ponerinae							
Gastrodieae							
Malaxideae							
Neottieae							
Nervilieae							
Podochileae							
Sobralieae	+						
Triphoreae							
Tropidieae							
Catasetinae							
Coeliopsidinae							
Cymbidiinae			+	+			
Cyrtopodiinae		+					
Eulophiinae			+	+			
Maxillariinae	+	+		+			
Oncidiinae		(+)					
Stanhopeinae							
Zygopetalinae		(in 1 var)					
Dendrobieae	(+)	+					(+)
Adrorhizinae							
Aeridinae & Angraecinae							
Agrostophyllinae							
Polystachyinae			+				

Table 6. Root. Protoxylem and pith

Taxon	Protoxylem poles	Pith parenchymatous	Pith sclerenchymatous	Pith present unspecified	Pith absent
Apostasioideae	poly	+			
Cypripedioideae	poly			+ (or abs)	+
Diseae	poly	+	(+)		
Acianthinae	3–10			+	
Caladeniinae	2–12			+	
Cryptostylidinae	8–11			+	
Diuridinae	4–9			+	
Drakaeinae	3–10		(+ in 1 sp.)	+	
Megastylidinae	2–18			+	
Prasophyllinae	3–8			+	
Thelymitrinae	3–9			+	
Orchideae	4–18(–65) (meristeles)	+		+	
Chloraeeae	poly				
Cranichideae	4–24	+			
Pogonieae	5–8				
Vanilleae	5–18(–32)	+	+		
Arethusinae	poly	+			
Coelogyninae	6–26	+	+		
Calypsoeae	3–9	+			(+ in 1 genus)
Collabieae	poly	+			
Bletiinae	8	+			
Chysinae	16	+			
Coeliinae	20				
Laeliinae	7–24	+	(+ in 1 gen)		
Pleurothallidinae	4–20	+	+		
Ponerinae	8–15	+			
Gastrodieae	3–6	+			
Malaxideae	5–16		+		
Neottieae	4–11			+	
Nervilieae	4–8(–15)			+	
Podochileae	6–22	+	(+ in 2 spp.)		
Sobralieae	6–40	+	+ (scl & par)		
Triphoreae	8, 20–22, 33	+			
Tropidieae	6–11, 23	+			
Catasetinae	8–17, 24	+			
Coeliopsidinae	7–15	+	+		
Cymbidiinae	10–24	+			
Cyrtopodiinae	15, 17	+			
Eulophiinae	8–19	+			
Maxillariinae	9–26	+	+		
Oncidiinae	4–27	+			
Stanhopeinae	5–20	+			
Zygopetalinae	6–18	+			
Dendrobieae	6–17(–66)	+	+		
Adrorhizinae					
Aeridinae & Angraecinae	6–28		+		
Agrostophyllinae	11–13				
Polystachyinae	11–14	+			

Notes on tables

GENERAL. ab = abaxial, abs = absent, ad = adaxial, gen = genus (genera), pres or + = present, sl. = slightly, & = and. Values in brackets are ones less common.

LEAF. Cuticle: patt = patterned. Hairs: multi = multicellular, uni = unicellular; 2, 3, etc. = 2-celled, 3-celled, etc; g = glandular, ng = non-glandular. Stomata: rais = raised, sunk = sunken; anom = anomocytic, cyclo = cyclocytic, para = paracytic, subs = subsidiaries, tet = tetracytic. Mesophyll: het = heterogeneous, ho = homogeneous. Silica: c = conical bodies, s = spherical bodies. Crystals: crys = unspecified or other crystal types, d = druses, epid = epidermis, ra = raphides, ro = rods.

ROOT. Tilosomes: var = variety. Protoxylem poles: poly = polyarch. Pith: par = parenchymatous, scl = sclerenchymatous.

BIBLIOGRAPHY

Abadie, J.-C., Püttsepp, U., Gebauer, G., Faccio, A., Bonfante, P., and Selosse, M.-A. (2006). *Cephalanthera longifolia* (Neottieae, Orchidaceae) is mixotrophic: a comparative study between green and nonphotosynthetic individuals. *Can. J. Bot.*, **84**, 1462–1477.

Abe, K. (1972). Contributions to the embryology of the family Orchidaceae. VI. Development of the embryo sac in 15 species of orchids. *Sci. Rep. Tohoku Univ. Ser. IV (Biol.)*, **36**, 135–178.

Ackerman, J., and Williams, N. (1981). Pollen morphology of the Chloraeinae (Orchidaceae: Diurideae) and related subtribes. *Am. J. Bot.*, **68**, 1392–1402.

Adams, P. B. (2011). Systematics of Dendrobiinae (Orchidaceae) with special reference to Australian taxa. *Bot. J. Linn. Soc.*, **166**, 105–126.

Aiton, W. T. (1810–13). *Hortus Kewensis* (2nd edn). Longman & Co., London.

Alconero, R. (1968a). Infection and development of *Fusarium oxysporum* f. sp. *vanillae* in vanilla roots. *Phytopathology*, **58**, 1281–1283.

Alconero, R. (1968b). *Vanilla* root anatomy. *Phyton (B. Aires)*, **25**, 103–110.

Alconero, R. (1969a). Mycorrhizal and parasitic infections of vanilla roots. *Phyton (B. Aires)*, **26**, 17–22.

Alconero, R. (1969b). Mycorrhizal synthesis and pathology of *Rhizoctonia solani* in *Vanilla* orchid roots. *Phytopathology*, **59**, 426–430.

Ames, O. (1922). Notes on New England orchids. II. The mycorrhiza of *Goodyera pubescens*. *Rhodora*, **24**, 37–46.

Andersen, T. F., Johansen, B., Lund, I., Rasmussen, F., Rasmussen, H., and Sorensen, I. (1988). Vegetative architecture of *Eria*. *Lindleyana*, **3**, 117–132.

Ando, T. (1987). Chimeral leaf structure found in Dendrobium moniliforme cultivars cultivated since 1835. pp. 84–91. In: *Proc. 12th World Orchid Conf., Tokyo, 1987* (Ed. Saito, K., Tanaka, R.). World Orchid Conference Trust, Tokyo.

Arber, A. (1925). *Monocotyledons: a morphological study*. Cambridge University Press, Cambridge.

Arditti, J., and Fisch, M. (1977). Anthocyanins of the Orchidaceae: distribution, heredity, functions, synthesis, and localization. pp. 117–155. In: *Orchid biology: Reviews and perspectives* (Ed. Arditti, J.). Comstock Publishing Associates, Ithaca, NY.

Arends, J., and Stewart, J. (1989). *Aerangis gracillima*: a definitive account of a rare African orchid of Cameroun and Gabon. *Lindleyana*, **4**, 23–29.

Ascensão, L., Francisco, A., Cotrim, H., and Pais, M. (2005). Comparative structure of the labellum in *Ophrys fusca* and *O. lutea* (Orchidaceae). *Am. J. Bot.*, **92**, 1059–1067.

Atwood, J. (1984). The relationships of the slipper orchids (subfamily Cypripedioideae, Orchidaceae). *Selbyana*, **7**, 129–247.

Atwood, J. (1986). The size of the Orchidaceae and the systematic distribution of epiphytic orchids. *Selbyana*, **9**, 171–186.

Atwood, J., and Williams, N. (1978). The utility of epidermal cell features in *Phragmipedium* and *Paphiopedilum* (Orchidaceae) for determining sterile specimens. *Selbyana*, **2**, 356–366.

Atwood, J., and Williams, N. (1979). Surface features of the adaxial epidermis in the conduplicate-leaved Cypripedioideae (Orchidaceae). *Bot. J. Linn. Soc.*, **78**, 141–156.

Aybeke, M. (2012). Comparative anatomy of selected rhizomatous and tuberous taxa of subfamilies Orchidoideae and Epidendroideae (Orchidaceae) as an aid to identification. *Plant Syst. Evol.*, **298**, 1643–1658.

Aybeke, M., Sezik, E., and Olgun, G. (2010). Vegetative anatomy of some *Ophrys*, *Orchis* and *Dactylorhiza* (Orchidaceae) taxa in Trakya region of Turkey. *Flora*, **205**, 73–89.

Ayensu, E., and Williams, N. (1972). Leaf anatomy of *Palumbina* and *Odontoglossum* subgenus *Osmoglossum*. *Am. Orchid Soc. Bull.*, **41**, 687–696.

Azevedo, N. F. dos S., D. (1970–1). Micorrizas de plantas espontâneas e cultivadas. *Mem. Soc. Brot.*, **21**, 329–342.

Baker, R. (1972). Foliar anatomy of the Laeliinae (Orchidaceae). Thesis, Department of Biology, Washington University.

Banerjee, A., and Rao, A. (1978). A preliminary epidermal study in a few taxa of *Coelogyne* (Orchidaceae). *Curr. Sci.*, **47**, 630–632.

Barroso, J., and Pais, M. (1985). Cytochimie. Caractérisation cytochimique de l'interface hôte/endophyte des endomycorrhizes d'*Ophrys lutea*. Role de l'hôte dans la synthèse des polysaccharides. *Ann. Sci. Nat. Bot.*, **7**, 237–244.

Barthlott, W. (1976). Struktur und Funktion des Velamen Radicum der Orchideen. pp. 438–443. In *Proc. 8th World Orchid Conf., Frankfurt, 1975* (Ed. Senghas, K.). World Orchid Conference Trust, Frankfurt.

Barthlott, W., and Capesius, I. (1975 [1976]). Mikromorphologische und funktionelle Untersuchungen am *Velamen radicum* der Orchideen. *Ber. Deutsch. Bot. Ges.*, **88**, 379–390.

Baruah, A. (1998). Vegetative anatomy of the endemic orchid *Vanilla pilifera* Holt. *Phytomorphology*, **48**, 101–105.

Baruah, A. (2001). Vegetative anatomical studies of *Vanda teres* Lindl. (Orchidaceae). *J. Econ. Taxon. Bot. Add. Ser.*, **19**, 161–164.

Baruah, A., and Saikia, N. (2002). Vegetative anatomy of the orchid *Vanilla planifolia* Andr. *J. Econ. Taxon. Bot.*, **26**, 161–165.

Bateman, R., Hollingsworth, P., Preston, J., Luo Y.-B., Pridgeon, A. M., and Chase, M. (2003). Molecular phylogenetics and evolution of Orchidinae and selected Habenariinae (Orchidaceae). *Bot. J. Linn. Soc.*, **142**, 1–40.

Bateman, R., Pridgeon, A. M., and Chase, M. (1997). Phylogenetics of subtribe Orchidinae (Orchidoideae, Orchidaceae) based on nuclear ITS sequences. 2. Infrageneric relationships and reclassification to achieve monophyly of *Orchis* sensu stricto. *Lindleyana*, **12**, 113–141.

Baytop, T. (1968). L'origine du Salep des Prés. *Istanbul Univ. Eczac. Fak. Mecmuasi*, **4**, 69–71.

Bechtel, H., Cribb, P., and Launert, E. (1992). *The manual of cultivated orchid species*. 3rd edn. Blandford, London.

Bell, A. (1991). *Plant form: an illustrated guide to flowering plant morphology*. Oxford University Press, Oxford.

Bell, A., Roberts, D., Hawkins, J., Rudall, P., Box, M., and Bateman, R. (2009). Comparative micromorphology of nectariferous and nectarless labellar spurs in selected clades of subtribe Orchidinae (Orchidaceae). *Bot. J. Linn. Soc.*, **160**, 369–387.

Bentham, G. (1881). Notes on Orchideae. *J. Linn. Soc. (Bot.)*, **18**, 281–360.

Bentham, G., and Hooker, J. D. (1883). *Genera plantarum*. Vol. 3. L. Reeve and Co., London.

Benzing, D. (1982). Mycorrhizal infections of epiphytic orchids in southern Florida. *Am. Orchid Soc. Bull.*, **51**, 618–622.

Benzing, D., and Friedman, W. (1981). Mycotrophy: its occurrence and possible significance among epiphytic Orchidaceae. *Selbyana*, **5**, 243–247.

Benzing, D., Friedman, W., Peterson, G., and Renfrow, A. (1983). Shootlessness, velamentous roots, and the pre-eminence of Orchidaceae in the epiphytic biotope. *Am. J. Bot.*, **70**, 121–133.

Benzing, D., Ott, D., and Friedman, W. (1982). Roots of *Sobralia macrantha* (Orchidaceae): structure and function of the velamen-exodermis complex. *Am. J. Bot.*, **69**, 608–614.

Benzing, D., and Pridgeon, A. M. (1983). Foliar trichomes of Pleurothallidinae (Orchidaceae): functional significance. *Am. J. Bot.*, **70**, 173–180.

Bernard, N. (1902). Etudes sur la tubérisation. *Rev. Gén. Bot.*, **14**, 5–25, 58–71, 101–119, 170–183, 269–179.

Bernard, N. (1903). La germination des Orchidées. *C. R. Acad. Sci., Paris*, **137**, 483–485.

Bernard, N. (1904). Recherches expérimentales sur les Orchidées. *Rev. Gén. Bot.*, **16**, 405–451, 458–476.

Bernard, N. (1909). L'évolution dans la symbiose. *Ann. Sci. Nat. Bot.*, **9**, 1–196.

Bernardos, S., Tyteca, D., and Amich, F. (2004b). Micromorphological study of some taxa of the genus *Epipactis* (Orchidaceae) from the Central-Western Iberian Peninsula. *Belg. J. Bot.*, **137**, 193–198.

Bernardos, S., Tyteca, D., Revuelta, J., and Amich, F. (2004a). A new endemic species of *Epipactis* (Orchidaceae) from northeast Portugal. *Bot. J. Linn. Soc.*, **145**, 239–249.

Bidartondo, M. (2005). The evolutionary ecology of mycoheterotrophy. *New Phytol.*, **167**, 332–352.

Birger, S. (1906–7). On Tuber Salep. *Arkiv Bot.*, **6** (13), 1–13.

Blackman, S., and Yeung, E. (1983). Comparative anatomy of pollinia and caudicles of an orchid (*Epidendrum*). *Bot. Gaz.*, **144**, 331–337.

Bloch, R. (1935). Observations on the relation of adventitious root formation to the structure of air-roots of orchids. *Proc. Leeds Phil. Lit. Soc. Sci. Sect.*, **3**, 92–101.

Bonates, L. C. de, M. (1993). Estudos ecofisiologicos de Orchidaceae da Amazonia. II. Anatomia ecologica foliar de especies com metabolismo CAM de uma campina da Amazonia Central. *Acta Amazonica*, **23**, 315–348.

Bonnier, G. (1903a). Sur les formations secondaires anormales du cylindre central dans les racines aériennes d'Orchidées. *Bull. Soc. Bot. Fr.*, **50**, 291–295.

Bonnier, G. (1903b). Influence de l'eau sur la structure des racines aériennes d'Orchidées. *C. R. Acad. Sci. Paris*, **137**, 505–510.

Borriss, H., Jeschke, E. N., and Bartsch, G. (1971). Elektronenmikroskopische Untersuchungen zur Ultrastruktur der Orchideen-Mykorrhiza. *Biol. Rdsch.*, **9**, 177–180.

Borsos, O. (1977). Magyarorszagi szabadfoldi orchideak anatomiai vizsgalata. I. Levelepidermisz vizsgalatok. [Blattanatomische Untersuchung der Wildorchideen Ungarns.] *Abstracta Bot. Budapest*, **5**, 1–13.

Borsos, O. (1980). Anatomy of wild orchids in Hungary. I. Tissue structure of leaf and floral axis. *Acta Agron. Acad. Sci. Hung.*, **29**, 369–389.

Borsos, O. (1982a). Magyaroszagi vadontermo orchideak anatomiai-hiszto-kemiai vizsgalata. [Anatomical–histological research of corms of the Hungarian native orchids.] *Orchidea*, **2**, 10–12, 16.

Borsos, O. (1982b). Anatomische Untersuchung der Wildorchideen Ungarns. II. Stengelanatomie. *Abstr. Bot. Inst. Taxon. Oikol. Plant., Univ. Sci. L. Eotvos, Budapest*, **7**, 117–129.

Borsos, O. (1983). Anatomisch-histochemische Untersuchung der Knollen der Wildorchideen Ungarns. *Orchidee (Sonderheft)*, 61–64.

Borsos, O. (1990). An anatomical-histochemical study of wild orchid tubers in Hungary. *Acta Univ. Wratislaviensis*, No. 1055, 25–31.

Bouriquet, G. (1954). *Le vanillier et la vanille dans le monde*. Paul Lechevalier, Paris.

Brandham, P. (1999). Cytogenetics. pp. 67–80. In: *Genera Orchidacearum*. Vol. 1. *General introduction, Apostasioideae, Cypripedioideae* (Ed. Pridgeon, A. M., and Cribb, P. J., Chase, M. W., and Rasmussen, F. N.). Oxford University Press, Oxford, UK.

Brown, R. (1810). *Prodromus floræ Novæ Hollandiæ et Insulæ Van-Diemen: exhibens characteres plantarum quas annis 1802–1805 per oras utriusque insulae collegit et descripsit Robertus Brown; insertis passim aliis speciebus auctori hucusque cognitis, seu evulgatis, seu ineditis, præsertim Banksianis, in primo itinere navarchi Cook detectis*. Vol. 1. Richard Taylor, London.

Brown, R. (1833). On the organs and mode of fecundation in Orchdieae and Asclepiadeae. *Trans. Linn. Soc., London*, **16**, 685–745.

Burgeff, H. (1909). *Die Wurzelpilze der Orchideen*. G. Fischer, Jena.

Burgeff, H. (1932). *Saprophytismus und Symbiose. Studien an tropischen Orchideen*. G. Fischer, Jena.

Burgeff, H. (1959). Mycorrhiza of orchids. pp. 361–395. In: *The orchids: a scientific survey* (Ed. Withner, C. L.). John Wiley and Sons, New York.

Burr, B., and Barthlott, W. (1991). On a velamen-like tissue in the root cortex of orchids. *Flora*, **185**, 313–323.

Cameron, K. M. (2001). An expanded phylogenetic analysis of Orchidaceae using three plastid genes: rbcL, atpB, and psbA. *Am. J. Bot.*, **88**, Suppl., abstract 2.

Cameron, K. M. (2003). Vanilloideae. pp. 281–334. In: *Genera Orchidacearum*. Vol. 3. *Orchidoideae (Part 2), Vanilloideae* (Ed. Pridgeon, A. M., Chase, M. W., and Rasmussen, F. N.). Oxford University Press, Oxford.

Cameron, K. M. (2009). On the value of nuclear and mitochondrial gene sequences for reconstructing the phylogeny of vanilloid orchids (Vanilloideae, Orchidaceae). *Ann. Bot.*, **104**, 377–385.

Cameron, K. M., and Chase, M. (1999). Phylogenetic relationships of Pogoniinae (Vanilloideae, Orchidaceae): an herbaceous example of the eastern North-America-Eastern Asia phytogeographic disjunction. *J. Plant Res.*, **112**, 317–329.

Cameron, K. M., and Chase, M. (2000). Nuclear 18s rDNA sequences of Orchidaceae confirm the subfamilial status and circumscription of Vanilloideae. pp. 457–464. In: *Monocots: systematics and evolution* (Ed. Wilson, K. L., and Morrison, D. A.). CSIRO, Melbourne, Australia.

Cameron, K. M., Chase, M., Whitten, W., Kores, P., Jarell, D., Albert, V., et al. (1999). A phylogenetic analysis of the Orchidaceae: evidence from rbcL nucleotide sequences. *Am. J. Bot.*, **86**, 208–224.

Cameron, K. M., and Dickison, W. (1998). Foliar architecture of vanilloid orchids: insights into the evolution of reticulate leaf venation in monocotyledons. *Bot. J. Linn. Soc.*, **128**, 45–70.

Campbell, E. (1962). The mycorrhiza of *Gastrodia cunninghamii* Hook. f. *Trans. R. Soc. N. Z. Bot.*, **1**, 289–296.

Campbell, E. (1970). The fungal association of *Yoania australis*. *Trans. R. Soc. N. Z. Biol. Sci.*, **12**, 5–12.

Campos Leite, V. M. de, and Oliveira, P. (1987). Morfo-anatomia foliar de *Cattleya intermedia* (Orchidaceae). *Napaea*, No. 2, 1–10.

Camus, E. (1908). *Monographie des Orchidées de l'Europe, de l'Afrique septentrionale, de l'Asie mineure et des provinces russes transcaspiennes*. P. Lechevalier, Paris.

Camus, E. (1929). *Iconographie des Orchidées d'Europe et du Bassin mediterraneen* (2 vols). P. Lechevalier, Paris.

Candolle, A. P. de (1827). *Organographie végétale* (2 vols). Deterville, Paris.

Capeder, E. (1898). Beiträge zur Entwicklungsgeschichte einiger Orchideen. *Flora*, **85**, 368–423.

Capesius, I., and Barthlott, W. (1975). Isotopen-Markierungen und Raster-elektronenmikroskopische Untersuchungen des Velamen radicum der Orchideen. *Z. Pflanzenphysiol.*, **75**, 436–448.

Carlquist, S. (2012). Monocot xylem revisited: new information, new paradigms. *Bot. Rev.*, **78**, 87–153.

Carlquist, S., and Schneider, E. (2006). Origins and nature of vessels in monocotyledons: 8. Orchidaceae. *Am. J. Bot.*, **93**, 963–971.

Carlson, M. (1938). Origin and development of shoots from the tips of roots of *Pogonia ophioglossoides*. *Bot. Gaz.*, **100**, 215–225.

Carlson, M. (1943). The morphology and anatomy of *Calopogon pulchellus*. *Bull. Torrey Bot. Cl.*, **70**, 349–368.

Carlsward, B. S. (2004). Molecular systematics and anatomy of Vandeae (Orchidaceae): the evolution of monopodial leaflessness. PhD thesis, University of Florida, Gainesville, FL.

Carlsward, B. S., and Stern, W. L. (2008a). *Corallorhiza*, a rootless, leafless, terrestrial. *Orchid Rev.*, **116**, 334–339.

Carlsward, B. S., and Stern, W. L. (2008b). Vegetative anatomy of Calypsoeae (Orchidaceae). *Lankesteriana*, **8**, 105–112.

Carlsward, B. S., and Stern, W. L. (2009). Vegetative anatomy and systematics of Triphorinae (Orchidaceae). *Bot. J. Linn. Soc.*, **159**, 203–210.

Carlsward, B. S., Stern, W. L., and Bytebier, B. (2006a). Comparative vegetative anatomy and systematics of the angraecoids (Vandeae, Orchidaceae) with an emphasis on the leafless habit. *Bot. J. Linn. Soc.*, **151**, 165–218.

Carlsward, B., Stern, W., Judd, W., and Lucansky, T. (1997). Comparative leaf anatomy and systematics in *Dendrobium*, sections *Aporum* and *Rhizobium* (Orchidaceae). *Int. J. Plant Sci.*, **158**, 332–342.

Carlsward, B. S., Whitten, W. M., Williams, N. H., and Bytebier, B. (2006b). Molecular phylogenetics of Vandeae (Orchidaceae) and the evolution of leaflessness. *Am. J. Bot.*, **93**, 770–786.

Carnevali, G. (1996). Systematics, phylogeny and twig epiphytism in *Cryptocentrum* (Orchidaceae). PhD thesis, University of Missouri St. Louis, MO.

Champagnat, M., and Loiseau, M. (1975). La différenciation des racines adventives sur le protocorme de *Neottia nidus-avis* Rich. *Ann. Sci. Nat. Bot.*, **16**, 1–16.

Champagnat, M., Morel, G., Chabut, P., and Cognet, A. (1966). Recherches morphologiques et histologiques sur la multiplication végétative de quelques Orchidées du genre *Cymbidium*. *Rev. Gén. Bot.*, **73**, 706–746.

Chang, Y.-Y., Kao, N.-H., Li, J.-Y., Hsu, W.-H., Liang, Y.-L., Wu, J.-W., and Yang, C.-H. (2010). Characterization of the possible roles for B class MADS box genes in regulation of perianth formation in orchids. *Plant Physiol.*, **152**, 837–853.

Chase, M. (1986). A monograph of *Leochilus* (Orchidaceae). *Syst. Bot. Mon.*, **14**, 1–97.

Chase, M. (1987). Systematic implications of pollinarium morphology in *Oncidium* Sw., *Odontoglossum* Kunth, and allied genera (Orchidaceae). *Lindleyana*, **2**, 8–28.

Chase, M. W. (2001). The origin and biogeography of Orchidaceae. pp. 1–5. In: *Genera Orchidacearum*. Vol. 2. *Orchidoideae* (Ed. Pridgeon, A. M., Chase, M. W., and Rasmussen, F. N.). Oxford University Press, Oxford.

Chase, M. W. (2005). Classification of Orchidaceae in the age of DNA data. *Curtis's Bot. Mag.*, **22**, 2–7.

Chase, M. W., Cameron, K. M., Barrett, R. L., and Freudenstein, J. V. (2003). DNA data and Orchidaceae systematics: a new phylogenetic classification. pp. 69–89. In: *Orchid conservation* (Ed. Dixon, K. W., Barrett, R. L., and Cribb, P. J.). Natural History Publications, Kota Kinabalo, Sabah.

Chase, M., Cameron, K., Hills, H., and Jarrell, D. (1994). DNA sequences and phylogenetics of the Orchidaceae and other lilioid monocots. pp. 61–73. In: *Proc. 14th World Orchid Conf., Glasgow, 1993* (Ed. Pridgeon, A. M.). pp. 84–91. World Orchid Conference Trust, Edinburgh.

Chase, M. W., Duvall, M. R., Hills, H. G., Conran, J. G., Cox, A. V., Eguiarte, L. E., et al. (1995). Molecular phylogenetics of Lilianae. pp. 109–137. In: *Monocotyledons: systematics and evolution* (Ed. Rudall, P. J., Cribb, P. J., Cutler, D. F., and Humphries, C. J.). Royal Botanic Gardens, Kew.

Chase, M. W., Hanson, L., Albert, V. A., Whitten, W. M., and Williams, N. H. (2005). Life history evolution and genome size in subtribe Oncidiinae. *Ann. Bot.*, **95**, 191–199.

Chase, M. W., and Olmstead, R. G. (1988). Isozyme number in subtribe Oncidiinae (Orchidaceae): an evaluation of polyploidy. *Am. J. Bot.*, **75**, 1080–1085.

Chase, M. W., and Palmer, J. (1989). Chloroplast DNA systematics of lilioid monocots: resources, feasibility, and an example from the Orchidaceae. *Am. J. Bot.*, **76**, 1720–1730.

Chase, M. W., and Pippen, J. (1988). Seed morphology in the Oncidiinae and related subtribes (Orchidaceae). *Syst. Bot.*, **13**, 313–323.

Chase, M. W., and Pippen, J. (1990). Seed morphology and phylogeny in subtribe Catasetinae (Orchidaceae). *Lindleyana*, **5**, 126–133.

Chase, M. W., Soltis, D. E., Soltis, P. S., Rudall, P. J., Fay, M. F., Hahn, W. H., *et al.* (2000). Higher level systematics of the monocotyledons: an assessment of current knowledge and a new classification. pp. 3–16. In: *Monocots: systematics and evolution* (Ed. Wilson, K. L., and Morrison, D. A.). CSIRO, Collingwood, Victoria, Australia.

Chatin, A. (1856; 1857). Anatomie des plantes aériennes de l'ordre des Orchidées, 1. Mém. Anatomie des racines. 2. Mém. Anatomie du rhizome, de la tige, et des feuilles. *Mém. Soc. Sci. Nat. Cherbourg*, 5–18 (1856); 1833–1869 (1857).

Cheadle, V. I. (1942). The occurrence and types of vessels in the various organs of the plant in the Monocotyledonae. *Am. J. Bot.*, **29**, 441–450.

Cheadle, V. I. (1943). The origin and certain trends of specialization of the vessel in the Monocotyledoneae. *Am. J. Bot.*, **30**, 11–17.

Cheadle, V. I. (1944). Specialization of vessels within the xylem of each organ in the Monocotyledoneae. *Am. J. Bot.*, **31**, 81–92.

Cheadle, V. I. (1968 (1969)). Vessels in Haemodorales. *Phytomorphology*, **18**, 412–420.

Cheadle, V. I. and Kosakai, H. (1982). The occurrence of kinds of vessels in Orchidaceae. *Phyta*, 45–57.

Chen, D., Guo, B., Hexige, S., Zhang, T., Shen, D., and Ming, F. (2007). SQUA-like genes in the orchid *Phalaenopsis* are expressed in both vegetative and reproductive tissues. *Planta*, **226**, 369–380.

Chen, Y. R. (1970). Observations on *Pleione formosana* Hayata. *Taiwania*, **15**, 253–270.

Chesselet, P. (1989). Systematic implications of leaf anatomy and palynology in the Disinae and Coryciinae (Orchidaceae). Master's thesis, University of Cape Town, South Africa.

Chiang, S. (1970). Development of the root of *Dendrobium kwashotense* Hay. with special reference to the cellular structure of its exodermis and velamen. *Taiwania*, **15**, 1–16.

Chiang, S. H., and Chou, T. (1971). Histological studies on the roots of orchids from Taiwan. *Taiwania*, **16**, 1–29.

Chiang, Y. L., and Chen, Y. R. (1968). Observations of *Pleione formosana* Hayata. *Taiwania*, **14**, 271–301.

Childers, N. F., Cibes, H. R., and Hernández-Medina, E. (1959). Vanilla—the orchid of commerce. pp. 477–508. In: *The orchids, a scientific survey* (Ed. Withner, C. L.). John Wiley and Sons, New York.

Chiron, G., and Guirad, J. (2008). Anatomie foliaire du genre *Baptistonia* Barb. Rodr. (Orchidaceae, Oncidiinae). *Candollea*, **63**, 101–113.

Chodat, R., and Lendner, A. (1896). Sur les mycorhizes du *Listera cordata*. *Bull. Herb. Boissier*, **4**, 256–272.

Cingel, N. A. van der (1995). *An atlas of orchid pollination: European orchids*. A. A. Balkema, Rotterdam.

Cingel, N. A. van der (2001). *An atlas of orchid pollination: America, Africa, Asia and Australia*. A. A. Balkema, Rotterdam.

Cisternas, M. A., Salazar, G. A., Verdugo, G., Novoa, P., Calderón, X., and Negritto, M. A. (2012). Phylogenetic analysis of Chloraeinae (Orchidaceae) based on plastid and nuclear DNA sequences. *Bot. J. Linn. Soc.*, **168**, 258–277.

Clements, M. (1988). Orchid mycorrhizal associations. *Lindleyana*, **3**, 73–86.

Clements, M. (1995). Reproductive biology in relation to phylogeny of the Orchidaceae especially the tribe Diurideae. PhD thesis, Australian National University.

Clements, M. (1999). Embryology. pp. 38–58. In: *Genera Orchidacearum*. Vol. 1. *General introduction, Apostasioideae, Cypripedioideae* (Ed. Pridgeon, A. M., Cribb, P. J., Chase, M. W., and Rasmussen, F. N.). Oxford University Press, Oxford.

Clifford, S., and Owens, S. (1990). The stigma, style, and ovarian transmitting tract in the Oncidiinae (Orchidaceae): morphology, developmental anatomy, and histochemistry. *Bot. Gaz.*, **151**, 440–451.

Colleta, R., and Silva, I. (2007). Morfoanatomia foliar de microorquideas de *Ornithocephalus* Hook. e *Psygmorchis* Dodson and Dressler. *Acta Bot. Bras.*, **21**, 1068–1076.

Conran, J., Bannister, J., and Lee, D. (2009). Earliest orchid macrofossils: Early Miocene *Dendrobium* and *Earina* (Orchidaceae, Epidendroideae) from New Zealand. *Am. J. Bot.*, **96**, 466–474.

Cordemoy, H. J. de (1904). Sur une fonction spéciale des mycorhizes des racines laterales de la vanille. *C. R. Acad. Sci. Paris*, **138**, 391–393.

Costantin, J. (1885). Recherches sur l'influence qu'exerce le milieu sur la structure des racines. *Ann. Sci. Nat. Bot.*, **1**, 135–182.

Cox, A., Pridgeon, A. M., Albert, V., and Chase, M. (1997). Phylogenetics of the slipper orchids (Cypripedioideae, Orchidaceae): nuclear rDNA ITS sequences. *Plant Syst. Evol.*, **208**, 197–223.

Cribb, P. J. (1978). Studies in the genus *Polystachya* (Orchidaceae) in Africa. *Kew Bull.*, **32**, 743–766.

Cribb, P. J. (1999). Morphology. pp. 13–23. In: *Genera Orchidacearum*. Vol. 1. *General introduction, Apostasioideae, Cypripedioideae* (Ed. Pridgeon, A. M., Chase, M. W., and Rasmussen, F. N.). Oxford University Press, Oxford.

Cribb, P. J. (2009). *Eulophia*. pp. 100–107. In: *Genera Orchidacearum*. Vol. 5. *Epidendroideae (Part two)* (Ed. Pridgeon, A. M., Cribb, P. J., Chase, M. W., and Rasmussen, F. N.). Oxford University Press, Oxford.

Cribb, P. J., and Gasson, P. (1982). Unusual asexual reproduction in the East African orchid, *Cynorkis uncata*. *Kew Bull.*, **36**, 661–663.

Cronquist, A. (1968). *The evolution and classification of flowering plants*. Houghton Mifflin, Boston, MA.

Cronquist, A. (1981). *An integrated system of classification of the flowering plants*. Columbia University Press, New York.

Currah, R., Hambleton, S., and Smreciu, A. (1988). Mycorrhizae and mycorrhizal fungi of *Calypso bulbosa*. *Am. J. Bot.*, **75**, 739–752.

Currah, R., Smreciu, E., and Hambleton, S. (1990). Mycorrhizae and mycorrhizal fungi of boreal species of *Platanthera* and *Coeloglossum* (Orchidaceae). *Can. J. Bot.*, **68**, 1171–1181.

Currah, R., Zelmer, C. D., Hambleton, S., and Richardson, K. A. (1997). Fungi from orchid mycorrhizas. pp. 117–170. In: *Orchid biology: Reviews and perspectives, VII* (Ed. Arditti, J., and Pridgeon, A. M.). Kluwer Academic Publishers, Dordrecht, The Netherlands.

Curry, K., McDowell, L., Judd, W., and Stern, W. (1991). Osmophores, floral features, and systematics of *Stanhopea* (Orchidaceae). *Am. J. Bot.*, **78**, 610–623.

Curry, K., Stern, W., and McDowell, L. (1989). Osmophore development in *Stanhopea anfracta* and *S. pulla* (Orchidaceae). *Lindleyana*, **3**, 212–220.

Curtis, K. (1917). The anatomy of the six epiphytic species of the New Zealand Orchidaceae. *Ann. Bot.*, **31**, 133–149.

Cutler, D. F. (1978). *Applied plant anatomy*. Longman, London.

Cyge, T. (1930). Etudes anatomiques et écologiques sur les feuilles d'Orchidées indigènes. *Mém. Acad. Polon. Sci. Lett. Cl. Sci. Math. Nat. B*, **4**, 1–73 + 6 plates.

Czapek, F. (1909). Beiträge zur Morphologie und Physiologie der epiphytischen Orchideen Indiens. *S. B. Akad. Wiss. Wien Math. Naturw. Kl.*, **118**, 1555–1580.

Dahlgren, R. M. T., and Clifford, H. T. (1982). *The monocotyledons: a comparative study*. Academic Press, London.

Dahlgren, R. M. T., Clifford, H. T., and Yeo, P. F. (1985). *The families of the monocotyledons: structure, evolution, and taxonomy*. Springer-Verlag, Berlin.

D'Amelio, E., and Zeiger, E. (1988). Diversity in guard cell plastids of the Orchidaceae: a structural and functional study. *Can. J. Bot.*, **66**, 257–271.

Dangeard, M., and Armand, L. (1898). Observations de biologie cellulaire. (Mycorrhizes d'*Ophrys aranifera*). *Rev. Mycol.*, **20**, 13–18.

Danilova, M., and Barmicheva, E. (1990). Root-phloem of the saprophytic orchid *Neottia nidus-avis*. *Agric. Ecosyst. Environ.*, **29**, 73–77.

Dannecker, E. (1898). Bau und Entwicklung hohler ameisenbewohnter Orchideenknollen nebst Beitrag zur Anatomie der Orchideenblätter. Dissertation, Freiburg i. Schweiz.

Darwin, C. (1859). *On the origin of species by means of natural selection, or the preservation of the favoured races in the struggle for life*. J. Murray, London.

Darwin, C. (1862). *On the contrivances by which British and foreign orchids are fertilized by insects, and on the good effects of intercrossing*. J. Murray, London.

Darwin, C. (1877). *The various contrivances by which orchids are fertilised by insects* (edn 2, revised). Appleton and Co., New York.

Das, S., and Paria, N. (1992). Stomatal structure of some Indian orchids with reference to taxonomy. *Bangladesh J. Bot.*, **21** (1), 65–72.

Davies, K. (1999). A preliminary survey of foliar anatomy in *Maxillaria*. *Lindleyana*, **14**, 126–135.

Davies, K., and Stpiczyńska, M. (2009). Comparative histology of floral elaiophores in the orchids *Rudolfiella picta* (Schltr.) Hoehne (Maxillariinae sensu lato) and *Oncidium ornithorhynchum* H.B.K. (Oncidiinae sensu lato). *Ann. Bot.*, **104**, 221–234.

D'Emerico, S. (2001). Cytogenetics (Orchideae). pp. 216–224. In: *Genera Orchidacearum*. Vol. 2. *Orchidoideae (Part one)* (Ed. Pridgeon, A. M., Cribb, P. J., Chase, M. W., and Rasmussen, F. N.). Oxford University Press, Oxford.

D'Emerico, S., Bianco, P., Medagli, P., and Ruggiero, L. (1990). Karyological studies of some taxa of the genera *Himantoglossum*, *Orchis*, *Serapias* and *Spiranthes* (Orchidaceae) from Apulia (Italy). *Caryologia*, **43**, 267–276.

D'Emerico, S., Pignone, D., and Bianco, P. (1996). Karyomorphological analyses and heterochromatin characteristic disclose phyletic relationships among 2n = 32 and 2n = 36 species of *Orchis* (Orchidaceae). *Plant Syst. Evol.*, **200**, 111–124.

Dexheimer, J., and Serrigny, J. (1983). Etude ultrastructurale des endomycorhizes d'une orchidée tropicale: *Epidendrum ibaguense* H. B. K. I. Localisation des activités phosphatasiques acides et alcalines. *Bull. Soc. Bot. Fr. Lett. Bot.* **130** (3), 187–194.

Dietz, J. (1930). Morphologisch-anatomische Untersuchungen der unterirdischen Organe tropischer Erdorchideen. *Ann. Jard. Bot. Buitenzorg*, **41**, 1–26.

Diskus, A., and Kiermayer, O. (1954). Die Raphidenzellen von *Haemaria discolor* bei Vitalfarbung. *Protoplasma*, **43**, 450–454.

Dixon, H. (1894). On the vegetative organs of *Vanda teres*. *Proc. R. Ir. Acad.*, **3**, 441–458.

Dixon, H., Pate, J., and Kuo, J. (1990). The Western Australian fully subterranean orchid *Rhizanthella gardneri*. pp. 37–62. In: *Orchid biology: Reviews and perspectives, V* (Ed. Arditti, J.). Timber Press, Portland, OR.

Dodson, C. H. (1962a). The importance of pollination in the evolution of the orchids of tropical America. *Am. Orchid Soc. Bull.*, **31**, 525–534, 641–649, 731–735.

Dodson, C. H. (1962b). Pollination and variation in the Subtribe Catasetinae (Orchidaceae). *Ann. Missouri Bot. Gard.*, **49**, 35–56.

Dodson, C. H., Dressler, R. L., Hills, H. G., Adams, R. M., and Williams, N. H. (1969). Biologically active compounds in orchid fragrances. *Science*, **164**, 1243–1249.

Dodson, C. H., and Hills, H. G. (1966). Gas chromatography of orchid fragrances. *Am. Orchid. Soc. Bull.*, **35**, 720–725.

Dong, Z.-B., and Zhang, W.-J. (1986). Studies on the characteristics of the structures in the cortical cells of *Gastrodia elata* after infection of *Armillaria mellea*. *Acta Bot. Sin.*, **28**, 349–354 + 2 plates. [In Chinese with English summary.]

Dörr, I., and Kollmann, R. (1969). Fine structure of mycorrhiza in *Neottia nidus-avis* (L.) L.C. Rich. (Orchidaceae). *Planta*, **89**, 372–375.

Dressler, R. (1974). Classification of the orchid family. pp. 259–279. In: *Proc. 7th World Orchid Conf., Medellin 1972*. World Orchid Conference Trust, Medellin, Redout, South Africa.

Dressler, R. (1981). *The orchids. Natural history and classification*. Harvard University Press, Cambridge, MA.

Dressler, R. (1986). Recent advances in orchid phylogeny. *Lindleyana*, **1**, 5–20.

Dressler, R. (1990). The Spiranthoideae: grade or subfamily. *Lindleyana*, **5**, 110–116.

Dressler, R. (1993). *Phylogeny and classification of the orchid family*. Dioscorides Press, Portland, OR.

Dressler, R., and Cook, S. (1988). Conical silica bodies in *Eria javanica*. *Lindleyana*, **3**, 224–225.

Dressler, R., and Dodson, C. (1960). Classification and phylogeny in the Orchidaceae. *Ann. Missouri Bot. Gard.*, **47**, 25–68.

Drude, O. (1873). *Die Biologie von Monotropa hypopitys L. und Neottia nidus avis L. unter vergleichender Hinzuziehung anderer Orchideen*. Preisschrift, Göttingen.

Duncan, R. E. (1959). List of chromosome numbers in orchids. pp. 529–587. In: *The orchids: a scientific survey* (Ed. Withner, C. L.). J. Wiley and Sons, New York.

Dunsterville, G. C. K. (1961). How many orchids on a tree? *Am. Orchid Soc. Bull.*, **30**, 362–363.

Du Puy, D. (1986). A taxonomic revision of the genus *Cymbidium* Sw. (Orchidaceae). Thesis, University of Birmingham, UK.

Du Puy, D., and Cribb, P. (1988). *The genus Cymbidium*. Christopher Helm and Timber Press, Portland, OR.

Duruz, A. (1960). Etude anatomique des stomates chez les Orchidées des régions tropicales. *Bull. Soc. Fribourg Sci. Nat.*, **50**, 207–240.

Dycus, A., and Knudson, L. (1957). The role of the velamen of the aerial roots of orchids. *Bot. Gaz.*, **119**, 78–87.

Ecott, T. (2004). *Vanilla: travels in search of the ice cream orchid*. Grove Press, New York.

Engard, C. (1944). Morphological identification of the velamen and exodermis in orchids. *Bot. Gaz.*, **105**, 457–462.

Erbar, C. (2007). Current opinions in flower development and the evo-devo approach in plant phylogeny. *Plant Syst. Evol.*, **269**, 107–132.

Ernst, R., and Rodriguez, E. (1984). Carbohydrates of the Orchidaceae. pp. 223–260. In *Orchid biology: Reviews and perspectives, III* (Ed. Arditti, J.). Cornell University Press, Ithaca, NY.

Faber, F. C. von (1904). *Beiträge zur vergleichenden Anatomie der Cypripedilinae*. C. Gruninger, Stuttgart.

Fabre, J. H. (1855). Recherches sur les tubercules de l'*Himantoglossum hircinum*. *Ann. Sci. Nat. Bot.*, **3**, 253–291.

Fabre, J. (1856). De la germination des Ophrydées et de la nature de leurs tubercules. *Ann. Sci. Nat. Bot.*, **5**, 163–186.

Falkenberg, P. (1876). *Vergleichende Untersuchungen über den Bau der Vegetationsorgane der Monocotyledonen*. F. Enke, Stuttgart.

Fan, L., Guo, S.-X., and Xiao, P.-G. (1999). Study on structure and localization of acid phosphatase of mycorrhizal root of *Cymbidium sinense* (Orchidaceae). *Acta Bot. Yunnan.*, **21**, 197–201 + 3 plates.

Fan, L., Guo, S.-X., and Xiao, P.-G. (2000). A study on the mycorrhizal microstructure of six orchids. *Chinese Bull. Bot.*, **17**, 73–79.

Feldman, A. R., and Alquini, Y. (1997). Anatomia de orquídeas nativas na região metropolitana de Curitiba (Paraná, Brasil). *Fontqueria (Spain)*, **48**, 11–23.

Feldman, A. R., and Alquini, Y. (1999). Análise estrutural de folhas e ramicaules de *Pleurothallis bicristata* Cogn. e *Pleurothallis mentigera* Kranz, micro-orquideas ocorrendo na região metropolitana de Curitiba, Paraná, Brasil. *Garcia de Orta Sér. Bot.*, **14** (2), 47–51.

Fellerer, C. (1892). Beiträge zur Anatomie und Systematik der Begoniaceen. Dissertation, Universität München, Germany.

Ferreira, J., Oliveira, P. L. de, and Mariath, J. E. de A. (1994). Anatomia foliar de especies do genero *Octomeria* (Orchidaceae). *Napaea*, **10**, 7–14.

Figueroa, C., Salazar, G., Zavaleta, H., and Engleman, M. (2008). Root character evolution and systematics in Cranichidinae, Prescottiinae and Spiranthinae (Orchidaceae, Cranichideae). *Ann. Bot.*, **101**, 509–520.

Filipello Marchisio, V., Berta, G., Fontana, A., and Marzetti Mannina, F. (1985). Endophytes of wild orchids native to Italy: their morphology, caryology, ultrastructure and cytochemical characterization. *New Phytol.*, **100**, 623–641.

Fockens, J. W. (1857). Über die Luftwurzeln der Gewächse. PhD thesis, University of Göttingen, Germany.

Foroughbakhch, R., Ferry, R., Hernandez-Pinero, J., Alvarado-Vazquez, M., and Rocha-Estrada, A. (2008). Quantitative measures of leaf epidermal cells as a taxonomic and phylogenetic tool for identification of *Stanhopea* species (Orchidaceae). *Phyton (B. Aires)*, **77**, 113–127.

Foster, A. S. (1956). Plant idioblasts: remarkable examples of cell specializations. *Protoplasma*, **46**, 184–193.

Freudenstein, J. (1991). A systematic study of endothecial thickenings in the Orchidaceae. *Am. J. Bot.*, **78**, 766–781.

Freudenstein, J. (1997). A monograph of *Corallorhiza* (Orchidaceae). *Harv. Pap. Bot.*, No. 10, 5–51.

Freudenstein, J. (2005). Tribe Calypsoeae. pp. 89–115. In: *Genera Orchidacearum*. Vol. 4. *Epidendroideae (Part one)* (Ed. Pridgeon, A. M., Cribb, P. J., Chase, M. W., and Rasmussen, F. N.). Oxford University Press, Oxford.

Freudenstein, J., and Chase, M. W. (2001). Analysis of mitochondrial nad1b-c intron sequences in Orchidaceae: utility and coding of length-change characters. *Syst. Bot.*, **26**, 643–657.

Freudenstein, J., Harris, E., and Rasmussen, F. (2002). The evolution of anther morphology in orchids: incumbent anthers, superposed pollinia, and the vandoid complex. *Am. J. Bot.*, **89**, 1747–1755.

Freudenstein, J., and Rasmussen, F. (1996). Pollinium development and number in the Orchidaceae. *Am. J. Bot.*, **83**, 813–824.

Freudenstein, J., and Rasmussen, F. (1999). What does morphology tell us about orchid relationships—a cladistic analysis. *Am. J. Bot.*, **86**, 225–248.

Freudenstein, J., Senyo, D., and Chase, M. (2000). Mitochondrial DNA and relationships in the Orchidaceae. pp. 421–429. In: *Monocots: systematics and evolution* (Ed. Wilson, K. L., and Morrison, D. A.). CSIRO, Melbourne, Australia.

Freudenstein, J., van den Berg, C., Goldman, D., Kores, P., Molvray, M., and Chase, M. (2004). An expanded plastid DNA phylogeny of Orchidaceae and analysis of jackknife branch support strategy. *Am. J. Bot.*, **91**, 149–157.

Freytag, K. (1956). Optik und Feinbau der Schleimzellen in den Knollen verschiedener *Orchis* Arten. *Protoplasma*, **47**, 237–241.

Fricke, G. (1926). Über die Beziehungen der Hochblätter zu den Laubblättern und Blüten. *Planta*, **2**, 249–294.

Fuchs, A., and Ziegenspeck, H. (1925). Bau und Form der Wurzeln der einheimischen Orchideen in Hinblick auf ihre Aufgaben. *Bot. Arch.*, **12**, 290–379.

Fuchs, A., and Ziegenspeck, H. (1926). Entwicklungsgeschichte der Axen der einheimischen Orchideen und ihre Physiologie und Biologie. I. *Cypripedium*, *Helleborine*, *Limodorum*, *Cephalanthera*. II. *Listera*, *Neottia*, *Goodyera*. *Bot. Arch.*, **14**, 165–260; **16**, 360–413.

Fuchs, A., and Ziegenspeck, H. (1927a). Die Entwicklungsgeschichte der einheimischen Orchideen und der Bau ihrer Axen. III. *Bot. Arch.*, **18**, 378–475.

Fuchs, A., and Ziegenspeck, H. (1927b). Die *Dactylorchis*-Gruppe der Ophrydineen. *Bot. Arch.*, **19**, 163–274.

Fuchs, A., and Ziegenspeck, H. (1927c). Entwickelung, Axen und Blätter einheimischer Orchideen. IV. *Bot. Arch.*, **20**, 275–422.

Gao, Q., Li, S.-Y., and Hong, H. (2009). Structure and annual changing pattern of mycorrhizae of four *Cypripedium* species. *Guihaia*, **29**, 187–191.

Garay, L. (1960). On the origin of the Orchidaceae. *Bot. Mus. Leafl.*, **19**, 57–96.

Garay, L. (1972). On the origin of the Orchidaceae, II. *J. Arnold Arbor.*, **53**, 202–215.

Garay, L. (1986). Olim Vanillaceae. *Bot. Mus. Leafl.*, **30**, 223–237.

Garay, L., and Christensen, E. (1995). *Danhatchia*: a new genus for *Yoania australis*. *Orchadian*, **11**, 469–471.

Garcia Cruz, C., and Hágsater, E. (1998). Revision of *Epidendrum anisatum* group (Orchidaceae). pp. 223–234. In: *Proc. 25th World Orchid Conf.*, Rio de Janeiro, 1996 (Ed. Pereira, C. E. de B.). Naturalia Publications, Turriers, France.

Gasson, P., and Cribb, P. (1986). The leaf anatomy of *Ossiculum aurantiacum* Cribb and van der Laan (Orchidaceae: Vandoideae). *Kew Bull.*, **41**, 827–832.

Gentry, A. H., and Dodson, C. H. (1987). Diversity and biogeography of neotropical vascular epiphytes. *Ann. Missouri Bot. Gard.*, **74**, 205–233.

Gerlach, G., and Schill, R. (1991). Composition of orchid scents attracting euglossine bees. *Bot. Acta*, **104**, 379–391.

Gillot, X. (1898). *Orchis alata* Fleury, morphologie et anatomie. *Monde Plantes*, **7**, 93–97.

Goebel, K. (1922). Erdwurzeln mit velamen. *Flora*, **115**, 1–26.

Goh, C. (1975). The anatomy of orchid leaves. *Malayan Orch. Rev.*, **12**, 14–23.

Goh, C., Arditti, J., and Avadhani, P. (1983). Carbon fixation in orchid aerial roots. *New Phytol.*, **95**, 367–374.

Goh, C., Sim, A., and Lim, G. (1992). Mycorrhizal associations in some tropical orchids. *Lindleyana*, **7**, 13–17.

Goldblatt, P. (1981). Index to plant chromosome numbers 1975–1978. *Mon. Syst. Bot.*, **5**. Missouri Botanical Garden, St. Louis, MO.

Goldblatt, P. (1984). Index to plant chromosome numbers 1979–1981. *Mon. Syst. Bot.*, **8**. Missouri Botanical Garden, St. Louis, MO.

Goldblatt, P. (1985). Index to plant chromosome numbers 1982–1983. *Mon. Syst. Bot.*, **13**. Missouri Botanical Garden, St. Louis, MO.

Goldblatt, P. (1988). Index to plant chromosome numbers 1984–1985. *Mon. Syst. Bot.*, **23**. Missouri Botanical Garden, St. Louis, MO.

Goldblatt, P., and Johnson, D. E. (1990). Index to plant chromosome numbers 1986–1987. *Mon. Syst. Bot.*, **30**. Missouri Botanical Garden, St. Louis, MO.

Goldblatt, P., and Johnson, D. E. (1991). Index to plant chromosome numbers 1988–1989. *Mon. Syst. Bot.*, **40**. Missouri Botanical Garden, St. Louis, MO.

Goldblatt, P., and Johnson, D. E. (1994). Index to plant chromosome numbers 1990–1991. *Mon. Syst. Bot.*, **51**. Missouri Botanical Garden, St. Louis, MO.

Goldblatt, P., and Johnson, D. E. (1996). Index to plant chromosome numbers 1992–1993. *Mon. Syst. Bot.*, **58**. Missouri Botanical Garden, St. Louis, MO.

Gonuz, A. (2001). Ultrastructural studies on stem and leaf of *Orchis anatolica* Boiss. and *Cyclamen hederifolium* Aiton. growing at different altitudes in West Anatolia. *Hacet. Bull. Nat. Sci. Eng. Ser. A*, **30**, 47–71.

Govaerts, R. (2003). Computer printout of the monocot checklist [21 February 2003]. Royal Botanic Gardens, Kew.

Grace, O. M. (1999). Velamen in African species of *Eulophia* R.Br. ex Lindl. (Orchidaceae). Honours thesis, University of Natal, Pietermaritzburg, South Africa.

Gravendeel, B. (2000). Reorganising the orchid genus Coelogyne—a phylogenetic classification based on molecules and morphology. PhD thesis, University of Leiden, The Netherlands.

Groom, P. (1893). On the velamen of orchids. *Ann. Bot.*, **7**, 143–151.

Groom, P. (1895–7). Contributions to the knowledge of monocotyledonous saprophytes. *J. Linn. Soc. (Bot.)*, **31**, 149–215.

Guan, Z.-J., Zhang, S.-B., Guan, K.-Y., Li, S.-Y., and Hu, H. (2011). Leaf anatomical structures of *Paphiopedilum* and *Cypripedium* and their adaptive significance. *J. Plant Res.*, **124**, 189–198.

Guillaud, A. (1878). *Recherches sur l'anatomie comparée et le développement des tissus de la tige des Monocotylédones*. G. Masson, Paris.

Gupta, R., Ansari, M., and Kapoor, L. (1970 [1972]). Pharmacognostical studies on Jivanti. I. *Desmotrichum fimbriatum* Blume. *Bull. Bot. Surv. India*, **12**, 29–36.

Gutiérrez, R. M. P. (2010). Orchids: a review of uses in traditional medicine, its phytochemistry and pharmacology. *J. Med. Plants Res.*, **4**, 592–638.

Guttenberg, H. von (1968). Der primäre Bau der Angiospermenwurzel. In *Handbuch der Pflanzenanatomie*. Vol. VIII (eds Zimmermann, W., Ozenda, P., and Wulff, H. D.). Borntraeger, Berlin.

Haberlandt, G. (1914). *Physiological plant anatomy*. Translated from 4th German edition by M. Drummond. Macmillan, London.

Häfliger, A. (1901). Beiträge zur Anatomie der Vanilla-Arten. Dissertation, Universität Basel, Switzerland.

Hágsater, E., and Soto Arenas, M. Á. (2004). *Epidendrum*. pp. 236–250. In: *Genera Orchidacearum*. Vol. 4. *Epidendroideae*

(Part one) (Ed. Pridgeon, A. M., Cribb, P. J., Chase, M. W., and Rasmussen, F. N.). Oxford University Press, Oxford.

Hágsater, E., Soto Arenas, M. Á., Salazar Chávez, G. A., Jiménez Machorro, R., López Rosas, M. A., and Dressler, R. L. (2005). *Orchids of Mexico*. Productos Farmacéuticos, Mexico City.

Hadley, G. (1982). Orchid mycorrhiza. pp. 83–118. In: *Orchid biology: Reviews and perspectives, II* (Ed. Arditti, J.). Cornell University Press, Ithaca, NY.

Hadley, G., Johnson, R., and John, D. (1971). Fine structure of the host-fungus interface in orchid mycorrhiza. *Planta*, **100**, 191–199.

Hadley, G., and Williamson, B. (1972). Features of mycorrhizal infection in some Malayan orchids. *New Phytol.*, **71**, 1111–1118.

Hall, I. (1976). Vesicular mycorrhizas in the orchid *Corybas macranthus*. *Trans. Br. Mycol. Soc.*, **66**, 160.

Hamada, M. (1939). Studien über die Mykorrhiza von *Galeola septentrionalis* Reichb. f. Ein neuer Fall der Mykorrhiza-Bildung durch intraradicale Rhizomorpha. *Jap. J. Bot.*, **10**, 151–211.

Harley, J., and Smith, S. (1983). *Mycorrhizal symbiosis*. Academic Press, London.

Harvais, G., and Hadley, G. (1967). The relation between host and endophyte in orchid mycorrhiza. *New Phytol.*, **66**, 205–215.

Hashimoto, T. (1990). A taxonomic review of the Japanese *Lecanorchis* (Orchidaceae). *Ann. Tsukuba Bot. Gard.*, **9**, 1–40.

Hayat, M. A. (1981a). *Principles and techniques of electron microscopy*. 2nd edn. Arnold, London.

Hayat, M. A. (1981b). *Fixation for electron microscopy*. Academic Press, New York.

Heaney, J. M. (2006). Comparative anatomy and systematics in *Polystachya* (Orchidaceae). MSc thesis, University of Florida, Gainesville, FL.

Heckel, E. (1899). Sur la structure des Vanilles aphylles. *C. R. Acad. Sci. Paris*, **129**, 347–349.

Heim, R. (1945). Sur les racines aériennes de *Phalaenopsis schilleriana* Rchb. *C. R. Acad. Sci. Paris*, **220**, 365–367.

Hering, L. (1900). Zur Anatomie der monopodialen Orchideen. *Bot. Zbl.*, **84**, 1–11, 35–45, 73–81, 113–122, 145–152, 177–184.

Hermans, J., Hermans, C., Du Puy, D., Cribb, P., and Bosser, J. (2007). *Orchids of Madagascar: annotated checklist*. 2nd edn. Royal Botanic Gardens, Kew.

Hirmer, M. (1920). Beiträge zur Organographie der Orchideenblüte. *Flora*, **113**, 213–310.

Hofmann, E. (1930). Über die Anatomie des Blattes von *Oncidium ascendens* Lindl. *Sber. Akad. Wiss. Wien Math. Nat. Kl.*, **I**, 189–193.

Hofsten A, v. (1973). The ultrastructure of mycorrhiza in *Ophrys insectifera*. *Zoon*, Suppl. 1, 93–96.

Ho, K.-K., Yeoh, H.-H., and Hew, C.-S. (1983). The presence of photosynthetic machinery in aerial roots of leafy orchids. *Plant Cell Physiol.*, **24**, 1317–1321.

Holm, T. (1900). *Pogonia ophioglossoides* Nutt. A morphological and anatomical study. *Am. J. Sci.*, **9**, 13–19.

Holm, T. (1904). The root-structure of North American terrestrial Orchideae. *Am. J. Sci.*, **18**, 197–212.

Holm, T. (1908). Medicinal plants of North America. 13. *Cypripedium pubescens* Willd. *Merck's Rep.*, **17**, 60–62.

Holm, T. (1913). Medicinal plants of North America. 72. *Corallorhiza odontorhiza* Nutt. *Merck's Rep.*, **22**, 120–122.

Holm, T. (1915). Medicinal plants of North America. 93. *Vanilla planifolia* Andrews. *Merck's Rep.*, **24**, 212–215.

Holm, T. (1927). Sciaphilous plant-types. *Beih. Bot. Centralbl.*, **44**, 1–89.

Holst, A. (1999). *The world of catasetums*. Timber Press, Portland, OR.

Holtermann, C. (1902). Anatomisch-physiologische Untersuchungen in den Tropen. *Sber. Akad. Wiss. Berlin*, **29–30**, 656–674.

Holtzmeier, M., Stern, W., and Judd, W. (1998). Comparative anatomy and systematics of Senghas's cushion species of *Maxillaria* (Orchidaceae). *Bot. J. Linn. Soc.*, **127**, 43–82.

Howard, R. (1969). The ecology of an elfin forest in Puerto Rico. 8. Studies of stem growth and form and of leaf structure. *J. Arnold Arbor.*, **50**, 225–267.

Hsiao, Y.-Y., Pan, Z.-J., Hsu, C.-C., Yang, Y.-P., Hsu, Y.-C., Chuang, Y.-C., et al. (2011). Research on orchid biology and biotechnology. *Plant Cell Physiol.*, **52**, 1467–1486.

Huber, B. (1921). Zur Biologie der Torfmoororchidee *Liparis loeselii* Rich. *Sber. Akad. Wiss. Wien Math. Nat. Kl.*, **130**, 307–328.

Hünecke, G. (1904). Zur Anatomie der Pleurothallidinae. Inaugural dissertation, Universität Heidelberg, Germany.

Hutchinson, J. (1973). *The families of flowering plants*. Clarendon Press, Oxford.

Huynh, T., McLean, C., Coates, F., and Lawrie, A. (2004). Effect of developmental stage and peloton morphology on success in isolation of mycorrhizal fungi in *Caladenia formosa* (Orchidaceae). *Aust. J. Bot.*, **52**, 231–241.

Inamdar, J. (1968). Stomatal ontogeny in *Habenaria marginata* Coleb. *Curr. Sci.*, **37**, 24–25.

Irmisch, T. (1850). *Zur Morphologie der monokotylischen Knollen-und Zwiebelgewächse*. Reimer, Berlin.

Irmisch, T. (1853). *Beiträge zur Biologie und Morphologie der Orchideen*. Ambrosius Abel, Leipzig.

Isaiah, J., Khasim, S., and Mohana Rao, P. (1990). Vegetative anatomy of two species of *Eria* (Orchidaceae). *Indian J. Bot.*, **13**, 16–22.

Isaiah, J., Khasim, S., and Mohana Rao, P. (1991). Vegetative anatomy of *Liparis duthiei* Hook. fil. (Orchidaceae). pp. 128–132. In: *Botanical researches in India* (Ed. Aery, N. C, and Chaudhary, B. L.). Himanshu Publishers, Udaipur.

Isaiah, J., and Mohana Rao, P. (1992). Vegetative anatomy of *Dendrobium jenkinsii* Wall. ex Lindl. *J. Orchid Soc. India*, **6**, 63–69.

Izaguirre de Artucio, P. (1972). El genero *Capanemia* (Orchidaceae) en el Uruguay. *Bol. Soc. Argent. Bot.*, **14**, 225–231.

Jakubska-Busse, A., and Gola, E. (2010). Morphological variability of hellebornines. I. Diagnostic significance of morphological features in *Epipactis helleborine* (L.) Crantz., *Epipactis atrorubens* (Hoffm.) Besser and their hybrid, *Epipactis x schmalhausenii* Richt. (Orchidaceae, Neottieae). *Acta Soc. Bot. Pol.*, **79**, 207–213.

Janczewski, E. de (1885). Organisation dorsiventrale dans les racines des Orchidées. *Ann. Sci. Nat. Bot.*, **2**, 55–81.

Janse, J. (1897). Les endophytes radicaux de quelques plantes javanaises. *Ann. Jard. Bot. Buitenzorg*, **14**, 53–201.

Jayeola, A., and Thorpe, J. (2000). A scanning electron microscope study of the genus *Calyptrochilum* Kraenzl. (Orchidaceae) in West Africa. *Feddes Rep.*, **111**, 315–320.

Jeffrey, D., Arditti, J., and Koopowitz, H. (1970). Sugar content in floral and extrafloral exudates of orchids: pollination, myrmecology, and chemotaxonomy implications. *New Phytol.*, **69**, 187–195.

Jennings, A., and Hanna, H. (1899). *Corallorhiza innata* R. Br., and its mycorrhiza. *Sci. Proc. Roy. Dublin Soc.*, **9**, 1–11.

Jeyamurthy, A., Dagar, J. C., and Rathore, R. K. S. (1990). Some morphological and anatomical observations on an interesting and rare orchid *Chiloschista lunifera* (Reichb. f.) J. J. Sm. from Andamans. *J. Indian Bot. Soc.*, **69**, 335–338.

Johansen, D. A. (1948). *Plant microtechnique*. McGraw-Hill, New York.

Johow, F. (1885). Die chlorophyllfreien Humusbewohner West-Indiens, biologisch-morphologisch dargestellt. *Jb. Wiss. Bot.*, **16**, 415–449.

Johow, F. (1889). Die chlorophyllfreien Humuspflanzen nach ihren biologischen und anatomisch-entwickelungsgeschichtlichen Verhältnissen. *Jb. Wiss. Bot.*, **20**, 475–525.

Jones, D. L. (1988). *Native orchids of Australia*. Reed Books, Frenchs Forest, NSW, Australia.

Jonsson, L. (1979). The African members of *Taeniophyllum* (Orchidaceae). *Bot. Notiser*, **132**, 511–519.

Jonsson, L. (1981). A monograph of the genus *Microcoelia* (Orchidaceae). *Symb. Bot. Upsal.*, **23** (4), 151 pp.

Jost, L. (1887). Ein Beitrag zur Kenntniss der Athmungsorgane der Pflanzen. *Bot. Zeitung (Berlin)*, **45**, 601–606, 617–628, 633–642 + 1 plate.

Judd, W., Stern, W., and Cheadle, V. (1993). Phylogenetic position of *Apostasia* and *Neuwiedia* (Orchidaceae). *Bot. J. Linn. Soc.*, **113**, 87–94.

Jurcak, J. (1999). Der innere Bau vegetativer Organe einiger europäischer Orchideen. *Orchidee*, **50**, 41–45.

Kaiser, R. (1993). *The scent of orchids: olfactory and chemical investigations*. Editiones Roche, Basle.

Karasawa, K., and Aoyama, M. (1981). Morphological studies on leaf of *Paphiopedilum*. *Bull. Hiroshima Bot. Gard.*, No. 4, 81–87.

Kasapligil, B. (1961). Foliar xeromorphy of certain geophytic monocotyledons. *Madroño*, **16**, 43–70.

Katiyar, R., Sharma, G., and Mishra, R. (1986). Mycorrhizal infections of epiphytic orchids in tropical forests of Meghalaya (India). *J. Indian Bot. Soc.*, **65**, 329–334.

Kausch, A., and Horner, H. (1983). Development of syncytial raphide crystal idioblasts in the cortex of adventitious roots of *Vanilla planifolia* L. (Orchidaceae). *Scan. Electron Microsc.*, **2**, 893–903.

Kaushik, P. (1982 [1983]). Anatomy of *Aerides* (Orchidaceae), and its ecological and taxonomical bearing. *Phytomorphology*, **32**, 157–166.

Kaushik, P. (1983). *Ecological and anatomical marvels of the Himalayan orchids*. Progress in Ecology 8. Today and Tomorrow's Printers and Publishers, New Delhi.

Khasim, S., and Mohana Rao, P. (1984). Structure and function of the velamen—exodermis complex in some epiphytic orchids. *Geobios New Rep.*, **3**, 133–136.

Khasim, S., and Mohana Rao, P. (1986). Anatomical studies in relation to habitat tolerance in some epiphytic orchids. pp. 49–57. In: *Biology, conservation, and culture of orchids* (Ed. Vij, S. P.). Affiliated East-West Press, New Delhi.

Khasim, S., and Mohana Rao, P. (1989). Anatomy of four species of *Dendrobium* (Orchidaceae). *J. Swamy Bot. Club*, **6**, 99–104.

Khayota, B. (1995). Systematics of the genus *Ansellia* Lindl. (Orchidaceae). PhD thesis, University of Reading, UK.

Kim, S., and Kim, Y. (1986). Morphological and cytological study on genus *Liparis* in Korea. *Korean J. Plant Taxon.*, **16** (1), 59–88.

Knudson, L. (1922). Non-symbiotic germination of orchid seeds. *Bot. Gaz.*, **73**, 1–25.

Kocyan, A., and Endress, P. (2001). Floral structure and development of *Apostasia* and *Neuwiedia* (Apostasioideae) and their relationships to other Orchidaceae. *Int. J. Plant Sci.*, **162**, 847–867.

Kocyan, A., Qiu, Y.-L., Endress, P. K., and Conti, E. (2004). A phylogenetic analysis of Apostasioideae based on ITS, trnL-F and matK sequences. *Plant Syst. Evol.*, **247**, 203–213.

Kohl, F. (1889). *Anatomisch-physiologische Untersuchung der Kalksalze und Kieselsäure in der Pflanze*. Marburg.

Kores, P. (1995). A systematic study of the genus *Acianthus* (Orchidaceae: Diurideae). *Allertonia*, **7**, 87–220.

Kottke, I., and Suárez, J. P. (2009). Mutualistic, root-inhabiting fungi of orchids: identification and functional types. pp. 84–99. In: *Proc. Second Scientific Conference on Andean Orchids* (Ed. Pridgeon, A. M., and Suárez, J. P.). Universidad Técnica Particular de Loja, Loja, Ecuador.

Kozhevnikova, A., and Vinogradova, T. (1999). Pseudobulb structure in some boreal terrestrial orchids. *Bull. Jard. Bot. Natn. Belg.*, **68**, 59–65.

Kraenzlin, F. (1910). Orchidaceae—Monandrae—Dendrobiinae. *Pflanzenreich*, No. 45, 5–6.

Kraenzlin, F. (1926). Monographiae der Gattung *Polystachya* Hook. *Feddes Repert. Beih.*, **39**, 1–136.

Kraft, M. M. (1949). Etude histologique de quelques racines aériennes d'Orchidées. *Bull. Soc. Vaud. Sci. Nat.*, **64**, 201–211.

Kroemer, K. (1903). Wurzelhaut, Hypodermis und Endodermis der Angiospermenwurzel. *Bibl. Bot.*, **59**, 1–151.

Krüger, P. (1883). Die oberirdischen Vegetationsorgane der Orchideen in ihren Beziehungen zu Clima und Standort. *Flora*, **66**, 435–443, 451–439, 467–477, 499–510, 515–424.

Kumar, M., and Manilal, K. (1985). Morphology and anatomy of *Satyrium nepalense* Don. *J. Plant Anat. Morph.*, **2** (2), 59–62.

Kumar, M., and Manilal, K. (1988). Floral anatomy of *Apostasia odorata* and the taxonomic status of apostasioids (Orchidaceae). *Phytomophology*, **38**, 159–162.

Kumar, M., and Manilal, K. (1992). Morphology and floral anatomy of some saprophytic orchids. *J. Orchid Soc. India*, **6**, 115–124.

Kumar, M., and Manilal, K. (1992 [1993]). Floral morphology and anatomy of *Paphiopedilum insigne* and the taxonomic

status of cypripedioids (Orchidaceae). *Phytomorphology*, **42**, 293–297.
Kumar, M., and Manilal, K. (1993). Anatomical studies and their impact on the understanding of floral morphology in orchids—a chronological review. *J. Orchid Soc. India*, **7**, 1–12.
Kumazawa, M. (1956). Morphology and development of the sinker in *Pecteilis radiata* (Orchidac.). *Bot. Mag. Tokyo*, **69**, 455–461.
Kumazawa, M. (1958). The sinker of *Platanthera* and *Perularia*—its morphology and development. *Phytomorphology*, **8**, 137–145.
Kurzweil, H. (1987a). Developmental studies in orchid flowers I: Epidendroid and vandoid species. *Nordic J. Bot.*, **7**, 427–442.
Kurzweil, H. (1987b). Developmental studies in orchid flowers II: Orchidoid species. *Nordic J. Bot.*, **7**, 443–451.
Kurzweil, H. (1988). Developmental studies in orchid flowers III: Neottioid species. *Nordic J. Bot.*, **8**, 271–282.
Kurzweil, H. (1990). Floral morphology and ontogeny in Orchidaceae subtribe Disinae. *Bot. J. Linn. Soc.*, **102**, 61–83.
Kurzweil, H. (1993). Developmental studies in orchid flowers IV. Cypripedioid species. *Nordic J. Bot.*, **13**, 423–430.
Kurzweil, H. (1996). Floral morphology and ontogeny in subtribe Satyriinae (Fam. Orchidaceae). *Flora*, **191**, 9–28.
Kurzweil, H. (1998 [1999]). Floral ontogeny of orchids: a review. *Beitr. Biol. Pfl.*, **71**, 45–100.
Kurzweil, H., Linder, H., Stern, W., and Pridgeon, A. M. (1995). Comparative vegetative anatomy and classification of Diseae (Orchidaceae). *Bot. J. Linn. Soc.*, **117**, 171–220.
Kurzweil, H., Weston, P., and Perkins, A. (2005). Morphological and ontogenetic studies on the gynostemium of some Australian members of Diurideae and Cranichideae (Orchidaceae). *Telopea*, **11**, 11–33.
Kusano, S. (1911). *Gastrodia elata* and its symbiotic association with *Armillaria mellea*. *J. Coll. Agric. Tokyo*, **4**, 1–66.
Kutschera, L., and Lichtenegger, E. (1982). *Wurzelatlas mitteleuropäischer Grunlandpflanzen. I. Monocotyledoneae*. Gustav Fischer Verlag, Stuttgart.
Kutschera, L., Lichtenegger, E., Sobotik, M., and Haas, D. (1997). *Die Wurzel das neue Organ, ihre Bedeutung für das Leben von Welwitschia mirabilis und anderer Arten der Namib sowie von Arten angrenzender Gebiete*. Pflanzensoziologische Institut, Klagenfurt.
Kuttelwascher, H. (1964). Entwicklungsanatomische und Vitalfarbe-Studien an Luftwurzeln einiger tropischer Orchideen. *Sber. Öst. Akad. Wiss. Math. Nat. Kl.*, **173**, 441–483.
Kuttelwascher, H. (1965 [1966]). Entwicklungsanatomische Untersuchungen an Orchideen-Luftwurzeln. *Ber. Deutsch. Bot. Ges.*, **78**, 307–313.
La Croix, I., and La Croix, E. (1997). *African orchids in the wild and in cultivation*. Timber Press, Portland, OR.
Lakshminarayana, S., and Venkateswarlu, V. (1950). Occurrence of velamen in *Eulophia graminea* R.Br. *Sci. Cult.*, **15**, 327–328.
Lal, V., Khosa, R., and Wahi, A. (1980). Astavarga: l. Pharmacognostic studies on *Habenaria edgeworthii* (Riddhi-Vriddhi). *Indian J. Bot.*, **3**, 18–23.

Lal, V., Khosa, R., and Wahi, A. (1982). Astavarga: II. Comparative pharmacognostical studies of *Habenaria marginata* Coleb and *H. edgeworthii* Hook. f. *Indian J. Bot.*, **5**, 87–91.
Lal, V., Wahi, A., and Khosa, R. (1979). A novel arrangement of vascular tissue in some orchids. *Curr. Sci.*, **48**, 64.
Lancaster, T. (1910 [1911]). Preliminary note on the fungi of the New Zealand epiphytic orchids. *Trans. Proc. N. Z. Inst.*, **43**, 186–191.
Latr, A., Curikova, M., Balaz, M., and Jurcak, J. (2008). Mycorrhizas of *Cephalanthera longifolia* and *Dactylorhiza majalis*, two terrestrial orchids. *Ann. Bot. Fenn.*, **45**, 281–289.
Lausi, D., Nimis, P., and Tretiach, M. (1989). Adaptive leaf structures in a *Myrica-Erica* stand on Tenerife (Canary Islands). *Vegetatio*, **79**, 133–142.
Lawler, L. (1984). Ethnobotany of the Orchidaceae. pp. 27–149. In: *Orchid biology: Reviews and perspectives, III* (Ed. Arditti, J.). Comstock Publishing Associated, Ithaca, NY.
Lawler, L., and Slaytor, M. (1969). The distribution of alkaloids in New South Wales and Queensland Orchidaceae. *Phytochemistry*, **8**, 1959–1962.
Lawler, L., and Slaytor, M. (1970). The distribution of alkaloids in orchids from the Territory of Papua and New Guinea. *Proc. Linn. Soc. N.S.W.*, **94**, 237–241.
Lawton, J., Hennessy, E., and Hedge, T. (1992). Morphology and ultrastructure of the leaf of three species of *Paphiopedilum* (Orchidaceae). *Lindleyana*, **7**, 199–205.
Lee, W., and Kim, S. (1986). A taxonomic study on genus *Cephalanthera* in Korea. *Korean J. Plant Taxon.*, **16** (1), 39–58.
Leitch, I., Kahandawala, I., Suda, J., Hanson, L., Ingrouille, M., Chase, M., and Fay, M. (2009). Genome size diversity in orchids: consequences and evolution. *Ann. Bot.*, **104**, 469–481.
Leitgeb, H. (1864a). Ueber kugelformige Zellverdickungen in der Wurzelhülle einiger Orchideen. *Sber. Akad. Wiss. Wien Math. Nat. Kl.*, **49**, 275–286.
Leitgeb, H. (1864b). Die Luftwurzeln der Orchideen. *Denkschr. Akad. Wiss. Wien Math. Naturw. Kl.*, **24**, 179–222.
Leitão, C. A. E., and Cortelazzo, A. L. (2010). Structure and histochemistry of the stigmatic and transmitting tissues of *Rodriguezia venusta* (Orchidaceae) during flower development. *Aust. J. Bot.*, **58**, 233–240.
Lemaire, C. (1856). *L'Illustration Horticole*, **3**, 30.
Liang, T. G., and Zheng, S. K. (1984). The morphological anatomy of the nutritive organs of three orchids in Wuyi. *J. Fujian Agric. Coll.*, **13**, 147–156.
Li, A., Ye, X., Chen, G., and Chen, Z. (2002). Anatomy of *Cymbidium sinense* (Andr.) Willd. *J. Trop. Subtrop. Bot.*, **10**, 295–300 + 2 plates.
Li, M., Xu, G., Xu, L., Jin, R., Sha, W., and Luo, J. (1989). Microscopical examination of leaf sheaths of the Chinese drug Shihu, herba dendrobii. *Acta Pharm. Sin.*, **24**, 139–146.
Lin, C.-C., and Namba, T. (1981a). Pharmacognostical studies on the crude drugs of Orchidaceae from Taiwan: VI. On Kimsoan-lian (1). *Shoyakugaku Zasshi*, **35**, 262–271.

Lin, C.-C., and Namba, T. (1981b). Pharmacognostical studies on the crude drugs of Orchidaceae from Taiwan: VII. On Kim-soan-lian (2). *Shoyakugaku Zasshi*, **35**, 272–286.

Lin, C.-C., and Namba, T. (1981c). Pharmacognostic studies on the crude drugs of Orchidaceae from Taiwan: 7.Chheng-chioh-nng. *Shoyakugaku Zasshi*, **35**, 303–315.

Lin, C.-C., and Namba, T. (1982a). Pharmacognostical studies on the crude drugs of Orchidaceae from Taiwan. IX. On 'Chheng-thian-liong-thiau'. *Shoyakugaku Zasshi*, **36**, 98–103.

Lin, C.-C., and Namba, T. (1982b). Pharmacognostical studies on the crude drugs of Orchidaceae from Taiwan: 10, On Koan-lan-hioh. *Shoyakugaku Zasshi*, **36**, 119–131. [In Japanese.]

Linder, H. P., and Kurzweil, H. (1990). Floral morphology and phylogeny of the Disinae (Orchidaceae). *Bot. J. Linn. Soc.*, **102**, 287–302.

Linder, H. P., and Kurzweil, H. (1994). the phylogeny and classification of the Diseae (Orchidoideae: Orchidaceae). *Ann. Missouri Bot. Gard.*, **81**, 687–713.

Linder, H. P., and Kurzweil, H. (1996). Ontogeny and phylogeny of *Brownleea* (Orchidoideae: Orchidaceae). *Nordic J. Bot.*, **16**, 345–357.

Linder, H. P., and Kurzweil, H. (1999). *Orchids of southern Africa*. A.A. Balkema, Rotterdam.

Lindley, J. (1830–40). *The genera and species of orchidaceous plants*. Ridgways, London.

Lindley, J. (1852–9). *Folia orchidacea*. J. Matthews, London.

Link, H. F. (1824). *Elementa philosophiae botanicae*. Haude and Spenersche, Berlin.

Link, H. (1849). Bemerkungen über den Bau der Orchideen, besonders der Vandeen. *Bot. Ztg.*, **7**, 745–750.

Link, H. (1849–50). Bemerkungen über den Bau der Orchideen, I,II. *Abhandl. Berliner Akad.* pp. 103–116 (1849); pp. 117–127 (1850) + 4 plates.

Linnaeus, C. (1753). *Species Plantarum*. Salvius, Stockholm.

Linnaeus, C. (1762–3). *Species Plantarum* (2nd edn). Salvius, Stockholm.

Liu, C., and Zhou, X. (1987). Studies on the endomycorrhiza of *Galeola faberi* Rolfe. *J. Wuhan Bot. Res.*, **5**, 104–111 + 4 plates. [In Chinese with English summary.]

Liu, H. X., Luo, Y., Liu, H., and Liu, H. (2010). Studies of mycorrhizal fungi of Chinese orchids. *Bot. Rev.*, **76**, 241–262.

Löv, L. (1926). Zur Kenntnis der Entfaltungszellen monokotyler Blätter. *Flora*, **120**, 283–343.

Luer, C. A. (1986). Icones Pleurothallidinarum: 1. Systematics of the Pleurothallidinae (Orchidaceae). *Mon. Syst. Bot.*, **15**, 1–81.

Lüning, B. (1964). Studies on Orchidaceae alkaloids. I. Screening of species for alkaloids I. *Acta Chem. Scand.*, **18**, 1507–1516.

Lüning, B. (1967). Studies on Orchidaceae alkaloids. IV. Screening of species for alkaloids 2. *Phytochemistry*, **6**, 857–861.

Lüning, B. (1974). Alkaloids of the Orchidaceae. pp. 349–382. In: *The orchids: scientific studies* (Ed. Withner, C. L.). John Wiley and Sons, New York.

Luo, Y.-B., and Chen, S.-C. (2000). The floral morphology and ontogeny of some Chinese representatives of orchid subtribe Orchidinae. *Bot. J. Linn. Soc.*, **134**, 529–548.

MacDougal, D. (1898). The mycorrhiza of *Aplectrum*. *Bull. Torrey Bot. Club*, **15**, 110–112.

MacDougal, D. (1899a). Symbiosis and saprophytism. *Bull. Torrey Bot. Club*, **26**, 511–530.

MacDougal, D. (1899b). Symbiotic saprophytism. *Ann. Bot.*, **13**, 1–47.

Macfarlane, J. (1892). A comparison of the minute structure of plant hybrids with that of their parents and its bearing on biological problems. *Trans. R. Soc. Edinb.*, **37**, 203–286.

Magnus, W. (1900). Studien an der endotrophen Mykorrhiza von *Neottia nidus avis* L. *Jb. Wiss. Bot.*, **35**, 205–272.

Maheshwari, P. (1950). *An introduction to the embryology of angiosperms*. McGraw-Hill, New York.

Mahyar, U. (1988). Variasi morfologi dan anatomi anggrek *Renanthera coccinea*. *Floribunda*, **1** (6), 23–24.

Malte, M. (1902). Untersuchungen über eigenartige Inhaltskörper bei den Orchideen. *Bih. Svenska Vetensk. Akad. Handl.*, **15**, 1–40.

Mangin, L. (1882). Origine et insertion des racines adventives et modifications corrélatives de la tige chez les monocotylédones. *Ann. Sci. Nat. Bot. Sér. 6*, **14**, 216–363 + plates 9–16.

Martos, F., Dulormne, M., Pailler, T., Bonfante, P., Faccio, A., Fournel, J., et al. (2009). Independent recruitment of saprotrophic fungi as mycorrhizal partners by tropical achlorophyllous orchids. *New Phytol.*, **184**, 668–681.

Masuhara, G., and Katsuya, K. (1992). Mycorrhizal differences between genuine roots and tuberous roots of adult plants of *Spiranthes sinensis* var. *amoena* (Orchidaceae). *Bot. Mag., Tokyo*, **105**, 453–460.

Mayer, J., Ribas, L., Bona, C., and Quoirin, M. (2008). Anatomia comparada das folhas e raizes de *Cymbidium* Hort. (Orchidaceae) cultivadas ex vitro e in vitro. *Acta Bot. Bras.*, **22**, 323–332.

McLennan, E. (1959). *Gastrodia sesamoides* R. Br. and its endophyte. *Aust. J. Bot.*, **7**, 225–229.

McLuckie, J. (1922). Studies in symboisis. I. The mycorrhiza of *Dipodium punctatum* R. Br. *Proc. Linn. Soc. N.S.W.*, **47**, 293–310.

McLuckie, J. (1923). Studies in symbiosis. V. A contribution to the physiology of *Gastrodia sesamoides* (R. Br.). *Proc. Linn. Soc. N.S.W.*, **48**, 436–448.

Meindl, U., and Kiermayer, O. (1987). Elektronenmikroskopischer Aspekt von Raphidenzellen der Orchidee *Haemaria discolor*. *Phyton (Austria)*, **26**, 157–164.

Meinecke, E. (1894). Beiträge zur Anatomie der Luftwurzeln der Orchideen. Dissertation, Universität Heidelberg.

Mejstrik, V. (1970). The anatomy of roots and mycorrhizae of the orchid *Denibrobium cunninghamii* Lindl. *Biol. Plant*, **12**, 105–109.

Mettenius, G. (1864). Ueber die Hymenophyllaceae. *Abhandl. Kon. Sachs. Ges. Wiss. Math. Phys. Cl.*, **7**, 403–504.

Metzler, W. (1924). Beiträge zur vergleichenden Anatomie blattsukkulenter Pflanzen. *Bot. Arch.*, **6**, 50–83.

Meyen, F. J. F. (1837). *Neues System der Pflanzenphysiologie*. Haude and Spenersche, Berlin.

Meyer, A. (1886). Ueber die Knollen der einheimischen Orchideen. VIII. Beitrag zur Kenntniss pharmaceutisch wichtiger Gewächse. *Arch. Pharm. Berl.*, **2**, 185, 233, 273, 321.

Milanez, F. R., Machado, R. D., and Costa, C. G. (1966). Observacoes sobre os elaioplastas da epiderme foliar da baunilha. *Anais Acad. Bras. Ci.*, **38**, 509–512 + 10 plates.

Möbius, M. (1886). Untersuchungen über die Stammanatomie einiger einheimischer Orchideen. *Ber. Deutsch. Bot. Ges.*, **4**, 284–292.

Möbius, M. (1887). Über den anatomischen Bau der Orchideenblätter und dessen Bedeutung für das System dieser Familie. *Jb. Wiss. Bot.*, **18**, 530–607.

Mohana Rao, P., Isaiah, J., and Khasim, S. (1988). Vegetative anatomy of *Panisea uniflora* Lindl. *J. Orchid Soc. India*, **2** (1–2), 1–7.

Mohana Rao, P., and Khasim, S. (1987a). Anatomy of three species of *Bulbophyllum* (Orchidaceae) with comments on their ecological adaptability and taxonomy. *Proc. Indian Acad. Sci. Plant Sci.*, **97**, 391–397.

Mohana Rao, P., and Khasim, S. (1987b). Anatomy of some members of Coelogyninae (Orchidaceae). *Phytomorphology*, **37**, 191–199.

Mohana Rao, P., Khasim, S., and Isaiah, J. (1991). Vegetative anatomy in some Orchidaceae: subfamily Epidendroideae. *J. Orchid Soc. India*, **5**, 85–91.

Mohana Rao, P., Kumari, S., Khasim, S., and Isaiah, J. (1989). Anatomy of some Sikkim Himalayan orchids with reference to their ecological adaptability. *Acta Bot. Indica*, **17**, 229–232.

Molisch, H. (1920). Aschenbild und Pflanzenverwandtschaft. *Sber. Akad. Wiss. Wien Math. Nat. Kl.*, **129**, 261–294.

Mollberg, A. (1884). Untersuchungen über Pilze in den Wurzeln der Orchideen. *Jena. Z. Naturwiss.*, **17**, 519–536.

Mollenhauer, H., and Larson, D. (1966). Developmental changes in raphide-forming cells of *Vanilla planifolia* and *Monstera deliciosa*. *J. Ultrastruct. Res.*, **16**, 55–70.

Møller, J., and Rasmussen, H. (1984). Stegmata in Orchidales: character state distribution and polarity. *Bot. J. Linn. Soc.*, **89**, 53–76.

Molvray, M., and Chase, M. (1999). Seed morphology. pp. 59–66. In: *Genera Orchidacearum*. Vol. 1. *General introduction, Apostasioideae, Cypripedioideae* (Ed. Pridgeon, A. M., Cribb, P. J., Chase, M. W., and Rasmussen, F. N.). Oxford University Press, Oxford.

Molvray, M., Kores, P. J., and Chase, M. W. (2000). Polyphyly of mycoheterotrophic orchids and functional influences on floral and molecular characters. pp. 441–448. In: *Monocots: sytematics and evolution* (Ed. Wilson, K. L., and Morrison, D. A.). CSIRO, Collingwood, Australia.

Moore, D. (1982). *Flora Europaea check-list and chromosome index*. Cambridge University Press, Cambridge.

Moore, R. J. (1970). *Index to Plant Chromosome Numbers for 1968*. Regnum Vegetabile, Vol. **68**. Oosthoek, Schelkema, and Holkema, Utrecht, The Netherlands.

Moore, R. J. (1971). *Index to Plant Chromosome Numbers for 1969*. Regnum Vegetabile, Vol. **77**. Oosthoek, Schelkema, and Holkema, Utrecht, The Netherlands.

Moore, R. J. (1972). *Index to Plant Chromosome Numbers for 1970*. Regnum Vegetabile, Vol. **84**. Oosthoek, Schelkema, and Holkema, Utrecht, The Netherlands.

Moore, R. J. (1973). *Index to Plant Chromosome Numbers for 1967–1971*. Regnum Vegetabile, Vol. **90**. Oosthoek, Schelkema, and Holkema, Utrecht, The Netherlands.

Moore, R. J. (1974). *Index to Plant Chromosome Numbers for 1972*. Regnum Vegetabile, Vol. **91**. Oosthoek, Schelkema, and Holkema, Utrecht, The Netherlands.

Moore, R. J. (1977). *Index to Plant Chromosome Numbers for 1973/74*. Regnum Vegetabile, Vol. **96**. Oosthoek, Schelkema, and Holkema, Utrecht, The Netherlands.

Moraes, C., Pedro, N., Diogo, J., and Marteline, M. (2007). Anatomia radicular de *Dendrobium* Stardust. *Bol. CAOB*, **68**.

Moreau, L. (1912). Etude de développement et de l'anatomie des *Pogonia* malgaches. *Rev. Gén. Bot.*, **24**, 97–112.

Moreau, L. (1913). Etude anatomique des Orchidées à pseudobulbes des pays chauds et de quelques autres espèces tropicales de plantes à tubercules. *Rev. Gén. Bot.*, **25**, 503–548.

Moreira, A., Filho, J., Zotz, G., and Isaias, R. (2009). Anatomy and photosynthetic parameters of roots and leaves of two shade-adapted orchids, *Dichaea cogniauxiana* Shltr. and *Epidendrum secundum* Jacq. *Flora*, **204**, 604–611.

Morisset, C. (1964). Structure et genèse du velamen dans les racines aériennes d'une orchidée epiphyte: le *Dendrobium nobile* Lindl. *Rev. Gén. Bot.*, **71**, 529–591.

Morris, M., Stern, W., and Judd, W. (1996). Vegetative anatomy and systematics of subtribe Dendrobiinae (Orchidaceae). *Bot. J. Linn. Soc.*, **120**, 89–144.

Mulay, B., Deshpande, B., and Williams, H. (1958). Study of velamen in some epiphytic and terrestrial orchids. *J. Indian Bot. Soc.*, **37**, 123–127.

Mulay, B., and Panikkar, T. (1956). Origin, development, and structure of velamen in the roots of some species of terrestrial orchids. *Proc. Rajasthan Acad. Sci.*, **6**, 31–48.

Mulay, B., Panikkar, T., and Prasad, M. (1954). *Curr. Sci.*, **24**, 416–417.

Mulay, B., Panikkar, T., and Prasad, M. (1956). Collateral vascular bundles in some orchid roots. *Proc. Rajasthan Acad. Sci.*, **6**, 70–73.

Müller, J. (1900). Ueber die Anatomie der Assimilationswurzeln von *Taeniophyllum zollingeri*. *Sber. Akad. Wiss. Wien*, **109**, 667–682.

Müller, R. (1908). Radix Senagae und ihre Substitutionen. *Pharm. Praxis*, **7**, 309–325.

Musampa Nseya, A., and Arends, J. (1995). Contribution à l'étude anatomique et caryologique des Orchidaceae: le genre *Cyrtorchis* Schltr. *Bull. Mus. Nat. Hist. Paris Sér. 4*, **17**, sect. B, Adansonia, 75–93.

Namba, T., and Lin, C. (1981a). Pharmacognostical studies on the crude drugs of Orchidaceae from Taiwan: 4. Chioh-hak, 1. *Shoyakugaku Zasshi*, **35**, 221–232.

Namba, T., and Lin, C. (1981b). Pharmacognostical studies on the crude drugs of Orchidaceae from Taiwan: 5. Chioh-hak, 2. *Shoyakugaku Zasshi*, **35**, 233–250.

Namba, T., Lin, C., and Kan, W. (1981b). Pharmacognostical studies on the crude drugs of Orchidaceae from Taiwan: II. On I-itam-hong. (2). *Shoyakugaku Zasshi*, **35**, 145–152.

Namba, T., Lin, C., and Kan, W. (1981c). Pharmacognostical studies on the crude drugs of Orchidaceae from Taiwan: III. On I-itam-hong (3). *Shoyakugaku Zasshi*, **35**, 153–156.

Namba, T., Lin, C., Kikuchi, T., and Kan, W. (1981a). Pharmacognostical studies on the crude drugs of Orchidaceae from Taiwan: I. On I-itam-hong (1). *Shoyakugaku Zasshi*, **35**, 138–144.

Napp-Zinn, K. (1953). Studien zur Anatomie einiger Luftwurzeln. *Ost. Bot. Z.*, **100**, 322–330.

Narayanaswamy, S. (1950). Occurrence of valamen and mycorrhiza in the subterranean roots of the orchid *Spiranthes australis* Lindl. *Curr. Sci.*, **19**, 250–251.

Nayar, B., Rai, R., and Vatsala, P. (1976). Dermal morphology of *Vanilla planifolia* Andr. and *V. wightii* Lind. *Proc. Indian Acad. Sci. B*, **84**, 173–179.

Nelson, E., and Sage, R. (2008). Functional constraints of CAM leaf anatomy: tight cell packing is associated with increased CAM function across a gradient of CAM expression. *J. Exp. Bot.*, **59**, 1841–1850.

Nelson, E., Sage, T., and Sage, R. (2005). Functional leaf anatomy of plants with crassulacean acid metabolism. *Funct. Plant Biol.*, **32**, 409–419.

Nelson, S., and Mayo, J. (1975). The occurrence of functional non-chlorophyllous guard cells in *Paphiopedilum* spp. *Can. J. Bot.*, **53**, 1–7.

Nestler, A. (1907). Das Sekret der Drusenhaare der Gattung *Cypripedium* mit besonderer Berücksichtigung seiner hautreizenden Wirkung. *Ber. Deutsch. Bot. Ges.*, **24**, 554–567 + plate XIV.

Neubauer, H. (1961a). Bau und Entwicklung der Luftwurzel von *Vanilla planifolia* Andr. *Beitr. Biol. Pfl.*, **36**, 239–253.

Neubauer, H. (1961b). Über Veränderungen im Blatte von *Vanilla planifolia* Andr., insbesondere über das Verschwinden der Elaoplasten und das Auftreten der Kristalle in den Epidermiszellen. *Beitr. Biol. Pfl.*, **36**, 255–271.

Neubauer, H. (1978). Über kortikale Idioblasten in Infloreszenzstielen von tropischen Orchideen. *Flora*, **167**, 121–125.

Neubig, K. M. (2012). Systematics of tribe Sobralieae (Orchidaceae): phylogenetics, pollination, anatomy and biogeography of a group of neotropical orchids. PhD thesis, University of Florida, Gainesville, FL.

Neubig, K. M., Whitten, W. M., Williams, N. H., Blanco, M. A., Endara, L., Burleigh, J. G., et al. (2012). Generic recircumscriptions of Oncidiinae (Orchidaceae: Cymbideae) based on maximum likelihood analysis of combined DNA datasets. *Bot. J. Linn. Soc.*, **168**, 117–146.

Neyland, R., Urbatsch, L., and Pridgeon, A. M. (1995). A phylogenetic analysis of subtribe Pleurothallidinae (Orchidaceae). *Bot. J. Linn. Soc.*, **117**, 13–28.

Nieuwdorp, P. (1972). Some observations with light and electron microscope on the endotrophic mycorrhiza of orchids. *Acta Bot. Neerl.*, **21**, 128–144.

Noack, F. (1892). Ueber Schleimranken in der Wurzelintercellularen einiger Orchideen. *Ber. Deutsch. Bot. Ges.*, **10**, 645–652.

Nobécourt, P. (1920). Sur la structure anatomique des tubercules des Ophrydées. *C. R. Acad. Sci. Paris*, **170**, 1593–1595.

Nobécourt, P. (1921). Les tubercules des Ophrydées. *Bull. Soc. Bot. Fr.*, **68**, 62–68.

Nobécourt, P. (1922). Etude sur les organes souterrains de quelques Ophrydées de Java. *Bull. Soc. Bot. Fr.*, **69**, 226–231.

Noel, A. (1974). Aspects of cell wall structure and the development of the velamen in *Ansellia gigantea* Reichb. f. *Ann. Bot.*, **38**, 495–504.

Noguera-Savelli, E., and Jauregui, D. (2011). Comparative leaf anatomy and phylogenetic relationships of 11 species of Laeliinae with emphasis on *Brassavola* (Orchidaceae). *Rev. Biol. Trop.*, **59**, 1047–1059.

Ogura, Y. (1953). Anatomy and morphology of the subterranean organs in some Orchidaceae. *J. Fac. Sci. Univ. Tokyo Sec. III*, **6**, 135–157.

Ogura, Y. (1964). Comparative morphology and classification of plants. *Phytomorphology*, **14**, 240–247.

Ojeda, I. (2003). Filogenia del complejo *Heterotaxis* Lindl. (Orchidaceae):evolución de la arquitectura vegetativa y los sindromes de polinización. Thesis, Centro de Invest. Ci. Yucatán, Mexico.

Olatunji, O., and Nengim, R. (1980). Occurrence and distribution of tracheoidal elements in the Orchidaceae. *Bot. J. Linn. Soc.*, **80**, 357–370.

Oliveira, V., and Sajo, M. (1999). Anatomia foliar de especies epifitas de Orchidaceae. *Rev. Bras. Bot.*, **22**, 365–374.

Oliveira, V., and Sajo, M. (2001). Morfo-anatomia caulinar de nove especies de Orchidaceae. *Acta Bot. Bras.*, **15**, 177–188.

Oliveira Pires, M. F. de, Semir, J., Mello Pinna, G., and Felix, L. (2003). Taxonomic separation of the genera *Prosthechea* and *Encyclia* (Laeliinae: Orchidaceae) using leaf and root anatomical features. *Bot. J. Linn. Soc.*, **143**, 293–303.

Oliver, W. (1930). New Zealand epiphytes. *J. Ecol.*, **18**, 1–50.

Olivier, L. (1881). Recherches sur l'appareil tégumentaire des racines. *Ann. Sci. Nat. Bot.*, **11**, 5–133.

Otero, J. T., Ackerman, J. D., and Bayman, P. (2002). Diversity and host specificity of endophytic *Rhizoctonia*-like fungi from tropical orchids. *Am. J. Bot.*, **89**, 1852–1858.

Oudemans, C. (1861). Ueber den Sitz der Oberhaut bei den Luftwurzeln der Orchideen. *Abhandl. K. Akad. Wiss. Amsterdam Math. Phys. Kl.*, **9**, 1–32.

Palla, E. (1889). Zur Anatomie der Orchideen-Luftwurzeln. *Sber. Akad. Wiss. Wien Math. Nat. Kl.*, **98**, 200–207.

Pan, Z.-J., Cheng, C.-C., Tsai, W., Chung, M.-C., Chen, W.-H., Hu, J.-M., and Chen, H.-H. (2011). The duplicated B-class MADS-box genes display dualistic characters in orchid floral organ identity and growth. *Plant Cell Physiol.*, **52**, 1515–1531.

Parrilla Diaz, A., and Ackerman, J. (1990). Epiphyte roots: anatomical correlates to environmental parameters in Puerto Rican orchids. *Orquidea (Mex.)*, **12** (1), 105–116.

Paschkis, H. (1880). Über zwei schleimliefernde Droguen. 1. Radix corniolae. 2. Tuber Aplectri hiemalis Nutt. *Pharm. Post.*, No. 16.

Payer, J.-B. (1857). *Traité d'organographie comparée de la fleur*. Victor Masson, Paris.

Pedersen, H. (1997). The genus *Dendrochilum* (Orchidaceae) in the Philippines—a taxonomic revision. *Opera Bot.*, **131**, 5–205.

Pedroso-de-Moraes, C., Souza-Leal, T. de, Brescansin, R. L., Pettini-Benelli, A., and Sajo, M. das G. (2012). Radicular anatomy of twelve representatives of the Catasetinae subtribe (Orchidaceae: Cymbideae). *Anais Acad. Bras. Cienc.*, **84**, 455–469.

Pfitzer, E. (1877). Beobachtungen über Bau und Entwicklung epiphytischer Orchideen. III. Über das Vorkommen von Kieselscheiben bei den Orchideen. *Flora*, **60**, 245–248.

Pfitzer, E. (1888). Untersuchungen über Bau und Entwicklung der Orchideenblüthe. I Theil: Cypripedilinae, Ophrydinae, Neottinae. *Jb. Wiss. Bot.*, **19**, 155–177.

Pfitzer, E. (1889). Orchidaceae. pp. 52–224. In: *Die natürlichen Pflanzenfamilien* (Ed. Engler, A., and Prantl, K.). Vol. 2, part 6. W. Engelmann, Leipzig.

Philipp, M. (1923). Über die verkorkte Abschlussgewebe der Monokotylen. *Bibl. Bot.*, **23** (92), 1–30.

Pirwitz, K. (1931). Physiologische und anatomische Untersuchungen an Speichertracheiden und Velamina. *Planta*, **14**, 19–76.

Pittman, H. (1929). Note on the morphology and endotrophic mycorrhiza of *Rhizanthella gardneri* Rogers and certain other Western Australian orchids. *J. R. Soc. West Aust.*, **15**, 71–79.

Podzorski, A. C., and Cribb, P. J. (1979). A revision of *Polystachya* sect. Cultriformes (Orchidaceae). *Kew Bull.*, **34**, 147–186.

Pompilian. (1883). *Contribution à l'étude des tiges de vanille.* Paris.

Porembski, S., and Barthlott, W. (1988). Velamen radicum micromorphology and classification of Orchidaceae. *Nordic J. Bot.*, **8**, 117–137.

Porsch, O. (1906). Orchidaceae. In: Wettstein, Ergebnisse der botanischen Expedition nach Süd-Brasilian 1901. *Denkschr. Akad. Wiss. Wien*, **79**, 1–75.

Poulsen, V. (1911). Bidrag til rodens anatomi. pp. 183–191. In: *Biologiske Arbejder tilegnede Eug. Warming* (Ed. Rosenvinge, L. K.). H. Hagerup, Copenhagen.

Prasad, M. (1960). Velamen in some terrestrial orchids. *J. Biol. Sci.*, **3** (1), 48–51.

Prete, C. D. and Miceli, P. (1999). Histoanatomical and taxonomical observations on some Central Mediterranean entities of *Orchis* Sect. Labellotrilobatae P. Vermeul, Subsections Masculae Newski and Provinciales Newski (Orchidee). *Caesiana Quaderno*, **12**, 21–44.

Pridgeon, A. M. (1978). Una revision de los generos *Coelia* y *Bothriochilus*. *Orquidea*, **7**, 57–94.

Pridgeon, A. M. (1981a). Absorbing trichomes in the Pleurothallidinae (Orchidaceae). *Am. J. Bot.*, **68**, 64–71.

Pridgeon, A. M. (1981b). Shoot anatomy of two additional species of *Dresslerella* (Orchidaceae). *Selbyana*, **5**, 274–278.

Pridgeon, A. M. (1982a). Diagnostic anatomical characters in the Pleurothallidinae (Orchidaceae). *Am. J. Bot.*, **69**, 921–938.

Pridgeon, A. M. (1982b). Numerical analyses in the classification of the Pleurothallidinae (Orchidaceae). *Bot. J. Linn. Soc.*, **85**, 103–131.

Pridgeon, A. M. (1984). On the integrity of *Restrepiopsis* (Orchidaceae). *Selbyana*, **7**, 312–314.

Pridgeon, A. M. (1987). The velamen and exodermis of orchid roots. pp. 139–192. In: *Orchid biology: Reviews and perspectives, IV* (Ed. Arditti, J.). Comstock Publishing Associates, Ithaca, NY.

Pridgeon, A. M. (1993). Systematic leaf anatomy of *Caladenia* (Orchidaceae). *Kew Bull.*, **48**, 533–543.

Pridgeon, A. M. (1994a). Systematic leaf anatomy of Caladeniinae (Orchidaceae). *Bot. J. Linn. Soc.*, **114**, 31–48.

Pridgeon, A. M. (1994b). Multicellular trichomes in tribe Diurideae (Orchidaceae): systematic and biological significance. *Kew Bull.*, **49**, 569–579.

Pridgeon, A. M. (1999). Anatomy. pp. 24–32. In: *Genera Orchidacearum.* Vol. 1. *General introduction, Apostasioideae, Cypripedioideae* (Ed. Pridgeon, A. M., Cribb, P. J., Chase, M. W., and Rasmussen, F. N.). Oxford University Press, Oxford.

Pridgeon, A. M. (2001). Anatomy of Diurideae. pp. 59–61, 65–56, 84–55, 115–116, 125, 134–115, 155–116, 175–116, 195, 198. In: *Genera Orchidacearum.* Vol. 2. *Orchidoideae (Part one)* (Ed. Pridgeon, A. M., Cribb, P. J., Chase, M. W., and Rasmussen, F. N.). Oxford University Press, Oxford.

Pridgeon, A. M., Bateman, R., Cox, A., Hapeman, J., and Chase, M. W. (1997). Phylogenetics of subtribe Orchidinae (Orchidoideae, Orchidaceae) based on nuclear its sequences. 1. Intergeneric relationships and polyphyly of *Orchis* sensu lato. *Lindleyana*, **12**, 89–109.

Pridgeon, A. M., and Chase, M. W. (1995). Subterranean axes in tribe Diurideae (Orchidaceae): morphology, anatomy, and systematic significance. *Am. J. Bot.*, **82**, 1473–1495.

Pridgeon, A. M., and Chase, M. W. (1998). Phylogenetics of subtribe Catasetinae (Orchidaceae) from nuclear and chloroplast DNA sequences. pp. 275–281. In: *Proc. 25th World Orchid Conf., Rio de Janeiro, 1996* (Ed. Pereira, C. E. de B.). Naturalia Publications, Turriers, France.

Pridgeon, A. M., Cribb, P. J., Chase, M. W., and Rasmussen, F. N. (1999). *Genera Orchidacearum.* Vol. 1. *General introduction, Apostasioideae, Cypripedioideae.* Oxford University Press, Oxford.

Pridgeon, A. M., Cribb, P. J., Chase, M. W., and Rasmussen, F. N. (2001). *Genera Orchidacearum.* Vol. 2. *Orchidoideae (Part one).* Oxford University Press, Oxford.

Pridgeon, A. M., Cribb, P. J., Chase, M. W., and Rasmussen, F. N. (2003). *Genera Orchidacearum.* Vol. 3. *Orchidoideae (Part two), Vanilloideae.* Oxford University Press, Oxford.

Pridgeon, A. M., Cribb, P. J., Chase, M. W., and Rasmussen, F. N. (2005). *Genera Orchidacearum.* Vol. 4. *Epidendroideae (Part one).* Oxford University Press, Oxford.

Pridgeon, A. M., Cribb, P. J., Chase, M. W., and Rasmussen, F. N. (2009). *Genera Orchidacearum.* Vol. 5. *Epidendroideae (Part two).* Oxford University Press, Oxford.

Pridgeon, A. M., Cribb, P. J., Chase, M. W., and Rasmussen, F. N. (2014). *Genera Orchidacearum.* Vol. 6. *Epidendroideae (Part three).* Oxford University Press, Oxford.

Pridgeon, A. M., Solano, R., and Chase, M. W. (2001). Phylogenetic relationships in subtribe Pleurothallidinae (Orchidaceae): combined evidence from nuclear and plastid DNA sequences. *Am. J. Bot.*, **88**, 2286–2308.

Pridgeon, A. M., and Stern, W. L. (1982). Vegetative anatomy of *Myoxanthus* (Orchidaceae). *Selbyana*, **7**, 55–63.

Pridgeon, A. M., and Stern, W. L. (1983). Ultrastructure of osmophores in *Restrepia* (Orchidaceae) *Am. J. Bot.*, **70**, 1233–1243.

Pridgeon, A. M., and Stern, W. L. (1985). Osmophores of *Scaphosepalum* (Orchidaceae). *Bot. Gaz.*, **146**, 115–123.

Pridgeon, A. M., Stern, W. L., and Benzing, D. H. (1983). Tilosomes in roots of Orchidaceae: morphology and systematic occurrence. *Am. J. Bot.*, **70**, 1365–1377.

Pridgeon, A. M., and Williams, N. H. (1979). Anatomical aspects of *Dresslerella* (Orchidaceae). *Selbyana*, **5**, 120–134.

Prillieux, E. (1856). Structure anatomique et mode de végétation du *Neottia nidus avis*. *Ann. Sci. Nat. Bot. Sér. 4*, **5**, 267–282.

Prillieux, E. (1865). Etude sur la nature, l'organisation et la structure des bulbes des Ophrydées. *Ann. Sci. Nat. Bot. Sér. 5*, **4**, 265–289.

Prillieux, E. (1879). Sur un détail de la structure de l'enveloppe des racines aériennes des Orchidées. *Bull. Soc. Bot. Fr.*, **26**, 275–282.

Prychid, C., Rudall, P., and Gregory, M. (2004). Systematics and biology of silica bodies in Monocotyledons. *Bot. Rev.*, **69**, 377–440.

Puri, H. (1971). Macro- and micromorphology of the tuber of *Eulophia hormusjii* Duthie. *Am. Orchid Soc. Bull.*, **40**, 704–706.

Queva, C. (1894; 1895). Anatomie de la tige de la vanille. *C. R. Ass. Fr. Av. Sci. Caen*, **161**, 162–163; **162**, 577–583.

Raciborski, M. (1898). Biologische Mittheilungen aus Java. *Flora*, **85**, 325–361.

Ramanujam, P., and Rao, T. A. (1998). Studies of foliar idioblasts and their systematic significance in a few taxa of Orchidaceae. pp. 23–27. In: *National seminar on in situ conservation and commercialisation of orchids of Western Ghats.*

Ramirez, S., Gravendeel, B., Singer, R., Marshall, C., and Pierce, N. (2007). Dating the origin of the Orchidaceae from a fossil orchid with its pollinator. *Nature*, **448**, 1042–1045.

Ramsay, R. R., Dixon, K. W., and Sivasithamparam, K. (1986). Patterns of infection and endophytes associated with Western Australian orchids. *Lindleyana*, **1**, 203–214.

Rao, T. A. (1998). *Conservation of wild orchids of Kodagu in the Western Ghats*. Centre for Technology Development and Agricultural Technologies and Services Private Ltd, Bangalore and New Delhi; Karnataka Association for Advancement of Science, Bangalore; World Wide Fund for Nature—India, New Delhi.

Rao, V. S. (1969 [1970]). The floral anatomy and relationship of the rare Apostasias. *J. Indian Bot. Soc.*, **48**, 374–386.

Rao, V. S. (1974). The relationship of the Apostasiaceae on the basis of floral anatomy. *Bot. J. Linn. Soc.*, **68**, 319–327.

Rasmussen, F. N. (1977). The genus *Corymborkis* Thou. (Orchidaceae): a taxonomic revision. *Bot. Tidsskr.*, **71**, 161–192.

Rasmussen, F. N. (1982). The gynostemium of the neottioid orchids. *Opera Bot.*, **65**, 1–96.

Rasmussen, F. N. (1985). The gynostemium of *Bulbophyllum ecornutum* (J. J. Smith) J. J. Smith (Orchidaceae). *Bot. J. Linn. Soc.*, **91**, 447–456.

Rasmussen, F. N. (1986). On the various contrivances by which pollinia are attached to viscidia. *Lindleyana*, **1**, 21–32.

Rasmussen, H. (1981a). The diversity of stomatal development in Orchidaceae subfamily Orchidoideae. *Bot. J. Linn. Soc.*, **82**, 381–393.

Rasmussen, H. N. (1981b). Terminology and classification of stomata and stomatal development. *Bot. J. Linn. Soc.*, **83**, 199–212.

Rasmussen, H. (1986). An aspect of orchid anatomy and adaptationism. *Lindleyana*, **1**, 102–107.

Rasmussen, H. (1987). Orchid stomata—structure, differentiation, function, and phylogeny. pp. 105–138. In: *Orchid biology: Reviews and perspectives, IV* (Ed. Arditti, J.). Comstock Publishing Associates, Ithaca, NY.

Rasmussen, H. N. (1990). Cell differentiation and mycorrhizal infection in *Dactylorhiza majalis* (Rchb. f.) Hunt and Summerh. (Orchidaceae) during germination in vitro. *New Phytol.*, **116**, 137–147.

Rasmussen, H. N. (1995). *Terrestrial orchids: from seed to mycotrophic plant*. Cambridge University Press, Cambridge.

Rasmussen, H., and Whigham, D. (2002). Phenology of roots and mycorrhiza in orchid species differing in phototrophic strategy. *New Phytol.*, **154**, 797–807.

Raunkiaer, C. (1895–9). *De danske blomsterplanters Naturhistorie*. Vol. I. H. Hagerup, Kjøbenhavn.

Ravnik, V., and Susnik, F. (1964). Contribution to the morphology and systematics of the genus *Nigritella* Rich. III. Morphology of the genera *Nigritella* and *Gymnadenia*. *Biol. Vestnik.*, **12**, 65–75.

Reiche, C. (1910). Orchidaceae Chilenses. Ensayo de una monografia de las orquideas de Chile. *Anales Mus. Nac. Chile Sec. Bot.*, **18**, 1–88.

Reichenbach, H. G. (1854–1900). *Xenia Orchidacea* (3 vols). Brockhaus, Leipzig.

Reinke, J. (1873). Zur Kenntniss des Rhizoms von *Corallorhiza* und *Epipogon*. *Flora*, **56**, 145–152, 161–167, 177–184, 209–224.

Richard, L. C. (1817). *De orchideis Europaeis annotationes, praesertim ad genera dilucidanda spectantes*. A. Belin, Paris.

Richter, A. (1901). Physiologisch-anatomische Untersuchungen über Luftwurzeln mit besonderer Berücksichtigung der Wurzelhaube. *Bibl. Bot.*, **54**, 1–50.

Rojas Leal, A. (1993). Anatomia foliar comparada de *Lemboglossum* Halbinger (Orchidaceae: Oncidiinae) y generos relacionados. Thesis, Universidad Nacional Autónoma de México, Mexico.

Rolfe, R. (1889). A morphological and systematic review of the Apostasieae. *J. Linn. Soc. Lond. (Bot.)*, **25**, 211–243.

Romero, G. (1990). Phylogenetic relationships in subtribe Catasetinae (Orchidaceae, Cymbidieae). *Lindleyana*, **5**, 160–181.

Rosinski, M. (1992). Untersuchungen zur funktionellen Anatomie der Laubblattstrukturen epiphytischer Coelogyninae und Eriinae (Orchidaceae). Dissertation, Universität des Saarlandes, Germany.

Rosso, S. (1966). The vegetative anatomy of the Cypripedioideae (Orchidaceae). *J. Linn. Soc. (Bot.)*, **59**, 309–334.

Roth, I., and Merida de Bifano, T. (1979). Morphological and anatomical studies of leaves of the plants of a Venezuelan cloud forest. II. Stomata density and stomatal patterns. *Acta Biol. Venez.*, **10**, 69–107.

Rotor, G., and MacDaniels, L. (1951). Flower bud differentiation and development in *Cattleya labiata* Lindl. *Am. J. Bot.*, **38**, 147–152.

Roux, P. (1954). Chapter 3. In: *Le vanillier et la vanille dans le monde* (by Bouriquet, G.). P. Lechevalier, Paris.

Roy, M., Yagame, T., Yamamoto, M., *et al.* (2009). Ectomycorrhizal *Inocybe* species associate with the mycoheterotrophic orchid *Epipogium aphyllum*. *Ann. Bot.*, **104**, 595–610.

Rŭgina, R., Lazar, M., and Neagu, G. (1987). Etude histo-anatomique de quelques Orchidées exotiques du Jardin Botanique de Iassy. *Anal. Stiint. Univ. Al. I. Cuza, Iasi*, **33**, 1–4.

Rüter, B., and Stern, W. (1994). An assessment of quantitative features of velamen stratification and protoxylem strands in roots of Orchidaceae. *Lindleyana*, **9**, 219–225.

Rutter, J., and Willmer, C. (1979). A light and electron microscopy study of the epidermis of *Paphiopedilum* spp. with emphasis on stomatal ultrastructure. *Plant Cell Environ.*, **2**, 211–219.

Sakharam Rao, J. (1953). Role of velamen tissue in roots of orchids. *Sci. Cult.*, **19**, 97–99.

Salazar, G. A., Cabrera, L. I., Madriñán, S., and Chase, M. W. (2009). Phylogenetic relationships of Cranichidinae and Prescottiinae (Orchidaceae, Cranichideae) inferred from plastid and nuclear DNA sequences. *Ann. Bot.*, **104**, 403–416.

Salmia, A. (1989a). Features of endomycorrhizal infection of chlorophyll-free and green forms of *Epipactis helleborine* (Orchidaceae). *Ann. Bot. Fenn.*, **26**, 15–26.

Salmia, A. (1989b). General morphology and anatomy of chlorophyll-free and green forms of *Epipactis helleborine* (Orchidaceae). *Ann. Bot. Fenn.*, **26**, 95–105.

Salokhin, A., Volkova, S., and Gorovoy, P. (2005). The leaf stomatography of short-rhizomatous species of *Cypripedium* (Orchidaceae) from Eastern Siberia and the Far East. *Turczaninowia*, **8** (2), 69–74.

Samuel, J., and Bhat, R. (1994). Epidermal structure, organographic distribution and ontogeny of stomata in vegetative and floral organs of *Stenoglottis fimbriata* (Orchidaceae). *S. Afr. J. Bot.*, **60**, 113–117.

Sandoval-Zapotitla, E. (1993). Anatomia foliar de *Cuitlauzina pendula*. *Orquidea*, **13**, 181–190.

Sandoval-Zapotitla, E., Garcia-Cruz, J., Terrazas, T., and Villaseñor, J. L. (2010a). Relaciones filogenéticas de la subtribu Oncidiinae (Orchidaceae) inferidas a partir de caracteres estructurales y secuencias de ADN (ITS y matK); un enfoque combinado. [Phylogenetic relationships of the subtribe Oncidiinae (Orchidaceae) inferred from structural and DNA sequences (matK, ITS): a combined approach.] *Rev. Mex. Biodiversidad*, **81**, 263–279.

Sandoval-Zapotitla, E., and Terrazas, T. (2001). Leaf anatomy of 16 taxa of the *Trichocentrum* clade (Orchidaceae: Oncidiinae). *Lindleyana*, **16**, 81–93.

Sandoval-Zapotitla, E., Terrazas, T., and Villaseñor, J. L. (2010b). Diversidad de inclusiones minerales en la subtribu Oncidiinae (Orchidaceae). *Rev. Biol. Trop.*, **58**, 733–755.

Sanford, W., and Adanlawo, I. (1973). Velamen and exodermis characters of West African epiphytic orchids in relation to taxonomic grouping and habitat tolerance. *Bot. J. Linn. Soc.*, **66**, 307–321.

Sarin, Y. (1960). Hetero-archic roots in *Eulophia herbacea* Lindl. *Proc. Rajasthan Acad. Sci.*, **7**, 48–51.

Sasikumar, B. (1973a). Morphological and anatomical peculiarities in some terrestrial orchids. *Glimpses Plant Res.*, **1**, 148–159.

Sasikumar, B. (1973b). A reinvestigation of the collateral vascular bundles in orchid roots. *Sci. Cult.*, **39**, 126–127.

Sasikumar, B. (1974). Anatomical peculiarities in the roots of some Orchidaceae. *J. Univ. Bombay Sci.*, **43**, 154–163.

Sasikumar, B. (1975a). Morphology and anatomy of the subterranean organs of some Orchidaceae. *J. Univ. Bombay Sci.*, **44**, 77–86.

Sasikumar, B. (1975b). Polystele in terrestrial orchids. *Sci. Cult.*, **41**, 216–217.

Sasikumar, B., and Navalkar, B. (1973 [1974]). Morphology and anatomy of the vegetative organs of *Microstylis rheedii* Wight and *Liparis walkeriae* Grah. *J. Univ. Bombay*, **42**, 93–103.

Scatena, V., and Nunes, A. (1996). Anatomia de *Pleurothallis rupestris* Lindl. (Orchidaceae) dos campos rupestres do Brasil. *Bol. Bot. Univ. São Paulo*, **15**, 35–43.

Schacht, H. (1856). *Lehrbuch der Anatomie und Physiologie der Gewächse*. G. W. F. Müller, Berlin.

Schelpe, S., and Stewart, J. (1990). *Dendrobiums: an introduction to the species in cultivation*. Orchid Sundries Ltd., Stour Provost, UK.

Schimper, A. (1884). Über Bau und Lebensweise der Epiphyten Westindiens. *Bot. Zbl.*, **17**, 192–195, 223–227, 253–258, 284–294, 319–326, 350–359, 381–389.

Schimper, A. (1888). Die epiphytische Vegetation Amerikas. *Bot. Mitt. Tropen Jena*, **2**, 1–162.

Schindler, H., and Toth, A. (1950). Zur Anatomie des Blattes von *Coelogyne flaccida*. *Phyton (Austria)*, **2**, 11–22.

Schlechter, R. (1918). Versuch einer natürlicher Neuordnung der afrikanischer angraekoiden Orchidaceen. *Beih. Bot. Centralbl.*, **36**, 62–181.

Schlechter, R. (1926). Das System der Orchidaceen. *Notizbl. Bot. Gart. Mus. Berlin Dahlem*, **9**, 563–591.

Schleiden, M. J. (1845–6). *Grundzüge der wissenschaftlichen Botanik* (2 vols). Wilhelm Englemann, Leipzig.

Schleiden, M. J. (1849). *Principles of scientific botany*. Translated from German 2nd edn by E. Lankester. Longman, Brown, Green, and Longmans, London.

Schmid, R., and Schmid, M. (1977). Fossil history of the Orchidaceae. pp. 25–45. In: *Orchid biology: Reviews and perspectives, 1* (Ed. Arditti, J.). Comstock Publishing Associates, Ithaca, NY.

Schmucker, T. (1927). Beiträge zur Kenntnis einer merkwurdigen Orchidee, *Haemaria discolor* Lindl. *Flora*, **121**, 157–171.

Schnepf, E., Deichgraber, G., and Barthlott, W. (1983). On the fine structure of the liquid producing floral gland of the orchid, *Coryanthes speciosa*. *Nordic J. Bot.*, **3**, 479–491.

Schuiteman, A. (1997). Revision of *Agrostophyllum* section Appendiculopsis (Orchidaceae), with notes on the systematics of *Agrostophyllum*. *Orchid Mon.*, **8**, 1–20, 175, 181–187 + plates 171–172.

Scrugli, A., Cogoni, A., and Riess, S. (1991). Beitrag zur Kenntnis der Mykorrhiza in der Gattung *Limodorum* Boehmer in C. G. Ludwig (Orchidaceae): *Limodorum trabutianum* Battand. *Orchidee*, **42**, 99–103.

Seberg, O., Petersen, G., Davis, J. I., Pires, J. C., Stevenson, D. W., Chase, M. W., et al. (2012). Phylogeny of the Asparagales based on three plastid and two mitochondrial genes. *Am. J. Bot.*, **99**, 875–889.

Segonzac, G. D. de (1958). L'ontogénie du phloème cheq *Vanilla planifolia* Andr. *Rev. Cytol. Biol. Veg.*, **19**, 153–184.

Seidenfaden, G. (1980). Orchid genera in Thailand: 9. *Flickingeria* Hawkes and *Epigeneium* Gagnep. *Dansk Bot. Ark.*, **34** (1), 1–104.

Seidenfaden, G. (1985). Orchid genera in Thailand: 12. *Dendrobium* Sw. *Opera Bot.*, **83**, 1–250.

Seidenfaden, G. (1992). The orchids of Indochina. *Opera Bot.*, **114**, 1–502.

Seidenfaden, G., and Wood, J. (1992). *The orchids of peninsular Malaysia and Singapore. A revision of R.E. Holttum Orchids of Malaya*. Olsen and Olsen, Fredensborg.

Senthilkumar, S., Krishnamurthy, K., and Vengadeshwari, G. (1998). Studies on the mycorrhizal association of the ornamental orchid, *Papilionanthe subulata* (J. Koenig) Garay. *Philipp. J. Sci.*, **127** (3), 189–199.

Sevgi, E., Altundag, E., and Karo, O. (2012). Studies on the morphology, anatomy and ecology of *Anacamptis pyramidalis* (L.) L.C.M.Richard (Orchidaceae) in Turkey. *Pakistan J. Bot.*, **44** (special issue 1), 135–141.

Sgarbi, E., and Del Prete, C. (2005). Histo-anatomical observations on some *Orchis* species (Orchidaceae) from the eastern Mediterranean. *Flora Medit.*, **15**, 321–329.

Sharman, B. (1939). The development of the sinker of *Orchis mascula* Linn. *J. Linn. Soc. (Bot.)*, **52**, 145–158.

Shershevskaya, E. Y. (1963). [On the anatomy of three *Cypripedium* species.] *Bot. Zh. SSSR*, **48**, 1692–1695. [Russian only.]

Shirley, J., and Lambert, C. (1918). The stems of climbing plants. *Proc. Linn. Soc. N.S.W.*, **43**, 600–609.

Siebe, M. (1903). Über den anatomischen Bau der Apostasiinae. Dissertation, Universität Heidelberg, Germany.

Silva, I., Meira, R., Azevedo, A., and Euclydes, R. (2006). Estratégias anatomicas foliares de treze espécies de Orchidaceae ocorrêntes em um campo de altitude no Parque Estadual de Serra do Brigadeiro (PESB)—MG, Brasil. [Strategies of anatomy from Orchidaceae species occurring in a 'high altitude grassland' in the State Park of Serra do Brigadeiro (PESB)—Minais Gerais State, Brazil.] *Acta Bot. Bras.*, **20**, 741–750.

Singh, A., and Duggal, S. (2009). Medicinal orchids: an overview. *Ethnobot. Leafl.*, **13**, 351–363.

Singh, H. (1981). Development and organisation of stomata in Orchidaceae. *Acta Bot. Indica*, **9**, 94–100.

Singh, H. (1983). Vessels in Orchidaceae. *Curr. Sci.*, **52**, 688–689.

Singh, H. (1986). Anatomy of root in some Orchidaceae. *Acta Bot. Indica*, **14**, 24–32.

Singh, V., and Singh, H. (1974). Organisation of stomatal complex in some Orchidaceae. *Curr. Sci.*, **43**, 490–491.

Slaytor, M. B. (1977). The distribution and chemistry of alkaloids in the Orchidaceae. pp. 95–115. In: *Orchid biology: Reviews and perspectives* (Ed. Arditti, J.). Comstock Publishing Associates, Ithaca, NY.

Sohma, K. (1954). Über die Mykorrhiza von *Sarcochilus japonicus*. *Ecol. Rev. (Japan)*, **13**, 257–261.

Solereder, H., and Meyer, F. (1930). *Systematische Anatomie der Monokotyledonen. VI. Microspermae*. Borntraeger, Berlin.

Solereder, H., and Meyer, F. (1969). *Systematic anatomy of the monocotyledons. VI. Microspermae*. Translated by A. Herzberg and B. Golek. Israel Program for Scientific Translations, Jerusalem.

Soto Arenas, M. A. (2003). Vanilla (in part). pp. 332–334. In: *Genera Orchidacearum*. Vol. 3. *Orchidoideae (Part two), Vanilloideae* (Ed. Pridgeon, A. M., Cribb, P. J., Chase, M. W., and Rasmussen, F. N.). Oxford University Press, Oxford.

Souèges, R. (1936–9). *Exposés d'embryologie et de morphologie végétale*; Vol. IV. *La segmentation. Les blastomères*; Vol. VIII. *Les lois du développement*; Vol. IX. *Embryogénie et classification. Essai d'un systéme embryogénique*. Hermann et Cie., Paris.

Sprenger, M. (1904). Über den anatomischen Bau von Bolbophyllinae. Dissertation, Universität Heidelberg, Germany.

St Arnaud, M., and Barabé, D. (1989). Comparative analysis of the flower vascularization of some *Cypripedium* species (Orchidaceae). *Lindleyana*, **4**, 146–153.

St George, I. (1994). The Pacific genus *Earina*: how many species are there? *N.Z. Native Orchid Group J.*, No. 50, 1–7.

Stancato, G., Mazzoni Viveiros, S., and Luchi, A. (1999). Stomatal characteristics in different habitat forms of Brazilian species of *Epidendrum* (Orchidaceae). *Nordic J. Bot.*, **19**, 271–275.

Staudermann, W. (1924). Die Haare der Monokotylen. *Bot. Arch.*, **8**, 105–184.

Stebbins, G., and Khush, G. (1961). Variation in the organization of the stomatal complex in the leaf epidermis of monocotyledons and its bearing on their phylogeny. *Am. J. Bot.*, **48**, 51–59.

Stern, W. L. (1997a). Vegetative anatomy of subtribe Orchidinae (Orchidaceae). *Bot. J. Linn. Soc.*, **124**, 121–136.

Stern, W. L. (1997b). Vegetative anatomy of subtribe Habenariinae (Orchidaceae). *Bot. J. Linn. Soc.*, **125**, 211–227.

Stern, W. L. (1999). Comparative vegetative anatomy of two saprophytic orchids from tropical America: *Wullschlaegelia* and *Uleiorchis*. *Lindleyana*, **14**, 136–146.

Stern, W. L., Aldrich, H. C., McDowell, L. M., Morris, M. W., and Pridgeon, A. M. (1993a). Amyloplasts from cortical root cells of Spiranthoideae (Orchidaceae). *Protoplasma*, **172**, 49–55.

Stern, W. L., and Carlsward, B. S. (2004). Vegetative constants in the anatomy of epiphytic orchids. *Orchid Rev.*, **112**, 119–122.

Stern, W. L., and Carlsward, B. S. (2006). Comparative vegetative anatomy and systematics of the Oncidiinae (Maxillarieae, Orchidaceae). *Bot. J. Linn. Soc.*, **152**, 91–107.

Stern, W. L., and Carlsward, B. S. (2008). Vegetative anatomy of Calypsoeae (Orchidaceae). *Lankesteriana*, **8**, 105–112.

Stern, W. L., and Carlsward, B. S. (2009). Comparative vegetative anatomy and systematics of Laeliinae (Orchidaceae). *Bot. J. Linn. Soc.*, **160**, 21–41.

Stern, W. L., Cheadle, V. I., and Thorsch, J. (1993b). Apostasiads, systematic anatomy, and the origins of Orchidaceae. *Bot. J. Linn. Soc.*, **111**, 411–455.

Stern, W. L., Curry, K., and Pridgeon, A. M. (1987). Osmophores of *Stanhopea* (Orchidaceae). *Am. J. Bot.*, **74**, 1323–1331.

Stern, W. L., and Judd, W. S. (1999). Comparative vegetative anatomy and systematics of *Vanilla* (Orchidaceae). *Bot. J. Linn. Soc.*, **131**, 353–382.

Stern, W. L., and Judd, W. S. (2000). Comparative anatomy and systematics of the orchid tribe Vanilleae excluding *Vanilla*. *Bot. J. Linn. Soc.*, **134**, 179–202.

Stern, W. L., and Judd, W. S. (2001). Comparative anatomy and systematics of Catasetinae (Orchidaceae). *Bot. J. Linn. Soc.*, **136**, 153–178.

Stern, W L., and Judd, W. S. (2002). Systematic and comparative anatomy of Cymbidieae (Orchidaceae). *Bot. J. Linn. Soc.*, **139**, 1–27.

Stern, W. L., Judd, W. S., and Carlsward, B. S. (2004). Systematic and comparative anatomy of Maxillarieae (Orchidaceae, sans Oncidiinae. *Bot. J. Linn. Soc.*, **144**, 251–274.

Stern, W. L., and Morris, M. W. (1992). Vegetative anatomy of *Stanhopea* (Orchidaceae) with special reference to pseudobulb water-storage cells. *Lindleyana*, **7**, 34–53.

Stern, W. L., Morris, M. W., and Judd, W. S. (1994). Anatomy of the thick leaves in *Dendrobium* section Rhizobium (Orchidaceae). *Int. J. Plant Sci.*, **155**, 716–729.

Stern, W. L., Morris, M. W., Judd, W. S., Pridgeon, A. M., and Dressler, R. (1993c). Comparative vegetative anatomy and systematics of Spiranthoideae (Orchidaceae). *Bot. J. Linn. Soc.*, **113**, 161–197.

Stern, W. L., and Pridgeon, A. M. (1984). Ramicaul, a better term for the pleurothallid 'secondary stem'. *Am. Orchid Soc. Bull.*, **53**, 397–401.

Stern, W L., Pridgeon, A. M., and Luer, C. (1985). Stem structure and its bearing on the systematics of Pleurothallidinae (Orchidaceae). *Bot. J. Linn. Soc.*, **91**, 457–471.

Stern, W L., and Warcup, J. (1994). Root tubercles in apostasiad orchids. *Am. J. Bot.*, **81**, 1571–1575.

Stern, W. L., and Whitten, W. M. (1999). Comparative vegetative anatomy of Stanhopeinae (Orchidaceae). *Bot. J. Linn. Soc.*, **129**, 87–103.

Stevens, M. (1990). The multifunctional aerial root of epiphytic orchids. pp. 285–288. In: *Proc. 13th World Orchid Conf., Auckland, 1990* (Ed. Kernohan, J.). World Orchid Conference Trust, Auckland.

Stojanow, N. (1916). Über die vegetative Fortpflanzung der Ophrydineen. *Flora*, **109**, 1–39 + plates 1 and 2.

Stpiczyńska, M. (1997). The structure of nectary of *Platanthera bifolia* L. Orchidaceae. *Acta Soc. Bot. Pol.*, **66**, 05–11.

Suárez, J. P., Weiss, M., Oberwinkler, F., and Kottke, I. (2009). Epiphytic orchids in a mountain rain forest in southern Ecuador harbour groups of mycorrhiza-forming Tulasnellales and Sebacinales subgroup B (Basidiomycota). pp. 184–196. In: *Proc. 2nd Sci. Conf. on Andean Orchids* (Ed. Pridgeon, A. M., and Suárez, J. P.). Universidad Técnica Particular de Loja, Loja, Ecuador.

Subrahmanyam, N. (1974). Velamen in pseudobulb. *Sci. Cult.*, **40**, 365–366.

Sulistiarini, D. (1986). Anatomi daua dan status kedudukan taksonoma *Luisia latipetala*. [Leaf anatomy and the taxonomic status of *Luisia latipetala*.] *Berita Biol.*, **3** (4), 143–145.

Sun, A. (1995). Investigations on leaves of *Cymbidium*, *Paphiopedilum*, *Dendrobium* under scanning electron microscope. *J. Wuhan Bot. Res.*, **13**, 289–294 + 3 plates.

Sun, T. X., Hu, Y. S., and Lang, K. Y. (1999). A study on micromorphological characters of leaf epidermis of *Neottianthe* in China. *Acta Bot. Yunnan.*, **21**, 57–62 + 2 plates.

Suryanarayana Raju, M. (1996). Morpho-anatomical studies of the endemic orchid *Vanilla wightiana* Lindl. (Orchidaceae). *Phytomorphology*, **46**, 371–375.

Swamy, B. (1948). Vascular anatomy of orchid flowers. *Bot. Mus. Leafl.*, **13**, 61–95.

Swamy, B. (1949). Embryological studies in the Orchidaceae. I. Gametophytes. II. Embryogeny. *Am. Midl. Nat.*, **41**, 184–201; 202–232.

Swartz, O. (1805). Genera et species orchidearum. *Neues J. Bot.*, **1**, 1–108.

Szlachetko, D. L. (1995). Systema orchidalium. *Frag. Florist. Geobot.*, Suppl. 3, 1–152.

Takahashi, S., Namba, T., and Hayashi, Y. (1965). Anatomical studies on 'chukanso'. *Syoyakugaku Zasshi*, **19** (1), 13–24.

Tanaka, R., and Kamemoto, H. (1974). List of chromosome numbers in species of the Orchidaceae. pp. 411–483. In: *The orchids: scientific studies* (Ed. Withner, C. L.). John Wiley and Sons, New York.

Tanaka, R., and Kamemoto, H. (1984). Chromosomes in orchids: counting and numbers. pp. 323–410. In: *Orchid biology: Reviews and perspectives, III* (Ed. Arditti, J.). Cornell University Press, Ithaca, NY.

Tatarenko, I. (1995). Mycorrhiza of orchids (Orchidaceae) of Primorye Territory. *Bot. Zh. Ross. Akad. Nauk*, **80** (8), 64–72 + 2 plates.

Teixeira, S. de P., Borba, E. L., and Semir, J. (2004). Lip anatomy and its implications for the pollination mechanisms of *Bulbophyllum* species (Orchidaceae). *Ann. Bot.*, **93**, 499–505.

Tham, F., Tan, R., and Haseloff, J. (2009). Mycorrhizal infection in the bamboo orchid (*Arundina graminifolia* (Don) Hochr.) as observed using confocal laser scanning microscopy. *Malayan Nat. J.*, **61**, 213–221.

Thomas, M. (1893). The genus *Corallorhiza*. *Bot. Gaz.*, **18**, 166–170.

Thorsch, J., and Stern, W. (1997). Tracheary studies and the terrestrial ancestry of Orchidaceae. *Int. J. Plant Sci.*, **158**, 222–227.

Tischler, G. (1910). Untersuchungen an Mangrove- und Orchideen-wurzeln mit spezieller Beziehung auf die Statolithentheorie des Geotropismus. *Ann. Jard. Bot. Buitenzorg*, Suppl., 3 (1), 131–186.

Tominski, P. (1905). Die Anatomie des Orchideenblattes in ihrer Abhängigkeit von Klima und Standort. Dissertation, Universität Berlin, Germany.

Toscano de Brito, A. L. V. (1994). Systematic studies in the subtribe Ornithocephalinae (Orchidaceae). PhD thesis, University of Reading, UK.

Toscano de Brito, A. L. V. (1998). Leaf anatomy of Ornithocephalinae (Orchidaceae) and related subtribes. *Lindleyana*, **13**, 234–258.

Toscano de Brito, A. L. V. (2001). Systematic review of the *Ornithocephalus* group (Oncidiinae; Orchidaceae) with comments on *Hofmeisterella*. *Lindleyana*, **16**, 157–217.

Treub, M. (1879). Note sur l'embryogénie de quelques Orchidées. *Verh. kon. Akad. Wet.*, Amsterdam.

Van den Berg, C., Goldman, D., Freudenstein, J. V., Pridgeon, A. M., Cameron, K. M., and Chase, M. (2005). An overview of the phylogenetic relationships within Epidendroideae inferred from multiple DNA regions and recircumscription of Epidendreae and Arethuseae (Orchidaceae). *Am. J. Bot.*, **92**, 613–624.

Van den Berg, C., Higgins, W., Dressler, R., Whitten, W., Soto-Arenas, M., and Chase, M. (2009). A phylogenetic study of Laeliinae (Orchidaceae) based on combined nuclear and plastid DNA sequences. *Ann. Bot.*, **104**, 417–430.

van der Pijl, L., and Dodson, C. H. (1966). *Orchid flowers: their pollination and evolution.* University of Miami Press, Coral Gables, FL.

Vermeulen, P. (1959). The different structure of the rostellum in Ophrydeae and Neottieae. *Acta Bot. Neerl.*, **8**, 338–355.

Veyret, Y. (1974). Development of the embryo and the young seedling stages of orchids. pp. 223–265. In: *The orchids: scientific studies* (Ed. Withner, C. L.). John Wiley and Sons, New York.

Vieira, M., Andrade, M., Bittencourt Jr, N., and de Carvalho-Okano, R. (2007). Flowering phenology, nectary structure and breeding system in *Corymborkis flava* (Spiranthoideae: Tropidieae), a terrestrial orchid from a Neotropical forest. *Aust. J. Bot.*, **55**, 635–642.

Vij, S., Kaushal, P., and Kaur, P. (1991). Observations on leaf epidermal features in some Indian orchids: taxonomic and ecological implications. *J. Orchid Soc. India*, **5**, 43–53.

Vij, S., Sharma, M., and Datta, S. (1985). Mycorrhizal endophyte of *Spiranthes lancea* (Sw.) Baker—white flowered taxon (Orchidaceae). *J. Indian Bot. Soc.*, **63**, 175–179.

Vinogradov, A. E. (2003). Selfish DNA is maladaptive: evidence from the plant Red List. *Trends Genet.*, **19**, 609–614.

Vogel, E. F. de (1969). Monograph of the tribe Apostasieae (Orchidaceae). *Blumea*, **17**, 313–350.

Vogel, S. (1962). Duftdrusen im Dienste der Bestaubung. Über Bau und Funktion der Osmophoren. *Akad. Wiss. Lit., Abh. Math. Nat. Kl. Mainz*, **10**, 601–763.

Vogel, S. (1974). Ölblumen und ölsammelnde Bienen. *Trop. Subtrop. Pflwelt*, **7**, 1–267.

Voigt, T. (1991). Untersuchungen der Mykorrhiza und Samenkeimung von Orchideen aus der Gattung *Epipactis* Zinn. *Orchidee*, **42**, 270–272.

Wahrlich, W. (1886). Beitrag zur Kenntniss der Orchideenwurzelpilze. *Bot. Zeitung*, **44**, 481–488, 497–505.

Wallach, A. (1938–9). Beiträge zur Kenntnis der Wasseraufnahme durch die Luftwurzeln tropischer Orchideen. *Z. Bot.*, **33**, 433–468.

Wang, H., and Wang, Y. S. (1992). Ultrastructural localization of acid phosphatase in the digesting cell of tuber of *Gastrodia elata* infected by *Armillaria mellea*. *Chin. J. Bot.*, **4**, 148–152 + 2 plates; **4**, 107–111 + 1 plate. [In English.]

Warcup, J. H. (1973). Symbiotic germination of some Australian terrestrial orchids. *New Phytol.*, **72**, 387–392.

Warcup, J. H. (1981). The mycorrhizal relationships of Australian orchids. *New Phytol.*, **87**, 371–381.

Warcup, J. H. (1985). *Rhizanthella gardneri* (Orchidaceae), its rhizoctonia endophyte and close association with *Melaleuca uncinata* (Myrtaceae) in Western Australia. *New Phytol.*, **99**, 273–280.

Watthana, S. (2007). The genus *Pomatocalpa* (Orchidaceae). A taxonomic monograph. *Harvard Pap. Bot.*, **11**, 207–256.

Weber, H. (1981). Orchideen auf dem Weg zum Parasitismus. Über die Möglichkeit einer phylogenetischen Umkonstruktion der Infektionsorgane von *Corallorhiza trifida* Chat. (Orchidaceae) zu Kontaktorganen parasitischer Blütenpflanzen. *Ber. Deutsch. Bot. Ges.*, **94**, 275–286.

Weiss, J. (1880). Anatomie und Physiologie fleischig verdickter Wurzeln. *Flora*, **63**, 81–89, 97–123.

Wejksnora, P. (1971). Anatomy of orchid leaves and stems, with special reference to *Oncidium sphacelatum*. Senior thesis, Brooklyn College, New York.

Weltz, M. (1897). *Zur Anatomie der monandrischen sympodialen Orchideen.* Inaugural-Dissertation Ruprecht-Karls-Universität Heidelberg. J. Horning, Heidelberg.

Went, F. (1895). Über Haft- und Nahrwurzeln bei Kletterpflanzen und Epiphyten. *Ann. Jard. Bot. Buitenzorg*, **12**, 1–72.

White, J. (1907). On polystely in roots of Orchidaceae. *Univ. Toronto Stud., Biol. Ser.*, No. 6, 1–20.

Whitten, W. M., Blanco, M. A., Williams, N. H., Koehler, S., Carnevali, G., Singer, R. B., *et al.* (2007). Molecular phylogenetics of *Maxillaria* and related genera (Orchidaceae: Cymbidieae) based on combined molecular data sets. *Am. J. Bot.*, **94**, 1860–1889.

Whitten, W. M., Williams, N. H., and Chase, M. W. (2000). Subtribal and generic relationships of Maxillarieae (Orchidaceae) with emphasis on Stanhopeinae: combined with molecular evidence. *Am. J. Bot.*, **87**, 1842–1856.

Whitten, W., Williams, N., Dressler, R., Gerlach, G., and Pupulin, F. (2005). Generic relationships of Zygopetalinae (Orchidaceae: Cymbidieae): combined molecular evidence. *Lankesteriana*, **5** (2), 87–107.

Widholzer, C., and Oliveira, P. L. de (1994). Tipo fotossintetico e anatomia foliar de *Sophronitis coccinea* (Orchidaceae). *Napaea*, **10**, 23–29.

Williams, C. A. (1979). The leaf flavonoids of the Orchidaceae. *Phytochemistry*, **18**, 803–813.

Williams, N. (1974). The value of plant anatomy in orchid taxonomy. pp. 281–298. In: *Proc. 7th World Orchid Conf., Medellin, 1972* (Ed. Ospina, M.). Bedout S.A., Medellin.

Williams, N. (1975). Stomatal development in *Ludisia discolor* (Orchidaceae): mesoperigenous subsidiary cells in the monocotyledons. *Taxon*, **24**, 281–288.

Williams, N. (1976). Subsidiary-cell development in the Catasetinae (Orchidaceae) and related groups. *Bot. J. Linn. Soc.*, **72**, 299–309.

Williams, N. (1979). Subsidiary cells in the Orchidaceae: their general distribution with special reference to development in the Oncidieae. *Bot. J. Linn. Soc.*, **78**, 41–66.

Williams, N. (1982). The biology of orchids and euglossine bees. pp. 119–171. In: *Orchid biology: Reviews and perspectives, II* (Ed. Arditti, J.). Cornell University Press, Ithaca, NY.

Williams, N., Chase, M., Fulcher, T., and Whitten, W. (2001). Molecular systematics of the Oncidiinae based on evidence from four DNA sequence regions: expanded circumscriptions of *Cyrtochilum, Erycina, Otoglossum,* and *Trichocentrum* and a new genus (Orchidaceae). *Lindleyana,* **16,** 113–139.

Williams, N., and Whitten, W. M. (1983). Orchid floral fragrances and male euglossine bees: methods and advances in the last sesquidecade. *Biol. Bull.,* **164,** 355–395.

Williams, N., Whitten, W., and Dressler, R. (2005). Molecular systematics of *Telipogon* (Orchidaceae: Oncidiinae) and its allies: nuclear and plastid DNA sequence data. *Lankesteriana,* **5** (3), 163–184.

Williamson, B. (1973). Acid phosphatase and esterase activity in orchid mycorrhiza. *Planta,* **112,** 149–158.

Williamson, B., and Hadley, G. (1970). Penetration and infection of orchid protocorms by *Thanatephorus cucumeris* and other *Rhizoctonia* isolates. *Phytopathology,* **60,** 1092–1096.

Withner, C., Nelson, P., and Wejksnora, P. (1974). The anatomy of orchids. pp. 267–347. In: *The orchids: scientific studies* (Ed. Withner, C. L.). John Wiley and Sons, New York.

Wolter, M., and Schill, R. (1986). Ontogenie von Pollen, Massulae und Pollinien bei den Orchideen. *Trop. Subtrop. Pflwelt,* **56,** 1–93.

World Checklist of Selected Plant Families (2010, 2012). Facilitated by the Royal Botanic Gardens, Kew. Available from: http://apps.kew.org/wcsp (retrieved 2013).

Wright, M., and Guest, D. (2005). Development of mycorrhizal associations in *Caldenia tentaculata. Selbyana,* **26,** 114–124.

Wu, J., Han, S., Zhu, Y., Lu, M., Wang, G., and Guo, W. (2005). Ultrastructure of mycorrhizal symbiosis between *Cymbidium goeringii* and *Rhizoctonia* sp. *J. Nanjing Norm. Univ. Nat. Sci. Ed.,* **29** (4), 105–108.

Wu, S.-J., Liu, Y.-S., Chen, T.-W., Ng, C.-C., Tzeng, W.-S., and Shyu, Y.-T. (2009). Differentiation of medicinal *Dendrobium* species (Orchidaceae) using molecular markers and scanning electron microscopy. *J. Food Drug Analysis,* **17,** 474–488.

Xu, J.-T., and Fan, L. (2001). Cytodifferentiation of the seeds (protocorm) and vegetative propagation corms colonized by mycorrhizal fungi. *Acta Bot. Sin.,* **43,** 1003–1010.

Xu, Y., Teo, L., Zhou, J., Kumar, P., and Yu, H. (2006). Floral organ identity genes in the orchid *Dendrobium crumenatum. Plant J.,* **46,** 54–68.

Xu, Z.-H., Zeng, Z.-Q., and Zhang, M.-T. (2009). Mycorrhizal microstructure of *Cymbidium goeringii* and its mycorrhizal fungi: isolation and identification. *J. Wuhan Bot. Res.,* **27,** 332–335.

Ye, Q.-S., Pan, R.-C., Qiu, C.-X., and Hew, C.-S. (1992). Study on leaf anatomy and photosynthesis of *Cymbidium sinense. Acta Bot. Sin.,* **34,** 771–776 + 1 plate.

Yeung, E. (1987a). The development and structure of the viscidium in *Epidendrum ibaguense* H. B. K. (Orchidaceae). *Bot. Gaz.,* **148,** 149–155.

Yeung, E. (1987b). Development of pollen and accessory structures in orchids. pp. 193–226. In: *Orchid biology: Reviews and perspectives, IV* (Ed. Arditti, J.). Comstock Publishing Associates, Ithaca, NY.

Yeung, E. (1988). The development and structure of the stigma of *Epidendrum ibaguense* (Orchidaceae). *Lindleyana,* **3,** 97–104.

Yu, H., and Goh, C. (2000). Identification and characterization of three orchid MADS-box genes of the AP1/AGL9 subfamily during floral transition. *Plant Physiol.,* **123,** 1325–1336.

Yukawa, T., Ando, T., Karasawa, K., and Hashimoto, K. (1990 [1991]). Leaf surface morphology in selected *Dendrobium* species. pp. 250–258. In: *Proc. 13th World Orchid Conf., Auckland, 1990* (Ed. Kernohan, J.). World Orchid Conference Trust, Auckland.

Yukawa, T., Ando, T., Karasawa, K., and Hashimoto, K. (1992). Existence of two stomatal shapes in the genus *Dendrobium* (Orchidaceae) and its systematic significance. *Am. J. Bot.,* **79,** 946–952.

Yukawa, T., and Stern, W. L. (2002). Comparative vegetative anatomy and systematics of *Cymbidium* (Cymbidieae: Orchidaceae). *Bot. J. Linn. Soc.,* **138,** 383–419.

Yukawa, T., and Uehara, K. (1996). Vegetative diversification and radiation in subtribe Dendrobiinae (Orchidaceae): evidence from chloroplast DNA phylogeny and anatomical characters. *Plant Syst. Evol.,* **201,** 1–14.

Zanenga Godoy, R., and Costa, C. (2003). Anatomia foliar de quatro espécies do genero *Cattleya* Lindl. (Orchidaceae) do Planalto Central Brasileiro. *Acta Bot. Bras.,* **17,** 101–118.

Zankowski, P., Fraser, D., and Reynolds, T. (1987). The developmental anatomy of velamen and exodermis in aerial roots of *Epidendrum ibaguense. Lindleyana,* **2,** 1–7.

Zeiger, E., Grivet, C., Assmann, S. M., Deitzer, G. F., and Hannegan, M. W. (1985). Stomatal limitation to carbon gain in *Paphiopedilum* sp. (Orchidaceae) and its reversal by blue light. *Plant Physiol.,* **77,** 456–460.

Zelmer, C., and Currah, R. (1995). Evidence for a fungal liaison between *Corallorhiza trifida* (Orchidaceae) and *Pinus contorta* (Pinaceae). *Can. J. Bot.,* **73,** 862–866.

Zhang, J.-X., Wu, K.-L., Zeng, S.-J., Duan, J., and Tian, L.-N. (2010). Characterization and expression analysis of PhalLFY, a homologue in *Phalaenopsis* of FLORICAULA/LEAFY genes. *Sci. Hortic.,* **124,** 482–489.

Zhang, S.-B., Guan, Z.-J., Sun, M., Zhang, J.-J., et al. (2012). Evolutionary association of stomatal traits with leaf vein density in *Paphiopedilum,* Orchidaceae. *Plos One,* **7** (6), e40080.

Zhang, Z.-J., Chen, Y., Lin, K.-R., and Pan, W.-X. (1992). The structure of nutritive body in *Anoectochilus roxburghii. Acta Bot. Yunnan.,* **14** (1), 45–48 + 1 plate. [In Chinese with English summary.]

Zhao, G.-X., and Wei, Z.-X. (1999). An anatomical study on the stems of two species of *Vanilla. Acta Bot. Yunnan.*, **21**, 65–67 + 2 plates.

Zhao, Z., Shimomura, H., Sashida, Y., Tujino, R., Okamoto, T., and Kazami, T. (1996). Identification of traditional patent Chinese medicines by a polariscope (1). Polariscope characteristics of starch grains and calcium oxalate crystals. *Natural Medicines*, **50**, 389–398.

Ziegler, B. (1981). Mikromorphologie der Orchideënsamen unter Berücksichtigung taxonomischer Aspekte. PhD thesis, Ruprecht Karls-Universität, Heidelberg, Germany.

Zimmerman, M. (1932). Über die extrafloralen Nectarien der Angiospermen. *Beih. Bot. Centralbl.*, **49**, 99–196.

Zörnig, H. (1903). Beiträge zur Anatomie der Coelogyninen. Dissertation, Universität Heidelberg, Germany.

INDEX

Note: Where genera have been specially studied, the page numbers are in bold type; italic page numbers refer to illustrations. Most cell and tissue types are not indexed for the tribal and subtribal descriptions.

A

Acampe 19, 37, **215–221**
Acanthephippium **112–116**
Aceras 7, 72
Acianthera 23, **130–135**, *131*
Acianthinae 8, 25, 29, 57–58, 226–230, 232
Acianthus **57–58**, 65
Acineta 28, **195–199**, *196*
Ackermania **199–202**
Acostaea **130–135**
Acranthae 10
Acriopsis **173–177**
Acrolophia 183
Acropera 197
Acrotonae 10
Ada **189–195**
Adenochilus **58–60**
Adrorhizinae 214–215, 221, 227–232
Adrorhizon **214**, 221
Aerangidinae 221
Aerangis **215–221**
Aeranthes 25, **215–221**
aerial roots *see* roots
Aerides 8, 25
Aeridinae 2, 4, 8, 19, 23, 25, 26, 32, 215–221, 227–232
Afrotrilepis 222
Aganisia **199–202**
Agrostophyllinae 21, 221–222, 227–232
Agrostophyllum **221–222**
Alamania 128, 129
algae 37
alkaloids 8
Alstroemeria 11
Altensteinia **74–79**
Amesiella **215–221**
amyloplasts 37, *77*
Anacamptis 7, **65–73**, *67*
Anathallis 18, 22, **120–135**
anatomical literature 15–40
Ancistrochilus **112–116**
Ancistrorhynchus **215–221**
Angraecinae *Pl. 3*, *Pl. 4*, 2, 5, 19, 25, 31, 32, 215–221, 227–232
Angraecopsis **215–221**
Angraecum 5, **215–221**, *216*, *218*
Anguloa **184–189**, *186*
Anoectochilus 3, 8, 12, **74–79**, *76*
Ansellia 3, 4, *20*, **173–177**, *174*, *175*
anthocyanins 3, 8
Anthogonium 3, *96*
Anthosiphon **184–189**

ants 3, 12, 27
Aphyllorchis 6, 12, **143–148**
Aplectrum 3, 21, 28, 35, **105–112**, *107*, *109*
Aporostylis 19, 57, **58–60**
Apostasia 2, 4, 5, 6, 9, 11, 12, 14, 17, 24, *36*, 37, **43–49**, *44*, *46–48*
Apostasieae 47, 48
Apostasiinae 47
Apostasioideae 2, 3, 4, 5, 6, 10, 11, 12, 16, 21, 24, 25, 27, 28, 30, 36, 43–49, 226–230, 232
Appendicula **150–157**, 222
Arabidopsis 14
Araceae 16
Arachnis 220
Arecaceae 11
Arethusa 3, **95–97**, *96*
Arethuseae 2, 24, 25, 30, 95–105
Arethusinae 3, 95–97, 226–228, 230–232
Arpophyllum **118–130**, *119–121*
Arthrochilus 18, 38, **61–62**
Artorima 128, 129
Arundina 3, 25, **95–97**, *97*
Ascocentrum 220
Aspasia **189–195**
Asphodelus 15
Asparagales 11
Asteraceae 1
Auxopus 6, 136

B

Baptistonia 194
Barbosella **130–135**
Barkeria **118–130**
Barlia 66, 71
Basiphyllaea 3, 28, **116**
Basitonae 10
Batemannia **199–202**
Beclardia **215–221**, *217*
bees 9, 11, 13
Benzingia 184, **199–202**
Bifrenaria 28, **184–189**
Bipinnula 3, 73
Bletia 3, **116**
Bletiinae 3, 28, 116, 226–228, 230–232
Bletilla **98–105**, *99*
Bollea **199–202**
Bolusiella **215–221**
Bonatea **65–73**
Bonniera **215–221**
Bothriochilus 118

Brachionidium **130–135**, *132*
Brachtia **189–195**
Brachycorythis 13, **65–73**
Braemia **195–199**
Brasiliorchis **184–189**, *184*, *187*, *188*
Brassavola 1, 3, **118–130**
Brassia 25, **189–195**
Bromheadia 6, **214–215**, 221
Bromheadiinae 221
Broughtonia **118–130**, *119*
Brownleea **52–56**
Brownleeinae 56
Bryobium **150–157**
Bulbophyllaria 213
Bulbophyllinae 16, 17
Bulbophyllum 2, 4, 5, 12, 13, 16, 19, 21, 25, 27, 28, *29*, 30, 31, *34*, **202–213**, *209–211*
Burlingtonia 194
Burmannia 194
Burmanniaceae 11
Burnettia **62–63**

C

Cadetia **202–214**
Caladenia 3, 4, 25, *26*, 38, *39*, *41*, 56, **58–60**, 65
Caladeniinae 3, 19, 21, 25, 29, 57, 58–60, 65, 226–229, 231, 232
Calanthe 3, 28, **112–116**
Calanthinae 112
Caleana 19, **61–62**
Callostylis 156
Calochilus **64–65**
Calopogon 3, 5, 24, **95–97**
Calypso 3, 19, 25, 28, **105–112**, *106*, *107*, *109*
Calypsoeae 3, 5, 8, 19, 21, 24, 25, 27, 28, 30, 35, 37, 105–112, 226–228, 230–232
Calyptrochilum **215–221**, *217*
Camaridium **184–189**
Camarotis 220
Campanulorchis **150–157**
Campylocentrum *Pl. 4*, 2, 4, 31, **215–221**, *219*
Capanemia 194, 195
casparian bands or strips 16, 27, 28, 36, 38, 40
Catasetinae *Pl. 1*, *Pl. 2*, 2, 3, 12, 19, 21, 23, 26, 28, 35, 168–172, *176*, *178*, 184, 226, 228–232
Catasetum *Pl. 2*, 3, 5, 7, 12, **168–172**, *169–171*

Cattleya 13, **118–130**, *124, 127*
Caucaea **189–195**
Caularthron 3, 12, 27, **118–130**, *119*
Cephalanthera 35, **143–148**, *144*
Cephalantheropsis **112–116**
Ceratandra 19, *40*, **52–56**
Ceratostylis **150–157**
Cestichis 142, 143
Chamaeangis 35, **215–221**, *219*
Chaseela 212
Chaubardia **199–202**
Chaubardiella **199–202**
Cheirostylis **74–79**
Chelonistele **98–105**
Chelyorchis **189–195**
Chiloglottis *20*, **61–62**
Chiloschista 2, 4, **215–221**
Chloraea 1, 3, **73**, 148
Chloraeeae 3, 30, 73, 226–228, 230–232
Chloraeinae 12, 78
chlorenchyma 22, 24, 25
chloroplasts 27, 35, 36
Chondrorhyncha **199–202**
Chondroscaphe **199–202**
chromosomes 7
Chrysoglossinae 112
Chrysoglossum **112–116**
Chysinae 19, 116–117, 226–228, 230–232
Chysis 19, 25, 29, **117**
Chytroglossa 194
Cirrhaea **195–199**
Cirrhopetalum 21, 25, **202–213**
Cischweinfia **189–195**
Cleisostoma 220
Cleistes 3, 4, **79–81**
Cleistesiopsis **79–81**, *80*
Clematepistephium 21, **81–95**, *93*
Clowesia *Pl. 1, Pl. 2*, **168–172**, *170, 171*
Cochleanthes **199–202**
Cochlioda **189–195**, *190, 192*
Codonorchideae 30, 73–74
Codonorchis 2, 4, **73**
Codonosiphon 212
Coelia 33, **117–118**
Coeliinae 31, 117–118, 226–227, 229–232
Coeliopsidinae 19, 172–173, 198, 226, 228–232
Coeliopsis **172–173**
Coeloglossum 66, 72
Coelogyne 3, 6, 25, 28, **98–105**
Coelogyninae 17, 19, 21, 23, 24, 25, 27, 28, 31, 98–105, 226–228, 230–232
Coilochilus **60**
Colax 202
Collabieae 2, 19, 30, 112–116, 226–228, 230–232
Collabiinae 2
'collar' 7, 25, *26*, 28
Commelinaceae 16
Comparettia **189–195**
Conchidium **150–157**
Condylago **130–135**

Constantia **118–130**
Corallorhiza 1, 3, 4, 6, **105–112**
Corallorhizinae 105
corm 3, 25, 28
Corsiaceae 11
cortex 3, 15, 17, 23, 25, 27, 28, *29*, 31, *32*, 33–36, *35*, *37*, 38, 41
Coryanthes 5, 13, 15, **195–199**
Corybas 1, 8, 16, **57–58**, 65
Coryciinae 19, 56
Corycium 4, 5, **52–56**
Corymborkis 2, 3, 4, 12, 37, **164–168**, *166, 167*
Corysanthes 65
Cottonia 220
cover cells 31, 35
Cranichideae *Pl. 1*, 2, 13, 15, 19, 21, 24, 25, 27, 28, 30, 31, 35, 36, 37, 38, 74–79, 110, 112, 168, 226–228, 230–232
Cranichidinae 5, 74, 77, 79
Cranichis **74–79**, *77*
Cremastra 24, 25, **105–112**, *107*
Crepidium **139–143**, *141*
Cribbia **215–221**, *216*
Cryptarrhena 193, **199–202**
Cryptarrheninae 202
Cryptocentrum **184–189**
Cryptochilus 150, *151*, 153
Cryptophoranthus **130–135**
Cryptopus **215–221**, *217*
Cryptostylidinae 57, 60, 226–229, 231, 232
Cryptostylis 8, 31, 37, 57, **60**, 65, *75*
crystals *24*, 25, 28, 37, 228–229
Cuitlauzina **189–195**, *192*
cuticle 12, 13, *15*, *18*, 19, 25, *26*, 28, 226–227
Cyanicula 25, **58–60**, 65
Cyclopogon 29, **74–79**
Cycnoches 3, *26*, 35, **168–172**, *169*, *170, 171*
Cymbidieae 4, 21, 22, 24, 25, 110, 168–202, 184, 189, 221
Cymbidiella **179–184**
Cymbidiinae 3, 19, 21, 22, 23, 173–177, 226, 228–232
Cymbidium 3, 6, 19, 21, 24, 28, 37, **173–177**, *174–177*
Cynorkis 15, **65–73**, *69*
Cyperaceae 222
Cypripedieae 9
Cypripedilinae 12
Cypripedioideae 2, 3, 4, 5, 6, 7, 10, 11, 12, 13, 14, 19, 20, 21, 22, 24, 25, 27, 30, 36, 49–52, 226–230, 232
Cypripedium 2, 3, 6, 8, 12, 17, *18*, 19, *20*, 25, 27, 28, 30, **49–52**, *50*, *51*
Cyrtidiorchis **184–189**
Cyrtochiloides 195
Cyrtochilum 194, 195
Cyrtopodiinae 3, 4, 19, 21, 177–178, 226, 228–232

Cyrtopodium 3, 4, 30, **177–178**
Cyrtorchis 35, **215–221**
Cyrtosia 2, 5, **81–95**, *92*
Cyrtostylis **57–58**
Cystorchis 78, 79
cytogenetics 7–8

D

Dactylorhiza 1, 3, 8, 13, 15, 38, **65–73**
Danhatchia 78, 79, 112
Dendrobieae 2, 4, 8, 19, 21, 23, 24, 25, 26, 27, 31, 32, 202–213, 222, 227–232
Dendrobiinae 22, 212, 213
Dendrobium 1, 3, 4, 8, 11, 12, 14, 19, 21, *24*, 25, 28, **202–213**, *203–207*, 222
Dendrochilum **98–105**
Dendrocolla 220
Dendrophylax *Pl. 3, Pl. 4*, 2, 4, **215–221**, *217*
Desmotrichum 213
Diacrium 129
Diandrae 10
Diaphananthe **215–221**
Diceratostele 2, 3, 10, 30, 37, **161–164**
Diceratosteleae 19, 79, 163, 168
Diceratostelinae 24, 79, 161, 168
Dichaea 2, **199–202**, *200*
Dictyophyllaria 3, 5, 81
Didymoplexiella 137
Didymoplexis **136–139**
Dienia **139–143**
Dilochia 105
Dilomilis 128, 135
Dimerandra **118–130**
Dinema 128, 129
Diodonopsis **130–135**
Diplocaulobium 213
Dipodium **173–177**
Dipteranthus **189–195**
Disa 8, 19, 25, *40*, **52–56**
Diseae 4, 5, 8, 13, 15, 16, 19, 25, 27, 28, 30, 38, 40, 52–56, 226–230, 232
Disinae 13, 19, 56
Disperis 19, 25, **52–56**, *53*
Dithonea 128
Diurideae 2, 3, 4, 5, 6, 8, 13, 15, 16, 19, 24, 25, 27, 28, 29, 30, 31, 37, 38, 39, 40, 56–65, 75, 79, 168
Diuridinae 3, 60–61, 226–229, 231, 232
Diuris 39, **60–61**
DNA sequencing 4, 6, 7, 10
Domingoa **118–130**
Doritis 220
Dossinia 8
Dracula **130–135**
Drakaea *18*, 19, 31, **61–62**
Drakaeinae 25, 61–62, 226–229, 231, 232
Drakonorchis 65
Dresslerella 3, 19, *20*, 21, **130–135**, *131*
Dressleria **168–172**, *171*

droppers 4, 16, 29, 38–40, *39*, *41*
druses 25, 228–229
Dryadella **130–135**
Drymoda 213
Duckeella **79–81**

E
Earina 11, 21, **221–222**
Echinosepala 20, 21, *23*, **130–135**, *131*
economic uses 9
Eggelingia **215–221**
elaiophores 5, 12, 13
Eleorchis 3, 96
Elleanthus 35, **157–161**, *159*
Elythranthera *18*, 25, **58–60**
Embreea **195–199**, *196*
embryos 6
Encyclia 12, **118–130**
endodermis 3, 15, 16, 27, 28, *29*, *35*, 36, 38, 40
endovelamen 31
Ephemerantha 212, 213
Epiblastus **150–157**
Epiblema 2, **64–65**
Epidanthus 128
Epidendreae 2, 4, 6, 8, 9, 19, 24, 25, 29, 116–136, 160, 221
Epidendroideae 2, 3, 5, 8, 10, 11, 13, 14, 19, 20, 21, 23, 24, 27, 28, 36, 37, 38, 49, 79, 95–225
Epidendrum 2, 4, 9, 12, 13, 15, 24, 28, 31, **118–130**
epidermis 16, 17, 19–21, *20*, *23*, 24, 25, *26*, 27, 28, 30, *41*
Epigeneium 212, 213
Epilyna **157–161**, *160*
Epipactis 6, 8, 17, 21, 28, 35, **143–148**, *145*
Epipogium 3, 4, 6, 12, **148–150**, *150*
Epistephium **81–95**, *94*
epivelamen 30–31
Eria 21, 24, **150–157**, *152*
Eriaxis 3, 21, **81–95**, *94*, *95*
Eriinae 4, 19, 23, 150, 157, 213
Eriochilus 19, **58–60**
Eriodes **112–116**
Eriopsidinae 178–179
Eriopsis **178–179**
Erycina 7, **189–195**
Erythrodes 8, 74, 75, 78
Erythrorchis 2, 4, **81–95**
Eulophia Pl. *2*, Pl. *3*, 1, 3, 6, 28, **179–184**, *180*, *181*
Eulophidium 184
Eulophiella **179–184**
Eulophiinae Pl. *2*, Pl. *3*, 19, 23, 28, 179–184, 226, 228–232
Eurychone **215–221**
Eurystyles **74–79**
'evo-devo' studies 14–15
Evotella **52–56**
exodermis 3, 4, 7, 16, 28, *29*, 30, 31, *32*, 33–35, *35*, 38, *41*

F
Fernandezia 195
fibre bundles 17, 22, 23, 24, 27, 228–229
flavonoid 8, *24*, 25, 37
Flickingeria 212, 213
floral anatomy 12–15
floral fragrances 8–9
floral vasculature 12
fossils 11, 19, 212, 222
fruits 5
fungal hyphae 6, 7, 25, 28, 37, 38
Fusarium 9

G
Galeandra **168–172**
Galearis **65–73**
Galeoglossum **74–79**
Galeola 2, 4, 6, 12, **81–95**, *92*
Galeolinae 81
Galeottia **199–202**
Galeottiella 3
Galeottiellinae 74, 79
Gastrochilus 220
Gastrodia 4, 6, **136–139**
Gastrodieae 2, 3, 28, 29, 37, 136–139, 230–232
Gastrorchis **112–116**
Gavilea 2, 73
Geesinkorchis **98–105**
genome size 7–8
Genoplesium **63–64**
Genyorchis 5, 213
Geoblasta 73
Geodorum 28, **179–184**, *179*
Glomera **98–105**
Glomerinae 221
Glossodia 17, 57, **58–60**
glucomannin 38
Gomesa **189–195**
Gonatostylis **74–79**
Gongora **195–199**, *197*
Goodyera Pl. *1*, 3, 8, 19, 35, **74–79**, *75*
Goodyerinae 3, 5, 8, 11, 74, 77, 79
Govenia 3, 19, 24, 28, 35, **105–112**, *107*
Grammangis **179–184**
Grammatophyllum 2, 3, **173–177**, *174*, *176*
Graphorkis 3, **173–177**, *176*
Grobya **168–172**, *170*
ground tissue 17, 27, 28, 38
Guarianthe **118–130**, *119*
Gymnadenia 15, 16, 66, 71, 72
Gyrostachys 79

H
Habenaria 1, 7, 15, 19, 35, **65–73**, *68–71*
Habenariinae 30, 36, 38, 65, 71, 72, 73
Haemaria 79
Hagsatera 128
hairs/trichomes 3, 7, 17, 18–19, *18*, 25, *26*, 28, 38, *41*, 226–227
Hammarbya 2, **139–143**
Hapalochilus 213

Harrisella 4, 218, 220
Hederorkis 221, 222
Helcia **187–195**
Helleborine 148
Helleriella **135–136**
Hemerocallis 15
Herminium 72
Herschelianthe 56
Hetaeria **74–79**
Heterotaxis 189
Hexalectris 3, 4, 6, 116
Hexisea 128
Himantoglossum **65–73**
Hintonella **189–195**
Hippeophyllum **139–143**
Hoffmannseggella 128
Hofmeisterella 194
Holothrix 21, 25, 27, **65–73**, *67*
Homalopetalum **118–130**
Horichia **195–199**
Hormidium 129
Houlletia **195–199**
Huntleya 24, **199–202**
Huttonaea **52–56**
Huttonaeinae 56
Hylaeorchis **184–189**
Hylophila **74–79**
hypodermis *20*, 21–22, *22*, *23*, 25, 27, 227–228
Hypoxidaceae 11, 24
Hypoxis 24

I
Ida 189
idioblasts 22, *23*, 25, *26*, 27, 28, 31, 35, 37, 38
Imerinaea 173
inflorescence axis 36
Ionopsis **189–195**, *191*
Ipsea **112–116**
Iridaceae 11
Isabelia *123*, 128, 129, 130
Isochilus 3, 19, **135–136**
Isotria 2, **79–81**, *80*

J
Jacquiniella **118–130**
Jumellea **215–221**, *217*, *219*
Juncaginaceae 15

K
Kefersteinia **199–202**
Kegeliella **195–199**
Kerosphaereae 10
Kingidium 220
Koellensteinia **199–202**

L
Laelia **118–130**
Laeliinae 5, 19, 22, 23, 25, 26, 27, 31, 32, 36, 37, 116, 118–130, 136, 226–227, 229–232
Laeliocattleya 129

leaf 2–3, 8, 11, 17–25, *18*, *20*, *22*, *23*, *24*, *26*, *36*, 226–229
Lecanorchidinae 81
Lecanorchis **81–95**, *95*
Lemboglossum 13, **189–195**
Lemurella **215–221**
Lemurorchis **215–221**
Leochilus **189–195**
Lepanthes **130–135**
Lepanthopsis **130–135**
Lepidogyne *37*, **74–79**, *77*
Leporella **62–63**
Leptoceras **58–60**, 65
Leptotes **118–130**
Ligeophila **74–79**, *76*
Liliaceae 16
Liliales 11
Limodorum 6, **143–148**
Liparis 6, 20, 35, **139–143**, *140*
Listera **143–148**
Listrostachys **215–221**, *218*
Lockhartia 4, **189–195**
Loroglossum 72
Ludisia 3, 8, **74–79**
Lueddemannia **195–199**
Luisia 2, 220
Lycaste 15, *30*, **184–189**
Lycastinae 189
Lycomormium **172–173**
Lyperanthus 19, 21, **62–63**, 65

M
Macodes 3, 79
Macradenia **189–195**
Macroclinium **189–195**
Macroplectrum 220
Malaxideae 2, 8, 24, 27, 30, 33, 35, 139–143, 226–227, 229–232
Malaxis 5, 6, 28, 33, 35, **139–143**
Manniella **74–79**
Manniellinae 74, 77, 79
Masdevallia 4, 25, **130–135**
Maxillaria 15, 25, **184–189**, *185–187*
Maxillarieae 19, 173, 177, 189, 193, 198, 202
Maxillariella 136
Maxillariinae 4, 13, 19, 22, 23, 24, 25, 27, 28, 31, 136, 184–189, 226, 228–232
medicinal uses 9
Mediocalcar **150–157**
medullary strands 36
Megaclinium **202–214**
Megastylidinae 62–63, 226–229, 231–232
Megastylis 2, **62–63**
Meiracyllium **118–130**, *125*
Meliorchis 11
meristeles 15, 36, 38
Mesoglossum **189–195**
mesophyll 17, *20*, 21, 22–23, *23*, 25, 228–229
Mesospinidium **189–195**
methods xi

Mexicoa **189–195**
Mexipedium 21, 49, 51
Microchilus **74–79**
Microcoelia Pl. 4, 4, **215–221**
Microepidendrum 128, 129
Microstylis **139–143**
Microtatorchis **215–221**
Microterangis **215–221**
Microtis 38, *39*, **63–64**
Miltonia **189–195**
Miltoniopsis **189–195**
molecular studies 7, 11
Monadenia 40, 56
Monandrae 10
Monomeria 211, 213
Monophyllorchis **161–164**, *161*
Monosepalum 212
Mormodes **168–172**
Mormolyca **184–189**, *184–188*
mucilage 13, 18, 19, 25, 27, 28, 35, 38
mucilaginous idioblasts 25, *26*, 27, 28, 35, 38
Mycaranthes **150–157**
mycorrhiza(e) 6–7, *37*
Myoxanthus 24, *35*, **130–135**, *134*
Myrmechis 78, 79
Myrmecophila 1, 3, 25, 27, 128, 129
Mystacidium **215–221**

N
Nabaluia **98–105**
Nageliella 118
nectaries 5, 12
Nemaconia 135, 136
Neobathiea **215–221**
Neobenthamia 221, **222–225**
Neoclemensia 137
Neocogniauxia **130–135**
Neofinetia **215–221**
Neogardneria **199–202**
Neogyna **98–105**
Neomoorea **184–189**
Neotinea 7, **65–73**
Neottia 6, 21, **143–148**, *147*
Neottianthe 72
Neottieae 2, 3, 5, 6, 8, 9, 10, 13, 21, 28, 35, 143–148, 225–227, 229–232
Neottioideae 10
Nephelaphyllum **112–116**, *113*
Nervilia 3, 28, 81, **148–150**
Nervilieae 3, 28, 29, 148–150, 226–227, 229–232
Neuwiedia 2, 5, 6, 9, 11, 12, 14, 17, 25, *26*, 28, *36*, 37, **43–49**, *44–46*, *48*, *49*
Nidema **118–130**
Nigritella 72
Notheria 150
Notylia 2, *186*, **189–195**

O
Oberonia 2, **139–143**
Octarrhena **150–157**
Octomeria 23, *26*, **130–135**, *132*

Odontoglossum **189–195**
Oeceoclades **179–184**, *182*
Oeonia **215–221**
Oeoniella **215–221**
oil 25, 37
Oncidieae 194
Oncidiinae 4, 7, 13, 19, 20, 22, 23, 24, 25, 27, 32, 186, 189–195, 226, 228–232
Oncidium 4, 7, 8, 9, 12, 15, **189–195**, *190*
Ophidion **130–135**
Ophrydeae 9
Ophrydinae 12
Ophrydoideae 10
Ophrys 8, 13, **65–73**
Orchidaceae
  classification 9–11, 17
  distribution 1
  'evo-devo' 14–15
  floral anatomy 12–15
  habitats 1
  inflorescence types 4
  origin and affinities 11–12
  phytogenetics 9–11, 18
  reproductive morphology 4–6
  size 1–2
  taxonomic history 9–11
  vegetative anatomical literature 15–40
  vegetative morphology 2–4
Orchidacites 11
Orchideae 4, 5, 7, 9, 12, 13, 15, 16, 21, 24, 25, 27, 35, 36, 38, 65–73, 226–229, 231, 232
Orchidinae 13, 15, 21, 29, 38, 65, 71, 72, 73
Orchidoideae 2, 3, 5, 8, 10, 11, 12, 14, 15, 16, 21, 24, 31, 36, 38, 49, 52–79, 168
Orchis 3, 7, 8, 13, 15, 16, **65–73**, *67*
Oreorchis 110
Orestias 142, 143
Orleanesia **118–130**
Ornithidium 189
Ornithocephalinae 194
Ornithocephalus **189–195**
Ornithophora **189–195**
Orthoceras **60–61**
Osmoglossum **189–195**
osmophores 5, 9, 12
Ossiculum **215–221**
Otochilus 3, **98–105**, *100*
Otoglossum **189–195**
Otostylis **199–202**

P
Pabstia **199–202**
Pabstiella **130–135**, *132*
Pachites **52–56**
Pachyphyllum **189–195**
Pachyplectron 2, **74–79**, *78*
Palaeorchis 11
palisade (cells) 19, 21, 22, *23*, 25, 228–229

Palmorchis 2, 3, 5, 6, 79
Palumbina **189–195**
Panisea **98–105**
Paphinia **195–199**
Paphiopedilum 2, 3, 4, 12, 17, 19, 22, 25, 27, **49–52**
Papilionanthe 215, 220
papillae 7, 12, 13, *18*, 19, *20*, 25, 226–227
Paracaleana **61–62**
Paradisanthus **199–202**
Paralophia 179
Paraphalaenopsis 2
parenchyma 15, 17, 18, 21, 25, 27, 34, 35, 36, 37, 38
passage cells 30, 31, *32–35*, 36, 38
Pecteilis 16, 72
Pedilochilus 213
Pelexia **74–79**
pelotons 4, 6, 7, 25, 28, *37*, 38, *41*
Peranium 79
pericycle 3, 28, 36, 37, 38, 40
Peristeria **172–173**
Perularia 72
Pescatoria **199–202**
petiole 3
Phaius 3, **112–116**
Phajeae 112
Phajinae 112
Phalaenopsis 8, 14, **215–221**
phloem 3,15, 16, 23, 24, 27, 36, 37, 38
Phloeophila 22, **130–135**
Pholidota 3, 21, 28, **98–105**, *101*
Phragmipedium 2, 5, 12, 19, 27, **49–52**
Phreatia **150–157**, *153*
phylogenetics 9–11
Phymatidium **189–195**
Physosiphon **130–135**
Physothallis **130–135**
phytochemistry 8–9
Pinalia 153, 157
pith 17, 27, *35*, 36, 38
Platanthera 12, 16, **65–73**
Platyclinis 104, 105
Platyglottis 128
Platyrhiza 194
Platystele 2, 25, **130–135**
Platythelys **74–79**, *77*
Plectrelminthus **215–221**
Plectrophora **189–195**
Pleione 21, **98–105**, *102*
Pleuranthae 10
Pleurothallidinae 1, 2, 3, 4, 5, 12, 17, 18, 19, 20, 21, 22, 23, 25, 27, 31, 130–135, 226–227, 229–232
Pleurothallis 8, 9, 25, 28, **130–135**, *134*
Pleurothallopsis 23, **130–135**
Plocoglottis **112–116**
pneumathodes 31–32, *35*
Podangis **215–221**
Podochileae 2, 6, 19, 24, 150–157, 204, 226–227, 229–232

Podochilinae 222
Podochilus **150–157**, 222
Pogonia 3, 7, 10, 73, **79–81**, 148, 149
Pogonieae 21, 79–81, 226–228, 230–232
Pogoniopsis 2, **79–81**
pollen 5, 11, 14
pollination 4, 5, 8, 12
pollinia 5, 9, 10, 11, 13, 14
Polychilos 220
Polycycnis **195–199**, *198*
Polyradicion 218, 220
Polyrrhiza 220
Polystachya 4, 8, *34*, 221, **222–225**, *223*, *224*
Polystachyinae 31, 221, 222–225, 227–232
Pomatocalpa 220
Ponera 20, **135–136**
Ponerinae 3, 23, 135–136, 226–227, 229–232
Ponerorchis **65–73**
Ponthieva **74–79**, *75*
Porpax **150–157**
Porphyroglottis **173–177**
Porroglossum **130–135**
Praecoxanthus 57, **58–60**
Prasophyllinae 2, 25, 27, 57, 63–64, 226–229, 231, 232
Prasophyllum *41*, **63–64**
Prescottia 28, **74–79**, *76*
Prescottiinae 74, 79
prisms 25
Pristiglottis **74–79**
Promenaea **199–202**, *201*
Proplebeia 11
Prosthechea 12, *33*, 37, **118–130**, *119*, *122*
protists 37, 43, 65, 83
protocorm 6, 7
Protorchis 11
Pseuderia **150–157**, *156*, *204*, 213
pseudobulb 2, 3, 11, 17, 23, 25, 26, 27, 28
Pseudocranichis 3
Pseudolaelia **118–130**
Pseudovanilla 4, 21, **81–95**, *92*, *93*
Psilochilus 3, 25, **161–164**, *161*, *164*
Psychilis **118–130**
Psychopsis 3, **189–195**
Psygmorchis 7, 194
Pteroglossaspis **179–184**
Pterostylidinae 4, 5, 25, 29, 38, 39, 40, 74, 78, 79
Pterostylis 4, 15, *26*, *39*, **74–79**
Pterygodium *40*, **52–56**
Pyrorchis 21, **62–63**, 65

Q
Quisqueya **118–130**

R
ramicaul 130, *132*, 133
Rangaeris **215–221**

raphides 24, 25, 28, 37, 38, 228–229
Rauhiella 194
Reichenbachanthus 128
Renanthera 4, 17, *30*, 220
reproductive morphology 4–6
Restrepia 12, **130–135**
Restrepiella 29, **130–135**, *134*
Restrepiopsis **130–135**
Rhipidoglossum **215–221**
Rhizanthella 3, 6, 28, 57, **64**, 65
Rhizanthellinae 64
rhizodermis 4, 28, 29, 30, 38
rhizome 2, 3, 6, 11, 17, 25, 26, 27, 28, 37
Rhyncholaelia **118–130**
Rhynchostele 13, **189–195**
Rhynchostylis 25, 220
Ridleyella **150–157**, *154*
Rimacola **62–63**
Risleya 139
Rodriguezia 13, **189–195**, *190*
rods (crystals) 228–229
root 2, 3–4, 6, 7, 8, 9, 11, 15, 17, 23, 24, 25, 28–37, *29*, *30*, *32–37*, *39*, 229–232
root hairs 4, 16, 28–29, 38
root tuber/tubercle 4, 7, 15, 16, 17, 28, 37, 38
roots, aerial 2, 3, 4, 15, 16, 37
Roscoea 15
Rossioglossum **189–195**
Rudolfiella 13, **184–189**

S
Saccoglossum 212
Saccolabium 220
salep 9, 38
Salpistele **130–135**
Sarcanthus 220
Sarcochilus 220
Sarcoglottis **74–79**
Sarcopodium 213
Sarracenella **130–135**
Satyridium 56
Satyriinae 13, 19, 56
Satyrium 2, 12, 19, *40*, **52–56**, *54*
Scaphosepalum 12, **130–135**
Scaphyglottis 24, **118–130**, *126*
Schizodium **52–56**
Schlimmia **195–199**
Schomburgkia 25, 128, 129
sclerenchyma 15, 17, 20, 21, 23, 24, 36
sclerenchyma band/ring/sheath 27, 28
sclerenchyma bridge 24, 27
sclerenchyma caps 23, 27
Scuticaria **184–189**
seeds 6, 7, 10, 11
Selenipedium 2, 5, 6, 19, 25, **49–52**
Serapias **65–73**
Sertifera **157–161**, *158*
Sievekingia **195–199**, *197*, *198*
Sigmatostalix 13, **189–195**
silica bodies 17, *24*, *26*, 28, 37, 228–229

sinker 16
Sirhookera **214–215**, 221
Sobennikoffia **215–221**
Sobralia 3, 16, 31, *32*, *33*, 35, **157–161**, *158*, *159*
Sobralieae 2, 19, 30, 31, 35, 157–161, 226–227, 229–232
Sobraliinae 160
Solenangis 35, **215–221**, *219*
Solenidium **189–195**
Sophronitella 128
Sophronitis **118–130**
Soterosanthus **195–199**
Spathoglottis **112–116**, *114*
Specklinia 20, **130–135**
Sphyrarhynchus **215–221**
Sphyrastylis **189–195**
Spiculaea **61–62**
Spiranthes 1, 28, **74–79**
Spiranthinae 28, 29, 31, 74, 77, 79
Spiranthoideae 10, 37, 65,79, 167, 168
spiranthosomes 17, 28, 35, *37*, 75, 76, 77, 78, 79, 83, *91*, 110, *111*, 137, *138*, *139*, 183
Stanhopea 12, 24, 36, **195–199**, *196–198*
Stanhopeinae 2, 3, 5, 19, 21, 27, 28, 29, 172, 173, 195–199, 226, 228–232
starch 12, 24, 27, 28, 35, 36, 37, 38
stegmata 10, 15, *24*, 27, 37
stele 3, 15, *29*, 36, 37, 38, 40 *see also* vascular cylinder
Stelis 33, **130–135**
Stellilabium **189–195**
stem 3, 7, 16, 17, 24, 25–28, *26*, *36*, 37, 40
stem tubers 3, 28
Stenia **199–202**
Stenocoryne **184–189**
Stenoglottis **65–73**, *68*, *70*
Stenorrhynchos 31, **74–79**, *77*
Stenotyla **199–202**
Stichorkis **139–143**
stolon 15, 16
stolonoid root 4, 16, 29, 38–40, *39*
Stolzia **150–157**, *155*
stomata 4, 8, 11, 16, *18*, *20*, 21, 27, 28, 38, 227–228
stomatal ontogeny 21
Sturmia 143
subsidiary cells 17, 20–21, *20*
Sudamerlycaste **184–189**
Summerhayesia **215–221**
Sunipia 5, 213
Symphyglossum **189–195**
Systeloglossum 195

T
Taeniophyllum 2, 4, **215–221**
Tainia **112–116**, 157
Tainiopsis 112, 115
tannin 25, 37
Tapeinoglossum 212
taxonomic history 9–11
Telipogon **189–195**
Telipogoninae 194
Tetramicra **118–130**
Teuscheria **184–189**
Thecostele **173–177**, *175*, *176*
Thelasinae 150, 157
Thelasis **150–157**
Thelymitra 2, 15, 38, *41*, **64–65**
Thelymitrinae 64–65, 226–229, 231, 232
Thelyschista 2
Thrixspermum 220
Thunia **98–105**, *103*
Thysanoglossa 194
tilosomes 17, 28, 31, *32*, *33*, *34*, 38, 230–231
Tipularia 3, 8, 19, 21, 24, 28, **105–112**, *107*, *109*
Tolumnia **189–195**, *191*
Townsonia 2, **57–58**
tracheids 2–4, 24, 27, *36*, 38, 40
Traunsteinera 72
Trichocentrum 7, *30*, **189–195**, *191*
Trichoceros **189–195**
Trichoglottis **215–221**
trichomes *see* hairs
Trichopilia **189–195**, *190*
Trichosalpinx *18*, **130–135**
Trichotosia 3, **150–157**
Tridactyle *30*, 31, **215–221**
Triglochin 15
Trigonidium 31, **184–189**
Triphora 2, 4, 38, **161–164**, *162*
Triphoreae 19, 24, 25, 30, 37, 38, 79, 161–164, 168, 226–227, 229–232
Triphorinae 24, 161, 163
Trisetella **130–135**
Trizeuxis **189–195**
Tropideae 167, 168
Tropidia 4, 10, 37, **164–168**, *165–167*
Tropidieae 4, 5, 13, 19, 24, 30, 37, 79, 164–168, 226, 228–232
tuber 3, 4, 9, 11, 15, 16, 25, 28, 38, *39*, *41*
tubercle 4, 15
Tulotis 72

U
Uleiorchis 28, 37, **136–139**, *138*, *139*

V
Vanda 8, 28, **215–221**
Vandeae 2, 4, 6, 9, 14, 24, 25, 28, 31, 35, 37, 213–225
Vandoideae 10, 20
Vanilla 2, 3, 4, 5, 9, 10, 11, 19, 21, 27, 28, 34, 35, 37, **81–95**, *84–91*
Vanillaceae 86
Vanilleae 2, 4, 21, 35, 36, 81–95, 226–228, 230–232
Vanillinae 81
Vanilloideae 3, 4, 5, 6, 10, 11, 14, 19, 21, 24, 27, 28, 30, 36, 37, 49, 79–95
Vargasiellinae 199
vascular bundles 23–24, *23*, 25, 27, 28, 40, 228–229
vascular cylinder 27, 36, 38, *see also* stele
vegetative anatomical literature 15–40
vegetative morphology 2–4
velamen 3, 4, 7, 15, 16, 17, 28–33, *29*, *30*, *32*, *33*, *35*, 37, 38, *41*, 229–230
Vellozia 222
Velloziaceae 222
vessel elements 24, 27, *36*, 45, 48, *49*
Vrydagzynea **74–79**

W
Warczewiczella **199–202**
Warmingia **189–195**
Warrea **199–202**
water-storage (cells) 2, 3, 17, 19, 22, 23, 26, 27, 28, 35
Wullschlaegelia 25, *37*, **105–112**, *106*, *111*, 137

X
Xanthorrhoeaceae 15
Xerorchideae 168
Xerorchis 168
xylem 3, 15, 16, 23, 24, 27, 36, 38, 40
Xylobium **184–189**, *184*

Y
Yoania 78, 79, 110, 112
Ypsilopus **215–221**, *216*

Z
Zelenkoa 195
Zeuxine 3, **74–79**
Zingiberaceae 15
Zootrophion 31, *34*, **130–135**, *132*
Zygopetalinae 2, 4, 19, 21, 23, 184, 199–202, 226, 228–232
Zygopetalum **199–202**
Zygosepalum **199–202**
Zygostates 13, **189–195**